Lecture Notes in Computer Science

Lecture Notes in Artificial Intelligence 15413

Founding Editor

Jörg Siekmann

The series Lecture Notes in Artificial Intelligence (LNAI) was established in 1988 as a topical subseries of LNCS devoted to artificial intelligence.

The series publishes state-of-the-art research results at a high level. As with the LNCS mother series, the mission of the series is to serve the international R & D community by providing an invaluable service, mainly focused on the publication of conference and workshop proceedings and postproceedings.

Aline Paes · Filipe A. N. Verri
Editors

Intelligent Systems

34th Brazilian Conference, BRACIS 2024
Belém do Pará, Brazil, November 17–21, 2024
Proceedings, Part II

Springer

Editors
Aline Paes
Universidade Federal Fluminense
Niterói, Brazil

Filipe A. N. Verri
Instituto Tecnológico de Aeronáutica
São José dos Campos, Brazil

ISSN 0302-9743 ISSN 1611-3349 (electronic)
Lecture Notes in Artificial Intelligence
ISBN 978-3-031-79031-7 ISBN 978-3-031-79032-4 (eBook)
https://doi.org/10.1007/978-3-031-79032-4

LNCS Sublibrary: SL7 – Artificial Intelligence

This Springer imprint is published by the registered company Springer Nature Switzerland AG
The registered company address is: Gewerbestrasse 11, 6330 Cham, Switzerland

Preface

The 34th Brazilian Conference on Intelligent Systems (BRACIS 2024) stands as one of the most significant events in Brazil in 2024 for researchers focused on publishing groundbreaking work in Artificial and Computational Intelligence. BRACIS was formed from the merger of Brazil's two leading scientific events in these fields: the Brazilian Symposium on Artificial Intelligence (SBIA, 21 editions) and the Brazilian Symposium on Neural Networks (SBRN, 12 editions). This year, the community decided to continue the numbering from the first SBIA edition to honor its history.

Supported by the Brazilian Computer Society (SBC), together with the Special Committees on Artificial Intelligence (CEIA) and Computational Intelligence (CEIC), BRACIS 2024 served as a platform to promote both theoretical advancements and practical applications in these areas. The conference promoted the exchange of innovative ideas among researchers, practitioners, and industry leaders.

In 2024, BRACIS was held in Belém do Pará, from November 17th to 21st, 2024, on the *Universidade Federal do Pará* Campus. The conference took place alongside three other events: the National Meeting on Artificial and Computational Intelligence (ENIAC), the Symposium on Information and Human Language Technology (STIL), and the Symposium on Knowledge Discovery, Mining and Learning (KDMiLe).

BRACIS 2024 received 307 paper submissions, each undergoing a rigorous review process conducted by a Program Committee of 198 experienced AI researchers, with the support of 71 additional reviewers, all within a double-anonymous format. This year, we accepted 116 (38%) of the submissions in three key tracks: 70 articles in the main track, showcasing cutting-edge AI methods and solid results; 10 articles in the AI for Social Good track, featuring innovative applications of AI for societal benefit using established methodologies; and 36 articles in other AI applications, presenting novel applications using established AI methods, naturally considering the ethical aspects of the application. In addition, seven top-tier papers published in leading AI conferences and journals were presented during the event.

The topics of interest included, but were not limited to, the following:

- Agent-Based and Multi-agent Systems
- AI for Social Good
- Bioinformatics and Biomedical Engineering using AI
- Cognitive Modeling and Human Interaction
- Combinatorial and Numerical Optimization
- Commonsense Reasoning
- Computer Vision
- Constraints and Search
- Foundations of AI
- Deep Learning and Neural Networks
- Distributed AI
- Education for AI and AI for Education

- Ethics in AI
- Evolutionary Computation and Metaheuristics
- Foundation Models
- Fuzzy Systems
- Game Playing and Intelligent Interactive Entertainment
- Generative AI
- Human-centric AI
- Information Retrieval, Integration, and Extraction
- Intelligent Robotics
- Knowledge Representation and Reasoning
- Large Language Models
- Logic-Based Knowledge Representation and Reasoning
- Machine Learning and Data Mining
- Meta-learning
- Model-Based Reasoning
- Multidisciplinary AI and CI
- Natural Language Processing
- Ontologies and the Semantic Web
- Pattern Recognition and Cluster Analysis
- Planning and Scheduling
- Probabilistic Reasoning and Approximate Reasoning
- Reinforcement Learning

We extend our heartfelt thanks to everyone who contributed to the success of BRACIS 2024. We are especially grateful to the Program Committee members and reviewers for their generous and voluntary efforts in the review process. Special thanks also go to all the authors who submitted their papers and worked tirelessly to refine them into the best possible versions. We also acknowledge the General Chairs, the Local Organization Committee for their unwavering support, the Brazilian Computing Society (SBC), and all our sponsors and supporters. We are confident that these proceedings showcase the exceptional work being done in the artificial and computational intelligence fields.

November 2024

Aline Paes
Filipe A. N. Verri

Organization

General Chairs

André Carlos Ponce de L. F. de Carvalho	Universidade de São Paulo, Brazil
Carlos Renato Lisboa Francês	Universidade Federal do Pará, Brazil
Evelin Helena Silva Cardoso	Universidade Federal do Pará, Brazil

Program Chairs

Aline Paes	Universidade Federal Fluminense, Brazil
Filipe A. N. Verri	Instituto Tecnológico da Aeronáutica, Brazil

Steering Committee (CEIA and CEIC)

Anna Helena Reali Costa	Universidade de São Paulo, Brazil
Anne Magaly de Paula Canuto	Universidade Federal do Rio Grande do Norte, Brazil
Denis Deratani Mauá	Universidade de São Paulo, Brazil
Felipe Rech Meneguzzi	University of Aberdeen, UK
Filipe De Oliveira Saraiva	Universiade Federal do Pará, Brazil
Gisele Lobo Pappa	Universidade Federal de Minas Gerais, Brazil
Leliane Nunes de Barros	Universidade de São Paulo, Brazil
Maria Viviane de Menezes	Universidade Federal do Ceará, Brazil
Murilo Coelho Naldi	Universidade Federal de São Carlos, Brazil
Renato Tinós	Universidade de São Paulo, Brazil
Ricardo M. Marcacin	Universidade de São Paulo, Brazil
Tatiane Nogueira Rios	Universidade Federal da Bahia, Brazil

Local Organizers

Jose Jailton Henrique Ferreira Junior	Universidade Federal do Pará, Brazil
Hugo Pereira Kuribayashi	Universidade do Sul e Sudeste do Pará, Brazil
Jorge Antonio Moraes de Souza	Universidade Federal Rural da Amazônia, Brazil

Program Committee

Adenilton da Silva	Universidade Federal de Pernambuco, Brazil
Adrião Duarte Dória Neto	Universidade Federal do Rio Grande do Norte, Brazil
Alex Fernandes de Souza	Universidade Federal de Itajubá, Brazil
Alexandre Ferreira	Universidade Estadual de Campinas, Brazil
Aline Neves	Universidade Federal do ABC, Brazil
Aline Paes	Universidade Federal Fluminense, Brazil
Alison R. Panisson	Universidade Federal de Santa Catarina, Brazil
Altigran Soares da Silva	Universidade Federal do Amazonas, Brazil
Alvaro Moreira	Universidade Federal do Rio Grande do Sul, Brazil
Amedeo Napoli	Centre Inria de l'Université de Lorraine, France
Ana Bazzan	Universidade Federal do Rio Grande do Sul, Brazil
Ana Carolina Lorena	Instituto Tecnológico de Aeronáutica, Brazil
Ana Cristina B. K. Vendramin	Universidade Tecnológica Federal do Paraná, Brazil
Ana Cristina Garcia	Universidade Federal do Estado do Rio de Janeiro, Brazil
André Britto	Universidade Federal do Sergipe, Brazil
André Carlos P. de L. F. de Carvalho	Universidade de São Paulo, Brazil
André Luis Debiaso Rossi	Universidade Estadual Paulista, Brazil
Andrés Eduardo Coca Salazar	Universidade Tecnológica Federal do Paraná, Brazil
Angelo Loula	Universidade Estadual de Feira de Santana, Brazil
Anna Helena Reali Costa	Universidade de São Paulo, Brazil
Anne Canuto	Universidade Federal do Rio Grande do Norte, Brazil
Antonio Parmezan	Universidade de São Paulo, Brazil
Araken Santos	Universidade Federal Rural do Semi-Árido, Brazil
Ariane Machado-Lima	Universidade de São Paulo, Brazil
Artur Jordão	Universidade de São Paulo, Brazil
Aurora Pozo	Universidade Federal do Paraná, Brazil
Bernardo Gonçalves	Universidade de São Paulo, Brazil
Bianca Zadrozny	IBM Research, Brazil
Bruno Nogueira	Universidade Federal de Mato Grosso do Sul, Brazil
Bruno Souza	Universidade Federal do Maranhão, Brazil
Carla Amor Divino M. Delgado	Universidade Federal do Rio de Janeiro, Brazil
Carlos Ferreira	Universidade Federal de Ouro Preto, Brazil

Carlos Silla	Pontifícia Universidade Católica do Paraná, Brazil
Carlos Thomaz	Centro Universitário FEI, Brazil
Carlos Henrique Ribeiro	Instituto Tecnológico de Aeronáutica, Brazil
Carolina Paula de Almeida	Universidade Estadual do Centro-Oeste, Brazil
Celia Ralha	Universidade de Brasília, Brazil
Cleber Zanchettin	Universidade Federal de Pernambuco, Brazil
Cristiano Torezzan	Universidade Estadual de Campinas, Brazil
Daniel Araújo	Universidade Federal do Rio Grande do Norte, Brazil
Daniel Oliveira Dantas	Universidade Federal de Sergipe, Brazil
Daniela Barreiro Claro	Universidade Federal da Bahia, Brazil
Danilo Sanches	Universidade Tecnológica Federal do Paraná, Brazil
Debora Medeiros	Universidade Federal do ABC, Brazil
Denis Fantinato	Universidade Estadual de Campinas, Brazil
Denis Mauá	Universidade de São Paulo, Brazil
Diana Adamatti	Universidade Federal do Rio Grande, Brazil
Diego Furtado Silva	Universidade de São Paulo, Brazil
Diego Mesquita	Fundação Getúlio Vargas, Brazil
Diego Pinheiro	Universidade de Pernambuco, Brazil
Eder Mateus Gonçalves	Universidade Federal do Rio Grande, Brazil
Edson Gomi	Universidade de São Paulo, Brazil
Edson Matsubara	Fundação Universidade Federal de Mato Grosso do Sul, Brazil
Eduardo Borges	Universidade Federal do Rio Grande, Brazil
Eduardo Goncalves	Escola Nacional de Ciências Estatísticas, Brazil
Eduardo Luz	Universidade Federal de Ouro Preto, Brazil
Eduardo Spinosa	Universidade Federal do Paraná, Brazil
Emerson Paraiso	Pontificia Universidade Católica do Paraná, Brazil
Evandro Costa	Universidade Federal de Alagoas, Brazil
Fabiano Silva	Universidade Federal do Paraná, Brazil
Fabio Faria	Universitdade Federal de São Paulo, Brazil
Fabiola Souza Fernandes Pereira	Universidade Federal de Uberlândia, Brazil
Fabrício Enembreck	Pontificia Universidade Católica do Paraná, Brazil
Fabrício A. Silva	Universidade Federal de Viçosa, Brazil
Fabricio Olivetti de França	Universidade Federal do ABC, Brazil
Fábio Cozman	Universidade de São Paulo, Brazil
Felipe Meneguzzi	University of Aberdeen, UK
Felix Antreich	Instituto Tecnológico de Aeronáutica, Brazil
Fernanda Maria da Cunha Santos	Universidade Federal de Uberlândia, Brazil
Fernando Maciano de Paula Neto	Universidade Federal de Pernambuco, Brazil
Fernando Osório	Universidade de São Paulo, Brazil

Filipe Saraiva	Universidade Federal do Pará, Brazil
Filipe Verri	Instituto Tecnológico da Aeronáutica, Brazil
Flavia Bernardini	Universidade Federal Fluminense, Brazil
Flavio Tonidandel	Centro Universitario da FEI, Brazil
Flavius Gorgônio	Universidade Federal do Rio Grande do Norte, Brazil
Flávio Soares Corrêa da Silva	Universidade de São Paulo, Brazil
Francisco De Carvalho	Universidade Federal de Pernambuco, Brazil
George D. da Cunha Cavalcanti	Universidade Federal de Pernambuco, Brazil
Geraldo Pereira Rocha Filho	Universidade Estadual do Sudoeste da Bahia, Brazil
Gerson Zaverucha	Universidade Federal do Rio de Janeiro, Brazil
Giancarlo Lucca	Universidade Católica de Pelotas, Brazil
Gisele Pappa	Universidade Federal de Minas Gerais, Brazil
Gleifer Vaz Alves	Universidade Tecnologica Federal do Paraná, Brazil
Guilherme Barreto	Universidade Federal do Ceará, Brazil
Guilherme Palermo Coelho	Universidade Estadual de Campinas, Brazil
Guilherme Derenievicz	Universidade Federal do Paraná, Brazil
Guilherme Dean Pelegrina	Universidade Presbiteriana Mackenzie, Brazil
Guillermo Simari	Universidad Nacional del Sur in Bahia Blanca, Argentina
Gustavo Giménez-Lugo	Universidade Tecnológica Federal do Paraná, Brazil
Heder Bernardino	Universidade Federal de Juiz de Fora, Brazil
Helen Senefonte	Universidade Estadual de Londrina, Brazil
Helena Caseli	Universidade Federal de São Carlos, Brazil
Helena Maia	Universidade Estadual de Campinas, Brazil
Helida Santos	Universidade Federal do Rio Grande, Brazil
Heloisa Camargo	Universidade Federal de São Carlos, Brazil
Islame Felipe da Costa Fernandes	Universidade Federal da Bahia, Brazil
Ivan Reis Filho	Universidade do Estado de Minas Gerais, Brazil
Ivette Luna	Universidade Estadual de Campinas, Brazil
Jadson Gertrudes	Universidade Federal de Ouro Preto, Brazil
Jaime Sichman	Universidade de São Paulo, Brazil
Jean Paul Barddal	Pontificia Universidade Católica do Paraná, Brazil
João Bertini	Universidade Estadual de Campinas, Brazil
João Papa	Universidade Estadual Paulista, Brazil
João C. Xavier-Júnior	Universidade Federal do Rio Grande do Norte, Brazil
João Paulo Canário	Stone Co., Brazil
Joniel Barreto	Instituto Tecnológico de Aeronáutica, Brazil

José Antonio Sanz	Universidad Pública de Navarra, Spain
José Augusto Baranauskas	Universidade de São Paulo, Brazil
Jose Eduardo Ochoa Luna	Universidad Católica San Pablo, Peru
Júlio Cesar Nievola	Pontifícia Universidade Católica do Paraná, Brazil
Julio C. S. Reis	Universidade Federal de Viçosa, Brazil
Karla Roberta Lima	Universidade de São Paulo, Brazil
Leonardo Emmendörfer	Universidade Federal de Santa Maria, Brazil
Leonardo Matos	Universidade Federal de Sergipe, Brazil
Leonardo Tomazeli Duarte	Universidade Estadual de Campinas, Brazil
Li Weigang	Universidade de Brasília, Brazil
Livy Real	B2W Digital/GLiC, Brazil
Lucelene Lopes	Universidade de São Paulo, Brazil
Luciano Digiampietri	Universidade de São Paulo, Brazil
Luiz Henrique de C. Merschmann	Universidade Federal de Lavras, Brazil
Luiza de Macedo Mourelle	Universidade do Estado do Rio de Janeiro, Brazil
Marcela Ribeiro	Universidade Federal de São Carlos, Brazil
Marcelo Bruno	Instituto Tecnológico de Aeronáutica, Brazil
Marcelo Finger	Universidade de São Paulo, Brazil
Marcilio de Souto	Université d'Orléans, France
Marco A. G. de Carvalho	Universidade Estadual de Campinas, Brazil
Marcos Domingues	Universidade Estadual de Maringá, Brazil
Marcos Quiles	Universidade Federal de São Paulo, Brazil
Marcos R. O. A. Maximo	Instituto Tecnológico de Aeronáutica, Brazil
Mariela Morveli-Espinoza	Universidade Tecnologica Federal do Paraná, Brazil
Marilton Aguiar	Universidade Federal de Pelotas, Brazil
Mariza Ferro	Universidade Federal Fluminense, Brazil
Marley M. B. R. Vellasco	Pontifícia Universidade Católica do Rio de Janeiro, Brazil
Marlo Souza	Universidade Federal da Bahia, Brazil
Marlon Mathias	Universidade de São Paulo, Brazil
Matheus Giovanni Pires	Universidade Estadual de Feira de Santana, Brazil
Mauri Ferrandin	Universidade Federal de Santa Catarina, Brazil
Murillo Carneiro	Universidade Federal de Uberlândia, Brazil
Murilo Loiola	Universidade Federal do ABC, Brazil
Murilo Naldi	Universidade Federal de São Carlos, Brazil
Myriam Delgado	Universidade Tecnológica Federal do Paraná, Brazil
Nádia Felix	Universidade Federal de Goiás, Brazil
Omar Andres Carmona Cortes	Instituto Federal do Maranhão, Brazil
Paula Costa	Universidade Estadual de Campinas, Brazil
Paulo Gabriel	Universidade Federal de Uberlândia, Brazil

Paulo Mann	Universidade do Estado do Rio de Janeiro, Brazil
Paulo Pirozelli	Universidade de São Paulo, Brazil
Paulo Quaresma	Universidade de Évora, Portugal
Paulo Henrique Pisani	Universidade Federal do ABC, Brazil
Paulo T. Guerra	Universidade Federal do Ceará, Brazil
Pedro Silva	Universidade Federal de Ouro Preto, Brazil
Petrucio Viana	Universidade Federal Fluminense, Brazil
Priscila Lima	Universidade Federal do Rio de Janeiro, Brazil
Rafael H. Bordini	Pontifícia Universidade Católica do Rio Grande do Sul, Brazil
Rafael Giusti	Universidade Federal do Amazonas, Brazil
Reinaldo Bianchi	Centro Universitario FEI, Brazil
Renato Tinos	Universidade de São Paulo, Brazil
Ricardo Cerri	Universidade de São Paulo, Brazil
Ricardo Dutra da Silva	Universidade Tecnológica Federal do Paraná, Brazil
Ricardo Marcacini	Universidade de São Paulo, Brazil
Ricardo Prudêncio	Universidade Federal de Pernambuco, Brazil
Ricardo Rios	Universidade Federal da Bahia, Brazil
Ricardo Suyama	Universidade Federal do ABC, Brazil
Ricardo Augusto Souza Fernandes	Universidade Federal de São Carlos, Brazil
Roberto Fray da Silva	Universidade de São Paulo, Brazil
Roberto Douglas G. de Aquino	Instituto Tecnológico de Aeronáutica, Brazil
Robespierre Pita	Universidade Federal da Bahia, Brazil
Rodrigo Lira	Instituto Federal de Pernambuco, Brazil
Rodrigo Silva	Universidade Federal de Ouro Preto, Brazil
Rodrigo Veras	Universidade Federal do Piauí, Brazil
Romis Attux	Universidade Estadual de Campinas, Brazil
Ronaldo Prati	Universidade Federal do ABC, Brazil
Roseli Ap. Francelin Romero	Universidade de São Paulo, Brazil
Rosiane de Freitas Rodrigues	Universidade Federal do Amazonas, Brazil
Rui Camacho	Universidade do Porto, Portugal
Sandro Fiorini	IBM Research, Brazil
Sarajane Marques Peres	Universidade de São Paulo, Brazil
Sílvia Maia	Universidade Federal do Rio Grande do Norte, Brazil
Sílvio César Cazella	Universidade Federal de Ciências da Saúde de Porto Alegre, Brazil
Silvia Botelho	Universidade Federal do Rio Grande, Brazil
Solange Rezende	Universidade de São Paulo, Brazil
Talles Medeiros	Universidade Federal de Ouro Preto, Brazil
Tatiane Nogueira	Universidade Federal da Bahia, Brazil

Thiago Covoes	Thomson Reuters Labs, USA
Thiago Pardo	Universidade de São Paulo, Brazil
Tiago Almeida	Universidade Federal de São Carlos, Brazil
Tiago Tavares	Instituto de Ensino e Pesquisa Insper, Brazil
Valéria Santos	Universidade Federal de Ouro Preto, Brazil
Valmir Macario	Universidade Federal Rural de Pernambuco, Brazil
Vander Freitas	Universidade Federal de Ouro Preto, Brazil
Vasco Furtado	Universidade de Fortaleza, Brazil
Vinicius de Carvalho	Universidade de São Paulo, Brazil
Vinicius Souza	Pontificia Universidade Católica do Paraná, Brazil
Vitor Curtis	Instituto Tecnológico de Aeronáutica, Brazil
Viviane Moreira	Universidade Federal do Rio Grande do Sul, Brazil
Viviane Torres da Silva	IBM Research, Brazil
Wagner Meira Jr.	Universidade Federal de Minas Gerais, Brazil

Additional Reviewers

Alessandra Aparecida Paulino
Alexandre Alcoforado
Aline Del Valle
Amanda Perez
Anderson Moraes
André De Oliveira
Annie Amorim
Arthur Torres
Arthur Valencio
Astrid Wiens
Bernardo Scapini Consoli
Caroline Pires Alavez Moraes
Débora Engelmann
Diedre Santos
Diego Minatel
Eduardo Silva
Elaine Gatto
Eliezer de Souza da Silva
Elineide Silva dos Santos
Eliton Perin
Estela Ribeiro
Eulanda Miranda dos Santos
Fabian Cardoso
Fabiano Silva
Felipe Serras
Filipe Cordeiro
Gabriel de Castro Michelassi
Gustavo Henrique do Nascimento
Hermon Faria Araújo
Israel Fama
Ivandro Sanches
João Barbon
Johny Moreira
Jose Jailton Henrique Ferreira Junior
Juan Colonna
Justino Santos
Karliane Vale
Keylly Santos
Leonardo da Silva Costa
Leonardo Sousa
Letícia Freire
Lucas Alegre
Lucas dos Santos
Lucas Rodrigues
Lucas Santos
Luciana Bencke
Luis Vogado
Luiz Felipe Vercosa

Miguel de Mello Carpi
Napoleao Galegale
Natanael Batista
Patrick Ferreira
Patrick Terrematte
Paulo Girardi
Rafael Berri
Renata Wassermann
Rodrigo Maia
Sarah Negreiros
Selma Carloto
Silvio Romero de Araujo Junior
Tamara Pereira
Thais Luca
Thiago Carvalho
Thiago Freire de Oliveira
Thiago Miranda
Thomas Palmeira Ferraz
Tiago Botelho
Tiago Pinho da Silva
Vítor Lourenço
Wandry Faria
Wesley Seidel

Contents – Part II

Main Track

Main Track

Going Bananas! - Unfolding Program Synthesis with Origami

Matheus Campos Fernandes(✉), Fabrício Olivetti de França, and Emilio Francesquini

Federal University of ABC (UFABC), São Paulo, Santo André, Brazil
{fernandes.matheus,folivetti,e.francesquini}@ufabc.edu.br

Abstract. Automatically creating a computer program using input-output examples can be a challenging task, especially when trying to synthesize computer programs that require loops or recursion. Even though the use of recursion can make the algorithmic description more succinct and declarative, this concept creates additional barriers to program synthesis algorithms such as the creation and the attempt to evaluate non-terminating programs. One reason is that the recursive function must define how to traverse (or generate) the data structure and, at the same time, how to process it. In functional programming, the concept of Recursion Schemes decouples these two tasks by putting a major focus on the data processing. This can also help to avoid some of the pitfalls of recursive functions during program synthesis, as argued in the paper that introduced the Origami technique. The authors showed how this technique can be effective in finding solutions for programs that require folding a list. In this work, we incorporate other Recursion Schemes into Origami, such as accumulated folding, unfolding, and the combination of unfolding and folding. We evaluated it on the 29 problems of the standard General Program Synthesis Benchmark Suite 1, obtaining favorable results against other well-known algorithms. Overall, Origami achieves the best results in 25% more problems than its predecessor (HOTGP) and an even higher increase when compared to other approaches. Not only that, but it can also consistently find a solution to problems for which other concurrent algorithms report a low success rate.

Keywords: Program Synthesis · Genetic Programming · Recursion

1 Introduction

Program Synthesis (PS) is the task of automatically generating a computer program written in a programming language of choice using some form of specification [6]. This can be framed as a search problem that explores the space of possible programs under a set of constraints, usually as a grammar and a maximum program size, that remove undesirable solutions from the search space.

A convenient format for the specification is a set of input-output examples that demonstrate the expected outputs for several different input cases.

A. Paes and F. A. N. Verri (Eds.): BRACIS 2024, LNAI 15413, pp. 3–18, 2025.
https://doi.org/10.1007/978-3-031-79032-4_1

In this particular case, the task is called *Inductive Synthesis* or *Programming-by-Example* (PBE). The main advantage of this approach is that sets of examples are easy to create, and often do not require deep knowledge of the problem in question. However, the lack of corner or special cases in the set allows for the generation of programs that do not follow the original intent of the user.

Among the vast selection of possible search algorithms, we highlight *Genetic Programming* (GP) [10]. GP is a search algorithm that tries to balance the exploration and exploitation of the search space using recombination and perturbation of a bag of solutions. It employs selective pressure at every step, inspired by the evolution of species, to favor the fittest solutions when replicating and applying such operators. Some benchmark problems extracted from common programming tasks have been successfully solved by recent variations of the original GP algorithm, such as PushGP [8], CBGP [15], GE [13], G3P [3], and HOTGP [1].

The representation of the solutions and the imposed constraints play a major role in search algorithms as they describe the navigability in this space and the coverage of possible programs. For example, the Push language [17] is a stack-based programming language in which the operations are executed sequentially storing and retrieving values from the stack corresponding to the types of the operation. This provides additional flexibility during the execution of a computer program as any invalid operation can be skipped and, thus, the operation at any given timestep is not constrained by output type of the previous step. This increases the possibilities of navigation of the search space. On the other hand, in Grammatical Evolution [13] the search space is defined by the grammar being evolved. This can limit the size of the search space and ensure that each step of the generated program is valid. Likewise, type-safe representations, such as in CBGP [15] and HOTGP [1], carry at each node the information of the input and output types, constraining the space to valid programs (*i.e.*, that execute without errors) and solutions that follow the type specification.

A common challenge to GP algorithms is how to handle recursion or loops. The problem here lies in the fact that it is possible to generate an unnecessarily long or even unbounded recursion/loop. This is alleviated by the use of programming patterns that hide the loop or recursion behind a declarative language. Popular examples of such patterns are the `map`, `filter`, and `fold` functions.

In particular, *fold* describes a recursive pattern capable of expressing recursive algorithms that traverse a structure aggregating the partial results. Examples of the use of this pattern are *sum*, *product*, and even *insertion sort*. A complementary pattern that describes the recursion that generates or builds a structure is known as *unfold*. For example, `unfold` can be used to generate the list of Fibonacci numbers, or reversing the order of the elements of a list.

Although many algorithms can be described using folds or unfolds, this approach has some limitations such as working only on lists, or the inability to store partial results. A more general set of recursive patterns that involves folding and unfolding is known as *Recursion Schemes* [11]. The Recursion Schemes extend the common folding and unfolding operations to work on any inductive type and include additional mechanisms to handle a wider variety of recursive function patterns[1]. These Recursion Schemes can be used to guide the synthesis of recursive programs, as their structure is well-defined and can be used as scaffolding, with variations only in certain parts of the program.

Origami [2] is an algorithm capable of synthesizing typed, pure, functional programs that support different Recursion Schemes. Origami's authors presented a proof-of-concept implementation, capable of partially synthesizing one single scheme. This implementation was evaluated, yielding promising preliminary results. In this paper, we provide and evaluate the first complete implementation of Origami, following the description given in the original paper. This represents a meaningful step in assessing the effectiveness of integrating Recursion Schemes into Program Synthesis. We evaluate the performance of Origami using the problems described in the General Program Synthesis Benchmark Suite 1 (PSB1) [9], comparing the obtained results to well-known GP-based program synthesis algorithms. When comparing the success rate of each problem, this new version of Origami achieves the best results in 25% more problems than its predecessor (HOTGP) and an even higher increase when compared to other approaches. Additionally, Origami was capable of achieving a high success rate (higher than 70%) in problems that most algorithms achieve less than 50%.

This text is organized as follows. Section 2 brings a brief literature review of Functional Program Synthesis and Recursion Schemes. Section 3 presents Origami and describes the details of our implementation. An analysis of the results and comparisons of Origami to HOTGP and other well-known methods is shown in Sect. 4. Finally, Sect. 5 concludes our work.

2 Related Work

There have been several attempts to use functional programming as the basis of program synthesis. One of the earliest attempts employed types to guide the search [12]. More recently, the use of recursion schemes [18] and higher-order functions [1,5,14] to represent recursion were proposed.

Out of these approaches, the one that is more closely related to this work is HOTGP [1], which is a GP algorithm that synthesizes pure, typed, and functional programs. Its approach to recursion includes support for higher-order functions, λ-functions, and parametric polymorphism.

Notably, in [18] the authors presented the first algorithm for synthesizing programs that exploit Recursion Schemes. Their work focuses only on catamorphisms over natural numbers using Peano representation (*i.e.*, the inductive type of natural numbers). The authors evaluate their approach with variations of the Fibonacci sequence, successfully obtaining the correct programs.

Origami was originally proposed in [2] where the authors evaluated the feasibility of using Recursion Schemes to synthesize recursive programs. In that work, they showed that the entire PSB1 benchmark can be solved by one of four different Recursion Schemes: catamorphism, accumulation, anamorphism, and

[1] The authors represent these recursive functions using a notation called bananas, lenses, envelopes, and barbed wires [11].

hylomorphism. It also described preliminary experiments with catamorphism showing that, for the problems that are solvable with this scheme, the use of scaffolding improved the success rate when compared to HOTGP. These results brought evidence that the synthesis process is simplified once the correct Recursion Scheme is determined, as it just needs to evolve non-recursive expressions.

Large Language Models as Code Assistants have also been employed for Program Synthesis. In [16], the authors use GitHub Copilot to synthesize programs from a textual description of the problem, obtaining correct programs more often than a selection of GP based synthesizers in 50% of the evaluated problems.

3 Origami Program Synthesis

Origami's implementation follows a Koza-style Genetic Programming [10] (tree representation). The main distinctions to traditional approaches are the introduction of immutable nodes (ensuring a certain Recursion Scheme); and the type-safety of the genetic operators (the same approach taken by HOTGP [1]).

The implementation is based on *patterns*, which are used to represent different Recursion Schemes. A pattern is composed of immutable nodes and a set of evolvable slots that, when replaced with expressions, can be evaluated. The immutable nodes describe the main definition of the Recursion Scheme (see Sect. 3.1) and are fixed once we choose the pattern, while the evolvable slots represent the inner mechanisms that need to be synthesized to correspond to the expected behavior described by the dataset. These slots have a well-defined output type (inferred from the problem description), and a well-defined set of bindings to which the expression has access.

In this work we focus on the six different patterns that comprehend the minimal set required to solve PSB1. Naturally, Origami is not limited to these patterns, and more could be included as needed. Section 3.1 details these six patterns. Due to space constraints, we assume the reader has a basic understanding of Recursion Schemes [11], how they can be used to solve PSB1 [2], as well as the Haskell language notation, which is similar to the ML notation.

3.1 Patterns

NoScheme. This is the simplest pattern in Origami, as it does not employ any recursion at all. It is represented by the following code:

```
f arg_0 ... arg_n = slot_1
```

This pattern has just a single slot, which has all the arguments in scope and returns a value of the same type as the output of the program. Its main use is to accommodate for problems that do not require any recursion.

Catamorphism over Indexed List. This pattern captures the most common Recursion Scheme observed in PSB1, and arguably in practical scenarios as well, *i.e.*, folding a list from the right. In Meijer-notation [11], this would be represented by the banana brackets $(\!|b, \oplus|\!)$, where b is the initial value and $\oplus$ is the combining function. In the context of Origami, it can be represented as:

```
f arg0 ... argn = cata alg arg0 where
      alg INil = slot1
      alg (ICons i x acc) = slot2
```

In a problem with arguments of type $\texttt{i}_0 \ldots \texttt{i}_n$ and of output type `o`, where $0 \equiv$ `[e]`[2], this pattern's slots are typed as follows:

- *slot₁* :: `o`, with nothing in scope;
- *slot₂* :: `o`, with scope { `i` :: `Int` ; `x` :: `e` ; `acc` :: `o` ; $\texttt{arg}_0$ :: $\texttt{i}_0 \ldots \texttt{arg}_n$ :: $\texttt{i}_n$}.

This pattern will be referred simply as *Cata* in the remainder of this paper.

Curried Catamorphism over Indexed List. This pattern captures a common variation of the Catamorphism, and can be represented by the following code:

```
f arg0 arg1 = cata alg arg0 arg1 where
      alg INil = \ys -> slot1
      alg (ICons i x f) = \ys -> slot2
```

As a problem of type $\texttt{i}_0$ `->` $\texttt{i}_1$ `-> o` can also be seen in its curried form as $\texttt{i}_0$ `-> (`$\texttt{i}_1$ `-> o)`, we can employ Catamorphism to accumulate a *function* over the first argument, and then apply this function to the second argument. This is useful when we need to apply a Catamorphism over the zip of two lists [2].

In a problem with arguments of type $\texttt{i}_0$, $\texttt{i}_1$[3] and of output type `o`, where $\texttt{i}_0 \equiv$ `[e]`, this pattern's slots are typed as follows:

- *slot₁* :: `o`, with scope { `ys` :: $\texttt{i}_1$ };
- *slot₂* :: `o`, with scope { `i` :: `Int` ; `x` :: `e` ; `f` :: $\texttt{i}_1$ `-> o` ; `ys` :: $\texttt{i}_1$ }.

For brevity, this will be referred to as simply *CurriedCata*. Note that both this and the previous pattern use Indexed Linked List as the data structure, allowing the program to access each element's index and value. For the remaining patterns we employ a regular list since it is enough to solve their problems (as shown by the canonical solutions presented in [2]).

---

[2] The notation $\texttt{i}_0 \equiv$ `[e]` is a restriction such that $\texttt{i}_0$ can be decomposed into the type `[e]`, which is the type of a list with elements of some type `e`.

[3] Notice that this pattern can only be applied to problems with exactly two arguments.

Anamorphism to a List. This pattern is commonly used in Haskell as `unfold`, which is used to generate a list. In Meijer-notation [11], this would be represented by the concave lenses $[\!(g, p)\!]$ where g is the generator function, and p is the predicate. In the context of Origami, it can be represented by the following code:

```
f arg0 ... argn = ana coalg slot1 where
      coalg seed = if slot2 then []
                   else slot3 : slot4
```

In a problem with arguments of type $\mathtt{i}_0 \ldots \mathtt{i}_\mathtt{n}$ and of output type `o`, where `o` $\equiv$ `[e]`, this pattern's slots are typed as follows:

- $slot_1$ :: $\mathtt{i}_0$, with scope { $\mathtt{arg}_0$ :: $\mathtt{i}_0 \ldots \mathtt{arg}_\mathtt{n}$ :: $\mathtt{i}_\mathtt{n}$};
- $slot_2$ :: `Bool`, with scope { `seed` :: $\mathtt{i}_0$; $\mathtt{arg}_0$ :: $\mathtt{i}_0 \ldots \mathtt{arg}_\mathtt{n}$ :: $\mathtt{i}_\mathtt{n}$};
- $slot_3$ :: `e`, with scope { `seed` :: $\mathtt{i}_0$; $\mathtt{arg}_0$ :: $\mathtt{i}_0 \ldots \mathtt{arg}_\mathtt{n}$ :: $\mathtt{i}_\mathtt{n}$};
- $slot_4$ :: $\mathtt{i}_0$, with scope { `seed` :: $\mathtt{i}_0$ }.

Note that while we do not enforce $\mathtt{arg}_0$ to be used in $slot_1$, it must be of the same type as $\mathtt{arg}_0$, as all of the solutions for PSB1 respected this constraint. For brevity, this will be referred to as simply *Ana* in the rest of this paper.

Accumulation over a List. This pattern captures using an accumulation strategy before using a `foldr`, and can be represented by the following code:

```
f arg0 ... argn = accu st alg arg0 slot1
      where
        st [] s = []
        st (x : xs) s = x : (xs, slot2)
        alg [] s = slot3
        alg (x : acc) s = slot4
```

In a problem with arguments of type $\mathtt{i}_0 \ldots \mathtt{i}_\mathtt{n}$ and of output type `o`, where $\mathtt{i}_0 \equiv$ `[e]`, **and given a type** `a`, this pattern's slots are typed as follows:

- $slot_1$ :: `a`, with scope { $\mathtt{arg}_0$ :: $\mathtt{i}_0 \ldots \mathtt{arg}_\mathtt{n}$ :: $\mathtt{i}_\mathtt{n}$};
- $slot_2$ :: `a`, with scope { `x` :: `e` ; `xs` :: `[e]` ; `s` :: `a` ; $\mathtt{arg}_0$ :: $\mathtt{i}_0 \ldots \mathtt{arg}_\mathtt{n}$ :: $\mathtt{i}_\mathtt{n}$};
- $slot_3$ :: `o`, with scope { `s` :: `a` ; $\mathtt{arg}_0$ :: $\mathtt{i}_0 \ldots \mathtt{arg}_\mathtt{n}$ :: $\mathtt{i}_\mathtt{n}$};
- $slot_4$ :: `o`, with scope { `x` :: `e` ; `acc` :: `o` ; `s` :: `a` ; $\mathtt{arg}_0$ :: $\mathtt{i}_0 \ldots \mathtt{arg}_\mathtt{n}$ :: $\mathtt{i}_\mathtt{n}$}.

This is the first pattern whose types are not fully determined by the type of the arguments and the expected output type: the accumulator type `a`. Types such as this will be referred to as *unbound types*. To keep the implementation simple, we assume unbound types are known and provided by the user. The exploration of different types is an interesting challenge that warrants dedicated research. This pattern will be referred to as simply *Accu* in the rest of this paper.

Hylomorphism Through a List. This pattern captures an Anamorphism followed by a Catamorphism, such as applying `foldr` to the result of `unfold`, in Haskell. In Meijer-notation [11], this would be represented by the envelopes $[\![(c, \oplus), (g, p)]\!]$. In Origami, it is represented by the following code:

```
f arg0 ... argn = hylo alg coalg arg0 where
    coalg seed = if slot1 then [] else slot2 : slot3
    alg [] = slot4
    alg (x : acc) = slot5
```

In a problem with arguments of type $\mathtt{i_0} \ldots \mathtt{i_n}$ and of output type `o`, **and given a type** `a`, this pattern's slots are typed as follows:

- $\underline{slot_1}$ :: `Bool`, with scope { `seed` :: $\mathtt{i_0}$; $\mathtt{arg_0}$ :: $\mathtt{i_0} \ldots \mathtt{arg_n}$ :: $\mathtt{i_n}$};
- $\underline{slot_2}$ :: `a`, with scope { `seed` :: $\mathtt{i_0}$; $\mathtt{arg_0}$ :: $\mathtt{i_0} \ldots \mathtt{arg_n}$ :: $\mathtt{i_n}$};
- $\underline{slot_3}$ :: $\mathtt{i_0}$, with scope { `seed` :: $\mathtt{i_0}$; $\mathtt{arg_0}$ :: $\mathtt{i_0} \ldots \mathtt{arg_n}$ :: $\mathtt{i_n}$};
- $\underline{slot_4}$ :: `o`, with nothing in scope;
- $\underline{slot_5}$ :: `o`, with scope { `x` :: `a` ; `acc` :: `o` ; $\mathtt{arg_0}$ :: $\mathtt{i_0} \ldots \mathtt{arg_n}$ :: $\mathtt{i_n}$}.

This pattern also contains an unbound type: the intermediary list has elements of type `a`. This pattern will be referred to as *Hylo*.

3.2 Genetic Programming

Origami synthesizes the evolvable slots using a Genetic Programming (GP) [10] algorithm. Since the patterns require more than a single slot, we represent each solution as a collection of programs represented as expression trees. Each element of this collection corresponds to one of the slots.

The GP starts with an initial random population of 1 000 individuals, and iterates by applying either crossover to a pair of parents, or mutation to a single parent, generating 1 000 new individuals in total. The entire population is replaced by the offspring population.

The initial population is generated using a ramped half-and-half where half of the individuals are generated using the full method and half using the grow method. The maximum depth for each method varies between 1 and 5. The parental selection is performed using a tournament selection of size 10.

Following a simple GP algorithm, in Origami the mutation randomly selects one of the evolvable slots, then picks one point in the tree at random to be replaced by a new subtree generated at random using the grow method, with a maximum depth of $5 - d_{current}$. Crossover also starts by picking one of the slots at random, then performing one of these two actions with equal probability: i) swap the entire slot of one parent with the same slot of the other parent; ii) swap two subtrees of the same output type from each parent.

3.3 Grammar

In HOTGP [1], the selected grammar focused on providing a minimal set of operations, including higher-order functions, that would enable the synthesis of programs under the functional programming paradigm.

With Origami, however, the main focus is assessing Recursion Schemes as the only means of synthesizing recursive programs. Therefore, we designed our grammar to avoid implicitly recursive functions, like `map`, `filter`, `sum`, and `product`. We acknowledge that this might remove shortcuts and potentially make the synthesis of certain problems harder. Notice that the recursion happens in the immutable nodes that describe the Recursion Schemes, so the recursion is provided rather than evolved. Additionally, the set of operations includes functions equivalent to those used by other methods, in particular those implemented by PushGP [8]. As a result, Origami has a larger set of operations than HOTGP. The full grammar is presented in Table 1. Once an execution is finished, the champion's slots are refined using the same procedure that was used in HOTGP [1]. To refine a tree, we pick the root node and check if replacing it with any of its children leads to a correctly-typed solution with an equal or better fitness. If so, we replace it with the best child; otherwise, we keep the original node. This process continues recursively, traversing the tree and greedily replacing nodes with their children when needed. This procedure applies Occam's Razor to choose a simpler solution [7], making sure the fitness in the training set is never worse.

4 Experimental Results

To evaluate our approach we conducted experiments to perform an automatic search for different patterns in the PSB1 [9] context. For each of the 29 datasets, we sequentially tried each pattern in increasing order of complexity: NoScheme; Cata, if $\texttt{arg}_0$ is a list; CurriedCata, if the problem has two arguments and $\texttt{arg}_0$ is a list; Ana, if the return type is a list; Accu, if $\texttt{arg}_0$ is a list; Hylo.

For each dataset, we executed 30 seeds of each pattern starting from the simplest and testing other patterns if none of the seeds succeeded in finding a solution (*i.e.*, the success rate was 0%). Each seed followed the instructions provided by PSB1, using the recommended number of training and test instances, and included the fixed edge cases in the training data, as well as using the fitness functions described in [9]. We also made the same adaptations to the benchmarks as in [2], and similar to [1] and [16]. Specifically, we changed the input of the grade benchmark from $(\texttt{arg}_0, \texttt{arg}_1, \texttt{arg}_2, \texttt{arg}_3, \texttt{arg}_4)$ to $([(\texttt{arg}_0, \texttt{'A'}), (\texttt{arg}_1, \texttt{'B'}), (\texttt{arg}_2, \texttt{'C'}), (\texttt{arg}_3, \texttt{'D'})], \texttt{arg}_4)$; and, since we only generate pure programs, we adapted the output to return the results instead of printing them on: checksum, digits, even-squares, for-loop-index, grade, pig-latin, replace-space-w-nl., string-differences, syllables, and word-stats.

Note that we deliberately placed the patterns with unbound types at the end of the sequence. Therefore, the unbound type in both Accu and Hylo is only decided after all other patterns have failed. For the benchmarks that Origami failed to find a solution with the other patterns, we applied one of these two patterns choosing the type that was known to be correct according to the canonical solutions [2]. For the cases in which the canonical solutions did not use Accu or Hylo, we chose a reasonable type as needed (see Table 2).

Table 1. The complete set of operations available for Origami. Each dataset only had access to the operations that involved its allowed types according to [9].

Operations	Types
`add, sub, mult, div, quot, mod, rem, min, max`	`Int -> Int -> Int`
`abs, succ, pred`	`Int -> Int`
`add, sub, mult, div, min, max`	`Float -> Float -> Float`
`abs, sqrt, sin, cos, succ, pred`	`Float -> Float`
`fromIntegral`	`Int -> Float`
`floor, ceiling, round`	`Float -> Int`
`lt, gt, gte, lte`	`Int -> Int -> Bool`
`lt, gt, gte, lte`	`Float -> Float -> Bool`
`and, or`	`Bool -> Bool -> Bool`
`not`	`Bool -> Bool`
`if`	`Bool -> a -> a -> a`
`eq, neq`	`a -> a -> Bool`
`show`	`a -> [Char]`
`charToInt`	`Char -> Int`
`intToChar`	`Int -> Char`
`isLetter, isSpace, isDigit`	`Char -> Bool`
`length`	`[a] -> Int`
`cons, snoc`	`a -> [a] -> [a]`
`mappend`	`[a] -> [a] -> [a]`
`elem`	`a -> [a] -> Bool`
`delete`	`a -> [a] -> [a]`
`null`	`[a] -> Bool`
`head, last`	`[a] -> a`
`tail, init`	`[a] -> [a]`
`zip`	`[a] -> [b] -> [(a, b)]`
`replicate`	`Int -> a -> [a]`
`enumFromThenTo`	`Int -> Int -> Int -> [Int]`
`reverse`	`[a] -> [a]`
`splitAt`	`Int -> [a] -> ([a], [a])`
`intercalate`	`[a] -> [a] -> [a]`
`fst`	`(a, b) -> a`
`snd`	`(a, b) -> b`
`mkPair`	`a -> b -> (a, b)`
`apply`	`(a -> b) -> a -> b`
`singleton`	`a -> b -> Map a b`
`insert`	`a -> b -> Map a b -> Map a b`
`insertWith`	`((b, b) -> b) -> a -> b -> Map a b -> Map a b`
`fromList`	`[(a, b)] -> Map a b`

Table 2. The chosen types for the unbound types in Accu and Hylo. The type is colored in blue when the decision was guided by the canonical solution.

Dataset	Accu	Hylo
checksum	`Int`	`Int`
collatz-numbers	–	`Int`
digits	–	`Int`
pig-latin	–	`[Char]`
string-differences	`Int`	`(Char, Char)`
sum-of-squares	–	`Int`
vector-average	`(Float, Int)`	`Int`
wallis-pi	–	`Float`
word-stats	`((Int, Int), (Int, Int))`	`[Char]`
x-word-lines	`Int`	`[Char]`

The maximum tree depth was set to 5 for each slot. As Origami is based in HOTGP, which was empirically shown to be robust to changes to the crossover rate, we set it to the same value as HOTGP (50%). We allowed a maximum of 300 000 evaluations with an early stop whenever the algorithm finds a perfectly accurate solution according to the training data. For patterns in which termination is not guaranteed, namely Ana and Hylo, a maximum number of iterations was imposed (empirically set to 10 000). Non-termination is also an issue that can occur with CurriedCata. Specifically, Origami was synthesizing solutions with the slot `alg (Cons i x f) = \ys -> f (f ys)`, essentially creating a "fork bomb". To tackle this, a maximum execution budget is enforced when the evaluation of a single iteration of a slot executes more than 10 000 operations. In this case, the program is assigned an infinitely bad fitness. This limit was reached by less than 0.5% of the individuals.

Table 3 shows the percentage of executions in which Origami was able to synthesize a solution that completely solved the test set (*i.e.*, success rate). Origami found a solution for all of the problems that were canonically solved by NoScheme as well as Cata. Surprisingly, it was also able to synthesize a solution for for-loop-index by using NoScheme, even though the canonical solution used Ana, and for grade by using Cata when the canonical solution used CurriedCata. Nonetheless, we also ran these problems with their canonical patterns and discovered Origami was also able to synthesize solutions, albeit less often. Moreover, Origami was able to find the solutions for 3 out of the 4 canonical CurriedCata problems, and 2 out of the 3 Ana problems. Accu and Hylo, however, appear to be the most difficult patterns to synthesize, as no solution for problems that canonically involve these patterns was found.

Considering the 4 canonical Accu problems, checksum and word-stats are historically hard, with few methods ever finding a solution. The same can be said for Hylo in the wallis-pi and collatz-numbers problems.

Table 3. Success rates obtained by Origami for each pattern in each benchmark. The "Best" column shows the highest success rate for that benchmark across all patterns, which is also underlined. We also show in blue the pattern of the canonical solution.

Dataset	NoScheme	Cata	CurriedCata	Ana	Accu	Hylo	**Best**
checksum	0	0	–	–	0	0	0
collatz-numbers	0	–	–	–	–	0	0
compare-string-lengths	90	–	–	–	–	–	90
count-odds	0	40	–	–	–	–	40
digits	0	–	–	0	–	0	0
double-letters	0	3	–	–	–	–	3
even-squares	0	–	–	3	–	–	3
for-loop-index	90	–	–	67	–	–	90
grade	0	100	10	–	–	–	100
last-index-of-zero	0	70	–	–	–	–	70
median	97	–	–	–	–	–	97
mirror-image	93	–	–	–	–	–	93
negative-to-zero	0	87	–	–	–	–	87
number-io	100	–	–	–	–	–	100
pig-latin	0	0	–	–	–	0	0
replace-space-w-nl.	0	3	–	–	–	–	3
scrabble-score	0	100	–	–	–	–	100
small-or-large	53	–	–	–	–	–	53
smallest	100	–	–	–	–	–	100
string-differences	0	0	0	–	0	0	0
string-lengths-back.	0	97	–	–	–	–	97
sum-of-squares	0	–	–	–	–	0	0
super-anagrams	0	0	73	–	–	–	73
syllables	0	7	–	–	–	–	7
vector-average	0	0	–	–	0	0	0
vectors-summed	0	0	20	–	–	–	20
wallis-pi	0	–	–	–	–	0	0
word-stats	0	0	–	–	0	0	0
x-word-lines	0	0	0	–	0	0	0

In vector-average, the canonical solution involved using Accu to compute both the sum and the count as a pair in the `st` slots, and using the `alg` slots to perform the division as a post-processing step, finally obtaining the average. The solution that got closer to the intended result was the following:

```
accu st alg arg0 (last arg0, length arg0)
  where
    st [] s = []
    st (x : xs) s = x : (xs, s)
    alg [] s = min 0 (last arg0)
    alg (x : acc) s = acc + ((max (x - acc) x) / (snd s))
```

Origami took a different approach from the canonical solution, by storing the length of the input in the second element of the tuple while having no use for the first element. The `st` section had no other purpose than to transmit this pre-processing step to the `alg` section. This solution got a perfect score during training but failed in testing for certain cases. If we were to replace `min 0 (last arg₀)` by `0` and `max (x - acc) x` by `x`, then this solution would be correct.

The Hylo solution for sum-of-squares employed `coalg` to generate a list of all the numbers from 0 to $\texttt{arg}_0$, and then used `alg` to square each number and accumulate the sum. Even though this was the simplest Hylo solution, as Hylo has 5 different slots, it has an increased search space in relation to other patterns, which seems to be a big challenge for the algorithm.

Table 4 compares Origami's results to HOTGP's. There was a substantial increase (> 30) in the success rate in 6 problems. In the 17 problems where the absolute difference is < 30, we highlight syllables, double-letters and even-squares problems, as those were problems for which HOTGP was not able to synthesize a solution, whereas Origami was successful at least once. The two problems with a more noticeable decrease are replace-space-w-nl. and vector-average. These can be explained by the change in grammar between the two algorithms, as HOTGP's solutions were arguably simpler due to having `map` and `filter` for replace-space-w-nl. and `sum` for vector-average. In a practical scenario, the inclusion of these functions would likely lead to a correct solution but, as previously noted, removing them was a conscious decision to enable the proper assessment of the impact of Recursion Schemes in PS. It would also allow for composite solutions, such as using Ana with a `map` inside instead of relying on Hylo to find the entire pattern, which might be easier to synthesize.

Table 4. Origami's success rates compared to HOTGP's on solved problems. The Δ column shows the relative success rate of Origami with respect to HOTGP.

Dataset	Origami	HOTGP	Δ	Dataset	Origami	HOTGP	Δ
scrabble-score	100	0	100	number-io	100	100	0
mirror-image	93	1	92	sum-of-squares	0	1	−1
super-anagrams	73	0	73	median	97	99	−2
last-index-of-zero	70	0	70	small-or-large	53	59	−6
grade	100	37	63	compare-string-lengths	90	100	−10
for-loop-index	90	59	31	count-odds	40	50	-10
string-lengths-back.	97	89	8	negative-to-zero	87	100	−13
syllables	7	0	7	vectors-summed	20	37	-17
double-letters	3	0	3	replace-space-w-nl.	3	38	−35
even-squares	3	0	3	vector-average	0	80	−80
smallest	100	100	0				

To assess how Origami fares with relation to the best methods found in the literature, we compare its results to those obtained by PushGP [9], Grammar-Guided Genetic Programming (G3P) [3], the extended grammar version of G3P (here called G3P+) [4], Code Building Genetic Programming (CBGP) [14], G3P with Haskell and Python grammars (G3Phs and G3Ppy) [5], as well as with Github Copilot [16]. In some of those works, only a subset of the problems was chosen, often avoiding the most difficult ones (*i.e.*, not previously solved by any other method), or problems not solvable by the proposed method itself. The results are reported in Table 5 ("–" indicates the authors did not test their method on that specific problem), and summarized in Table 6.

Table 5. Success rate with the best values underlined. The last row displays the ratio of victories of each algorithm against Origami by the amount of tested problems.

Dataset	Origami	HOTGP	DSLS	UMAD	PushGP	CBGP	G3P	G3P+	G3Phs	G3Ppy	Copilot
checksum	0	0	18	5	0	–	0	0	–	–	89
collatz-numbers	0	–	–	–	0	–	0	0	–	–	73
compare-string-lengths	90	100	51	42	7	22	2	0	5	0	70
count-odds	40	50	11	12	8	0	12	3	–	–	98
digits	0	0	28	11	7	0	0	0	–	–	0
double-letters	3	0	50	20	6	–	0	0	–	–	88
even-squares	3	0	2	0	2	–	1	0	–	–	11
for-loop-index	90	59	5	1	1	0	8	6	–	–	72
grade	100	37	2	0	4	–	31	31	–	–	84
last-index-of-zero	70	0	65	56	21	10	22	44	0	2	61
median	97	99	69	48	45	98	79	59	96	21	79
mirror-image	93	1	99	100	78	100	0	25	–	–	70
negative-to-zero	87	100	82	82	45	99	63	13	0	66	99
number-io	100	100	99	100	98	100	94	83	99	100	93
pig-latin	0	0	0	0	0	–	0	3	–	–	54
replace-space-w-nl.	3	38	100	87	51	0	0	16	–	–	87
scrabble-score	100	0	31	20	2	–	2	1	–	–	35
small-or-large	53	59	22	4	5	0	7	9	4	0	51
smallest	100	100	98	100	81	100	94	73	100	89	66
string-lengths-back.	97	89	95	86	66	–	68	18	0	34	60
sum-of-squares	0	1	25	26	6	–	3	5	–	–	90
super-anagrams	73	0	4	0	0	–	21	0	5	38	55
syllables	7	0	64	48	18	–	0	39	–	–	96
vector-average	0	80	97	92	16	88	5	0	4	0	92
vectors-summed	20	37	21	9	1	100	91	21	68	0	87
x-word-lines	0	0	91	59	8	–	0	0	–	–	1
Win ratio vs. Origami		$\frac{9}{25}$	$\frac{10}{25}$	$\frac{9}{25}$	$\frac{7}{29}$	$\frac{5}{14}$	$\frac{3}{29}$	$\frac{5}{29}$	$\frac{2}{11}$	$\frac{0}{11}$	$\frac{14}{29}$

Table 6. Number of problems with a success rate ≥ a certain % for each method.

	Origami	HOTGP	DSLS	UMAD	PushGP	CBGP	G3P	G3P+	G3Phs	G3Ppy	Copilot
= 100%	4	4	1	3	0	4	0	0	1	1	0
≥ 75%	10	7	8	7	3	7	4	1	3	2	12
≥ 50%	13	10	13	9	5	7	6	3	4	3	22
≥ 25%	14	13	16	13	7	7	7	7	4	5	23
> 0%	19	15	24	21	22	9	17	17	8	7	26

Among the other GP algorithms, the best performer (DSLS) achieves a higher success rate than Origami in only 10 problems. Considering that they both get a 0 in pig-latin, Origami outperforms DSLS in 14 problems. Notably, Origami frequently outperforms CBGP and the G3P variants. It also has the highest number of problems solved with 100%, ≥ 75%, and ≥ 50%, and is second-place in ≥ 25%. When we consider problems to which Origami found at least one solution, we note that it outperforms HOTGP, CBGP, and all the G3P variations, placing Origami at the fourth place. It is also worth noting that Origami outperforms HOTGP in both the number of best results and amount of problems above all thresholds, which demonstrates it is a substantial improvement over HOTGP.

We also compare with the results obtained using Copilot on PSB1, as reported by [16]. In that paper, the authors tested Copilot with a different formulation of the program synthesis problem: instead of receiving example of input-outputs, they made the problem description and the function signature available to Copilot. This input format can be more difficult to process, as it requires the extraction of useful information from a textual description, but contains additional information that may be implicit in the input-output format. Out of the 29 problems, Copilot had better results in 14 of them, and Origami is equivalent or outperforms Copilot in 15 problems. While Copilot solves more problems (at least once) than Origami, it struggles with consistency, and does not achieve 100% success rate on any of the tested problems.

5 Conclusion

This work is the first full implementation of Origami, a GP algorithm proposed in [2], and builds on its previous work, HOTGP [1]. Origami's main differential is the use of Recursion Schemes, well-known constructs in functional programming that enable recursive algorithms to be defined in a unified manner. The main motivation for using these in the PS context is enabling recursive programs to be synthesized in a controlled manner, without sacrificing expressiveness.

We evaluate our approach in the 29 problems in the PSB1 dataset, which is known to be solvable by just a handful of Recursion Schemes. In general, Origami performs better than other similar methods, synthesizing the correct solution more often than other methods in most problems. It was also able to obtain the highest count of problems with success rate = 100%, ≥ 75% and ≥ 50% among the GP methods. Furthermore, Origami achieved comparable results to Github Copilot, solving some problems that the LLM achieved 0% score. We should stress that the problem formulation is different for both approaches, indicating that combining LLM with GP and Recursion Schemes could be beneficial to improve the results. These experimental results suggest that using Recursion Schemes to guide the search is a promising research avenue. Currently, the main challenge of Origami appears to be dealing with harder Recursion Schemes, such as Accumulation and Hylomorphism. Different evolutionary mechanisms, such as other selection methods and mutation/crossover operators, should be evaluated in this context to understand if they can positively impact the search process.

Acknowledgments. This research was partially funded by Fundação de Amparo à Pesquisa do Estado de São Paulo (FAPESP), grant numbers #2021/12706-1, #2019/26702-8, #2021/06867-2, #2023/00811-0, #2024/01115-0, CNPq #301596/2022-0, CNPq-AWS #421787/2022-8 and Coordenação de Aperfeiçoamento de Pessoal de Nível Superior - Brasil (CAPES).

References

1. Fernandes, M.C., De França, F.O., Francesquini, E.: HOTGP - Higher-Order Typed Genetic Programming. In: Proceedings of the Genetic and Evolutionary Computation Conference, GECCO 2023, pp. 1091-1099. Association for Computing Machinery, New York (2023). https://doi.org/10.1145/3583131.3590464
2. Fernandes, M.C., de Franca, F.O., Francesquini, E.: Origami: (un)folding the Abstraction of Recursion Schemes for Program Synthesis, pp. 263–281. Springer Nature Singapore, Singapore (2024). https://doi.org/10.1007/978-981-99-8413-8_14
3. Forstenlechner, S., Fagan, D., Nicolau, M., O'Neill, M.: A grammar design pattern for arbitrary program synthesis problems in genetic programming. In: McDermott, J., Castelli, M., Sekanina, L., Haasdijk, E., García-Sánchez, P. (eds.) EuroGP 2017. LNCS, vol. 10196, pp. 262–277. Springer, Cham (2017). https://doi.org/10.1007/978-3-319-55696-3_17
4. Forstenlechner, S., Fagan, D., Nicolau, M., O'Neill, M.: Extending program synthesis grammars for grammar-guided genetic programming. In: Auger, A., Fonseca, C.M., Lourenço, N., Machado, P., Paquete, L., Whitley, D. (eds.) PPSN 2018. LNCS, vol. 11101, pp. 197–208. Springer, Cham (2018). https://doi.org/10.1007/978-3-319-99253-2_16
5. Garrow, F., Lones, M.A., Stewart, R.: Why functional program synthesis matters (in the realm of genetic programming). In: Proceedings of the Genetic and Evolutionary Computation Conference Companion, GECCO 2022, pp. 1844–1853. Association for Computing Machinery, New York (2022). https://doi.org/10.1145/3520304.3534045
6. Gulwani, S.: Dimensions in program synthesis. In: Proceedings of the 12th International ACM SIGPLAN Symposium on Principles and Practice of Declarative Programming, PPDP 2010, pp. 13–24. Association for Computing Machinery, New York (2010). https://doi.org/10.1145/1836089.1836091, https://doi.org/10.1145/1836089.1836091

7. Helmuth, T., McPhee, N.F., Pantridge, E., Spector, L.: Improving generalization of evolved programs through automatic simplification. In: Proceedings of the Genetic and Evolutionary Computation Conference, GECCO 2017, pp. 937–944. Association for Computing Machinery, New York(2017). https://doi.org/10.1145/3071178.3071330
8. Helmuth, T., McPhee, N.F., Spector, L.: Program synthesis using uniform mutation by addition and deletion. In: Proceedings of the Genetic and Evolutionary Computation Conference, pp. 1127–1134 (2018)
9. Helmuth, T., Spector, L.: General program synthesis benchmark suite. In: Proceedings of the 2015 Annual Conference on Genetic and Evolutionary Computation, pp. 1039–1046 (2015)
10. Koza, J.R.: Genetic Programming: On the Programming of Computers by Means of Natural Selection. Bradford Books, Cambridge, MA (Dec, Complex Adaptive Systems (1992)
11. Meijer, E., Fokkinga, M., Paterson, R.: Functional programming with bananas, lenses, envelopes and barbed wire. In: Hughes, J. (ed.) FPCA 1991. LNCS, vol. 523, pp. 124–144. Springer, Heidelberg (1991). https://doi.org/10.1007/3540543961_7
12. Montana, D.J.: Strongly typed genetic programming. Evol. Comput. **3**(2), 199–230 (1995)
13. O'Neill, M., Ryan, C.: Grammatical evolution. IEEE Trans. Evol. Comput. **5**(4), 349–358 (2001)
14. Pantridge, E., Helmuth, T., Spector, L.: Functional code building genetic programming. In: Proceedings of the Genetic and Evolutionary Computation Conference, GECCO 2022, pp. 1000–1008. Association for Computing Machinery, New York (2022). https://doi.org/10.1145/3512290.3528866
15. Pantridge, E., Spector, L.: Code building genetic programming. In: Proceedings of the 2020 Genetic and Evolutionary Computation Conference, pp. 994–1002 (2020)
16. Sobania, D., Petke, J., Briesch, M., Rothlauf, F.: A comparison of large language models and genetic programming for program synthesis. IEEE Trans. Evolutionary Comput. (2024). https://doi.org/10.1109/TEVC.2024.3410873
17. Spector, L., Robinson, A.: Genetic programming and autoconstructive evolution with the push programming language. Genet. Program Evolvable Mach. **3**, 7–40 (2002)
18. Swan, J., Krawiec, K., Kocsis, Z.A.: Stochastic synthesis of recursive functions made easy with bananas, lenses, envelopes and barbed wire. Genetic Program. Evolvable Mach. **20**(3), 327–350 (2019). https://doi.org/10.1007/s10710-019-09347-3

GovBERT-BR: A BERT-Based Language Model for Brazilian Portuguese Governmental Data

Mariana O. Silva, Gabriel P. Oliveira, Lucas G. L. Costa, and Gisele L. Pappa(✉)

Universidade Federal de Minas Gerais (UFMG), Belo Horizonte, MG, Brazil
{mariana.santos,gabrielpoliveira,glpappa}@dcc.ufmg.br, lucas-lage@ufmg.br

Abstract. Given the growing interest in natural language processing (NLP) for governmental applications, particularly in Brazil, where vast amounts of governmental data are processed daily, the need for specialized NLP models tailored to the nuances of Brazilian Portuguese and the legal and administrative domains has become increasingly apparent. However, existing models may struggle to accurately interpret the complexities of governmental texts, often leading to suboptimal performance in document classification and analysis tasks. To address these challenges, we introduce GovBERT-BR, a pre-trained language model tailored to the Brazilian governmental context, covering legal and administrative domains. Leveraging insights from diverse governmental texts, GovBERT-BR addresses the challenges of accurately interpreting Brazilian Portuguese and the unique legal and bureaucratic terminology prevalent in governmental documents. We present the pre-training process and experimental evaluation of GovBERT-BR, comparing its performance against baseline models across various text classification tasks relevant to the Brazilian public sector. Our findings demonstrate that GovBERT-BR outperforms existing models in document and short-text classification tasks, showcasing its efficacy in accurately analyzing governmental text data. Furthermore, our analysis reveals insights into the convergence behavior of GovBERT-BR during fine-tuning, highlighting its rapid adaptation to downstream tasks.

Keywords: Natural Language Processing · Brazilian Governmental Data · Legal Texts · BERT · Language Models

1 Introduction

In recent years, the application of natural language processing (NLP) tools to government data has surged significantly. This trend is largely driven by the promise of NLP to enhance public administration's efficiency, transparency, and accountability. In Brazil, where a vast volume of governmental data is generated and processed daily [4], NLP tools can streamline various administrative processes by automating the analysis and interpretation of textual data, thereby facilitating improved decision-making [8,10,20].

A. Paes and F. A. N. Verri (Eds.): BRACIS 2024, LNAI 15413, pp. 19–32, 2025.
https://doi.org/10.1007/978-3-031-79032-4_2

However, the effective utilization of this data is often constrained by the need for specialized NLP models capable of accurately interpreting the specificities of Brazilian Portuguese and the unique legal and bureaucratic terminology. Existing NLP models are predominantly generic [19], trained on datasets that do not adequately reflect the particularities of Brazilian governmental documents. This mismatch usually leads to suboptimal performance when these models are applied to the Brazilian public sector [20].

In this context, some efforts have been made to develop domain-specific models focused on tackling legal [8,18,20] and administrative [10] tasks. Although these endeavors represent important steps forward, they are often limited in scope, addressing only specific sub-contexts of governmental data. The interdisciplinary nature of governmental documents requires models capable of seamlessly integrating insights from multiple domains. Legal documents, for example, frequently intersect with administrative procedures, and understanding these interrelations is crucial for effective decision-making within public administration.

Therefore, we introduce GovBERT-BR, a pre-trained language model specifically tailored to the Brazilian governmental context. It integrates both administrative and legal domains, providing a multifaceted perspective to better align with the diverse nature of governmental text data. In this paper, we present the pre-training process and the experimental setup of GovBERT-BR. Our evaluation involves a series of text classification tasks relevant to the Brazilian governmental context, comparing the performance of GovBERT-BR against several baseline models. Our main contributions are summarized as follows.

- We introduce a novel pre-trained language model designed explicitly for the Brazilian governmental domain, incorporating insights from both administrative and legal domains to enhance its adaptability and effectiveness.
- We provide a comprehensive evaluation framework, including a diverse set of text classification tasks pertinent to the Brazilian governmental context. This allows for a thorough assessment of GovBERT-BR's performance across different types of governmental text data.
- Our experimental results demonstrate GovBERT-BR's superior or competitive performance compared to baseline models across various datasets. This underscores its efficacy in accurately classifying governmental text, highlighting its potential for improving text understanding and decision-making within the Brazilian public sector.

2 Related Work

Using language models in government applications has changed how governments interact with citizens and manage information. Indeed, such models offer advanced capabilities in NLP, enabling automation in understanding public inquiries, analyzing policies, and other applications using Open Government Data (OGD) [7,14]. They can be applied in several areas, including consulting services [9], topic modeling [10], and sentiment analysis related to government policies [23]. In addition, language is crucial in these models, as the model's language must align with the languages used in the target domain to ensure more efficient adaptation. Therefore, recent works have proposed distinct models to tackle government-related tasks in several languages [1,3,11,15,22].

In the Brazilian context, most models that address government tasks are built from BERTimbau, a general-purpose model pre-trained on a large corpus of Portuguese texts scraped from the Web [19]. Such government-related models can be divided into two distinct categories according to their purpose: legal domain and administrative domain models. The first group includes models focusing mainly on juridical and legislative tasks, including texts from the Brazilian legal system into their pre-training phase to handle legal language and concepts effectively.

For example, LegalBERTPT-br [18] is a model for mining topics within comments on Brazilian draft laws. LegalBert-pt [20], JurisBERT [21] and CLSJUR.BR [12] use texts exclusively from the Brazilian judiciary branch for tasks such as named entity recognition (NER), semantic textual similarity (STS), and document summarization. Finally, RoBERTaLexPT [8] uses a large corpus from the judiciary and legislative bodies over NER and classification tasks. The main difference between such models is that RoBERTaLexPT's corpus contains texts in both European and Brazilian varieties of the Portuguese language, whereas the others consider only texts in Brazilian Portuguese.

On the other hand, administrative domain models are designed to support various governmental functions outside the legal sphere, such as managing bidding processes and processing information from the official gazettes. The development of solutions for this specific domain is relatively recent, with few models dedicated exclusively to this context. One notable example is LiBERT-SE [10], a model developed using a corpus of Brazilian official gazette segments and evaluated in topic detection tasks within bidding documents.

In light of the aforementioned related work, our goal is to propose a model that addresses the legal and administrative domains in Brazilian Portuguese. GovBERT-BR bridges the gap between these two domains, providing a comprehensive solution for handling diverse governmental text data. By integrating insights from both legal and administrative contexts, GovBERT-BR aims to enhance the efficiency and effectiveness of NLP applications within the Brazilian public sector, ultimately contributing to improved decision-making.

3 GovBERT-BR

Governmental data is a rich and diverse source of information that contains many documents, including public bidding documents, legal texts, legislative bills, administrative reports, and more. However, the complexity and specialized nature of this data pose significant challenges for generic NLP models, which often fail to capture the nuances and specific terminology used in governmental texts. Moreover, existing domain-specific models tend to focus on narrow subcontexts, failing to address the interdisciplinary nature inherent in governmental documents comprehensively.

To overcome these challenges, we introduce GovBERT-BR, a pre-trained language model specifically tailored to the Brazilian governmental context.[1] It considers both legal and administrative domains, providing a multifaceted perspective that can better align with the diverse nature of governmental text data. In this section, we describe the pre-trained model used as the foundation for GovBERT-BR (Sect. 3.1) and detail the methodologies employed in its development (Sect. 3.2).

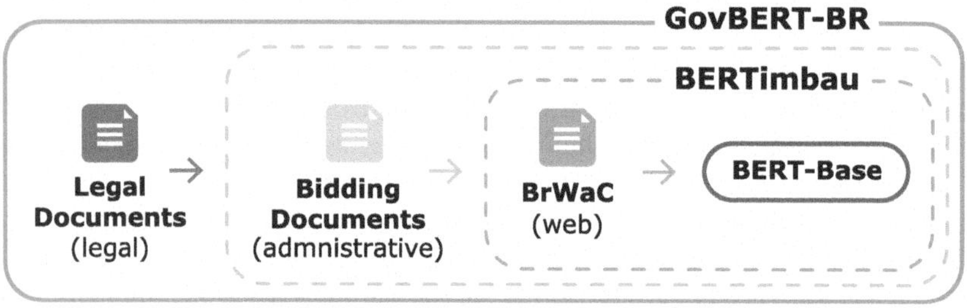

Fig. 1. Overview of GovBERT-BR pre-training methodology

3.1 Pre-trained Model

GovBERT-BR is built upon the architecture of BERTimbau [19], the first large-scale Portuguese language model. BERTimbau, based on the BERT (Bidirectional Encoder Representations from Transformers) [6] architecture, has been fine-tuned on various tasks to enhance its performance in understanding Portuguese text. GovBERT-BR inherits this architecture and extends it to focus specifically on the nuances of Brazilian governmental and legal documents.

3.2 Pre-training

The GovBERT-BR pre-training process involves further training BERTimbau using two distinct corpora, each tailored to capture the intricacies of governmental and legal texts in Brazilian Portuguese. We believe such a multi-domain approach allows GovBERT-BR to provide a comprehensive perspective that captures the interplay between legal and administrative language, offering a powerful resource for interpreting Brazilian governmental documents.

Figure 1 shows the overview of the GovBERT-BR pre-training methodology, which involves a two-phase approach. Initially, the model is pre-trained using data from the administrative domain to equip the model with an understanding of bureaucratic terminology and administrative procedures commonly found in governmental texts. Next, the model undergoes further pre-training using a dataset comprised of legal documents related to extraordinary appeals received by the Brazilian Supreme Court (STF).

Administrative Data. The initial pre-training corpus includes public bidding documents and segments from the official gazette, which contain summarized

[1] Available at https://huggingface.co/dccmpmgfinalisticas/GovBERT-BR.

Table 1. Overview of the baseline models

Model	Domain	Corpus	Size
[19] BERTimbau	General	brWac	3.5M
[10] LiBERT-SE	Admin	Official Gazette Segments	300K
[20] LegalBERT-PT	Legal	Legal Documents	1.5M
GovBERT-BR	Admin Legal	Public Bidding Documents Official Gazette Segments VICTOR dataset	3.4M

texts published to provide information regarding the bidding process [5]. It comprises 1,578,107 sentences and encompasses various document types and administrative data, including announcements, notices, and bidding terms.

Legal Data. The second pre-training corpus includes legal documents related to extraordinary appeals received by the STF. It contains a subset of 1,760,862 pages sourced from the VICTOR dataset [13].

Task. The pre-training task employed for GovBERT-BR is the Masked Language Model (MLM) task. In this task, a percentage of words in a sequence (15%) are masked, and the model is trained to predict these masked tokens based on the surrounding context. This method enables the model to learn contextual relationships and develop a deep understanding of the language patterns specific to the domain.

Setup. Each pre-training session is limited to 10 epochs, with checkpoints saved at the end of each epoch. We use a batch size of 64 sequences, each containing a maximum of 300 tokens. Both pre-training steps involved over 245,000 steps, ensuring thorough exposure to the diverse dataset.

4 Experimental Design

To evaluate the effectiveness of the pre-trained GovBERT-BR language model, we conduct a series of experiments across various text classification tasks relevant to the Brazilian governmental context. These experiments aim to assess GovBERT-BR's performance compared to baseline models and existing NLP approaches, demonstrating its ability to analyze and interpret governmental documents accurately. In this section, we describe the baselines considered (Sect. 4.1), the downstream tasks (Sect. 4.2) and datasets (Sect. 4.3), and the evaluation metrics used (Sect. 4.4).

4.1 Baselines

Our main objective is to assess whether integrating legal and administrative pre-training data into GovBERT-BR leads to better performance on governmental text classification tasks. Therefore, we compare our model against several baseline models, including **general-purpose** and **domain-specific** models (Table 1). Regarding the general domain, we consider the BERTimbau-Base model, a widely used general-purpose language model pre-trained on a large-scale Portuguese corpus and used as the basis for the GovBERT-BR model.

In addition to BERTimbau, we include domain-specific models trained on legal and administrative data separately. For the **legal domain**, we consider the LegalBERT-PT model [20], which is also built upon BERTimbau and pre-trained on a vast corpus of legal documents, enabling it to capture the nuances of legal language and terminology. As GovBERT-BR builds upon BERTimbau, comparing both models allows us to evaluate the impact of additional pre-training on legal and administrative texts.

For the **administrative domain**, we consider LiBERT-SE [10], a pre-trained language model tailored for the public bidding context. LiBERT-SE is built upon BERTimbau using the MLM task and a dataset of 300,000 official gazette segments sourced from various municipalities in Minas Gerais, as previously extracted in [5]. As it focuses on administrative language and procedures, such a model is a relevant benchmark for evaluating GovBERT-BR's performance in administrative tasks.

4.2 Downstream Tasks

As downstream tasks, we focus on two distinct text classification tasks within the governmental domain: document classification and short text classification.

Document Classification. This task involves categorizing documents, such as legal texts, legislative bills, administrative reports, and public bidding documents, into predefined classes. Document classification provides a high-level understanding of the content and purpose of each document, facilitating efficient document retrieval and organization within governmental systems.

Short Text Classification. Unlike document classification, this task focuses on classifying shorter text segments, such as item descriptions, summaries, or motions. Short text classification is particularly useful for extracting relevant information from large volumes of textual data, enabling quick decision-making and analysis in governmental processes.

Table 2. Datasets used for the downstream tasks

Ref.	Dataset	Year	Domain	Downstream Task	Size
[17]	LiPSet	2022	Administrative	Document Classification	9,761
-	NaPEx	2024	Administrative	Short text Classification	583,174
-	ProdServ	2024	Administrative	Short text Classification	583,174
[13]	SVic	2020	Legal	Document Classification	339,478
[16]	Motions	2021	Legal	Short text Classification	6,449
[2]	RRIoP	2021	Legal	Short text Classification	10,784

4.3 Datasets

For each domain, three datasets are selected to evaluate the performance of GovBERT-BR and the baseline models. Table 2 summarizes each dataset's key characteristics, further described as follows.

LiPSet [17]. Contains 9,761 labeled public bidding documents from Minas Gerais, Brazil. Such documents are classified into 12 classes: *Ata Registro Preços* (price registration), *Ata Dispensa* (minutes of waiver), *Ata Pregão Presencial* (face-to-face auction), *Outras Atas* (other minutes), *Edital* (public notice), *Contrato* (contract), *Aditamento* (amendment), *Aviso* (notice), *Ratificação* (ratification), *Errata* (erratum), *Homologação* (homologation), and *Outros* (others).

NaPEx. Contains 583,174 labeled public expenditure items from Minas Gerais, Brazil.[2] Each item presents a short text description and is classified into five expenditure nature classes: *Obras* (construction), *Serviços* (services), *Material de Consumo* (consumables), *Material Permanente* (permanent materials), and *Locação* (rent).

ProdServ. Represents the same public expenditure items as NaPEx, but in this dataset, each item description is further classified as either *products* or *services.*

SVic [13]. It is a subset of the VICTOR dataset, containing 94,267 legal documents with 339,478 labeled pages. Instances are labeled as *Acórdão* (lower court decisions under review), *Recurso Extraordinário* (appeal petitions), *Agravo de Recurso Extraordinário* (motions against the appeal petition), *Sentença* (judgments), *Despacho* (court orders), and Others.

Motions [16]. Contains 6,449 legal proceedings, each with an individual and variable number of motions, labeled by lawyers. Legal proceedings refer to formal legal actions or cases brought before a court of law. Motions, in turn, are formal requests made to a judge during legal proceedings, typically seeking a decision or ruling on a specific aspect of the case. Instances in this dataset are labeled as *Arquivado* (archived), *Ativo* (active), or *Suspenso* (suspended), indicating the status of the legal proceedings.

RRIoP [2]. Contains rhetorical annotations within the legal domain, focusing on sentences extracted from judicial sentences from the Court of Justice of Mato Grosso do Sul (TJMS), Brazil. It comprises 10,784 petitions from 70 civil lawsuits filed in TJMS court and was manually labeled with rhetorical roles specifically tailored for petitions. Instances in this dataset are labeled as *Identificação das partes* (identification), *Fatos* (facts), *Argumentos* (arguments), *Fundamentação Legal* (legal basis), *Jurisprudência* (precedents), *Pedidos* (requests), *Valor da causa* (remedy), and Others.

4.4 Evaluation Setup

In our experimental setup, each dataset is partitioned into distinct subsets for training, validation, and testing, maintaining the distribution of label values

[2] https://dadosabertos.tce.mg.gov.br/.

Table 3. Performance comparison on document and short text classification tasks in the **administrative** domain. F1-Macro scores are reported with standard deviations

Model	Document	Short Text	
	LiPSet	NaPEx	ProdServ
BERTimbau	0.902 ± 0.026	0.839 ± 0.003	0.816 ± 0.003
LiBERT-SE	0.937 ± 0.006	0.836 ± 0.002	0.815 ± 0.003
LegalBERT-PT	0.945 ± 0.009	0.831 ± 0.008	0.814 ± 0.004
GovBERT-BR	0.948 ± 0.008	0.839 ± 0.003	0.814 ± 0.004

across partitions. Given that all datasets exhibit class imbalance, we perform the split in a stratified manner by class. This stratified partitioning scheme allocates 70% of the data to training, 20% to validation, and the remaining 10% to testing. However, it is important to note that we do not use cross-validation in this study, which may introduce bias into the results as the model's performance is evaluated on a single split rather than multiple folds. This choice limits the ability to generalize the findings across different data partitions.

Moreover, the class imbalance present across all datasets poses a challenge. While we refrain from using oversampling or undersampling techniques to reflect real-world scenarios, balancing techniques could be applied during training without compromising the integrity of the imbalanced test scenario. Such techniques might help mitigate the impact of class imbalance on model performance, leading to more reliable and interpretable outcomes. We leave the exploration of these balancing strategies for future work, as it could further enhance the robustness of the models in handling skewed data distributions.

All evaluated language models are trained for a fixed number of five epochs without extensive hyperparameter search. Models are trained until convergence on the validation set loss is achieved, ensuring stable performance. Following convergence, or at the last epoch, model performance is evaluated using the F1-Macro metric. To ensure the robustness of our evaluations, we conduct five experiments for each model, varying the random seeds. Next, the results are averaged to produce a more reliable estimation of model performance.

5 Experimental Results

In this section, we present the experimental results obtained from evaluating GovBERT-BR across various text classification tasks relevant to the Brazilian governmental domain. We analyze GovBERT-BR's performance in terms of overall effectiveness (Sect. 5.1), the impact of multi-domain pre-training (Sect. 5.2), and its convergence across different governmental domains (Sect. 5.3).

5.1 Overall Performance

We first assess how GovBERT-BR's performance compares to other state-of-the-art NLP models. Tables 3 and 4 present the results of document and short text

Table 4. Performance comparison on document and short text classification tasks in the **legal** domain. F1-Macro scores are reported with standard deviations

Model	Document	Short Text	
	SVic	Motions	RRIoP
BERTimbau	0.779 ± 0.008	0.782 ± 0.004	0.825 ± 0.010
LiBERT-SE	0.774 ± 0.010	0.777 ± 0.005	0.814 ± 0.006
LegalBERT-PT	0.787 ± 0.007	0.781 ± 0.002	0.839 ± 0.005
GovBERT-BR	0.794 ± 0.012	0.784 ± 0.003	0.838 ± 0.005

classification tasks in the administrative and legal domains, respectively. The statistical significance is reported using the Wilcoxon rank-sum test to compare the distributions of F1-Macro scores. In document classification, GovBERT-BR achieves the highest F1-Macro on the LiPSet dataset, with an average score of 0.948, outperforming significantly the general-purpose BERTimbau ($W = 25$, $p < 0.05$) and the domain-specific LiBERT-SE models ($W = 22$, $p \leq 0.05$). However, regarding LegalBERT-PT, the difference in performance is not statistically significant ($W = 15$, $p > 0.05$).

Similarly, in the SVic dataset, GovBERT-BR achieves an F1-Macro score of 0.794, demonstrating its competitive performance. It significantly outperforms LiBERT-SE ($W = 23$, $p < 0.05$), but the difference in performance compared to BERTimbau ($W = 21$, $p > 0.05$) and LegalBERT-PT ($W = 20$, $p > 0.05$) is not statistically significant. These results demonstrate that pre-training on both legal and administrative data provides GovBERT-BR with a solid understanding of the nuances of Brazilian governmental text, enabling it to perform well across diverse datasets and domains.

Regarding short text classification, GovBERT-BR shows competitive performance across all datasets. It achieves the highest F1-Macro score on the NaPEx dataset, tying with BERTimbau at 0.839, and performs competitively on the Motions and RRIoP datasets, indicating its robustness in handling varied short-text data within governmental contexts. Although LegalBERT-PT slightly outperforms GovBERT-BR on the RRIoP dataset, GovBERT-BR's balanced performance across all datasets highlights its versatility. Finally, our results reveal no statistical difference between all models in the ProdServ dataset.

Overall, GovBERT-BR demonstrates superior or competitive performance compared to baseline models across multiple datasets, underscoring its efficacy in accurately classifying both document-level and short-text data within the governmental domain. Its robust performance highlights its versatility and effectiveness in handling various types of textual governmental data.

5.2 Multi-domain Pre-training

To assess the impact of incorporating both administrative and legal domain pre-training data, we compare GovBERT-BR's performance to models pre-trained

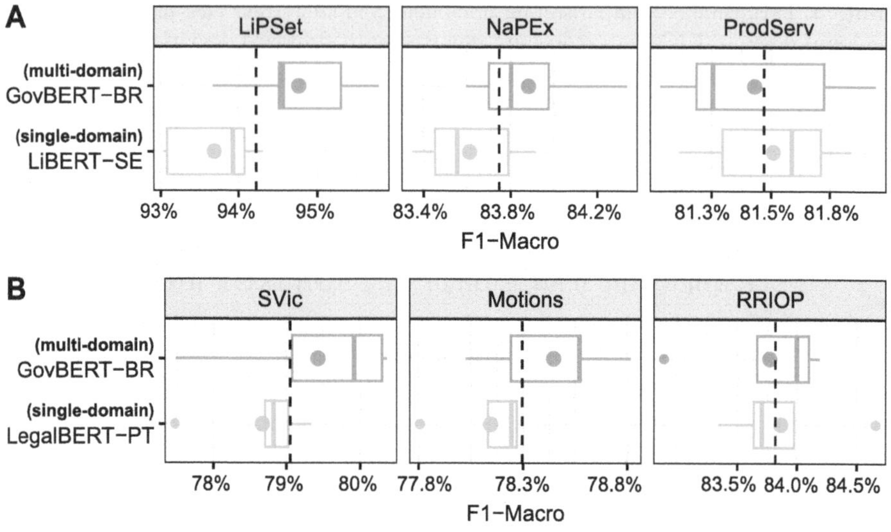

Fig. 2. Distribution of F1-Macro scores on single domains (LiBERT-SE and LegalBERT-PT) and multi-domains (GovBERT-BR) across administrative and legal text classification tasks. (**A**) Administrative and (**B**) Legal domains. The horizontal dashed line represents the mean F1-Macro score for each dataset

on single domains. Figure 2 shows the distribution of F1-Macro scores for models pre-trained on a single domain only (LiBERT-SE and LegalBERT-PT) and multi-domains (GovBERT-BR), for text classification tasks in (**A**) administrative and (**B**) legal domains. The horizontal dashed line represents the mean F1-Macro score for each dataset.

In the administrative domain (Fig. 2**A**), for most datasets, GovBERT-BR achieves a high mean F1-Macro score compared to the single-domain LiBERT-SE model. The exception is the ProdServ dataset, where the performance difference between GovBERT-BR and LiBERT-SE is minimal, indicating that the benefit of multi-domain pre-training may vary depending on the dataset. This suggests that while multi-domain pre-training generally enhances performance, its impact can be less pronounced in certain administrative contexts.

Similarly, in the legal domain (Fig. 2**B**), GovBERT-BR outperforms the single-domain LegalBERT-PT model across most datasets, except for the RRIoP. In this particular dataset, LegalBERT-PT shows a slight edge, possibly due to its specialized training in legal texts, which are more aligned with the content of RRIoP. However, the difference in performance is not statistically significant, indicating that GovBERT-BR's inclusion of administrative domain data does not detract from its legal domain capabilities.

Overall, the results highlight the advantages of multi-domain pre-training. GovBERT-BR's consistent performance across various datasets, covering both administrative and legal texts, demonstrates the robustness and versatility of multi-domain pre-training. This approach leverages a broader contextual understanding, making the model more adaptable to different types of text classification tasks within governmental contexts. This versatility underscores the potential of GovBERT-BR as a valuable resource for interpreting diverse governmental documents, ultimately supporting more efficient and accurate information processing in governmental operations.

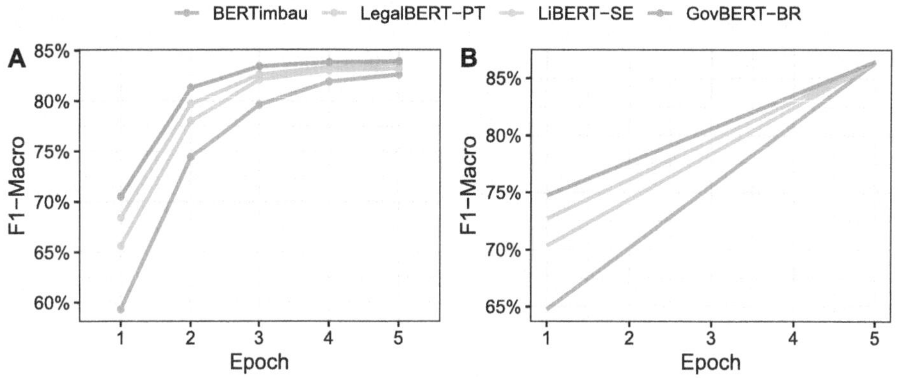

Fig. 3. Convergence analysis. (**A**) Average performance, in terms of F1-Macro, of each model across all datasets, highlighting convergence over training epochs. (**B**) Regression lines of the average performance trajectory during fine-tuning sessions

5.3 Convergence Analysis

To further evaluate the models' performance and stability, we analyzed the F1-Macro scores across epochs during fine-tuning on all downstream tasks. Figure 3**A** presents the average performance, in terms of F1-Macro, of each model across the six datasets, highlighting how each model converges over time. Figure 3**B** shows the regression lines for the average performance of each model across all datasets during fine-tuning sessions. The linear regression lines provide insights into the overall performance improvement or degradation trend throughout the fine-tuning process over epochs.

In Fig. 3**A**, we observe that BERTimbau initially exhibits lower average performance than the domain-specific models, indicating a slower convergence rate in the early epochs. Such a result is expected as the model undergoes significant adjustments in the initial epochs, transitioning from broad pre-training to more specialized tasks. As training progresses, BERTimbau's performance steadily increases, eventually reaching comparable levels to the domain-specific models, albeit with a slower rate of improvement. This suggests that BERTimbau may need more epochs to achieve full adaptation.

Overall, all domain-specific models demonstrate rapid convergence compared to the BERTimbau baseline, with GovBERT-BR notably achieving the highest F1-Macro scores early in the training process. LiBERT-SE and LegalBERT-PT also converge quickly, demonstrating the effectiveness of domain-specific pre-training. This allows the models to quickly adjust to the nuances of the target tasks and achieve high-performance levels early in the training process. However, GovBERT-BR consistently maintains a higher F1-Macro score, suggesting that multi-domain pre-training provides a significant advantage across various governmental text classification tasks.

Figure 3**B** corroborates these observations, providing a straightforward visual representation of the trajectory of performance improvement over epochs. The steeper the slope of the line, the faster the model's performance improves during fine-tuning. Conversely, a shallower slope indicates a slower rate of improvement. Although BERTimbau shows the steepest slope among the models, indicating relatively rapid improvement, it starts from a lower performance level compared to the domain-specific models. Conversely, GovBERT-BR, LiBERT-SE, and LegalBERT-PT, while exhibiting slightly gentler slopes, start from higher initial performance levels, reflecting the benefits of domain-specific pre-training.

6 Conclusions and Future Work

This paper introduces GovBERT-BR, a specialized pre-trained language model tailored explicitly for the Brazilian governmental context, encompassing legal and administrative domains. Through pre-training on diverse governmental texts, GovBERT-BR effectively addresses the challenges of interpreting Brazilian Portuguese and governmental documents' intricate legal bureaucratic terminology. Our experimental evaluation demonstrates GovBERT-BR's superior performance compared to existing models across various document and short-text classification tasks relevant to the Brazilian public sector, highlighting its efficacy in accurately analyzing governmental text data.

Moreover, our analysis of GovBERT-BR's convergence behavior during fine-tuning reveals its rapid adaptation to downstream tasks, highlighting its versatility and effectiveness in handling diverse text classification challenges within governmental contexts. By offering a comprehensive solution for processing governmental text data, GovBERT-BR significantly improves the efficiency and effectiveness of NLP applications within the Brazilian public sector, directly contributing to more informed decision-making and governance processes.

In summary, GovBERT-BR represents a significant advancement in applying natural language processing to the Brazilian governmental context. Our model bridges the gap left by more generic models by addressing the specific linguistic and terminological nuances of Brazilian Portuguese and the distinct domains of legal and administrative documents. Our comprehensive pre-training approach, incorporating diverse governmental texts, ensures that GovBERT-BR can handle the complexities inherent in public administration tasks more effectively than previously available models.

Looking ahead, future research could explore the potential extensions and applications of GovBERT-BR in other domains within the Brazilian governmental landscape. Moreover, ongoing efforts to refine and optimize GovBERT-BR's performance could further enhance its utility and impact in real-world governmental applications. Finally, exploring ways to integrate GovBERT-BR with other existing NLP models or frameworks could lead to synergistic effects, potentially improving its overall performance and expanding its capabilities in addressing complex governmental challenges.

Acknowledgments. This work was funded by the Prosecution Service of State of Minas Gerais (in Portuguese, *Ministério Público do Estado de Minas Gerais*, or simply MPMG) through its Analytical Capabilities Project and by CNPq, CAPES, FAPEMIG and the partnership project between Amazon Web Services (AWS) and CNPq.

References

1. Al-Qurishi, M., AlQaseemi, S., Souissi, R.: Aralegal-bert: a pretrained language model for arabic legal text. In: NLLP@EMNLP, pp. 338–344. Association for Computational Linguistics (2022)
2. Aragy, R., Fernandes, E.R., Caceres, E.N.: Rhetorical role identification for portuguese legal documents. In: Britto, A., Valdivia Delgado, K. (eds.) BRACIS 2021. LNCS (LNAI), vol. 13074, pp. 557–571. Springer, Cham (2021). https://doi.org/10.1007/978-3-030-91699-2_38
3. Bogdanovic, M., Kocic, J., Stoimenov, L.: Srberta - a transformer language model for serbian cyrillic legal texts. Information **15**(2), 74 (2024)
4. Brandão, M.A., et al.: PLUS: a semi-automated pipeline for fraud detection in public bids. Digit. Gov. Res. Pract. **5**(1), 5:1–5:16 (2024). https://doi.org/10.1145/3616396
5. Constantino, K., et al.: Segmentação e Classificação Semântica de Trechos de Diários Oficiais Usando Aprendizado Ativo. In: SBBD, pp. 304–316. SBC (2022).https://doi.org/10.5753/sbbd.2022.224656
6. Devlin, J., Chang, M., Lee, K., Toutanova, K.: BERT: pre-training of deep bidirectional transformers for language understanding. In: NAACL-HLT, pp. 4171–4186. Association for Computational Linguistics (2019).https://doi.org/10.18653/v1/n19-1423
7. Gao, S., Gao, L., Li, Q., Xu, J.: Application of large language model in intelligent Q&A of digital government. In: CNCIT, pp. 24–27. ACM (2023)
8. Garcia, E.A.S., et al.: Robertalexpt: a legal roberta model pretrained with deduplication for portuguese. In: PROPOR, pp. 374–383. ACL (2024)
9. Han, J., Lu, J., Xu, Y., You, J., Wu, B.: Intelligent practices of large language models in digital government services. IEEE Access **12**, 8633–8640 (2024)
10. Hott, H.R., Silva, M.O., Oliveira, G.P., Brandão, M.A., Lacerda, A., Pappa, G.L.: Evaluating contextualized embeddings for topic modeling in public bidding domain. In: BRACIS. LNCS, vol. 14197, pp. 410–426. Springer, Cham (2023). https://doi.org/10.1007/978-3-031-45392-2_27
11. Licari, D., Comandé, G.: ITALIAN-LEGAL-BERT models for improving natural language processing tasks in the italian legal domain. Comput. Law Secur. Rev. **52**, 105908 (2024)
12. Lins, A.A., Carvalho, C.S., das Chagas Jucá Bomfim, F., de Carvalho Bentes, D., Pinheiro, V.: CLSJUR.BR - a model for abstractive summarization of legal documents in portuguese language based on contrastive learning. In: PROPOR, pp. 321–331. ACL (2024)
13. Luz de Araujo, P.H., de Campos, T.E., Braz, F.A., da Silva, N.C.: VICTOR: a dataset for Brazilian legal documents classification. In: LREC, pp. 1449–1458. ELRA (2020)

14. Mamalis, M.E., Kalampokis, E., Karamanou, A., Brimos, P., Tarabanis, K.A.: Can large language models revolutionalize open government data portals? A case of using chatgpt in statistics.gov.scot. In: PCI, pp. 53–59. ACM (2023)
15. Miyazaki, K., Yamada, H., Tokunaga, T.: Cross-domain analysis on japanese legal pretrained language models. In: AACL/IJCNLP (Findings), pp. 274–281. Association for Computational Linguistics (2022)
16. Polo, F.M., Ciochetti, I., Bertolo, E.: Predicting legal proceedings status: approaches based on sequential text data. In: ICAIL, pp. 264–265. ACM (2021). https://doi.org/10.1145/3462757.3466138
17. Silva, M.O., et al.: LiPSet: Um Conjunto de Dados com Documentos Rotulados de Licitações Públicas. In: DSW, pp. 13–24. SBC (2022). https://doi.org/10.5753/dsw.2022.224925
18. Silva, N.F.F., et al.: Evaluating topic models in portuguese political comments about bills from brazil's chamber of deputies. In: Britto, A., Valdivia Delgado, K. (eds.) BRACIS 2021. LNCS (LNAI), vol. 13074, pp. 104–120. Springer, Cham (2021). https://doi.org/10.1007/978-3-030-91699-2_8
19. Souza, F., Nogueira, R., Lotufo, R.: BERTimbau: pretrained BERT models for Brazilian Portuguese. In: Cerri, R., Prati, R.C. (eds.) BRACIS 2020. LNCS (LNAI), vol. 12319, pp. 403–417. Springer, Cham (2020). https://doi.org/10.1007/978-3-030-61377-8_28
20. de V. Silveira, R., Ponte, C., Almeida, V., Pinheiro, V., Furtado, V.: LegalBert-pt: a pretrained language model for the Brazilian Portuguese legal domain. In: BRACIS. LNCS, vol. 14197, pp. 268–282. Springer, Cham (2023). https://doi.org/10.1007/978-3-031-45392-2_18
21. Viegas, C.F.O., Costa, B.C., Ishii, R.P.: Jurisbert: a new approach that converts a classification corpus into an STS one. In: ICCSA. LNCS, vol. 13956, pp. 349–365. Springer, Cham (2023)
22. Xiao, C., Hu, X., Liu, Z., Tu, C., Sun, M.: Lawformer: a pre-trained language model for chinese legal long documents. AI Open **2**, 79–84 (2021)
23. Yulita, I.N., Wijaya, V., Rosadi, R., Sarathan, I., Djuyandi, Y., Prabuwono, A.S.: Analysis of government policy sentiment regarding vacation during the COVID-19 pandemic using the bidirectional encoder representation from transformers (BERT). Data **8**(3), 46 (2023)

Growing Self-Organizing Maps for Multi-label Classification

Pedro Henrique Casarotto[1] and Ricardo Cerri[2(✉)]

[1] Department of Computing, Federal University of São Carlos, São Carlos, SP, Brazil
pedro.casarotto@estudante.ufscar.br
[2] Institute of Mathematics and Computer Science, University of São Paulo, São Carlos, SP, Brazil
cerri@icmc.usp.br

Abstract. In Machine Learning, multi-label classification is the problem of simultaneously classifying an instance into two or more labels. It is a challenging problem since each label has its specialty, and correlations between them must be considered. A Self-Organizing Map (SOM) is a Neural Network where neurons organized in a grid are tuned to represent the input instances in self-organization. After tuning, similar instances in the input space are mapped to closer neurons in the grid. SOMs have already been used for multi-label problems, obtaining competitive results with other methods. However, the static nature of their grid of neurons is a disadvantage since it is difficult to define the optimized grid size for each problem. The Growing Self-Organizing Maps (GSOM) extends the SOMs, allowing the network to grow during execution based on the data characteristics. This paper proposes a GSOM to predict multi-label data. The experiments showed that GSOM obtained better or more competitive results in most of the datasets investigated compared to SOM and had a competitive performance compared to other methods.

Keywords: Growing Self-Organizing Maps · Multi-label Classification

1 Introduction

In the Machine Learning context, the most common classification problem is called single-label. This problem involves classifying a data input in a single class l from a set of classes L, where $|L| > 1$. If the data input is expected to be classified over two or more classes simultaneously, then the problem is considered a multi-label classification problem. In the literature, this problem is presented over diverse areas, such image classification and recognition [8,22], text and document classification [9,12], music classification [33,40], in biomedicine, such in genome classification [39,42] and protein function prediction [29,30].

There are many strategies to deal with multi-label problems, divided into two main approaches. The first is called the Local Approach, in which traditional classification algorithms independently predict each label. The other strategy is

A. Paes and F. A. N. Verri (Eds.): BRACIS 2024, LNAI 15413, pp. 33–48, 2025.
https://doi.org/10.1007/978-3-031-79032-4_3

Global, which considers label correlations by creating specific algorithms to deal with the problem and all labels simultaneously. A hybrid of these two approaches is also possible [19].

Colombini et al. [17] proposed using Self-Organizing Maps [21] to solve the multi-label problems. Self-Organizing Maps (SOM), or Kohonen Maps, is an unsupervised neural network model containing a one-layer grid of neurons. Each neuron has a discriminant function measuring the similarity of the neuron weights to an input instance. In addition, Colombini et al. associated a *prototype vector* to each neuron, which is the mean of the input label vectors of the instances mapped to the neuron. When the model receives a new instance, it is compared with all weights of the neurons of the grid. The neuron that shows more similarity with the instance is selected as the *winner neuron*, and his prototype vector is returned as the prediction for that instance.

The Growing Self-Organizing Maps (GSOM) [6] are an extension of SOMs in which the number of neurons in the grid increases during training. The increase rate is established by the *Growth Threshold*, a parameter that compares the total error of the neuron with the input data. A new neuron is created if the error value exceeds the *Growth Threshold*. The benefit of this adaptation is that the number of neurons is more precise, and thus, the results are better.

In this paper, we propose a method for multi-label classification using Growing Self-Organizing Maps. The method is called Growing Self-Organizing Maps Multi-label Learning (GSOM-MLL), and improves the SOM-MLL proposed in the work of Colombini et al. [17].

The remainder of this paper is organized as follows. Section 2 contains a literature review of the most common algorithms for multi-label problems. Section 3 presents an overview of the Kohonen Maps and introduces the SOM-MLL proposed by Colombini et al. [17]. Our proposal, GSOM-MLL, is presented in Sect. 4, while our methodology is presented in Sect. 5. Section 6 presents and discusses our experiments and results. Finally, Sect. 7 concludes the paper and points to future work.

2 Multi-label Classification

This section presents the two main approaches to classifying multi-label data: Local (Problem Transformation) and Global (Algorithm Adaptation).

2.1 Local Approach

The local approach reduces the original problem into many different single-label problems. Each problem is individually solved, and the corresponding predictions are combined to form the final multi-label classification. The simplest local-based strategy is called Binary-Relevance (BR) [35] in which L binary classifiers are trained, each associated with one of the L labels. One advantage of this strategy is that the classifiers can be run independently from each other (in parallel, for example), and the peculiarities of each label independently influence its corresponding classifier. Another advantage is that traditional algorithms can be used. However, the disadvantage is that this strategy ignores the correlations between labels, a fundamental characteristic of multi-label problems.

Many proposed improvements in the BR algorithm have been explored over the last few years. *Classifier Chains* (CC) [27] modifies BR, adding a dependency mechanism in the labels. This is done by a chain that randomly connects the binary classifiers, considering the previous labels/predictions as input for training a classifier in the chain. It has the advantage of maintaining BR's simplicity while learning the labels' relationships. A similar strategy is Meta Staking (MS) [20], which indirectly considers these relations by adding the number of examples based on the previous classification. In Cherman et al. [15], BR was explored using decision trees and Naïve Bayes to consider label relations. Dembczyński et al. [14] is another approach that also uses the Naïve Bayes algorithm to optimize predictions. The BRkNN [32] is another approach that uses the kNN algorithm conceptually equivalent to BR, with the advantage of being faster.

Another common local-based strategy is called Label-Powerset (LP) [11] in which all classes assigned to each instance are turned into a new and unique class. This is a form to consider the correlations between classes, something BR could not do. However, this method can escalate quickly and create many classes with few positive instances. A modified version called RAndom k-LabELsets (RAKEL) [37] creates combinations of Label-Powerset classifiers iteratively, which considers label correlations without escalating the problem.

Pruned Sets (PS) [26] is a strategy that identifies the relation of the labels and eliminates the ones with less usage, grouping them based on their density and frequency while trying not to lose data. The Ensemble of Pruned Sets (EPS) is also presented as a version of the algorithm that avoids the model overfitting.

2.2 Global Approach

The global approach creates classifiers that take into account label dependency in a direct way, dealing with all labels simultaneously. It uses only one classifier to predict the L labels of the problem. The advantage is that it considers the relationships between labels and is usually faster to execute. However, the approach does not give too much weight to each individual label's peculiarity. Additionally, the algorithm is more difficult to parallelize, bringing more computational cost to the algorithms and spending more training time.

One of the first global-based strategies is C4.5M [16], a decision-tree-based method created by adapting the C4.5 algorithm to multi-label problems. This was done by altering the entropy formula. The algorithm was also changed to represent a set of labels in their tree leaf nodes.

The MLkNN method [41,43] is based on the kNN algorithm where, for each instance, the K closest neighbors classes are compared. Based on this, the instance is labeled. In Zhang and Zhou [42], the Back-propagation Multi-Label Learning was proposed, in which a multi-label error measure is proposed and used in the output layer of an artificial neural network.

In Tsoumakas et al. [34], the Hierarchy Of Multi-label classifiERs (HOMER) is proposed to deal with large label sets, focusing on being effective and computationally efficient. This method uses a "divide and conquer" label hierarchy strategy. It was able to have linear training and logarithmic testing complexity.

Colombini et al. [17] proposed SOM-MLL (Self-Organizing Maps Multi-label Learning), implementing Kohonen Maps to predict multi-label data. This algorithm was also used by Alshanqiti and Namoun [7] to compare the method with other baseline models, like linear regression and matrix factorization. In Saini et al. [28], a different structure of SOM for multi-label data was presented and called ML-SOM, where the label vector is determined using the closest and neighboring neurons, obtaining better results compared to SOM-MLL in some cases.

3 Multi-label Classification with Self-Organizing Maps

This section provides an overview of Self-Organizing Maps (Kohonen Maps) and the Self-Organizing Map Multi-Label Learning (SOM-MLL) [17].

3.1 Self-Organizing Maps

Self-organizing maps (SOM) [21] are unsupervised neural networks with a fixed-size bi-dimensional grid of neurons. When an instance is mapped to the grid, its best-matching neuron is selected as the winning neuron, which strengthens its proximity to the instance. The mapping process is illustrated in Fig. 1.

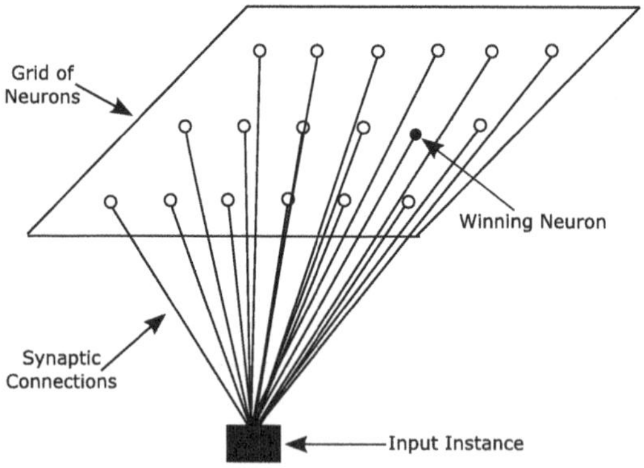

Fig. 1. Mapping an input instance in the Kohonen Map

The training starts by calculating the Euclidean distance between the input instance and all neuron weight vectors. The winning neuron is the one with the minimum distance according to Eq. 1. In this equation, d_j is the distance between the input data $\mathbf{x}$ and the neuron j, with A the dimension of the input instance. Each neuron weight vector is initialized with small random values.

$$d_j(x) = \sqrt{\sum_{i=1}^{A} (x_i - w_{j,i})^2} \tag{1}$$

As the *winner neuron* is defined, the value of its weights is updated based on the input data and the weights of the neurons in its neighborhood, according to Eq. 2. The hyperparameter α is the learning rate of the model, which typically decreases with training time.

$$w_j(t+1) = w_j(t) + \alpha(t) \times h_{j,i} \times (\mathbf{x} - w_j(t)) \tag{2}$$

The $h_{j,i}$ is the influence that the winning neuron j has in its neighborhood of neurons. The most commonly used formula is the Gaussian, which can be seen in Eq. 3. The σ hyperparameter defines the influence's size, i.e., the neighborhood width.

$$h_{j,i} = \exp\left(-\frac{d_j^2(x)}{2\sigma^2}\right) \tag{3}$$

The Self-Organizing Map Multi-label Learning (SOM-MLL) is trained similarly to the common SOM. The difference is that in the prediction process, multiple labels are considered instead of only one label.

3.2 Predicting New Data

In the SOM-MLL, every instance $\mathbf{x}_i$ is associated with a binary vector $\boldsymbol{v}_i$, which has the value 1 in position j if the instance is classified in class c_j, and 0 otherwise. After training, these binary vectors are used to construct a prototype vector $\overline{\boldsymbol{v}}_i$ for each neuron. The prototype of a neuron n is the column-wise average of the matrix formed by all binary label vectors associated with the training instances that had the neuron n as their winning neuron. This calculation is presented in Eq. 4, where $S_{n,j}$ is the set of training instances mapped to the neuron n classified in the class c_j and S_n is the total instances mapped to neuron n.

$$\overline{\mathbf{v}}_{i,j} = \frac{|S_{n,j}|}{|S_n|} \tag{4}$$

The prototype vector represents the probability that the mapped instance x_i will be classified in each of the problem's L labels. A threshold is applied to the prototype vector to obtain a final binary prediction. The common threshold used is 0.5, so that if $\overline{\mathbf{v}}_{i,j}$ is greater or equal to 0.5, x_i is classified in class c_j. Figure 2 shows an example of applying a threshold to a prototype vector.

4 Growing Self-Organizing Maps for Multi-label Classification

This section introduces the Growing Self-Organizing Maps (GSOM) and our proposal for a multi-label classification method called Growing Self-Organizing Maps Multi-Label Learning (GSOM-MLL).

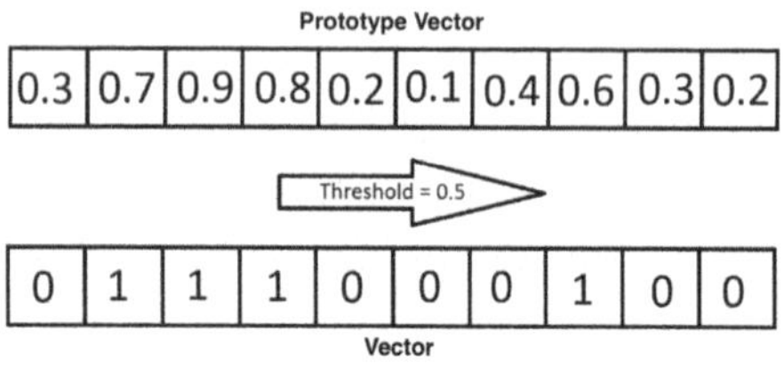

Fig. 2. Obtaining a final prediction by applying a threshold to a prototype vector

4.1 Growing Self-Organizing Maps

The Growing Self-Organizing Maps [6] is an extension of the Kohonen Maps able to increase its number of neurons during training based on the characteristics of the input data. The algorithm has two phases: the Growing phase, where new neurons are added to the map, and the Smooth phase, where the weights of neurons are faceted by their neighbors. The algorithm has a hyperparameter called *Spread Factor* SF, $0 \leq SF \leq 1$, which controls the number of grid nodes. Lower SF values create more populated maps, while higher values create more concentrated maps.

4.2 Growing Phase

The growing phase is responsible for creating new nodes on the map by means of a *Growth Threshold* (GT), a *Spread Factor* (SF), and the dimension D of the training data. This can be seen in Eq. 5.

$$GT = -D \times \log SF \tag{5}$$

The Euclidean Distance (Eq. 1) is used for each input training instance to find the corresponding closest neuron. After this winning neuron i is found, the weights are adjusted for all neurons j of the grid in the Neighborhood Influence ($h_{j,i}$) of neuron i. This adjustment procedure is the same one used in the original SOM (Eqs. 1, 2 and 3).

As an extension of SOMs, each neuron in GSOM has an associated *error value* E used to compute the accumulated error in the neuron. This is shown in Eq. 6, where $\mathbf{x}$ is the input sample, and A is the number of attributes.

$$E_j(t+1) = E_j(t) + \sqrt{\sum_{i=1}^{A} (x_i - w_{j,i})^2} \tag{6}$$

The largest *error value* E is called $H_{E_{rr}}$, and is updated every time a new input instance is given to the model. If $H_{E_{rr}} > GT$, the model creates a new node and sets $H_{E_{rr}} = GT/2$.

4.3 Node Generation

A new node will always be connected to a boundary node, a neuron with at least one of its four immediate neighboring positions free. Figure 3 shows the process of a node generation, being (i) the original grid, (ii) the grid where a neuron has an error larger than GT, and (iii) the two possible positions where a new neuron can be inserted in the grid.

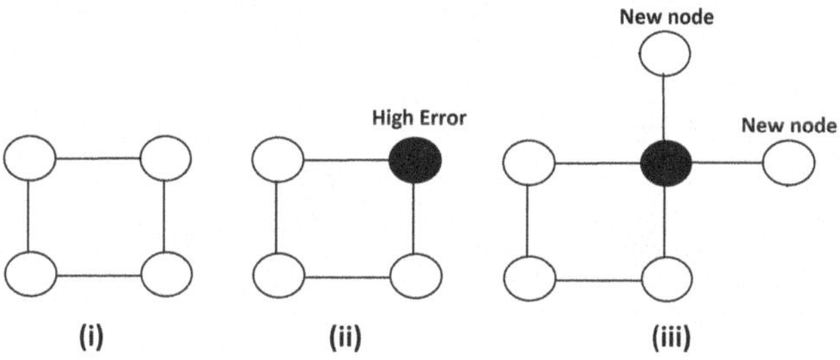

Fig. 3. Insertion of a new neuron in a grid. Adapted from Alahakoon et al. [6]

The weights vector $\mathbf{W}_{new}$ of a new neuron is based on the weight vectors $\mathbf{W}1$ and $\mathbf{W}2$ of its neighbors so that certain regions have similar characteristics. Figure 4 illustrates the four cases that dictate the weight initialization of a new neuron. All comparisons and operations in the four cases are executed individually for each weight of the vectors $\mathbf{W}1$ and $\mathbf{W}2$.

1. The new neuron has two consecutive older nodes on one of its sides (Fig. 4(a)). In this case, the following rules are applied:
   - If $W2 > W1$, then $W_{new} = W1 - (W2 - W1)$;
   - If $W1 > W2$, then $W_{new} = W1 + (W1 - W2)$.
2. The new neuron is in between two older nodes (Fig. 4(b)). In this case:
   - $W_{new} = \frac{W1+W2}{2}$.
3. The new neuron has only one direct neighbor: an older neuron (Fig. 4(c)). In this case, the following rules are applied:
   - If $W2 > W1$, then $W_{new} = W1 - (W2 - W1)$;
   - If $W1 > W2$, then $W_{new} = W1 + (W1 - W2)$.

4. The new neuron has only one direct neighbor, which is an old isolated neuron (Fig. 4(d)). In this case:

   - $W_{new} = (r1 + r2)/2$, where $r1$ and $r2$ are the lower and upper values of the range of the weight vector distribution.

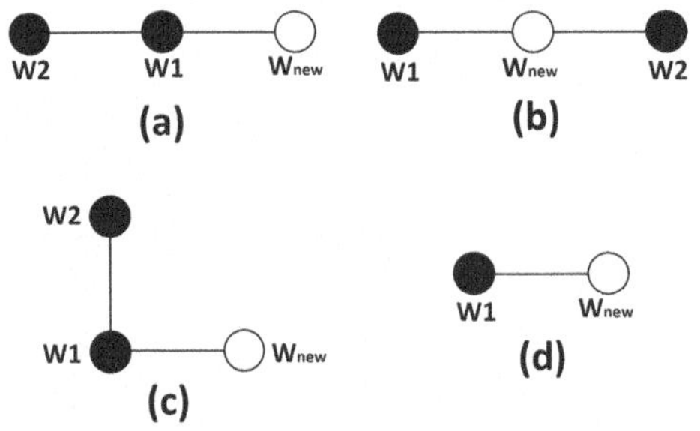

Fig. 4. Initializing weights of a new neuron in a grid. Adapted from Alahakoon et al. [6]

4.4 Smoothing Phase

This phase adjusts the neuron's weights, tuning the grid smoother. At each epoch, the learning rate α is decreased according to Eq. 7.

$$\alpha(e+1) = \alpha \times v(e) \times \alpha(e) \tag{7}$$

In Alahakoon et al. [6], v is defined as shown in Eq. 8, where R is experimentally set to 3.8 in cases of a four nodes initialization grid, and $N(e)$ the number of nodes in the beginning of that epoch.

$$v(e) = 1 - \frac{R}{N(e)} \tag{8}$$

The neighborhood around the winning neuron must also be reduced with training time. This is guaranteed decreasing the neighborhood width according to Eq. 9. In the equation, σ_0 is the neighborhood width in the algorithm's initialization, while τ is a time constant.

$$\sigma(t) = \sigma_0 \exp\left(-\frac{t}{\tau}\right) \tag{9}$$

4.5 Predicting New Data

After finishing training, each neuron is associated with a prototype vector $\overline{\boldsymbol{v}}_i$ calculated the same way as SOM-MLL using Eq. 4. A threshold is then applied to obtain a binary final prediction vector (Fig. 2).

5 Materials and Methods

This section presents the datasets used in the experiments, the baseline algorithms used for comparison, the evaluation measures employed, and the tuning procedure employed to define the hyperparameters of GSOM-MLL.

5.1 Datasets

The datasets used are available in the Mulan Repository[1]. Table 1 shows some characteristics of each dataset: domain, the number of instances, the number of labels and distinct combinations of labels, and the label cardinality and label density. Label Cardinality (LC) is a metric of the average number of labels per instance, while Label Density (LD) is the LC metric over the total number of labels. Equations (10) and (11) show LD and LC, where Y_i is the set of positive labels for instance i, m is the total number of instances, and q is the total number of labels.

Table 1. Characteristics of the Datasets

Name	Domain	#Instances	#Labels	#Distinct	Cardinality	Density
cal500	music	502	174	502	26.044	0.150
birds	audio	645	19	133	1.014	0.053
emotions	music	593	6	27	1.869	0.311
flags	image	194	7	54	3.392	0.485
scene	image	2407	6	15	1.074	0.179
yeast	biology	2417	14	198	4.237	0.303

$$LC = \frac{1}{m} \sum_{i=1}^{m} |Y_i| \tag{10}$$

$$LD = \frac{1}{m} \sum_{i=1}^{m} \frac{|Y_i|}{q} \tag{11}$$

5.2 Baseline Classifiers

The proposed GSOM-MLL was compared with the original SOM-MLL proposed by Colombini et al. [17], where a hexagonal grid of 25 neurons was used, the learning rate was set to decrease linearly from 0.05 to 0.01 at each epoch, and the Gaussian function (Eq. 3) was used as the neighborhood with $\sigma = 2/3$. GSOM-MLL was also compared with different local and global-based methods. They are listed below.

[1] https://mulan.sourceforge.net/datasets-mlc.html.

- Binary Relevance (BR) and Label Powerset (LP) [35], with the following algorithms as base classifiers: Support Vector Machine (SVM) [38], J48 decision tree induction [24] and k-Nearest Neighbors (KNN) [5];
- Random k-Labelsets (Rakel) [37] and Classifiers Chains (CC) [27], both also using SVM, J48 and KNN as base classifiers;
- Hierarchy Of Multi-label classifiERs (HOMER) [23], using LP and BR, with SVM, J48 and KNN as base classifiers;
- Back-Propagation Multi-Label Learning (BPMLL) [42], Multi-Label k-Nearest Neighbors (MLKNN) [43], Instance-Based Learning by Logistic Regression (IBLR-ML and IBLR-ML+) [13], and Predictive Clustering Trees (Clus) [10].

All the baseline classifiers were executed within Mulan, a library for multi-label learning [36], using their default hyperparameter values. The KNN-based methods used $K = 3$. BPMLL was executed with one hidden layer with its number of neurons equal to 80% of the number of input features.

5.3 Evaluation Measures

We evaluated all classifiers using the well-known F1-Score multi-label measure [20], which is the harmonic mean of Precision and Recall. Its calculation is presented in Eq. 12, with Z_i the set of predicted labels, Y_i the set of true labels, and m the number of instances.

$$\textit{F1-Score} = \frac{1}{m}\sum_{i=1}^{m}\frac{2 \times |Y_i \cap Z_i|}{|Y_i| + |Z_i|} \tag{12}$$

5.4 Hyperparameters and Tuning

There are many GSOM implementations available in the literature. Adeu et al. [4], for example, evaluated three of them: PyGSOM Python package [3], GSOM Python package [2] and the GrowingSOM R package [1]. These are all implementations based on Alahakoon et al. [6], and according to Adeu et al., all have different implementation details focusing on specific applications that can slightly change the results. Because of this, they chose to implement their own package[2] focusing on their specific application. In the same direction, we created a new Python package to meet our needs here. GSOM-MLL[3] is an adaptation of Adeu et al. for multi-label classification.

We used the Area Under the Precision-Recall Curve (AUPRC) to find the best hyperparameter values for our method. A Precision-Recall curve is obtained by applying threshold values in the interval $[0, 1]$ to the classifier's outputs. After interpolation all the precision-recall points (PR-points), and posterior connecting them, we can calculate the area under the PR-curve [18].

[2] https://github.com/rodrigosales/GSOM.
[3] https://github.com/PedroHCasarotto/GSOM-MLL.

We performed a grid search to investigate different ranges of values for the hyperparameters. For the initial learning rate, we tested 20 values ranging from 0.005 to 0.195; for the spread factor, we evaluated 20 values ranging from 0.05 to 0.95; and for the smoothing and growing factors, we tested values ranging from 1 to 5 and 1 to 10, respectively. We chose the combination of values that led to the best AUPRC value.

The Gaussian function with $\sigma = 1$ for both growing and smoothing phases was used as the neighborhood influence. After obtaining the best hyperparameter values, the threshold of 0.5 was applied to the prototype vectors in order to obtain the final binary predictions. All grids were initialized with 4 neurons.

6 Experiments and Discussion

All the experiments were performed using the iterative stratified 10-fold cross-validation proposed for multi-label data [31]. Tables 2, 3, and 4 present, respectively, the average F1-score values and standard deviations over the 10 executions for each dataset and method. The best results are highlighted in boldface.

Table 2. Comparing GSOM-MLL with the Strategies BR, LP, and CC

Method	cal500	birds	emotions	flags	scene	yeast
GSOM-MLL	0.32 ± 0.020	0.16 ± 0.060	0.60 ± 0.040	0.67 ± 0.040	0.64 ± 0.030	0.60 ± 0.010
SVM-BR	0.33 ± 0.013	0.65 ± 0.037	0.59 ± 0.056	**0.72 ± 0.037**	0.62 ± 0.037	0.61 ± 0.013
J48-BR	0.34 ± 0.023	0.62 ± 0.033	0.54 ± 0.052	0.71 ± 0.051	0.57 ± 0.030	0.55 ± 0.018
KNN-BR	0.35 ± 0.017	**0.68 ± 0.041**	0.61 ± 0.053	0.67 ± 0.052	0.69 ± 0.033	0.61 ± 0.021
SVM-LP	0.33 ± 0.010	0.67 ± 0.036	**0.67 ± 0.047**	0.69 ± 0.044	**0.75 ± 0.030**	**0.62 ± 0.023**
J48-LP	0.33 ± 0.021	0.58 ± 0.030	0.54 ± 0.053	0.67 ± 0.067	0.61 ± 0.020	0.51 ± 0.017
KNN-LP	0.33 ± 0.015	0.65 ± 0.037	0.60 ± 0.070	0.66 ± 0.058	0.71 ± 0.032	0.57 ± 0.020
CC-SVM	0.31 ± 0.016	0.66 ± 0.028	0.62 ± 0.037	0.70 ± 0.033	0.70 ± 0.035	0.60 ± 0.012
CC-J48	0.35 ± 0.026	0.60 ± 0.034	0.55 ± 0.053	0.69 ± 0.056	0.62 ± 0.035	0.52 ± 0.019
CC-KNN	**0.36 ± 0.025**	0.67 ± 0.053	0.62 ± 0.046	0.68 ± 0.048	0.71 ± 0.029	0.60 ± 0.011

The overall results show that GSOM-MLL is superior to SOM-MLL and competitive with the performances of the other methods investigated. The GSOM-MLL F1-score was superior to SOM-MLL, especially in the dataset flags and scene.

We can observe that GSOM-MLL performed surprisingly badly in the birds dataset. This dataset has the smallest label density, meaning it has a very sparse distribution of labels assigned for each dataset instance. Our tuning procedure may not have been effective in maximizing performance in these circumstances.

Overall, the experiments show us that dynamically increasing the neuron grid dimension is better than working with a fixed grid size. Although the F1-score results were not as satisfactory as expected, GSOM-MLL is competitive with the other methods and is a promising strategy, especially for data visualization,

Table 3. Comparing GSOM-MLL with the Methods RAKEL and HOMER

Method	cal500	birds	emotions	flags	scene	yeast
GSOM-MLL	0.32 ± 0.020	0.16 ± 0.060	0.60 ± 0.040	0.67 ± 0.040	0.64 ± 0.030	0.60 ± 0.010
Rakel-SVM	0.34 ± 0.010	0.66 ± 0.035	**0.67 ± 0.047**	**0.74 ± 0.031**	0.70 ± 0.030	0.63 ± 0.015
Rakel-J48	0.38 ± 0.010	0.62 ± 0.051	0.59 ± 0.056	0.72 ± 0.050	0.65 ± 0.022	0.60 ± 0.011
Rakel-KNN	0.36 ± 0.020	**0.68 ± 0.041**	0.62 ± 0.048	0.69 ± 0.053	0.70 ± 0.031	0.60 ± 0.017
HBR-SVM	**0.45 ± 0.010**	0.67 ± 0.023	0.65 ± 0.038	0.74 ± 0.032	0.60 ± 0.026	0.64 ± 0.015
HBR-J48	0.39 ± 0.010	0.61 ± 0.039	0.53 ± 0.041	0.73 ± 0.040	0.55 ± 0.020	0.57 ± 0.015
HBR-KNN	0.39 ± 0.010	0.66 ± 0.033	0.63 ± 0.057	0.70 ± 0.049	0.71 ± 0.029	0.63 ± 0.018
HLP-SVM	0.44 ± 0.020	0.67 ± 0.027	0.65 ± 0.043	0.72 ± 0.023	0.66 ± 0.018	**0.64 ± 0.014**
HLP-J48	0.33 ± 0.020	0.55 ± 0.029	0.56 ± 0.043	0.69 ± 0.070	0.58 ± 0.022	0.51 ± 0.018
HLP-KNN	0.36 ± 0.010	0.63 ± 0.034	0.61 ± 0.063	0.66 ± 0.045	**0.72 ± 0.026**	0.60 ± 0.017

Table 4. Comparing GSOM-MLL with SOM-MLL, MLKNN, BPMLL, IBLR and Clus

Dataset	cal500	birds	emotions	flags	scene	yeast
GSOM-MLL	0.32 ± 0.020	0.16 ± 0.060	0.60 ± 0.040	0.67 ± 0.040	0.64 ± 0.030	0.60 ± 0.010
SOM-MLL	0.32 ± 0.010	0.56 ± 0.030	0.60 ± 0.060	0.64 ± 0.050	0.52 ± 0.040	0.59 ± 0.010
MLKNN	0.32 ± 0.013	**0.64 ± 0.048**	0.58 ± 0.062	0.70 ± 0.033	**0.65 ± 0.027**	0.62 ± 0.018
BPMLL	**0.46 ± 0.016**	0.40 ± 0.075	**0.66 ± 0.044**	0.69 ± 0.044	0.39 ± 0.109	**0.63 ± 0.025**
IBLR-ML	0.32 ± 0.015	0.63 ± 0.053	0.59 ± 0.057	0.70 ± 0.032	0.64 ± 0.031	0.62 ± 0.014
IBLR-ML+	0.32 ± 0.019	0.53 ± 0.041	0.60 ± 0.051	0.67 ± 0.059	0.62 ± 0.034	0.62 ± 0.015
Clus-Original	0.36 ± 0.023	0.59 ± 0.033	0.57 ± 0.063	0.71 ± 0.043	0.60 ± 0.033	0.53 ± 0.021
Clus-Pruned	0.31 ± 0.012	0.46 ± 0.019	0.57 ± 0.055	**0.72 ± 0.039**	0.61 ± 0.034	0.56 ± 0.018

since we can observe how instances are grouped in the grid. We can have a 2D visualization, keeping the original properties of the input space.

Table 5 presents the mean values for each hyperparameter after the tuning over the 10-fold cross-validation procedure: the number of growing and smooth epochs, the spread factors, the learning rates, the time in seconds to run the model, and the AUPRC values.

Table 5. Mean Values of the GSOM Hyperparameters after Tuning

Dataset	Growing Epoch	Smooth Epoch	Spread Factor	Learning Rate	Time (s)	AUPRC
cal500	1.00 ± 0.00	1.70 ± 0.64	0.05 ± 0.00	0.10 ± 0.03	0.48 ± 0.13	0.48 ± 0.01
birds	4.10 ± 1.37	3.30 ± 1.27	0.95 ± 0.02	0.18 ± 0.02	89.70 ± 40.83	0.31 ± 0.01
emotions	2.40 ± 0.49	2.60 ± 1.69	0.86 ± 0.13	0.10 ± 0.06	8.57 ± 6.70	0.69 ± 0.01
flags	2.20 ± 0.87	2.20 ± 1.25	0.59 ± 0.17	0.14 ± 0.04	0.19 ± 0.12	0.76 ± 0.01
scene	3.67 ± 0.94	3.00 ± 1.41	0.95 ± 0.00	0.14 ± 0.05	549.07 ± 120.61	0.61 ± 0.00
yeast	3.20 ± 1.83	3.00 ± 1.10	0.88 ± 0.15	0.15 ± 0.03	67.13 ± 27.67	0.66 ± 0.01

Table 5 also shows that a high spread factor is preferred for most datasets, indicating a large grid instead of a grid with few neurons. Another observation is that the learning rate is well-suited to most datasets, except for birds, which is very close to the maximum value used during the tuning stage (0.195).

Table 6 shows each dataset's final number of neurons in the grid. The results of the GSOM-MLL experiment suggest that the number of neurons in the grid significantly influenced the outcomes. This is because all the datasets in the original SOM-MLL experiment had a fixed grid of 25 neurons, while in GSOM-MLL experiment, all datasets started with a grid of 4 neurons.

Table 6. Final Number of Neurons in the GSOM-MLL Grid

Dataset	Number of Neurons
cal500	4.00 ± 0.00
birds	68.10 ± 17.06
emotions	31.90 ± 15.92
flags	9.40 ± 3.80
scene	156.20 ± 4.17
yeast	59.40 ± 14.28

Another point to consider is that the time spent on each iteration of the algorithm correlates with the number of neurons on the final grid. This is because more neurons can lead to increased computational costs.

7 Conclusion and Future Work

In this work, we presented GSOM-MLL, a growing self-organizing map tailored for multi-label classification. The model's overall performance proved to be an improvement over the original SOM-MLL. The high values of the spread factor show that the datasets benefit from a higher number of neurons in a dynamic grid. Unfortunately, GSOM-MLL performs poorly in datasets with balanced label cardinality and small label density with many distinct label sets. However, it shows consistent and competitive results in other scenarios.

In future works, we can explore the neighborhood's influence on the creation of the prototype vector. Exploring different learning rate ranges also allows for analyzing the model's performance, as many datasets have their optimal value near the limits tested when considering error rates.

Another way to expand the research would be to utilize Hierarchical Growing Self-Organizing Maps [25]. This involves creating a hierarchy of GSOMs based on a hierarchical threshold as a direct extension of SOM and GSOM.

Acknowledgment. This study was financed by the National Council for Scientific and Technological Development (CNPq) and the São Paulo Research Foundation (FAPESP) grant #2022/02981-8.

References

1. GrowingSOM R package. https://github.com/alexhunziker/growingsom
2. GSOM - the growing self organizing map implementation on Python. https://github.com/anantadata/gsom
3. Pygsom - a GSOM (growing self-organizing map) implementation for Python. https://github.com/philippludwig/pygsom
4. Adeu, R.S., Ferreira, K.R., Andrade, P.R., Santos, L.: Evaluating growing self-organizing maps for satellite image time series clustering. In: GEOINFO, 20 Years After! p. 243 (2019)
5. Aha, D.W., Kibler, D., Albert, M.K.: Instance-based learning algorithms. Mach. Learn. **6**(1), 37–66 (1991)
6. Alahakoon, D., Halgamuge, S.K., Srinivasan, B.: Dynamic self-organizing maps with controlled growth for knowledge discovery. IEEE Trans. Neural Netw. **11**(3), 601–614 (2000)
7. Alshanqiti, A., Namoun, A.: Predicting student performance and its influential factors using hybrid regression and multi-label classification. IEEE Access **8**, 203827–203844 (2020)
8. Baltruschat, I.M., Nickisch, H., Grass, M., Knopp, T., Saalbach, A.: Comparison of deep learning approaches for multi-label chest X-ray classification. Sci. Rep. **9**(1), 1–10 (2019)
9. Banerjee, S., Akkaya, C., Perez-Sorrosal, F., Tsioutsiouliklis, K.: Hierarchical transfer learning for multi-label text classification. In: Proceedings of the 57th Annual Meeting of the Association for Computational Linguistics, pp. 6295–6300 (2019)
10. Blockeel, H., Raedt, L.D., Ramon, J.: Top-down induction of clustering trees. In: Proceedings of the Fifteenth International Conference on Machine Learning, ICML 1998, pp. 55–63. Morgan Kaufmann Publishers Inc., San Francisco, CA, USA (1998)
11. Boutell, M.R., Luo, J., Shen, X., Brown, C.M.: Learning multi-label scene classification. Pattern Recogn. **37**(9), 1757–1771 (2004)
12. Chen, Q., et al.: Multi-label classification for biomedical literature: an overview of the BioCreative VII LitCovid Track for COVID-19 literature topic annotations. Database **2022**, baac069 (2022)
13. Cheng, W., Hüllermeier, E.: Combining instance-based learning and logistic regression for multilabel classification. Mach. Learn. **76**(2–3), 211–225 (2009)
14. Cheng, W., Hüllermeier, E., Dembczynski, K.J.: Bayes optimal multilabel classification via probabilistic classifier chains. In: Proceedings of the 27th International Conference on Machine Learning, ICML-10, pp. 279–286 (2010)
15. Cherman, E.A.: Aprendizado de máquina multirrótulo: explorando a dependência de rótulos e o aprendizado ativo. Ph.D. thesis, Universidade de São Paulo (2013)
16. Clare, A., King, R.D.: Knowledge discovery in multi-label phenotype data. In: De Raedt, L., Siebes, A. (eds.) PKDD 2001. LNCS (LNAI), vol. 2168, pp. 42–53. Springer, Heidelberg (2001). https://doi.org/10.1007/3-540-44794-6_4
17. Colombini, G.G., de Abreu, I.B.M., Cerri, R.: A self-organizing map-based method for multi-label classification. In: 2017 International Joint Conference on Neural Networks (IJCNN), pp. 4291–4298. IEEE (2017)
18. Davis, J., Goadrich, M.: The relationship between precision-recall and ROC curves. In: International Conference on Machine Learning, pp. 233–240 (2006)

19. Gatto, E.C., Valejo, A.D.B., Ferrandin, M., Cerri, R.: Community detection for multi-label classification. In: Naldi, M.C., Bianchi, R.A.C. (eds.) Intelligent Systems, BRACIS 2023. LNCS, vol. 14195, pp. 78–93. Springer, Cham. https://doi.org/10.1007/978-3-031-45368-7_6
20. Godbole, S., Sarawagi, S.: Discriminative methods for multi-labeled classification. In: Dai, H., Srikant, R., Zhang, C. (eds.) PAKDD 2004. LNCS (LNAI), vol. 3056, pp. 22–30. Springer, Heidelberg (2004). https://doi.org/10.1007/978-3-540-24775-3_5
21. Kohonen, T.K.: The self-organizing map. Proc. IEEE **78**(9), 1464–1480 (1990)
22. Li, Q., Peng, X., Qiao, Y., Peng, Q.: Learning category correlations for multi-label image recognition with graph networks. arXiv preprint arXiv:1909.13005 (2019)
23. Papanikolaou, Y., Tsoumakas, G., Katakis, I.: Hierarchical partitioning of the output space in multi-label data. Data Knowl. Eng. **116**, 42–60 (2018)
24. Quinlan, J.R.: C4.5: Programs for Machine Learning. Morgan Kaufmann Publishers Inc., San Francisco, CA, USA (1993)
25. Rauber, A., Merkl, D., Dittenbach, M.: The growing hierarchical self-organizing map: exploratory analysis of high-dimensional data. IEEE Trans. Neural Netw. **13**(6), 1331–1341 (2002)
26. Read, J., Pfahringer, B., Holmes, G.: Multi-label classification using ensembles of pruned sets. In: 2008 8th IEEE International Conference on Data Mining, pp. 995–1000. IEEE (2008)
27. Read, J., Pfahringer, B., Holmes, G., Frank, E.: Classifier chains for multi-label classification. In: Buntine, W., Grobelnik, M., Mladenić, D., Shawe-Taylor, J. (eds.) ECML PKDD 2009. LNCS (LNAI), vol. 5782, pp. 254–269. Springer, Heidelberg (2009). https://doi.org/10.1007/978-3-642-04174-7_17
28. Saini, N., Saha, S., Bhattacharyya, P.: Incorporation of neighborhood concept in enhancing SOM based multi-label classification. In: Deka, B., Maji, P., Mitra, S., Bhattacharyya, D.K., Bora, P.K., Pal, S.K. (eds.) PReMI 2019. LNCS, vol. 11941, pp. 91–99. Springer, Cham (2019). https://doi.org/10.1007/978-3-030-34869-4_11
29. Santos, B.Z., Nakano, F.K., Cerri, R., Vens, C.: Predictive bi-clustering trees for hierarchical multi-label classification. In: Hutter, F., Kersting, K., Lijffijt, J., Valera, I. (eds.) ECML PKDD 2020. LNCS (LNAI), vol. 12459, pp. 701–718. Springer, Cham (2021). https://doi.org/10.1007/978-3-030-67664-3_42
30. Schietgat, L., Vens, C., Struyf, J., Blockeel, H., Kocev, D., Džeroski, S.: Predicting gene function using hierarchical multi-label decision tree ensembles. BMC Bioinf. **11**(1), 1–14 (2010)
31. Sechidis, K., Tsoumakas, G., Vlahavas, I.: On the stratification of multi-label data. In: Gunopulos, D., Hofmann, T., Malerba, D., Vazirgiannis, M. (eds.) ECML PKDD 2011, Part III. LNCS (LNAI), vol. 6913, pp. 145–158. Springer, Heidelberg (2011). https://doi.org/10.1007/978-3-642-23808-6_10
32. Spyromitros, E., Tsoumakas, G., Vlahavas, I.: An empirical study of lazy multilabel classification algorithms. In: Darzentas, J., Vouros, G.A., Vosinakis, S., Arnellos, A. (eds.) SETN 2008. LNCS (LNAI), vol. 5138, pp. 401–406. Springer, Heidelberg (2008). https://doi.org/10.1007/978-3-540-87881-0_40
33. Trohidis, K., Tsoumakas, G., Kalliris, G., Vlahavas, I.P., et al.: Multi-label classification of music into emotions. In: ISMIR, vol. 8, pp. 325–330 (2008)
34. Tsoumakas, G., Katakis, I., Vlahavas, I.: Effective and efficient multilabel classification in domains with large number of labels. In: Proceeding of the ECML/PKDD 2008 Workshop on Mining Multidimensional Data, MMD 2008, vol. 21, pp. 53–59 (2008)

35. Tsoumakas, G., Katakis, I., Vlahavas, I.: Mining multi-label data. In: Data Mining and Knowledge Discovery Handbook, pp. 667–685 (2010)
36. Tsoumakas, G., Spyromitros-Xioufis, E., Vilcek, J., Vlahavas, I.: MULAN: a Java library for multi-label learning. J. Mach. Learn. Res. **12**, 2411–2414 (2011)
37. Tsoumakas, G., Vlahavas, I.: Random *k*-Labelsets: an ensemble method for multilabel classification. In: Kok, J.N., Koronacki, J., Mantaras, R.L., Matwin, S., Mladenič, D., Skowron, A. (eds.) ECML 2007. LNCS (LNAI), vol. 4701, pp. 406–417. Springer, Heidelberg (2007). https://doi.org/10.1007/978-3-540-74958-5_38
38. Vapnik, V.N.: The Nature of Statistical Learning Theory. Information Science and Statistics, Springer, New York (1999). https://doi.org/10.1007/978-1-4757-3264-1
39. Vens, C., Struyf, J., Schietgat, L., Džeroski, S., Blockeel, H.: Decision trees for hierarchical multi-label classification. Mach. Learn. **73**(2), 185–214 (2008)
40. Wu, B., Zhong, E., Horner, A., Yang, Q.: Music emotion recognition by multi-label multi-layer multi-instance multi-view learning. In: Proceedings of the 22nd ACM International Conference on Multimedia, pp. 117–126 (2014)
41. Zhang, M.L., Zhou, Z.H.: A k-nearest neighbor based algorithm for multi-label classification. In: 2005 IEEE International Conference on Granular Computing, vol. 2, pp. 718–721. IEEE (2005)
42. Zhang, M.L., Zhou, Z.H.: Multilabel neural networks with applications to functional genomics and text categorization. IEEE Trans. Knowl. Data Eng. **18**(10), 1338–1351 (2006)
43. Zhang, M.L., Zhou, Z.H.: ML-KNN: a lazy learning approach to multi-label learning. Pattern Recogn. **40**(7), 2038–2048 (2007)

HEACT: Hybrid Evolutionary Algorithm for the Multi-region Multi-objective Cloud Task Scheduling Problem. A Study of Workflow Scheduling in AWS EC2

Vinicius Renan de Carvalho(✉) and Jaime Simão Sichman

Escola Politécnica, Universidade de São Paulo (USP), São Paulo, Brazil
{vrcarvalho,jaime.sichman}@usp.br

Abstract. Cloud computing has revolutionized the provisioning and access of computing resources, offering scalable and flexible alternatives to traditional infrastructure. However, defining how to use these computational resources may be challenging. This paper addresses the challenge of workflow scheduling in cloud environments, focusing on Amazon Web Services (AWS) Elastic Compute Cloud (EC2). We present HEACT, a novel approach that integrates a multi-objective evolutionary algorithm with a specialist scheduling heuristic. The evolutionary algorithm is responsible for generating an initial set of machines (with their performance capability and cost information). The set is sent to the specialist scheduling heuristic for efficient task assignment in these machines. Our approach considers fourteen AWS regions, accurate pricing information from AWS, and employs SimGrid to simulate task execution. The proposed method was benchmarked considering established heuristics (HEFT, PEFT, HSIP, MPEFT) and meta-heuristics (NSGA-II, AGEMOEA2). Results demonstrated that the combinations of AGEMOEA2 with MPEFT and AGEMOEA2 with HEFT yield the best performance, indicating AGEMOEA2's efficacy as a state-of-the-art meta-heuristic for workflow scheduling.

Keywords: Cloud Scheduling · Multi-objective optimization · Evolutionary Algorithms · Scheduling · Heuristics

1 Introduction

Cloud computing changed how computational resources are provisioned and accessed, offering a scalable and flexible alternative to traditional on-premises infrastructure. Moreover, elasticity allows users to dynamically adjust their resource usage based on demand, ensuring that users can scale their applications up or down, paying only for the resources they consume.

Amazon Web Services (AWS) is an Infrastructure as a Service (IaaS) that provides virtualized computing resources over the Internet. In this context, with

A. Paes and F. A. N. Verri (Eds.): BRACIS 2024, LNAI 15413, pp. 49–64, 2025.
https://doi.org/10.1007/978-3-031-79032-4_4

AWS EC2[1], users can provision virtual machines (VMs) on demand according to their requirements. AWS EC2 VMs come in various instance types, from smaller to high-performance computing instances. Their computational power correlates with their capabilities and costs.

Effective resource management and scheduling are important in maximizing the benefits of cloud computing, especially when processing workflows. Workflow scheduling in cloud computing, which is known as NP-complete, consists of orchestrating complex sequences of tasks to be processed. Workflows are commonly represented as directed acyclic graphs (DAGs), where nodes represent tasks and edges denote dependencies between tasks.

Several approaches have been proposed over the years in order to schedule tasks to optimize performance. The list-based scheduling algorithm is the most popular scheduling paradigm for the computation of complex workflow applications [1]. Among them, we can cite HEFT [26], PEFT [2], HSIP [27] and MPEFT [20] (all described in Sect. 2) as heuristic specialized in assigning tasks to processors (or machines) focusing on reducing the total time for executing all tasks. This total time is named as *makespan*.

Over the years, genetic algorithms (GA) were also considered for scheduling tasks in this domain. In [15], a single objective genetic algorithm combined with HEFT. In this approach, the GA considers single-point crossover and mutation operators. The chromosome was defined as an ordered list of tasks. As commonly performed in heuristics for this problem, a fixed set of processors was considered.

Nasonov et al. [17] proposed Linewise Earliest Finish Time (LEFT) by combining the HEFT heuristic and the genetic algorithm to optimize the makespan for cloud workload optimization.

In [24], the authors introduced two hybrid heuristics based on genetic algorithms for task scheduling. Both algorithms enhance the initial population of the GA by incorporating guided chromosomes using novel prioritization equations and improved mutation strategies. Similar to [15], these algorithms consider as chromosomes an ordered list of tasks, use two-point crossover and proposed mutation, use makespan as a single-objective function, and consider a specialist heuristic, in that case, HLTSD [25].

These mentioned works are, in fact, effective. However, an overall literature review reveals an evident lack of workflow scheduling algorithms that optimize equally important makespan and monetary cost functions [13]. Despite of that, Durillo et al. [11] introduced MOHEFT, a multi-objective workflow scheduling algorithm based on HEFT. MOHEFT aims to optimize two key objectives simultaneously: minimizing makespan and enhancing energy efficiency.

However, there are significant challenges in directly applying these algorithms to Cloud environments because most of them are still based on traditional heterogeneous environments such as Grid [30].

In order to tackle this challenge, Zhu et al. [30] proposed using an evolutionary multi-objective optimization algorithm to address the Cloud workflow scheduling problem as a multi-objective problem. This approach generates a

[1] AWS Elastic Compute 2, the virtual machine product of AWS.

series of schedules with different tradeoffs between cost and time so that users can choose acceptable schedules that fit their preferences. The chromosome is defined using three arrays: one responsible for ordering the tasks, the other responsible for assigning the tasks to a machine, and the other for defining the type of the machine. They applied NSGA-II [9], SPEA2 [31], and MOEA/D [29]. The case study revolves around Amazon EC2, utilizing instance specifications and pricing schemes, focusing on the US East region, considering only a single region and a small number of machine types.

Chen and Menzies [7] proposed RIOT (Randomized Instance Order Types). This heuristic method considers aspects from HEFT to group tasks in the workflow, considering a probability model and a surrogate-based method. This paper used the exact chromosome representation considered by [30] and a similar setup.

Qin et al. [21] approach the same problem, focusing on three key objectives: minimizing makespan and cost while maximizing reliability. This problem is situated within the context of multi-cloud systems, where machines from both Azure and AWS were utilized. To address this task, they proposed a variation of the memetic algorithm [16] for multi-objective problems. The studied case study considered the same workflow scheduling problem but in the multi-cloud scenario.

Motivated by research that combines genetic algorithms with specialist heuristics and approaches that consider the problem as a multi-objective problem, this work proposes HEACT, **H**ybrid **E**volutionary **A**lgorithm for the Multi-region Multi-objective **C**loud **T**ask Scheduling Problem.

Contributions: This research has as contributions 1. Proposes a hybrid algorithm that considers a multi-objective evolutionary algorithm and a specialist heuristic in order to solve the multi-objective task scheduling problem considering makespan and cost as objectives; 2. Considers multi-region: instances located in different places in the world have different associated costs. Moreover, regions may have a different set of machine types; 3. Uses realistic AWS information to instantiate the simulator (Simgrid); 4. Considers state-of-the-art scheduling heuristic (specially HSIP and MPEFT) with state-of-art meta-heuristic (especially AGEMOEA2 [19]). As far as we know, MPEFT was never studied together with a GA, while AGEMOEA was never applied to the cloud scheduling problem.

This paper is organized as follows: First, in Sect. 2, we provide some background necessary for the understanding of our proposal. Then, in Sect. 3, we present our proposal. Then, in Sect. 4, we detail our experiments and provide results to demonstrate the effectiveness of our approach. Finally, in Sect. 5, we present our conclusions and further work.

2 Background

In this section, we provide the background necessary for understanding our proposal. First, we define the cloud scheduling problem and then detail the specialized heuristics used to solve it. Finally, we introduce meta-heuristics briefly.

2.1 Scheduling Problem Formulation

The cloud task scheduling problem can be modeled as a DAG $G = (V, E)$ composed by a set of nodes $t_i \in V$, which represents an application task to be executed, and a set of edges E, which represents dependency between tasks and have an associated cost. Thus, given $e(i, j) \in E$, a task t_i should be wholly executed before starting task t_j.

The scheduling problem consists of assigning tasks to processors considering dependency between n tasks presented in the *DAG* and information regarding the processors. In this paper, we commonly use the name processor and, in our context, consider a processor to be the same as a virtual machine. Equation 1 calculates the objective function of the scheduling problem.

$$makespan = \max\{AFT(t_{exit})\} \tag{1}$$

The Actual Finish Time (AFT) represents when all the tasks and dependencies were executed and considers the last task t_{exit}. In order to calculate an AFT, we have to calculate the *Earliest Starting Time* (EST) and the *Earliest Finish Time* (EFT) for all the tasks.

$$EFT(t_i, p_j) = w_{t_i,p_j} + EST(t_i, p_j) \tag{2}$$

Thus, to calculate when a task will finish, we sum the starting time with the computational cost w_{t_i,p_j}. This is performed for a given task t_i considering a given processor p_j. The computation cost matrix W of size $nt \times np$, where nt is the number of tasks, and np is the number of processors. Each $w_{i,j} \in W$ provides the estimated execution time to complete task t_i on processor p_j.

The *Earliest Starting Time* (EST in the previous equation) is calculated according to Eq. 3:

$$EST(t_i, p_j) = \max\{T_{avaliable}(p_j), \max_{t_m \in pred(t_i)} \{EFT(t_m) + c_{m,i}\}\} \tag{3}$$

In this equation, $T_{avaliable}(p_j)$ is the time at which processor p_j is available and $pred(t_i)$ represents all tasks that t_i is dependent. We first calculate the latest time we can get all dependencies executed plus communication time. Thus, we guarantee that all dependencies will be respected.

This scheduling problem considers that a given task t_m could depend on another task t_i, which is located in another processor. Hence, it considers how much time will be necessary for the data generated by t_i, which is an input in t_m, arrives at the processor that t_m will be processed. This is what we call as a *communication cost* and it is represented by $c_{m,i}$, which is the communication cost for tasks t_i and t_j and it is defined by:

$$c_{m,i} = L_{t_m} + \frac{data_{i,m}}{B_{t_m,t_i}} \tag{4}$$

In this equation, B represents the transfer rates between the processors. This matrix has a size of $np \times np$, where np is the number of processors. L represents the processors' communication startup costs. $data_{i,m}$ represents the amount of data to be transferred and processed. If t_i and t_m are at the same machine, then $c_{m,i} = 0$.

The mono-objective scheduling problem considers only the *makespan* as an objective to be minimized. This is a realistic scenario only when we have a static set of processors waiting to receive tasks to run, so in this case, we are just trying to use the resources we already have (on-premises servers) better. Recently, with the advent of cloud providers, we can use the machines we need. By doing this, costs are reduced when using other companies' computational resources.

In the context of task scheduling, we can also consider the cost to be optimized and then treat this problem as a multi-objective optimization problem (MOP). Equation 5 presents this function; there, we convert the makespan into a fraction of an hour and then multiply it by the cost of every machine (processor) we are considering. Each machine has a type (p_{type}), which defines how powerful the machine is and how much it will cost. The same happens for the region where the machine is located (p_{region}). Usually, the same machine type in different regions may have different prices.

$$TotalCost = \sum_{p \in P} \frac{Makespan}{3600} * Cost_{(p_{type}, p_{region})} \tag{5}$$

2.2 Heuristics to Deal with the Problem

Several of the specialized heuristics in the scheduling problem consist of sorting the tasks according to a criterion and then iterating through this ranking in order to assign tasks to processors considering the shorter starting time available. For example, in the Heterogeneous Earliest Finish Time algorithm (HEFT) [26], tasks are ranked decreasingly according to the *upward rank value* (Eq. 6), which is based on mean computation and mean communication costs.

$$rank_u(t_i) = \bar{w_i} + \max_{t_j \in succ(t_i)} (\bar{c_{i,j}} + rank_u(t_j)) \tag{6}$$

In this equation, *succ* is the set of immediate successors of task t_i $\bar{c_{i,j}}$ is the average communication cost of an edge (i, j) and $\bar{w_i}$ is the average computation cost of task t_i among the processors.

Similarly, the Predict Earliest Finish Time algorithm (PEFT) [2] also ranks tasks but considers the Optimistic Cost Table (OCT) for the ranking phase. The OCT (Eq. 7) is a matrix of size $np \times nt$ where each element represents, for a given task t_i and processor p_k, the maximum of the shortest paths of t_i dependencies tasks to the exit node. $\bar{c_{i,j}} = 0$ if $p_w = p_k$.

$$OCT(t_i, p_k) = \max_{t_j \in succ(t_i)} \left[\min_{pw \in P} \{OCT(t_j, p_w) + w(t_j, p_w) + c_{i,j}\} \right] \tag{7}$$

The OCT represents the maximum optimistic processing time of the children of task t_i because it considers that dependencies tasks are executed in the processor that minimizes processing time (communications and execution) independently of processor availability [2].

With the values of OCT calculated, we can then rank the values decreasingly, according to Eq. 8.

$$rank_{oct}(t_i) = \frac{\sum_{k=1}^{qp} OCT(t_i, p_k)}{qp} \tag{8}$$

In order to select a processor for a task, the algorithm considers the lowest Optimistic EFT ($OEFT$ Eq. 9), which sums to the Estimated Finish Time of a given task to the computation time of the longest path to the exit node (OCT). The aim is to guarantee that the tasks ahead will finish earlier, which is the purpose of the OCT table.

$$OEFT(t_i, p_j) = EFT(t_i, p_j) + OCT(t_i, p_j) \tag{9}$$

In [27], the authors proposed the heterogeneous scheduling algorithm with improved task priority (HSIP). This algorithm is considered a prioritization strategy based on computational cost standard deviation and total communication cost for a task (Eq. 10). σ_i is the standard deviation of the computational cost of the given task t_i all over the processors while $\bar{w}_i$.

$$rank_u(t_i) = \max_{t_j \in succ(t_i)} \{\bar{w}_i \times \sigma_i + occw(t_i) + rank_u(t_j)\} \tag{10}$$

Moreover, the $occw$ term is defined by Eq. 11, where $occw(t_{exit}) = 0$.

$$occw(t_i) = \sum_{t_j \in succ(t_i)} c_{i,j} \tag{11}$$

HSIP also implements an entry task duplication in order to improve the makespan and uses an improved idle time slot insertion-based optimizing policy to make the task scheduling more efficient.

Recently, the Modified Predict Earliest Finish Time algorithm (MPEFT) [20] was proposed. Equation 12 presents how this algorithm ranks tasks.

$$rank_{AP}(t_i) = DCT(t_i) + \sum_{t_j \in OS(t_i)} DCT(t_j) \tag{12}$$

The MPEFT ranking equation considers all direct and indirect successor tasks. Thus, instead of only using $succ$ as the other algorithms, MPEFT also considers all indirect tasks according to Eq. 13.

$$OS(t_i) = \begin{cases} \bigcup_{t_j \in succ(t_i)} OS(t_j) \iff t_j \not t_{exit} \\ \emptyset \text{ otherwise} \end{cases} \tag{13}$$

MPEFT ranking considers DCT, where the minimal computational time for a given task t_i (represented by $w^*(t_i)$) is the sum with the summation of the average transfer costs $c_{i,j}$. This is expressed by Eq. 14.

$$DCT(t_i) = w^*(t_i) + \sum_{t_j \in succ(t_i)} c_{i,j} \tag{14}$$

Similarly to PEFT, MPEFT does not directly consider the Estimated Finish Time (EFT) but incorporates it into another task's estimated finish, represented by MEFT (Eq. 15).

$$MEFT(t_i, p_k) = EFT(t_i, p_k) + OCT(t_j, p_k) \times k(t_i, p_k) \tag{15}$$

The MEFT equation also considers the OCT value (Eq. 7) as performed in PEFT but also uses a weight for the OCT value. This weight is calculated according to Eq. 16.

$$k(t_i, p_k) = \begin{cases} \frac{rank_{AP}(CPS(t_i, p_k))}{\sum_{t_j \in succ(t_i)} rank_{AP}(t_j) + c_{i,j}} & \iff |succ(t_i)| > |P| + 1 \\ 1 \text{ otherwise} \end{cases} \tag{16}$$

Where the CPS is essentially the argmax of Eq. 7, this is given by Equation 17, which is pretty similar to Eq. 7.

$$CPS(t_i, p_k) = \text{argmax}_{t_j \in succ(t_i)} \left[\min_{pw \in P} \{OCT(t_j, p_w) + w(t_j, p_w) + c_{i,j}\} \right] \tag{17}$$

In all presented heuristics, tasks are ordered according to the heuristic criterion. Based on that order, tasks are assigned to the processor that is the most early available. The best available processor is found by considering the time that the processor will be available (last task executing in the processor + task duration) + the amount of time to send the task to that processor.

2.3 Meta-heuristics

Meta-heuristics are algorithms, solution methods that orchestrate an interaction between local improvement procedures (heuristics) and higher-level strategies to create a process capable of escaping from local optima and performing a robust search of a solution space [12]. Thus, heuristics usually specializes in solving problems for one particular domain, while meta-heuristics are more generic and adaptive in several domains.

One of the most used meta-heuristic is the Evolutionary Algorithm (EAs) [8]. EAs are meta-heuristics that employ Darwin's theory of the survival of the fittest as their inspiration. This kind of algorithm generates solutions using the crossover and mutation heuristic operators. It employs an objective function to choose which individuals (solutions for the problem) will compose the next population.

Multi-Objective Evolutionary Algorithm (MOEA) has been successfully applied to the solution of MOPs (Multi-objective optimization problems) [5]. These algorithms are heuristic techniques that allow a flexible representation of the solutions and do not impose continuity conditions on the functions to be optimized. Moreover, MOEAs are extensions of EAs for multi-objective problems that usually apply the concepts of Pareto dominance. These algorithms typically differ in their replacement strategies. In a replacement strategy, a population of current solutions P that was used to generate offspring solutions O is combined, generating $P \cup O$. So, a replacement strategy (RS) to generate a new population of solutions P' by $P' \leftarrow RS(P \cup O)$.

The Non-dominated Sorting Genetic Algorithm-II [9] (NSGA-II) performs the replacement strategy considering Pareto Dominance and Crowding Distance selection. This selection evaluates how close solutions are to their neighbors, giving a better evaluation of large values allows a better diversity in the population. Thus, NSGA-II selects surviving solutions from $P \cup O$, it first takes non-dominated solutions to compose P', and if the set size is smaller than the maximum population size, it iteratively adds dominated solutions with higher Crowding Distance values until P' is complete. If the set size exceeds the maximum population size, NSGA-II discards the solutions with lower Crowding values until P' is complete.

The Adaptive Geometry Estimation based MOEA (AGEMOEA) [18] inherits the overall framework of NSGA-II. However, it replaces the crowding distance with a survival score that combines diversity and proximity of the non-dominated fronts. First, this algorithm uses a fast heuristic with $O(M \times N)$ computational complexity to estimate the geometry of the front in each generation. Then, proximity is computed as the distance between each population member and the ideal point, while the diversity is measured using the distance among population members. The distance used to calculate proximity and diversity corresponds to the Lp norm that is associated with the estimated geometry.

A second version of the AGE Algorithm (AGEMOEA2) [19] incorporated the Newton-Raphson iterative method and the geodesic distance into the AGE-MOEA framework. This was performed by proposing a novel method for non-dominated front modeling using the Newton-Raphson iterative method. Second, they compute the distance (diversity) between each pair of non-dominated solutions using geodesics, which are generalizations of the distance on Riemann manifolds (curved topological spaces).

In this paper, we considered NSGA-II because this algorithm is probably the most popular MOEA in the literature and has had good results in recent works on cloud scheduling. AGEMOEA2 is one state-of-the-art algorithm for continuous multi-objective optimization benchmarks and was never applied to the cloud scheduling problem.

3 Proposal

Figure 1 presents the HEACT workflow. The workflow to be scheduled is sent to the *Data preparation* process. This process is also responsible for getting real data about EC2 instance types from AWS (using the boto3 library). Then, with that information, for each region, the *Data preparation* calculates how long each task $t \in T$ demands to run in each machine. This creates the table W with size $nr \times np \times nt$, where nr is the number of regions, np is the number of machines (this number varies in each region), and nt is the number of tasks in T.

To create W, we considered two pieces of information about EC2 instances taken from boto3: (1) the network performance and (2) the *Estimated Compute Unit* (ECU) of the instance type. According to AWS, "ECU provides the equivalent CPU capacity of a 1.0–1.2 GHz 2007 Opteron or 2007 Xeon processor" [22]. In order to use Simulation of Distributed Computer Systems (SimGrid) [6] as the simulator, we follow [14] and convert ECU to GigaFlops using the formula $GigaFlops = ECU * 4.2$.

Once W is created, the 'Data preparation' process disseminates this information to nr distinct processes, each tasked with finding solutions for a specific region. This is where the MOEA and a specialist scheduling heuristic come into play, running in parallel. MOEAs (e.g. NSGA-II) are responsible for defining the set of machines to be used and heuristics (e.g. HEFT) are responsible for assigning tasks to the selected machine. After completion, each process sends its final population to the 'Pareto dominance' analysis, eliminating dominated and repeated (same decision variables) solutions. The resulting solution is then stored as a JSON file.

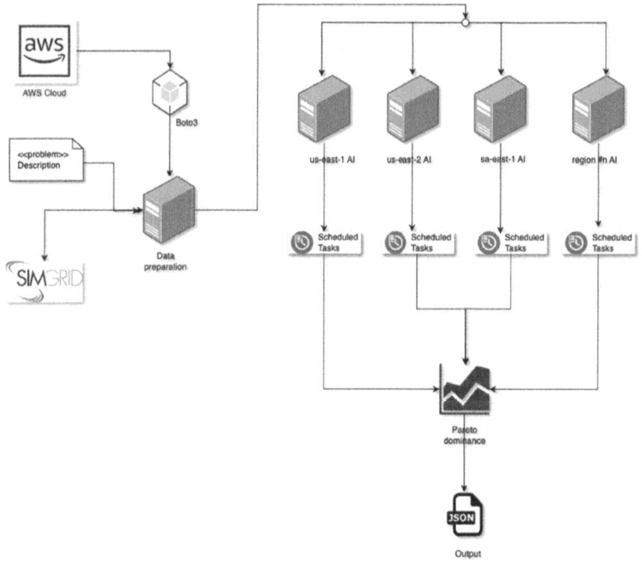

Fig. 1. HEACT Architectural representation

An essential part of an MOEA is how chromosomes are designed. In our approach, we used a fixed-size length integer array. Each array element represents an instance type, numbered from zero to np. When a number is negative, it means that this slot will not be used. That is a logical way to have a variable-size array. This is important because it allows us to use more straightforward crossover operators in the MOEA. Figure 2 presents two examples of how the solution is defined.

After generating a solution through crossover and mutation, the partial solution is sent to a specialized heuristic, which will be responsible for allocating tasks in machines specified by the previous partial solution. Thus, every task $t \in T$ must be assigned to a machine considering the performance characteris-

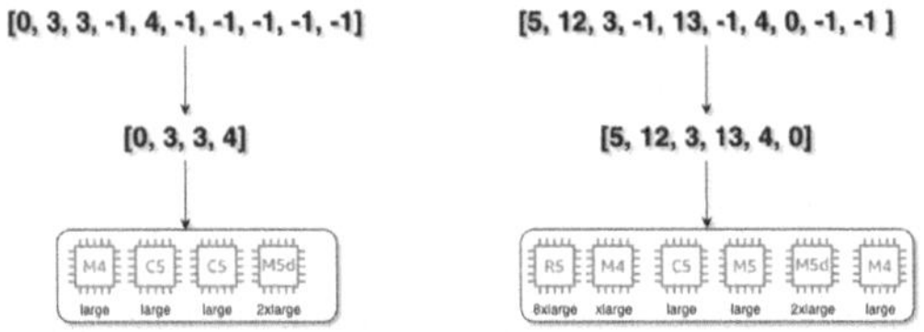

Fig. 2. MOEA chromosomes definition illustration

Table 1. Solution #1

Machine	Task	Begin	End
M4 large	T4 T6 T2		
C5 large	T1 T3 T5 T7 T8	0	10
C5 large	-		
M5d 2xlarge	-		

Table 2. Solution #2

Machine	Task	Begin	End
R5 8xlarge	T4 T6 T2		
M4 xlarge	T1 T3 T8	0	10
C5 large	-		
M5 large	-		
M5d 2xlarge	T5 T7		
M4 large	-		

tics of each EC2 instance type, such as network performance and how long it takes to solve a task (available in table W).

Tables 1 and 2 show an example of eight tasks being assigned to the solutions represented in Fig. 2. Notice that not every machine will be used; this is something particular to the heuristic because it considers the cost of transferring gigabytes of data through the network or just waiting for a dependency task at the same machine to finish and having a network transfer cost of zero.

Finally, the solution is updated by taking all the scheduling information (which task will run in each machine) and which machine will, in fact, be used. Thus, the solution [0,3,3,−1,4,−1,−1,−1,−1,−1] ([0,3,3,4]) becomes [0,3,−1,−1,−1,−1,−1,−1,−1,−1] ([0,3]) and [5,12,3,−1,13,−1,4,0,−1,−1] ([5,12,3,13,4,0]) becomes [5,12,−1,−1,−1,−1,4,−1,−1,−1] ([5,12,4]). With the complete solution, scheduling + machine types, we are able to calculate the objectives makespan and cost.

4 Experiments

We run experiments to evaluate how HEACT performs. For this purpose, we considered several real-world workflows, such as Montage, CyberShake, Epigenomics, Inspiral, and Sipht [3]. We named the problems in the format problem_nodes. So, Montage_25 is an instance of Montage, which has 25 tasks. Figure 3 shows the different workflows.

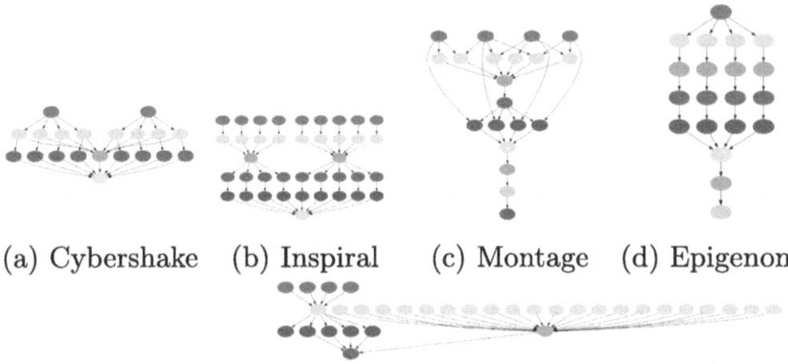

(a) Cybershake (b) Inspiral (c) Montage (d) Epigenomics

(e) Sipht

Fig. 3. Real-world workflows used to evaluate HEACT

We created eight different instances by combining the meta-heuristics NSGA-II and AGEMOEA2 with scheduling heuristics HEFT, PEFT, HSIP, and MPEFT. The population size was set as 50. The number of generations was set to 700, totaling 35000 heuristic execution and solution evaluation, defined empirically. We considered 14 AWS regions. These are the regions that have EC2 instances with complete data about ECU. Each AWS region was assigned to a process that would run for 50 generations (or 2500 solution evaluations). For NSGA-II and AGEMOEA2, we considered a two-point crossover with a probability of 90% and flip mutation with a probability of 10%.

All algorithms were coded using Python 3.11. We used pymoo [4] for meta-heuristics, pysimgrid [23] and SimGrid 3.24 for the cloud simulation, and AWS boto3 1.28 to get AWS pricing data.

After running each of the eight different algorithm instances twenty times (for each problem), we took the resulting population and removed all the dominated solutions. After that, as commonly performed in the literature, we calculated the normalized Hypervolume [28] to get the Pareto Front's quality for each execution. After getting twenty hypervolume values (for each problem and each algorithm instance), we calculated the average and submitted them to the Kruskal-Wallis statistical test with 99% confidence, as commonly performed in the literature.

Table 3. Hypervolume Averages

	AGE+HEFT	AGE+HSIP	AGE+MPEFT	AGE+PEFT	NS+HEFT	NS+HSIP	NS+MPEFT	NS+PEFT
Epigenomics_24	**0.9798**	0.9820	0.9790	0.9783	0.9796	**0.9821**	0.9800	0.9782
Montage_25	**0.9963**	**0.9961**	0.9959	0.9962	**0.9964**	0.9962	0.9955	0.9961
CyberShake_30	**0.9984**	0.9981	0.9980	0.9970	**0.9984**	0.9981	0.9980	0.9968
Inspiral_30	0.9980	0.9956	**0.9980**	0.9977	0.9979	0.9941	**0.9981**	0.9977
Sipht_30	0.9499	**0.9511**	0.9499	0.7736	0.9499	0.9511	0.9498	0.8126
Epigenomics_46	**0.9807**	**0.9800**	**0.9794**	0.6488	**0.9801**	0.9776	0.9778	0.6170
CyberShake_50	**0.9972**	**0.9975**	0.9975	**0.9974**	0.9974	0.9973	**0.9975**	0.9974
Inspiral_50	**0.9888**	0.9877	**0.9901**	0.5024	**0.9895**	0.9863	0.9897	0.5024
Montage_50	0.9827	0.9920	**0.9978**	0.9976	0.9923	0.9925	**0.9977**	0.9976
Sipht_60	**0.9805**	0.9761	0.9771	0.9732	**0.9805**	0.9760	0.9773	0.9717
CyberShake_100	0.9894	0.9939	0.9972	**0.9981**	0.9952	0.9920	0.9972	**0.9981**
Epigenomics_100	**0.9894**	**0.9909**	**0.9926**	0.9822	**0.9765**	**0.9885**	**0.9923**	0.9844
Inspiral_100	**0.9905**	0.9744	0.9693	0.0771	**0.9902**	0.9449	**0.9846**	0.1542
Montage_100	0.9790	**0.9966**	**0.9986**	0.9794	**0.9976**	0.9974	0.9978	0.9809
Sipht_100	**0.9862**	0.9852	**0.9850**	0.9801	**0.9856**	0.9836	0.9841	0.9767

Table 3 presents Hypervolume results; highlighted values mean the highest hypervolume average, while bold values mean that the value is statistically tied with the best value. NS means NSGA-II while AGE means AGEMOEA2. We can see that no single algorithm statistically overcomes all the others. We can also see that instances that use PEFT as a heuristic have, in general, the poorest performance. However, regarding CyberShake_100, AGE+PEFT and NS+PEFT were indeed the best-performing algorithms. NS+HSIP have similar performance compared with PEFT instances by being the best at one problem and have statistically tied results with the best algorithm in one problem. AGE+HSIP is a bit better because it has the best average at one problem, but it is tied with the best in five problems and is overcome in nine problems.

Based on the statistical results, we sum how many problems each algorithm instance got the best hypervolume average, tied with the best, and got the averages statistically worse when compared with the best algorithm. Table 4 presents this analysis. Moreover, we follow [10] and perform the ranking of algorithms considering the Friedman ranking.

According to Friedman Ranking, AGE+MPEFT is the best algorithm. This happens especially because it is the algorithm that has been the best more times. The ranking states that NS+HEFT, AGE+HEFT, and NS+MPEFT are in sequence. However, an interesting aspect is that NS+HEFT is the fourth algorithm in the ranking, but this is the algorithm with fewer overcomes (Qtd worse column). By this analysis, we can conclude that AGE+MPEFT and AGE+HEFT are the most suitable algorithms in general.

Another interesting aspect to evaluate is how solutions are spreading across the world. Due to this fact, we looked for solutions for each region independently and then selected the surviving ones using the Pareto non-dominance criterion. One region may have more solutions than another. Figure 4 presents this percentual. We can see that always us-east-2 (US East (Ohio)) have more solutions. The second best is us-west-2 (US West (Oregon)) followed by ap-south-1 (Asia

Table 4. Average Rankings of the algorithms (Friedman)

Algorithm	Qtd. Best	Qtd. tied	Qtd worse	Friedman Ranking
AGE+MPEFT	4	3	8	3.4334
NS+HEFT	3	6	6	3.4667
AGE+HEFT	3	7	5	3.5335
NS+MPEFT	2	3	10	3.9
AGE+HSIP	1	5	9	4.2667
NS+HSIP	1	1	13	5.0667
AGE+PEFT	1	1	13	6.0334
NS+PEFT	0	1	14	6.3001

Pacific (Mumbai)) and us-east-1 (US East (Virginia)). Of course, these values can change depending on the price AWS assigns to instances.

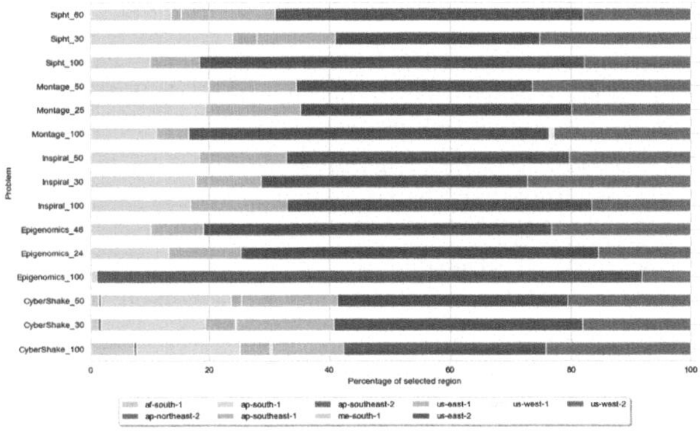

Fig. 4. Percentual of non-dominated solutions considering regions

5 Conclusions and Further Work

Planning the execution of a workflow is a challenging task. Over the years, several approaches have been proposed in order to tackle this problem. This present paper introduced HEACT, a novel approach that finds solutions for workflow scheduling by first running a multi-objective evolutionary algorithm in order to create the set of machines and then submitting this setup to a specialist heuristic responsible for the task assigning. This process was performed considering different regions, each one with its own set of instance types. Our proposal considered accurate pricing information from AWS and simulated the workflow execution considering the simgrid simulator. Moreover, fourteen AWS regions were considered.

We instanced our approach by considering heuristics, such as HEFT, PEFT, HSIP, and MPEFT, as well as meta-heuristic NSGA-II and AGEMOEA2. Results showed that AGEMOEA2 + MPEFT and AGEMOEA2 +HEFT were the best combinations. Thus, AGEMOEA2, a recent state-of-art meta-heuristic, seems promising for this kind of problem. Moreover, MPEFT, a recent work, seems a really interesting heuristic for scientific workload problems, as also verified in [20]. Further work may focus on combining these heuristics and meta-heuristics to create a hyper-heuristic aiming for the best overall performance. This is important because no single algorithm was the best across all the problems. Moreover, information about carbon emissions should also be considered in future work.

Acknowledgments. We gratefully acknowledge financial support from the São Paulo Research Foundation (FAPESP) grant 2023/15949-8.

References

1. Ahmad, D., Gautam, G., Alam, B., Bhati, B.: An analytical review and performance measures of state-of-art scheduling algorithms in heterogenous computing enviornment. Arch. Comput. Meth. Eng. (2024)
2. Arabnejad, H., Barbosa, J.G.: List scheduling algorithm for heterogeneous systems by an optimistic cost table. IEEE Trans. Parallel Distrib. Syst. **25**(3), 682–694 (2014)
3. Bharathi, S., Chervenak, A., Deelman, E., Mehta, G., Su, M.H., Vahi, K.: Characterization of scientific workflows. In: 2008 3rd Workshop on Workflows in Support of Large-Scale Science, pp. 1–10 (2008)
4. Blank, J., Deb, K.: pymoo: multi-objective optimization in Python. IEEE Access **8**, 89497–89509 (2020)
5. Boussaid, I., Lepagnot, J., Siarry, P.: A survey on optimization metaheuristics. Inf. Sci. **237**, 82–117 (2013)
6. Casanova, H., Giersch, A., Legrand, A., Quinson, M., Suter, F.: Versatile, scalable, and accurate simulation of distributed applications and platforms. J. Parallel Distrib. Comput. **74**(10), 2899–2917 (2014)
7. Chen, J., Menzies, T.: RIOT: a stochastic-based method for workflow scheduling in the cloud. In: IEEE International Conference on Cloud Computing, CLOUD 2018-July, pp. 318–325 (2018)
8. Coello, C.: Evolutionary Algorithms for Solving Multi-objective Problems. Springer, New York (2007). https://doi.org/10.1007/978-0-387-36797-2
9. Deb, K., Pratap, A., Agarwal, S., Meyarivan, T.: A fast and elitist multiobjective genetic algorithm: NSGA-II. IEEE Trans. Evol. Comput. **6**(2), 182–197 (2002)
10. Derrac, J., García, S., Molina, D., Herrera, F.: A practical tutorial on the use of nonparametric statistical tests as a methodology for comparing evolutionary and swarm intelligence algorithms. Swarm Evol. Comput. **1**(1), 3–18 (2011)
11. Durillo, J.J., Fard, H.M., Prodan, R.: MOHEFT: a multi-objective list-based method for workflow scheduling. In: 4th IEEE International Conference on Cloud Computing Technology and Science Proceedings, pp. 185–192 (2012)

12. Gendreau, M., Potvin, J.Y.: Handbook of Metaheuristics, 2nd edn. Springer, Heidelberg (2010)
13. Hosseini Shirvani, M., Noorian Talouki, R.: Bi-objective scheduling algorithm for scientific workflows on cloud computing platform with makespan and monetary cost minimization approach. Complex Intell. Syst. **8** (2021)
14. Iosup, A., Ostermann, S., Yigitbasi, M.N., Prodan, R., Fahringer, T., Epema, D.: Performance analysis of cloud computing services for many-tasks scientific computing. IEEE Trans. Parallel Distrib. Syst. **22**(6), 931–945 (2011)
15. Keshanchi, B., Souri, A., Navimipour, N.J.: An improved genetic algorithm for task scheduling in the cloud environments using the priority queues: Formal verification, simulation, and statistical testing. J. Syst. Softw. **124**, 1–21 (2017)
16. Moscato, P.: On evolution, search, optimization, genetic algorithms and martial arts: towards memetic algorithms. Technical report 826, Caltech Concurrent Computation Program, C3P (1989)
17. Nasonov, D., Visheratin, A., Butakov, N., Shindyapina, N., Melnik, M., Boukhanovsky, A.: Hybrid evolutionary workflow scheduling algorithm for dynamic heterogeneous distributed computational environment. J. Appl. Logic **24**, 50–61 (2017)
18. Panichella, A.: An adaptive evolutionary algorithm based on non-Euclidean geometry for many-objective optimization. In: Proceedings of the Genetic and Evolutionary Computation Conference, GECCO 2019, pp. 595–603. ACM, New York (2019)
19. Panichella, A.: An improved pareto front modeling algorithm for large-scale many-objective optimization. In: Proceedings of the Genetic and Evolutionary Computation Conference, GECCO 2022, pp. 565–573. ACM, New York, NY, USA (2022)
20. Pu, J., Meng, Q., Chen, Y., Sheng, H.: MPEFT: a novel task scheduling method for workflows. Front. Environ. Sci. **10** (2023). https://doi.org/10.3389/fenvs.2022.996483
21. Qin, S., Pi, D., Shao, Z., Xu, Y., Chen, Y.: Reliability-aware multi-objective memetic algorithm for workflow scheduling problem in multi-cloud system. IEEE Trans. Parallel Distrib. Syst. **34**(4), 1343–1361 (2023)
22. Amazon Web Services: Amazon EC2 instance-types (2024). https://www.amazonaws.cn/en/ec2/instance-types/. Accessed 21 May 2024
23. Sukhoroslov, O.: Toward efficient execution of data-intensive workflows. J. Supercomput. **77**(8), 7989–8012 (2021)
24. Sulaiman, M., Halim, Z., Lebbah, M., Waqas, M., Tu, S.: An evolutionary computing-based efficient hybrid task scheduling approach for heterogeneous computing environment. J. Grid Comput. **19** (2021)
25. Sulaiman, M., Halim, Z., Waqas, M.M., Aydm, D.: A hybrid list-based task scheduling scheme for heterogeneous computing. J. Supercomput. **77**, 10252–10288 (2021)
26. Topcuoglu, H., Hariri, S., Wu, M.Y.: Performance-effective and low-complexity task scheduling for heterogeneous computing. IEEE Trans. Parallel Distrib. Syst. **13**, 260–274 (2002)
27. Wang, G., Wang, Y., Liu, H., Guo, H.: HSIP: a novel task scheduling algorithm for heterogeneous computing. Sci. Program. **2016**, 1–11 (2016)
28. While, L., Bradstreet, L., Barone, L.: A fast way of calculating exact hypervolumes. IEEE Trans. Evol. Comput. **16**(1), 86–95 (2012)
29. Zhang, Q., Liu, W., Li, H.: The performance of a new version of MOEA/D on CEC09 unconstrained MOP test instances. Technical report, CES-491, School of CS & EE, University of Essex, February 2009

30. Zhu, Z., Zhang, G., Li, M., Liu, X.: Evolutionary multi-objective workflow scheduling in cloud. IEEE Trans. Parallel Distrib. Syst. **27**(5), 1344–1357 (2016)
31. Zitzler, E., Laumanns, M., Thiele, L.: SPEA2: improving the strength pareto evolutionary algorithm for multiobjective optimization. In: Evolutionary Methods for Design Optimization and Control with Applications to Industrial Problems. pp. 95–100. International Center for Numerical Methods in Engineering, Barcelona, Spain (2001)

Heuristic Solutions for the 2D Bin-Packing Problem with Varied Size

Eduardo Augusto Militão Fernandes[1(✉)], Thiago Ferreira de Noronha[1], and Amadeu Almeida Coco[2]

[1] Department of Computer Science, Universidade Federal de Minas Gerais, Belo Horizonte, MG 31270-901, Brazil
{eduardofernandes,tfn}@dcc.ufmg.br

[2] ORKAD, CRIStAL, UMR 9189, CNRS, Centrale Lille, Université de Lille, Lille, France
amadeu.almeidacoco@univ-lille.fr

Abstract. This work addresses the 2D Bin-Packing Problem with Varied Size and proposes heuristic solutions for it. An extensive literature review on the problem is carried out, and the state-of-the-art literature algorithm for its variation with guillotine constraint is reproduced in order to validate the reported results. It is also used to build a baseline which is compared with a proposed improvement for this algorithm in order to enable the addition of a new constraint to the problem that meets an industry demand for its practical application.

Keywords: Bin-Packing Problem · Heuristics · Metaheuristics

1 Introduction

Given one-dimensional boxes and items of varying lengths, the Bin Packing Problem (BPP) seeks to determine who to allocate items to each box in order to pack all items while minimizing the number of boxes used. Such a problem is famously NP-Hard, which means that a solution to the BPP is also a solution to every problem in the NP-Hard class, and, going further, there is no known algorithm of polynomial complexity to solve it. The Bin-Packing Problem is one of the most studied classical problems in computing, with several heuristic algorithms having already been proposed to quickly find good, but not necessarily optimal, solutions for it.

In this context, the 2D Bin-Packing Problem with Varied Size (2DVSBPP) is a variation of the original BPP, with rectangular two-dimensional boxes and items, with such boxes having different sizes, and the objective is to pack all items minimizing the total area spent for this (or, in the same way, maximizing the use of the total area). More formally, given N rectangular items of widths W_i and heights H_i, with $i \in [1, N]$, and M rectangular boxes of widths W_j and heights H_j, with $j \in [1, M]$, one wants to define the position of each item in exactly one box such that the items are entirely within a box, there is no

A. Paes and F. A. N. Verri (Eds.): BRACIS 2024, LNAI 15413, pp. 65–80, 2025.
https://doi.org/10.1007/978-3-031-79032-4_5

overlapping of items and the sides of an item are parallel to the sides of its respective box, in order to minimize the sum of the areas of the boxes used.

As it is a generalization of BPP, 2DVSBPP is also in the class of NP-Hard problems, and methods to solve it become even more complex and elaborate, given the exponentially greater number of possible combinations of items with boxes due to the addition of a extra dimension. Furthermore, 2DVSBPP is of great relevance to the industry [2,4], having as its immediate application the determination of the layout for cutting metal sheets in order to minimize material waste. In this context, another way of describing the problem (which will be used throughout the work) is to determine a sequence of vertical and horizontal cuts in the base material in order to obtain rectangular items from it. Since these cuts have to be from one end of the base material to the opposite one, this version of the problem is known as **2DVSBPP with Guillotine Constraint** [9]. Due to its direct applicability in the industry, it is of great interest to study this problem and develop more efficient solutions for it, including ones that fit in strict industry requirements [2,4,5].

The objectives of this work are, thus, three-fold: to carry out an extensive literature review for the 2D Bin-Packing Problem, focusing on its version with Varied Size and Guillotine Constraint; reproduce the state-of-the-art algorithm for the 2DVSBPP with Guillotine Constraint and analyse the results obtained by [3]; propose an improvement to this algorithm, taking into account not only the use of the total area spent but also the real practicality of the generated solutions - by reducing the number of changes of cut orientations in the solution - and enable the introduction of a new constraint to the problem in order to increase its applicability in the industry, by setting a hard limit on such number of cut orientation changes.

The remainder of this paper is organized as follows: Sect. 2 presents the literature review carried out for the 2DVSBPP. Section 3 introduces the methodology used in the work, specifying the process of collecting test instances, implementing the baseline algorithm, proposing an improvement to the algorithm, adding a new constraint to the problem and the corresponding results obtained. Finally, Sect. 4 concludes the work.

2 Literature Review

In [9], the authors proposed a notation to define 4 different variations of the 2D Bin Packing Problem:

- $2BP|O|G$: items cannot be rotated and must be positioned to allow guillotine cutting
- $2BP|R|G$: items can be rotated and must be positioned to allow guillotine cutting
- $2BP|O|F$: items cannot be rotated and the cutting shape is free
- $2BP|R|F$: items can be rotated and the cutting shape is free

A guillotine cut is a straight, vertical or horizontal cut that must be made from one side of the box to the other. Thus, the guillotine cutting constraint implies finding a configuration for the arrangement of items in such a way as to allow such items to be extracted only through cuts of this type. Such constraint is very common in industrial applications. Still in this work, a different heuristic is presented for each of the four variations of the problem, and an algorithm based on Tabu Search that allows using the four heuristics to solve the corresponding variation was also presented, and became the state-of-the-art of the era.

In the work of [12], an exact solution for the 2DVSBPP is presented. The authors used an Integer Programming (IP) formulation for the problem which, according to the text, is very difficult to be solved by standard solvers. They then presented four *lower-bounds* for the problem, and then implement a "branch-and-price" algorithm that applies these bounds to optimally solve the problem. A new set of instances for the problem, with 5 bin sizes and number of items equal to 20, 40, 60, 80 and 100, was also proposed. It is shown that one of the presented lower-bounds, which makes use of the Dantzig-Wolfe decomposition, is extremely accurate, capable of providing a bound of 3.89% on average at the root node of the branch-and-cut tree. Such lower-bound is found through a Mixed-Integer Programming model.

In [11], the $2DVSBP|O|F$ is discussed, that is, $2BP|O|F$ with items and boxes of varying sizes. The authors presented a two-phase algorithm for the problem. The first step consists of solving consecutive Strip Packing problems, using the width of the current box as the width of the strip, until all items have been packed. In the second step, a repacking process is carried out on each box, in the hope of repacking all of its items into a box of smaller area, through successive attempts on increasingly larger boxes. The algorithm was tested on instances from the literature and new proposed instances, whose generation process is described in detail.

In [8], the authors presented two algorithms based on Dynamic Programming (DP) to solve the 2DVSBPP with a limited number of each bin size. The first algorithm, H1, only solves the 2DVSBPP without guillotine cuts and with orientation constraints. The second, H2, is capable of incorporating or not each of these constraints. The DP does not solve this problem optimally, since it has an exponential number of states. In this way, the DP is used to solve the problem approximately, by uniting several states into just one, thus losing information but gaining time and space. On average, the H1 algorithm is able to obtain considerably better solutions than H2, when the constraints that H2 can incorporate are not present.

Alvarez-Valdés et al. [1] tackled the 2D and 3D Variable Sized and Variable Cost Bin Packing Problems. The authors proposed a two-phase algorithm to solve the problem. The first phase consists of a GRASP, which generates a set of local optimal solutions. The second phase is a Path Relinking algorithm, responsible for combining the solutions from the first phase into a better solution.

Hong et al. [6] addressed the $2DVSBP|O|G$ variant and introduced a heuristic called BTVS, which selects the next box size using a procedure based on Backtracking. For each size chosen, a semi-greedy algorithm is used to try to fill the box with the largest area of items possible, considering different item orderings. Such heuristic makes use of two new algorithms, one semi-greedy and the other greedy, described by the authors, and in the end the best solution returned by one of the two methods is chosen. The authors also presented a heuristic based on Simulated Annealing, where the neighborhood function simply consists of selecting two random items and swapping their order. This method is designed to improve the solution found by BTVS.

In [2], the authors proposed the Two-Dimensional Three-Stage Guillotine Cutting Problem with Batch Precedence Constraints, which is similar to 2DVS-BPP, but with an additional constraint that solutions can have a maximum depth of three - that is, they must start with a set of vertical cuts, followed by a set of horizontal cuts, followed by a final set of vertical cuts - and the order of the items being obtained follows a pre-defined sequence. Such restrictions are motivated by demands from the glass cutting industry. Three heuristics are presented for this problem, and experiments are carried out on publicly available instances. Similarly, in [4] the same authors propose constructive heuristic methods, whereas in [5], a Biased Random-key Genetic algorithm is proposed for the same problem.

In [13], the authors presented a *matheuristic* approach for $2DVSBP|O|G$. In this algorithm, feasible solutions are successively constructed by solving entire schedules iteratively, one sub-problem each time. Three different Integer Programming models were presented, each one better but more expensive than the other. The developed algorithm surpasses the then state of the art, both in the problem in question and in the 2D Single Size BPP problem. In the latter, the procedure was able to prove the optimality of the solution in 415 of the 500 instances tested.

Recently, Gardeyn and Wauters [3] proposed a method for $2DVSBP| * |G$, that is, the orientation constraint can be applied or not. The authors used a rooted tree to represent a solution, where leaf nodes are empty items/spaces, and inner nodes define which direction child nodes are cut (vertical or horizontal). The proposed algorithm is based on the Ruin-and-Recreate paradigm, and computational experiments show that the algorithm is the new state-of-the-art for $2DVSBP| * |G$, and equals the state-of-the-art for $2DBP| * |G$, although it has considerably longer execution times.

Finally, [10] carried out an extensive bibliographical review of the original Bin Packing Problem, exposing its formal notation, applications and, mainly, proposes a notation note to catalog all the variations and restrictions of the BPP. Furthermore, it also presents statistics that illustrate how the problem has been more or less researched over the years.

3 Methodology and Contribution

Based on the literature review, it was decided to collect and use the test instances proposed in [7,12] and [11], for a total of 855 instances. The first set contains 500 instances separated into 10 groups, namely MC_B1, up to MC_B10. The second set consists of 340 instances divided into two groups of 170 each, where the first, Nice, contains instances of items of homogeneous size, while the second, Path, contains items of vastly different and extreme sizes and dimensions. The last set contains 15 instances divided into 3 groups: M1, M2 and M3.

3.1 Literature Baseline Implementation

The algorithm chosen for reproduction and implementation was the one proposed in [3], for the $2DVSBP| * |G$, as it is an extremely recent work, is the current state of the art for this problem and is well specified in the original article. The entire description of the algorithm presented by the authors was meticulously followed, and the parts of the approach not detailed in the original article were implemented with a focus on minimizing the time and memory complexity of the algorithm, while maintaining the expected behavior.

The algorithm is based on a *goal-driven* heuristic of *ruin and recreate.* A rooted tree represents a solution, where leaf nodes are empty items/spaces, and internal nodes define in which direction child nodes are cut. More specifically, if an internal node represents cuts in a certain direction, its child node, which is also internal, represents cuts in the opposite direction to that of its parent (that is, an internal node represents a sequence of cuts in the same direction). The procedure uses two values to guide the search for better solutions: the total area of boxes used and the total area of unpacked items. This is because the procedure addresses invalid solutions rather than discarding them. The Ruin phase consists of deleting random nodes (item or internal) from the solution tree until a certain maximum number of nodes have been deleted and as long as the area of the modified solution is greater than an area limit. The Recreate phase tries to insert each unpacked item into the solution, whether or not it may open new boxes to do so. Each item is inserted in the best empty space for it, considering both orientations for it and for the cut that will generate it. However, to introduce a stochastic element, there is a probability that one of the options is ignored, which could lead to insertion in a sub-optimal space. Finally, to decide whether the new generated solution will be accepted or not, the Late-Acceptance Hill-Climbing metaheuristic is used. In this metaheuristic, a new solution is accepted or not if, and only if, it is better than the current solution X iterations ago. Algorithm 1 presents a high-level overview of the algorithm.

Algorithm 1. Gardeyn and Wauters 2021 - Ruin and Recreate Heuristic

```
S, S*, S' ← ∅
while There is time left do
    S' ← RuinSolution(S) //randomly remove sub-trees from the solution
    S' ← RecreateSolution(S') //select items to fill empty spaces
    S'_val ← EvaluateSolution(S')
    if EvaluateSolution(S') < EvaluateSolution(S) then
        S ← S'
        if SolutionIsFeasible(S) then
            if EvaluateSolution(S') < EvaluateSolution(S*) then
                S* ← S'
            end if
        end if
    end if
end while
```

Results. Our implementation of the algorithm was run for all 855 instances described in the beginning of this section. The code was developed in C++11 and compiled with GCC v9.4.0. The tests were run on an Ubuntu 18.04 machine with an Intel Core i7 980 and 24 GB of RAM. In the original article, each instance ran for 10 min with 8 threads. As multi-threading was not implemented in this work and to analyze the impact of running time on the algorithm's performance, it was decided to run our version for 25 min, then reporting the final result and the result obtained in the first 10 min of each execution.

The Tables 1, 2 and 3 display the results obtained in the instances of [7,11] and [12], respectively. Each line presents the average value of utilization, in percentage, obtained in each instance subset, for our algorithm with 10 and 25 min of execution, the result reported in [3] and the difference between the result in the article original and our 25-minute performance. The final line presents the averages for each column. The use is defined as the sum of the area of all items divided by the sum of the area of all boxes used in the solution found. Therefore, its maximum value is 1 (100%).

Table 1. Obtained vs original results on instances of [7] on $2VSBP|R|G$

Class Name	Sol Value Mean (%)			
	Ours (10 m)	Ours (25 m)	Gardeyn	Diff (G - O(25 m))
M1	98.38	98.38	98.38	0.00
M2	96.53	97.43	97.87	0.44
M3	97.98	98.26	98.85	0.59
Average	97.63	98.02	98.37	0.34

Table 2. Obtained vs original results in instances of [11] in $2VSBP|R|G$

Class Name	Sol Value Mean (%)			
	Ours (10 m)	Ours (25 m)	Gardeyn	Diff (G - O(25 m))
Nice25i	91.70	91.89	96.82	4.93
Nice50i	90.31	91.25	92.97	1.72
Nice100i	87.41	88.76	94.49	5.73
Nice200i	89.38	90.64	94.21	3.57
Nice300i	88.24	89.99	93.10	3.11
Nice400i	89.89	91.51	93.74	2.23
Nice500i	88.29	90.22	93.54	3.32
Path25i	91.72	91.89	98.88	6.99
Path50i	94.34	95.27	96.49	1.22
Path100i	92.85	93.89	97.17	3.28
Path200i	92.58	93.73	96.56	2.83
Path300i	93.29	94.15	95.85	1.70
Path400i	93.47	94.44	97.00	2.56
Path500i	91.60	94.08	96.56	2.48
Average	91.08	92.27	95.53	3.26

Table 3. Obtained vs original results in instances of [12] on $2VSBP|R|G$

Class Name	Sol Value Mean (%)			
	Ours (10 m)	Ours (25 m)	Gardeyn	Diff (G - O(25 m))
MB_C1	94.03	95.19	97.20	2.01
MB_C2	96.45	96.65	96.75	0.10
MB_C3	90.64	92.09	95.32	3.23
MB_C4	94.39	95.33	96.54	1.21
MB_C5	88.87	90.11	93.69	3.58
MB_C6	91.94	92.93	94.68	1.75
MB_C7	90.30	91.46	95.01	3.55
MB_C8	89.74	91.18	95.25	4.07
MB_C9	75.22	76.51	79.99	3.48
MB_C10	91.84	93.27	96.01	2.74
Average	90.34	91.47	94.04	2.57

Through an analysis of the results, it can be seen that the results obtained by our implementation are close to those reported in the original article, and, therefore, it is a faithful reproduction of the original algorithm, in addition to a positive validation of such reported results in [3]. This implementation, then, served as a baseline for the following stages of this work, since it is possible to compare the performance between the algorithm and proposed modifications.

3.2 Proposed Improvement to the Baseline Algorithm

Although the main objective of the 2DVSBPP and, consequently, of the Gardeyn algorithm, is to maximize the percentage of box area used by the solution, in practical applications this is not the only priority taken into account when one needs to produce good solutions to the problem. More specifically, in industrial applications where 2DVSBPP is used to define a sequence of guillotine cuts in metal/glass sheets in order to obtain smaller pieces, one of the criteria applied to find a feasible solution is the number of cuts to be made or, in more particular scenarios, the number of exchanges between a sequence of cuts in one orientation (for example, vertical) to a sequence of cuts in the other orientation (for example, horizontal), as addressed in [2]. This is mainly due to the fact that, in such scenarios, changing the orientation of machine cuts requires performing slow and costly procedures, such as rotating all the base material or parts of the machine itself. So, a solution that minimizes such number of exchanges of cut orientation can drastically save a lot of time and money for a manufacture. In this scenario, this last criterion can be measured as the maximum depth among the trees that represent a solution constructed by the baseline algorithm, since, as already described, an internal node of a solution tree represents a sequence of cuts in the same direction, and the children nodes represent cuts in the opposite direction. Therefore, it was decided to measure the maximum depth of the solutions generated by our reproduction of the algorithm described above. The results can be seen in Tables 4, 5 and 6.

It was noticed that in the instances of [12] and, mainly, of [11], many solutions had depths considered impractical by industry standards (in [2], the industry requirement is a max depth of 3). This result motivated the elaboration of a modification to the original algorithm, which not only aims at improving the use of the area used in the solution, but also focus on generating solutions of smaller depths, less variable and with no extreme values (in certain instances, the original method generated solutions with depth above 20).

Description of the Proposed Algorithm. While analysing the solutions generated by reproducing the original algorithm, it was noticed that the high depth of the solutions was ofter concentrated in some specific regions of the box where many items were grouped with cutting directions that alternated with high frequency. Therefore, the proposed solution to mitigate this problem, in order to generate solutions of less depth but maintaining or possibly increasing the use of boxes, is to use a Dynamic Programming (DP) algorithm. Given an empty space in the box, choose a set of items to be arranged side by side in this space in order to maximize the sum of the area of the selected items. Such DP is equivalent to solving a Knapsack Problem, where items with weight p_i and value v_i must be chosen in order to maximize the sum of their values without

Table 4. Maximum depth of solutions obtained in instances of [11] in $2VSBP|R|G$

Class Name	Deepest Level			
	Mean	Max	Min	Std
Nice25i	4.90	7	3	1.21
Nice50i	6.60	12	5	1.71
Nice100i	7.56	13	5	2.14
Nice200i	8.68	16	6	2.14
Nice300i	9.84	18	6	2.82
Nice400i	10.28	17	7	2.61
Nice500i	11.24	25	8	3.53
Path25i	5.35	10	3	1.69
Path50i	7.20	10	5	1.41
Path100i	8.52	12	7	1.48
Path200i	10.64	14	7	2.06
Path300i	11.72	16	9	1.65
Path400i	11.96	19	8	2.73
Path500i	12.76	19	9	2.20
Average	9.09	14.86	6.29	2.10

Table 5. Maximum depth of solutions obtained in instances of [12] in $2VSBP|R|G$

Class Name	Deepest Level			
	Mean	Max	Min	Std
MB_C1	4.56	7	2	1.32
MB_C2	5.70	8	4	1.02
MB_C3	4.70	6	4	0.61
MB_C4	6.62	9	4	1.05
MB_C5	4.96	6	4	0.70
MB_C6	7.12	10	6	1.12
MB_C7	4.88	7	4	0.80
MB_C8	4.80	6	3	0.76
MB_C9	4.16	6	3	0.74
MB_C10	5.68	7	4	0.71
Average	5.32	7.20	3.80	0.88

Table 6. Maximum depth of solutions obtained in instances of [7] in $2VSBP|R|G$

Class Name	Deepest Level			
	Mean	Max	Min	Std
M1	6.00	7	5	0.71
M2	5.00	5	5	0.00
M3	5.40	7	4	1.14
Average	5.47	6.33	4.67	0.62

the sum of their weights exceeding a pre-defined limit. In this case, v_i is the area of the item i and p_i is either the width or the height of the item, depending on how it is arranged in the empty area.

More formally, let I be the set of items not yet added to the solution such that each item would fit individually in the empty space to be filled, H be the height of the space and W its width. Let h_i also be the height of item $i \in I$ and w_i be the width of $i \in I$. If $W > H$, then the items will be arranged from top to bottom in the empty space. Otherwise, left to right, side by side. Without loss of generalization, considering that $W > H$, the items are chosen according to the following recurrence equation:

$$F(k,l) = \begin{cases} 0, & \text{if } k = 0. \\ -\infty, & \text{if } l < 0. \\ max\{F(k-1,l), F(k-1,l-w_k) + (w_k \cdot h_k)\}, & \text{if } w_k \leq h_k \cup w_k > H. \\ max\{F(k-1,l), F(k-1,l-h_k) + (w_k \cdot h_k)\}, & \text{if } h_k \leq w_k \cup h_k > H. \end{cases} \tag{1}$$

The initial call of the equation is $F(\|I\|, W)$. The asymptotic time and memory complexity of this algorithm are equal to $O(\|I\| \cdot W)$, since the values of $F(k,l)$ are computed and stored for every pair of values $(k \in \{1...\|I\|\}, l \in \{1...W\})$.

Therefore, in order to preserve the core functioning of the original algorithm and maintain its stochastic character, it was decided to introduce such DP filling as part of the *Recreate* step of the original algorithm. This way, after each insertion of an item during this step, one of the new empty spaces generated by such inclusion (randomly selected with equal chances) is filled with the new proposed method with probability $p = 1 - \gamma^d$, where d is the depth of the node corresponding to the empty space in question and $\gamma \in \mathbb{R}$ is a parameter between 0 and 1. This means that this method is more likely to be called when the current branch of the tree is deep. After experimentation, it was decided to set $\gamma = 0.6$. Furthermore, after DP defines which items should be inserted, they are cut from highest to lowest, from left to right, and the direction of their cuts is defined using the same criteria as the original article.

Results. Since solutions with excessive depth were observed in test instances of the [11] and [12] sets, the algorithm with the proposed modification was executed in these instances, on the same machine and same parameters described in Sect. 3.2.1, for 25 min each. The results are displayed in Tables 7 and 8. Furthermore, in order to visualize the difference in the variation in the depth of the solutions found within the same set of instances, *boxplots*, available in Figs. 1 and 2, were generated with the distribution of such depth.

It can be seen that in both Ortmann and Pisinger instances, no significant difference was observed between the percentage of use of the solutions generated by the two methods. However, when observing the boxplot with the distribution of solution depths, it is evident that, especially in the most challenging sets of instances in Ortmann's dataset - Nice300i, Nice400i, Nice500i, Path300i,

Path400i and Path500i -, the solutions found by the proposed algorithm have depths smaller, less varied and without extreme values than those displayed by the original algorithm. Such a difference, however, is not present in the Pisinger set, where the results displayed by the two methods are extremely close.

Table 7. Results obtained vs original in instances of [11] at $2VSBO|R|G$

Class Name	Sol Value Mean (%)			Deepest Level Mean		
	Gardeyn	Proposed	Diff (G - P)	Gardeyn	Proposed	Diff (G - P)
Nice25i	91.62	92.24	−0.61	4.90	5.20	−0.30
Nice50i	91.25	90.92	0.32	6.60	6.04	0.56
Nice100i	88.54	88.27	0.27	7.56	6.68	0.88
Nice200i	90.60	90.41	0.19	8.68	8.20	0.48
Nice300i	89.88	90.04	−0.16	9.84	8.72	1.12
Nice400i	91.14	91.68	−0.54	10.28	8.72	1.56
Nice500i	90.18	91.31	−1.13	11.24	8.84	2.40
Path25i	92.20	92.89	−0.69	5.35	5.40	−0.05
Path50i	94.80	94.81	−0.01	7.20	7.72	−0.52
Path100i	93.80	93.21	0.58	8.52	8.88	−0.36
Path200i	94.79	92.94	1.85	10.64	10.52	0.12
Path300i	94.11	93.45	0.66	11.72	10.40	1.32
Path400i	94.59	94.11	0.48	11.96	10.76	1.20
Path500i	93.80	91.99	1.81	12.76	11.24	1.52
Average	92.24	92.02	0.22	9.09	8.38	0.71

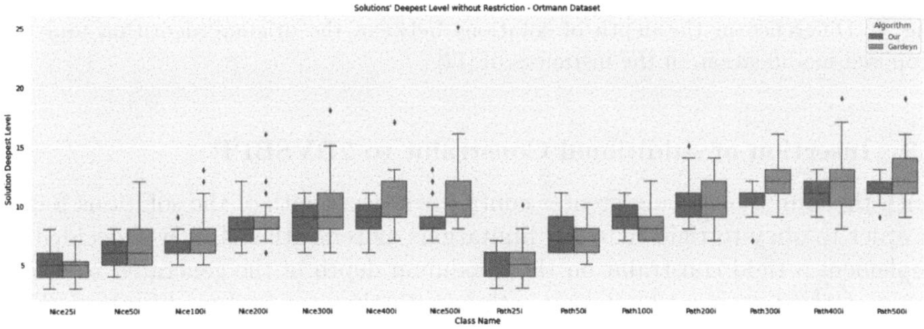

Fig. 1. Difference in the depth of solutions between the original algorithm and the proposed modification, in the instances of [11]

Table 8. Results obtained vs original in instances of [12] in $2DVSBP|R|G$

Class Name	Sol Value Mean (%)			Deepest Level Mean		
	Gardeyn	Proposed	Diff (G - P)	Gardeyn	Proposed	Diff (G - P)
MB_C1	95.18	95.43	−0.25	4.56	4.62	−0.06
MB_C2	96.66	96.70	−0.03	5.70	5.46	0.24
MB_C3	91.97	92.30	−0.33	4.70	4.78	−0.08
MB_C4	95.23	95.35	−0.12	6.62	6.38	0.24
MB_C5	89.98	90.48	−0.51	4.96	5.20	−0.24
MB_C6	92.67	92.52	0.15	7.12	7.08	0.04
MB_C7	91.40	90.95	0.45	4.88	4.98	−0.10
MB_C8	91.04	91.11	−0.07	4.80	4.84	−0.04
MB_C9	76.60	76.41	0.19	4.16	4.10	0.06
MB_C10	93.22	93.44	−0.22	5.68	5.72	−0.04
Average	91.39	91.47	−0.07	5.32	5.32	0.00

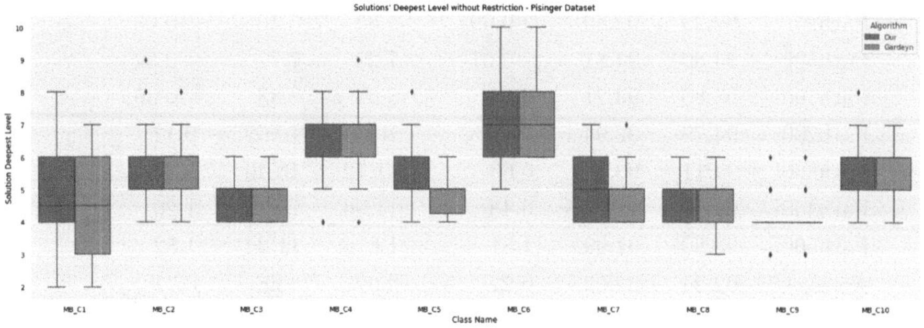

Fig. 2. Difference in the depth of solutions between the original algorithm and the proposed modification, in the instances of [12]

3.3 Insertion of Additional Constraint to 2DVSBPP

With the aim of obtaining greater control over the depth of the solutions found in order to obey to rigid industry limitations aforementioned, it was decided to implement a rigid constraint on the maximum depth of the generated solution, parameterized as *input* to the algorithm, with the proposed modification. This constraint was implemented in order to minimize the impact on the quality of solutions found without it. Thus, the DP filling method is only run in empty spaces whose corresponding node has a depth less than the maximum minus 1, in order to prevent the choice of an item whose cut increases the depth of the solution beyond what is allowed. In other steps of the algorithm, when selecting the order of cuts to be added and the items to be inserted, these choices are also restricted to avoid breaking the constraint.

Results. The proposed algorithm with the strict maximum depth constraint was executed for the instances of [11] and [12], on the same machine and parameters described in Sect. 3.2.1, with the same value of *gamma* mentioned in Sect. 3.3.1, and maximum depth of 7. The results comparing the performance of the algorithm with and without the additional constraint are in Tables 9 and 10. Furthermore, Figs. 9 and 10 show the distribution of the solution maximum depths of both proposed methods.

It can thus be seen that it was possible to add the new restriction with minimal impact on the quality of the solutions found, since the average difference between the use of the area of solutions with and without the constraint is less than 1%. Even in the most challenging sets of Ortmann's dataset, where the original solutions are quite deep, the constrained algorithm was able to find solutions of practically the same quality as those without the constraint (Fig. 3 and 4).

Table 9. Results obtained vs original in instances of [11] at $2DVSBP|R|G$

Class Name	Sol Value Mean (%)		
	Without Rest.	With Rest.	Diff (WO - W)
Nice25i	92.24	91.61	0.62
Nice50i	90.92	89.49	1.43
Nice100i	88.27	87.48	0.80
Nice200i	90.41	89.95	0.46
Nice300i	90.04	89.46	0.57
Nice400i	91.68	91.12	0.57
Nice500i	91.31	90.24	1.07
Path25i	92.89	92.01	0.88
Path50i	94.81	94.35	0.46
Path100i	93.21	92.72	0.49
Path200i	92.94	92.42	0.52
Path300i	93.45	92.86	0.59
Path400i	94.11	93.29	0.82
Path500i	91.99	91.08	0.91
Average	92.02	91.29	0.73

Table 10. Results obtained vs original in instances of [12] in $2DVSBP|R|G$

Class Name	Sol Value Mean (%)		
	Without Rest.	With Rest.	Diff (WO - W)
MB_C1	95.43	94.74	0.69
MB_C2	96.70	96.62	0.08
MB_C3	92.30	91.45	0.85
MB_C4	95.35	94.56	0.78
MB_C5	90.48	89.63	0.85
MB_C6	92.52	92.05	0.47
MB_C7	90.95	89.61	1.34
MB_C8	91.11	88.84	2.27
MB_C9	76.41	75.70	0.71
MB_C10	93.44	92.53	0.91
Average	91.47	90.57	0.89

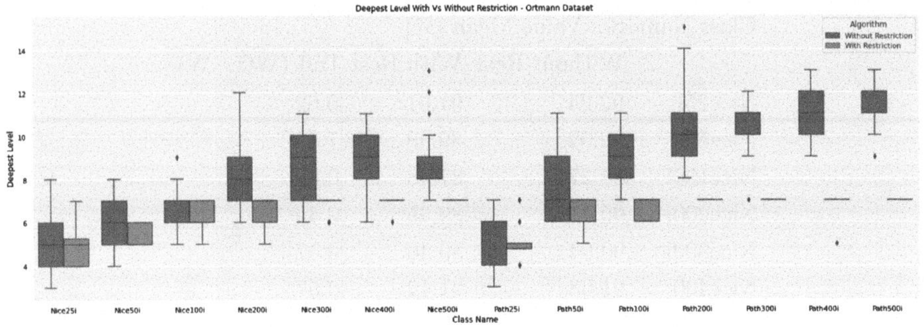

Fig. 3. Difference in the depth of solutions between the proposed algorithm without restrictions and with restrictions, in the instances of [11]

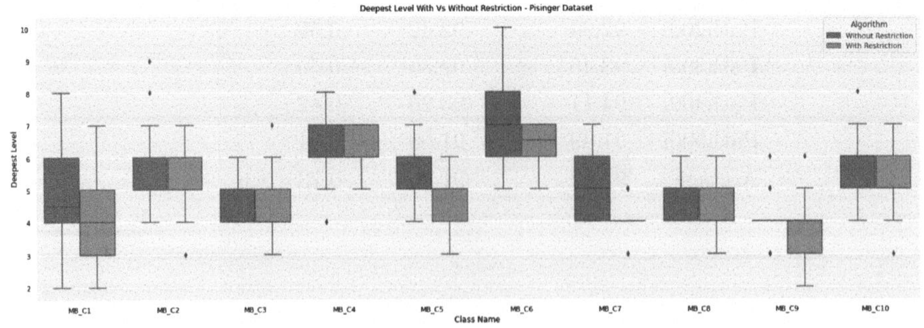

Fig. 4. Difference in the depth of solutions between the proposed algorithm without restrictions and with restrictions, in the instances of [12]

4 Conclusion

This work presents an extensive review of the literature on the 2D Bin Packing Problem, focusing on its variable size version, the re-implementation of the state-of-the-art algorithm for the 2D Bin Packing Problem with guillotine cuts constraint, the proposal of a modification to this method in order to improve the real practicality of the solutions found by it and the insertion of an additional constraint to the 2DVSBPP and, correspondingly, to the proposed algorithm in order to deal with this.

Through the results obtained by the reproduced code, it was possible to verify the results reported in [3]. Furthermore, the proposal for a modification to the original algorithm by using a DP to more optimally choose the items that fill empty spaces in the boxes in order to reduce the final depth of the solution - and, so, make it more practical in industrial applications - made this method more stable under this criterion, meaning it can find solutions without extreme maximum depth values and with smaller depth variation between similar instances. Finally, the addition of a hard constraint on the maximum depth of solutions gave more control over the algorithm and made it even more applicable in practical situations that require such constraint.

References

1. Alvarez-Valdés, R., Parreño, F., Tamarit, J.M.: A grasp/path relinking algorithm for two-and three-dimensional multiple bin-size bin packing problems. Comput. Oper. Res. **40**(12), 3081–3090 (2013)
2. Bogue, E.T., Guimarães, M.V., Noronha, T.F., Pereira, A.H., Carvalho, I.A., Urrutia, S.: The two-dimensional guillotine cutting stock problem with stack constraints. In: 2021 XLVII Latin American Computing Conference (CLEI), pp. 1–9. IEEE (2021)
3. Gardeyn, J., Wauters, T.: A goal-driven ruin and recreate heuristic for the 2D variable-sized bin packing problem with guillotine constraints. Eur. J. Oper. Res. **301**(2), 432–444 (2022)
4. Guimaraes, M., Bogue, E., Pereira, A., Carvalho, I., Noronha, T., Urrutia, S.: Heurısticas construtivas para o problema de corte guilhotinado bidimensional em 3 estágios com restriçoes de precedência. Anais do LI Simpósio Brasileiro de Pesquisa Operacional **2** (2019)
5. Guimarães, M.V., Bogue, E.T., Carvalho, I.A., Pereira, A.H., Noronha, T.F., Urrutia, S.: A biased random-key genetic algorithm for the 2-dimensional guillotine cutting stock problem with stack constraints. In: Dorronsoro, B., Yalaoui, F., Talbi, EG., Danoy, G. (eds.) International Conference on Metaheuristics and Nature Inspired Computing, pp. 155–169. Springer, Heidelberg (2021). https://doi.org/10.1007/978-3-030-94216-8_12
6. Hong, S., Zhang, D., Lau, H.C., Zeng, X., Si, Y.W.: A hybrid heuristic algorithm for the 2D variable-sized bin packing problem. Eur. J. Oper. Res. **238**(1), 95–103 (2014)
7. Hopper, E., Turton, B.: Problem generators for rectangular packing problems. Stud. Inform. Univ. **2**(1), 123–136 (2002)

8. Liu, Y., Chu, C., Wang, K.: A dynamic programming-based heuristic for the variable sized two-dimensional bin packing problem. Int. J. Prod. Res. **49**(13), 3815–3831 (2011)
9. Lodi, A., Martello, S., Vigo, D.: Heuristic and metaheuristic approaches for a class of two-dimensional bin packing problems. INFORMS J. Comput. **11**(4), 345–357 (1999)
10. Mezghani, S., Haddar, B., Chabchoub, H.: The evolution of rectangular bin packing problem—a review of research topics, applications, and cited papers (2023)
11. Ortmann, F.G., Ntene, N., Van Vuuren, J.H.: New and improved level heuristics for the rectangular strip packing and variable-sized bin packing problems. Eur. J. Oper. Res. **203**(2), 306–315 (2010)
12. Pisinger, D., Sigurd, M.: The two-dimensional bin packing problem with variable bin sizes and costs. Discret. Optim. **2**(2), 154–167 (2005)
13. Polyakovskiy, S., M'Hallah, R.: A lookahead matheuristic for the unweighed variable-sized two-dimensional bin packing problem. Eur. J. Oper. Res. **299**(1), 104–117 (2022)

Humanities and AI: Ethical Education in Technology Careers

Laura Simões Camargo[1,2](✉), Enio Alterman Blay[2](✉), and Gabriela Soares Schmidt[2](✉)

[1] José Luiz Egydio Setúbal Foundation, Av. Angélica, 2071, São Paulo 01227-200, Brazil
lscamargo27@gmail.com
[2] University of São Paulo, R. da Reitoria, 374, São Paulo 05508-220, Brazil
{enio.alterman.blay,gabriela.soares.schmidt}@usp.br

Abstract. This paper presents a quantitative and qualitative study of the teaching of ethics and related subjects in undergraduate courses in Computing of 17 top-ranked universities evaluated in three different ranking systems. The analysis was performed using documents from the selected universities and semi-structured interviews with professors and a head of department. This article discusses the relevance of ethical and social-oriented teaching and classifies the syllabuses of the sampled disciplines, assessing the type of focus that is apparently given. The overall results show considerable variability in the availability of data, ranging from courses whose content is very detailed to those that don't present their syllabus at all. Considering that the development of critical reasoning based on ethics is extremely relevant to the education of future Artificial Intelligence developers, the authors concluded that universities are not fulfilling their role adequately.

Keywords: Artificial Intelligence · Ethics · Human Rights · Computing · Education

1 Introduction

The world has changed since the release of OpenAI's generative conversational tool at the end of 2022 [5]. The consequent concern about regulation and ethical behavior in the face of Artificial Intelligence (AI) has become the order of the day worldwide [7]. However, little attention seems to be paid to critical education for AI developers, especially in the Global South. UNESCO itself [9,10] has published reports on the impacts of AI on education at basic levels and on society at large but has not addressed the professional education of those who create and employ the algorithms themselves. In addition, studies such as [4,6,14,16] and [11] are at the frontier of the discussion on ethics, education and AI, but address issues outside the Brazilian context.

In the various areas of Computing, higher education is the natural path for both scientists and technical professionals who work in AI. In order to focus on

A. Paes and F. A. N. Verri (Eds.): BRACIS 2024, LNAI 15413, pp. 81–93, 2025.
https://doi.org/10.1007/978-3-031-79032-4_6

education as a possible way of mitigating the risks and negative social impacts of AI, this article maps out the humanistic subjects offered in bachelor's degree programs in the field of Computing at main Brazilian universities and classifies their approaches.

Despite the worldwide impact of the social challenges posed by recent generative technologies, one might consider that the discussion of ethics in the field of Computing has been going on for more than 20 years. Not only was the Association of Computing Machinery's (ACM) code of ethics [1] published in the 1990s, but some academics have been dedicated to teaching this subject long before the contemporary phenomenon of AI [8], even presenting the same type of dilemmas, such as the reproduction of racism, the misappropriation of intellectual property, among others. However, through interviews conducted in this research, it was possible to gather evidence that these efforts have been and continue to be individual, rather than systemic or structural. Although there are groups concerned with the social impacts of AI within Computing, they are still small and their initiatives are insufficient in the face of current challenges.

There are more than 50 million places offered by around 22,000 bachelor's courses in Computer Science, Information Systems, Software Engineering, and Computer Engineering registered with the Ministry of Education (MEC). This data demonstrates the magnitude of the area and reinforces the relevance of the discussion on the critical education of these students.

Considering the size of the universe to be researched, it would be beyond the scope of this work to analyze all the courses. the authors, therefore, opted for a subset of 17 universities, chosen for their importance in terms of publications in the area of AI, verified on the Scopus platform and in a Fapesp publication [12], and a ranking system, presented by the specialized publication Folha de São Paulo University Ranking 2023, as detailed in the Sect. 3 section.

In addition to this introduction, this article is composed of five other sections. The second section presents aspects of the field of Computing related to ethics. The third section presents the method used to collect the data. The next section discusses the quantitative and qualitative results of the subjects. The fifth section systematizes the syllabuses according to their classifications. Finally, the final section presents some of the authors' conclusions.

2 The Construction of the Idea of Ethics in Computing

Professional activities have always involved ethical issues. This goes back to the most ancient civilizations whose history is documented, as well as the rules of native peoples whose traditions are lost in time. Of today's professions, the medical field has the oldest tradition of ethical values. This can be traced back to the Hippocratic Oath, for example, which deals with commitments such as secrecy and beneficence in the doctor-patient relationship. Created more than 2.5 millennia ago and introduced into Brazilian medical practice in 1771, the oath is still taken at the end of undergraduate medical courses. Perhaps that's why the biological sciences have such well-structured practices and regulations, as well as a network of 890 Research Ethics Councils (CEP) throughout Brazil.

In the case of Computing, although some discussions about the ethical use of technology goes way back, systematic discussions in congresses and seminars date to end of the 1980s and begin of the 1990s [15]. In addition to the fact that there are no Research Ethics Councils in Computing, it is not customary for professionals in the field to submit their projects to other councils, which should be the case, especially when involving human beings. This means that there is no formal impediment to computer professionals carrying out their work in an unethical manner.

3 Data Collection and Analysis

Having explained the context of Computer Science education and its dimension in Brazil, this section details how the data was collected and analyzed in this research.

In order to build the sample, the universities analyzed were chosen from three sources: i) the article "Scientific publications on artificial intelligence" by Fapesp [12], ii) Folha de São Paulo University Ranking 2023 (RUF) and iii) searches carried out by the authors on the Scopus platform. These sources were chosen to combine the relevance of the academic output of these universities with their other characteristics, such as teaching, innovation, and internationalization.

After defining the sources, the ten institutions with the greatest relevance in each of them were selected. Once the data had been cross-referenced and overlaps excluded, the final sample was consolidated in Table 1, presented below. A detailed explanation of the selection process is described in the following subsections.

3.1 Fapesp and Scopus Rankings

According to the article "Scientific publications on artificial intelligence" [12], the 10 leading institutions in AI publications between 2014 and 2018 were, respectively: USP, Unicamp, UFPE, UFMG, UFRN, UFRGS, UFPR, UFSCar, Unesp, and UTFPR.

Although international rankings like Edurank [17] and Times Higher Education [18] evaluate more metrics to achieve universities classification, they provide a generic and global ranking on all areas. As this article focuses on AI, the authors simplified the criteria aiming at this specific area in Brazil.

Based on Scopus searches using the keywords "artificial intelligence", "machine learning", "deep learning" and "natural language processing", the 10 Brazilian institutions with the largest number of publications on each topic were listed. After eliminating overlaps, 11 universities were added. They were: USP, Unicamp, UFPE, UFMG, UFRGS, UFRJ, UFSC, UnB, PUC-RJ, UFF, and PUC-RS.

It is worth noting that just by searching for "artificial intelligence" as a keyword, there were more than 65,000 articles published in Brazil between 1974 and 2024, i.e. in the last 50 years. As shown in Fig. 1, the universities with the

Table 1. Sample Composition. Source: Author's elaboration, 2024.

	Name	
1	USP	University of São Paulo
2	Unicamp	State University of Campinas
3	UFPE	Federal University of Pernambuco
4	UFMG	Federal University of Minas Gerais
5	UFRN	Federal University of Rio Grande do Norte
6	UFRGS	Federal University of Rio Grande do Sul
7	UFPR	Federal University of Paraná
8	UFSCar	Federal University of São Carlos
9	Unesp	São Paulo State University
10	UTFPR	Federal University of Technology - Paraná
11	UFRJ	Federal University of Rio de Janeiro
12	UFSC	Federal University of Santa Catarina
13	UnB	University of Brasília
14	Unifesp	Federal University of São Paulo
15	PUC-RJ	Pontifical Catholic University of Rio de Janeiro
16	UFF	Fluminense Federal University
17	PUC-RS	Pontifical Catholic University of Rio Grande do Sul

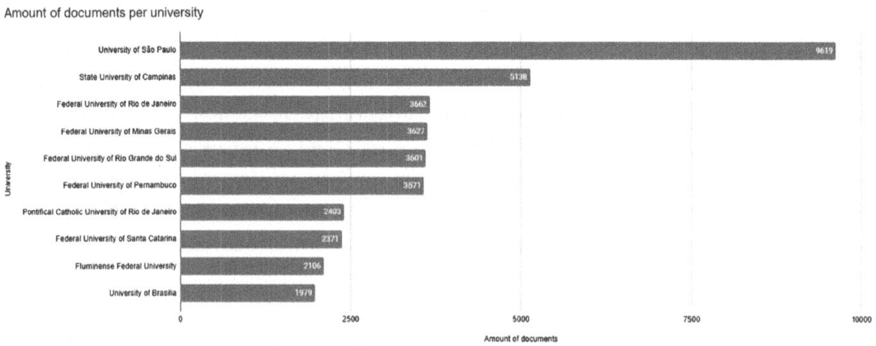

Fig. 1. AI publications by author affiliation. Source: Plataform Scopus, 2024.

most publications in the area are, respectively: USP, Unicamp, UFRJ, UFMG, UFRGS, UFPE, PUC-RJ, UFSC, UFF, and UnB.

3.2 RUF

According to the Folha de São Paulo University Ranking 2023 (RUF) [13], an annual assessment of higher education in Brazil that considers teaching, research, market orientation, innovation and internationalization indexes, the best-placed universities are, respectively: USP, Unicamp, UFRGS, UFRJ, UFMG, Unesp, UFSC, UFPR, UnB and UFPE.

3.3 Consolidating the Field and Building the Database

Once the sample of universities to be analyzed had been selected, the bachelor's degree courses were chosen. During the initial phase of the research, the authors held conversations with researchers from USP's Center for Artificial Intelligence (C4AI) about the AI field. A university degree in Computing was presented as the main qualification among AI developers in the private sector. In addition, Resolution No. 5 of November 16, 2016 [3], which establishes the National Curriculum Guidelines for undergraduate courses in the area of Computing, states that the area is composed of the following courses: Computer Science, Information Systems, Computer Engineering and Software Engineering. These were the bachelor's degree courses chosen by the authors for analysis.

In order to identify and analyze the subjects related to ethics and social sciences offered by these 17 universities in the four bachelor's degree courses described above, the authors created a database containing the following variables for each subject: university where the course was offered, institute or faculty, course, name of the subject, whether it is obligatory or optional course, and syllabus.

To search for the subjects, the authors used the Curricular Matrices available on the official websites of the faculties and institutes or the Pedagogical Political Projects, depending on which document was more up-to-date. The subjects were searched for using the following keywords: ethics, society, social, Sociology, diversity, human rights, humanism and humanities. The terms were chosen from the bibliography, Resolution No. 5 of November 16, 2016, and a pre-analysis of the curricula.

3.4 Categorization of the Syllabuses

After creating the categories for analysis, adapted from the literature review by [2], each syllabus was read and classified by at least two authors independently. Both classifications were then checked, and, in cases of disagreement, the authors debated until a consensus was reached.

3.5 Semi-structured Interviews

To complement the understanding of ethics teaching in the context of higher education, three semi-structured interviews were carried out, one with a professor of ethics in Computer Science, one with a professor of ethics in Engineering, and one with a professor of ethics and head of a Computer Science Department. In all cases, the interviewees were part of the teaching staff of the universities that make up the sample studied.

4 Analysis of Ethics-Related Subjects in Computing Courses

This section presents two analyses. Firstly, considerations on the survey of ethics-related subjects offered in the courses evaluated. These are quantitative assessments. In the next subsection, qualitative analyses are made. The authors define "ethics-related subjects" as the set of subjects identified using the selected keywords: ethics, society, social, Sociology, diversity, human rights, humanism, and humanities.

4.1 Descriptive Analysis of Ethics-Related Disciplines

This section deals with the research analysis into the quantitative evaluation of ethics-related disciplines in the Computer Science, Computer Engineering, Software Engineering, and Information Systems courses of the 17 universities studied. To the assessment, quantity and the compulsory nature of those disciplines were considered. This analysis made it possible to understand the interviewees' accounts of the ideological disputes in the field, which prioritize certain areas of knowledge to the detriment of others, and prompt a broader debate about the need for critical education of future AI professionals.

Although the research sample included 17 universities, not all have the four elected bachelor's degree courses. Table 2 below shows the relationship between the courses and the universities where they are offered. It is important to take this information into account when analyzing the number of ethics-related subjects below, since the number of disciplines was calculated in absolute numbers and not proportional to the number of courses.

Table 2. Courses offered by each University. Source: Author's elaboration, 2024.

Bachelor course	Offering
Computer Science	All 17 universities
Computer Engineering	Unicamp, UFPE, UFRGS, USP, UFRN, UFSCar, UTFPR, UFRJ, UFSC, UnB, Unifesp, PUC-RJ, PUC-RS
Software Engineering	UFRN, UTFPR, UnB, PUC-RS
Information Systems	Unicamp, UFPE, UFMG, USP, UFRN, Unesp, UFSCar, UTFPR, UFSC, PUC-RJ, UFF, PUC-RS

Based on a survey of ethics subjects and related topics (such as social sciences and human rights) in curricular matrices and Political Pedagogical Plans, a total of 204 subjects were found. Of these, 83 were part of the bachelor's degree courses in Computer Science, 67 belonged to the Computer Engineering courses, 12 to the Software Engineering courses, and 42 to the Information Systems courses.

As seen in Fig. 2, UTFPR is the university with the largest number of ethics-related disciplines. It offers all four degrees, having 55 disciplines, with Computer Engineering having the most (19) and Software Engineering the least (4). At the other end, Unicamp and Unesp have only one discipline each, and none was found at UFPR.

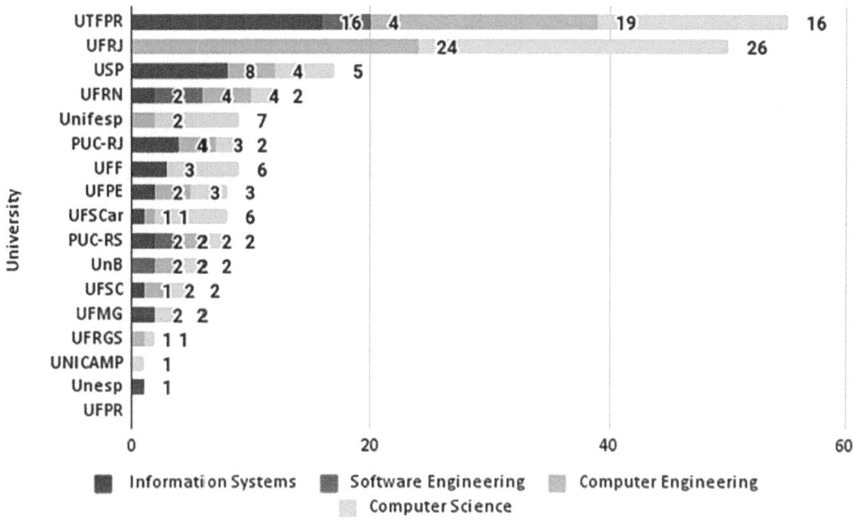

Fig. 2. Number of disciplines related to ethics, per course and university. Source: Author's elaboration, 2024.

The disciplines were classified as compulsory or optional/ elective/ complementary by analyzing the above-mentioned documents. The optional/ elective/ complementary disciplines were identified in the Curricular Matrices and Political Pedagogical Plans and were considered as a single block here treated as "optional disciplines". Furthermore, as non-compulsory subjects can be elective (compulsorily offered by the university, regardless of the number of applicants) or optional (they may not be offered), it was not possible to determine their classification in the surveyed data and whether they have been, in fact, offered or not. In the interviews, the authors found out that there are registered disciplines in university systems that are not, and have never been, offered.

As shown in Fig. 3, 132 disciplines are optional, which represents a large majority compared to the 62 compulsory ones. Thus, from the total of 204 ethics-related disciplines, only around 30.4% are obligatory, and it is up to the students individually to choose from the remaining 69.6%.

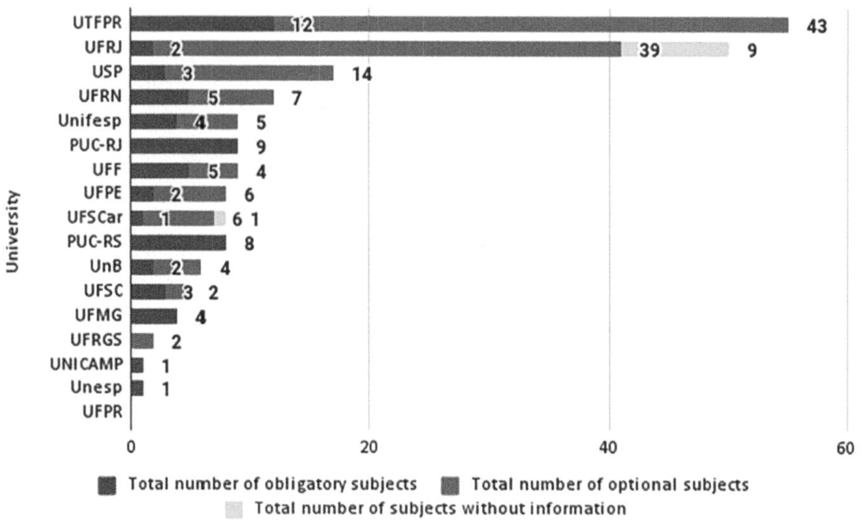

Fig. 3. Number of obligatory vs. optional disciplines per university. Source: Author's elaboration, 2024.

4.2 Analysis of the Interviews

The findings about the disparity between compulsory and optional subjects are aligned with the information from the interviews conducted with Computer Science and Engineering professors. According to what was reported, disciplines related to ethics, social responsibility, and social issues linked to technology are usually the result of personal initiatives by teachers who are interested in the subject and who, often in a self-taught way, begin to study philosophy in particular, incorporating these themes into their disciplines. They usually do this informally in technical classes, discussing real cases involving ethical issues, for example. Less often, they opt to formalize specific subjects on the topic.

When asked about the legal Resolution No. 5 of November 16, 2016, that makes humanistic teaching compulsory, the teachers and the head of department interviewed said they had no prior knowledge of it. In addition, the head of department said that, during a visit to inspect the course, the MEC did not analyze compliance with the aforementioned Resolution. In addition, the interviewees said that there was a certain amount of demise or even resistance from colleagues to the introduction of this kind of disciplines. According to them, there is an ideological dispute in the field about the essential nature of higher education for professionals in the field of Computing. Although not explicit, the neglect of ethics-related in favor of technical disciplines is based on a traditional and market-oriented discourse.

As for student interest in ethical issues, the interviewees described the profile of Engineering and Computing students as more introspective and less participative in class. For one of the interviewees, this attitude is due to the context of what is valued by companies as well as a lack of interest in social impact. However, the other interviewees attributed this behavior to typical characteristics of the social group and a more traditional hierarchy between teachers and students that is ingrained in these bachelor's degrees, and not necessarily due to the discipline itself. Furthermore, one of the interviewees considered quota students to be more politicized and interested in discussions about ethics.

5 Analysis of the Syllabuses

5.1 Creation of Categories

Ethics teaching covers a wide range of concepts and has roots in various knowledge fields. Reading the syllabuses of the collected subjects showed themes related to the fields of Sociology, Philosophy, and Law. The categories used to analyze the syllabuses were adapted from the review by [2]. In this review, the articles were identified according to the presence or absence of an explicit description of the conception of ethics, and how these concepts appeared (exclusively or mixed) in them.

[2] present three main conceptions of ethics found in the literature review: Relative to Philosophical Frameworks, Computing Applications, and Social Impacts of Computing. In general, the authors state that these conceptions used to mention "normative ethical theories or situate ethics in relation to concepts of Computing, social impact and professionalism" [2] (p. 12).

About the philosophical conception, [2] ethics is identified as a Western philosophical theory, including issues such as consequentialism, virtue, and duty, as well as a discussion of decision-making processes. On the other hand, the articles that dealt with ethics from a computational perspective did so from debates close to the legal field, addressing issues of security, privacy, and responsibility, for example. The third approach focused on the relationship between technology design and its social impacts, as well as issues such as accessibility.

As mentioned in the Sect. 3 section, the selection of subjects from Computing courses was broader than just those specific to ethics. Sociological and humanistic subjects were also taken into account. For this methodological reason, an adapted version of the three conceptions of ethics by [2] was adopted. The Computing Applications and Social Impacts of Computing categories were partly merged and partly fragmented. On the one hand, the issue of social impacts was expanded to cover not only the relationship between Computing and society but also typical Sociology topics that go beyond technology, such as Social Science theories, racism, gender inequalities and others. On the other hand, all the topics related to the profession of future Computing professionals, including both legal and responsibility debates, were compiled as a second block.

Based on the inspiration provided by the work of [2] and the particular reading of the set of syllabuses in this research, three ways of classifying the issues linked to ethics and society found in Computing subjects were drawn up:

a) Philosophical framework: Syllabuses that refer to theories, concepts, thoughts, and thinkers from the philosophical tradition; epistemology of ethics; human values; principles of justice, truth, good, freedom, etc.; mention of the notions of consequentialism (ethics based on consequences); deontology (ethics based on duties), and ethics based on virtues.
b) Applied to Computing: Syllabuses that refer to ethics in relation to the concepts and areas of application of Computing; everyday dilemmas faced by professionals; responsibility of professionals towards the community and society; ethics in decision-making; best practices and procedures; quality and security of computer systems; quality of scientific knowledge production; topics related to privacy, intellectual property, legislation, and codes of professional ethics.
c) Social issues: Syllabuses introduce the concept of ethics through the foundations of sociological thought and address the social impacts of Computing; reference to the social and historical character of the technological artifact; social repercussions and changes in how society is organized, such as in work, motivated by technology; the problem of reproducing inequalities, and stereotypes (of gender, race, class, sexuality, etc.); notions of risks and threats to the well-being of society; notions of citizenship.

5.2 Systematization of Syllabuses

All the syllabuses found in the Curricular Matrices and Political Pedagogical Projects of the four types of courses considered were analyzed based on the three categories mentioned in Subsect. 5.1. This showed 180 non-repeated disciplines. It is important to note that, of these 180 subjects, 48 did not have their syllabus available. Among those that did, there was a great disparity in the amount of information available. While some had detailed content in several descriptive paragraphs, others provided simplified generalizations in just one line. Thus, although there was one example with more than 200 words, the average was 50 words. With this in mind, about the classification, each discipline was placed in one to three categories, depending on the syllabus. Table 3 shows the following scenario:

Table 3. Courses offered by each University. Source: Author's elaboration, 2024.

Category	Total of disciplines
a. Philosophical framework	59
b. Applied to Computing	69
c. Social issues	104

As can be seen in Table 3, all three categories have a significant presence in all disciplines. However, disciplines that deal with social issues are more frequent and those with a philosophical framework make up a smaller group.

Furthermore, as shown in Table 4, the category "Social issues" appeared more often alone than the others. When combined, it is usually linked to the "Applied to Computing" category. Only 10 disciplines were categorized under "Philosophical framework" and "Applied to Computing", and only eight had all three approaches.

Table 4. Classification. Source: Author's elaboration, 2024.

One Classification	a	15
	b	15
	c	36
Two Classifications	ab and ba	10
	bc and cb	30
	ac and ca	18
Three Classifications	abc	8
No Classification	no syllabus	48
TOTAL		180

6 Conclusions

The emergence of Generative AI and the widespread use of AI-based technologies have intensified the debate on ethical issues, both nationally and internationally. The establishment of professional ethical values through education is an important step towards mitigating the negative social impacts generated by the use of AI tools.

The topic's relevance is also demonstrated by the size of the field of Computing, which in Brazil offers thousands of courses and millions of bachelor's degree places. For this reason, this article presents an introductory overview of ethics-related teaching in bachelor's degree courses in Computer Science, Computer Engineering, Software Engineering, and Information Systems at the 17 main Brazilian universities.

Based on the bibliography, an analysis of university documents and semi-structured interviews, the authors found an important disparity in the provision of ethics-related subjects/disciplines between universities. If, on the one hand, the Federal Technological University of Paraná (UTFPR) and the Federal University of Rio de Janeiro (UFRJ) were the only institutions to have more than 50 humanistic subjects on their courses, 13 universities out of the 17 evaluated offered less than 10 subjects of this type; the State University of Campinas (Unicamp) and the São Paulo State University (Unesp) with just one and the case at the Federal University of Paraná (UFPR) with none.

A significant number of subjects had no syllabus, around 24%. However, among those that did, at least one reference to some philosophical framework, applications in Computing or social issues was identified. As in the review work by [2] this does not mean that clear conceptualizations of ethics were found in the syllabuses. It just shows that, in terms of the delimitation suggested in this work, it was possible to analyze the distribution of references to fields of knowledge and to some sets of themes that the disciplines refer to in order to address the problem of ethics and social dilemmas in the context of Computing education. The balanced distribution of categories suggests no predominance of one approach. One possible explanation is that, since there is still no "canon" for the teaching of ethics in the area, there is a plurality of approaches depending on the proposal of each subject. The semi-structured interviews, although only three, showed a tendency toward relative pessimism and frustration, with little belief in the possibility of greater academic space for such ethical discussions and devaluation of the subject in relation to technological aspects.

The authors recognize the limitations of the data, which does not allow them to draw any major conclusions about the state of ethics teaching in Brazil. This article was intended to record the first findings of a more extensive research agenda, which could expand both to a bibliographical analysis of the disciplines offered by universities and to an understanding of the way they are taught by lecturers and understood by students. In addition, it is proposed to deepen the gathering and interpretation of discourses on the importance of critical education in Computing and its relationship with the development of AI for the social good. Finally, encouraging scientific debate on the ethical education of future AI developers could stimulate the application of the Resolution No. 5 of November 16, 2016 and suggest approaches to make this education effective.

Acknowledgments. The authors would like to thank the interviewees for their important insights.

References

1. Anderson, R.E.: ACM code of ethics and professional conduct. Commun. ACM **35**(5), 94–99 (1992)
2. Brown, N., Xie, B., Sarder, E., Fiesler, C., Wiese, E.S.: Teaching ethics in computing: a systematic literature review of ACM computer science education publications. ACM Trans. Comput. Educ. **24**(1), 36 (2024). https://doi.org/10.1145/3634685
3. Câmara de Educação Superior, Conselho Nacional de Educação, Ministério da Educação. Resolução CNE/CES 5/2016. Diário Oficial da União, Brasília, 17 de novembro de 2016, Seção 1, pp. 22–24 (2016)
4. Fiesler, C., Garrett, N., Beard, N.: What do we teach when we teach tech ethics? A syllabi analysis. In: Proceedings of the 51st ACM Technical Symposium on Computer Science Education (SIGCSE 2020), pp. 289–295. Association for Computing Machinery, New York (2020). https://doi.org/10.1145/3328778.3366825
5. George, A.S., George, A.H.: A review of ChatGPT AI's impact on several business sectors. Partn. Univ. Int. Innov. J. **1**(1), 9–23 (2023)

6. Gorur, R., Hoon, L., Kowal, E.: Computer science ethics education in Australia – a work in progress. In: 2020 IEEE International Conference on Teaching, Assessment, and Learning for Engineering (TALE), Takamatsu, Japan, 2020, pp. 945–947 (2020). https://doi.org/10.1109/TALE48869.2020.9368375
7. Hacker, P., Engel, A., Mauer, M.: Regulating ChatGPT and other large generative AI models. In: Proceedings of the 2023 ACM Conference on Fairness, Accountability, and Transparency, pp. 1112–1123 (2023)
8. Masiero, P.C.: Ética em computação. São Paulo, Edusp (2000)
9. Organização das Nações Unidas para a Educação, a Ciência e a Cultura. Currículos de IA para a educação básica: um mapeamento de currículos de IA aprovados pelos governos. Paris (2022). Author Proof Humanities and AI: Ethical Education in Technology Careers 13
10. Organização das Nações Unidas para a Educação, a Ciência e a Cultura. Recomendações sobre a Ética da Inteligência Artificial. Tradução de Tikinet Edições Ltda. Paris (2022)
11. Paltiel, M., Cheong, M., Coghlan, S., Lederman, R.: Approaches and models for teaching digital ethics in information systems courses – a review of the literature. Australas. J. Inf. Syst. 27 (2023). https://doi.org/10.3127/ajis.v27i0.4517
12. "Publicações científicas sobre inteligência artificial". Pesquisa Fapesp, mai. (Edição 291) (2020). https://revistapesquisa.fapesp.br/publicacoes-cientificassobre-inteligencia-artificial1/. Accessed 10 May 2024
13. Ranking de universidades. https://ruf.folha.uol.com.br/2023/ranking-deuniversidades/principal/. Accessed 10 May 2024
14. Saltz, J.S., Dewar, N.I., Heckman, R.: Key concepts for a data science ethics curriculum. In: Proceedings of the 49th ACM Technical Symposium on Computer Science Education (SIGCSE 2018), pp. 952–957. Association for Computing Machinery, New York (2018). https://doi.org/10.1145/3159450.3159483
15. Stahl, B.C., Ess, C.M.: 20 years of ETHICOMP: time to celebrate? J. Inf. Commun. Ethics Soc. **13**(3/4), 166–175 (2015)
16. Stavrakakis, I., et al.: The teaching of computer ethics on computer science and related degree programmes. a European survey. Int. J. Ethics Educ. **7**, 101–129 (2022). https://doi.org/10.1007/s40889-021-00135-1
17. University Ranking. https://edurank.org/methodology/. Accessed 28 Aug 2024
18. University Ranking. https://www.timeshighereducation.com/. Accessed 28 Aug 2024

Impact of Parent Selection Operator on the FDEA Algorithm

João Pedro A. de Oliveira(✉) and Olacir R. Castro Jr.

Universidade Federal do Acre, Rio Branco, AC, Brazil
contato@jpedro.com.br

Abstract. Many-Objective Optimization Problems (MaOPs) present a significant challenge due to the increased complexity associated with optimizing more than three conflicting objectives. Traditional Pareto-based algorithms often struggle to scale effectively with the number of objectives, leading researchers to explore alternative approaches. This study investigates the impact of parent selection strategies on the performance of the Fuzzy Decomposition-Based Multi/Many-Objective Evolutionary Algorithm (FDEA). The FDEA uses a fuzzy decomposition approach to divide a MaOP into subproblems, each solved individually to improve the adaptability and precision of solutions. The study compares the FDEA with three selection methods: binary tournament based on Pareto dominance, binary tournament using crowding distance, and a combination of Pareto dominance and crowding distance. Empirical evaluations were conducted using the DTLZ and WFG benchmark problem sets, with 2, 3, 5, 8, 10, and 15 objectives. The results indicate that incorporating quality indicators into the selection process improves the efficiency and effectiveness of the FDEA, especially in high-dimensional optimization scenarios. However, no variant was superior in all contexts. The study highlights the potential for methodological advances in Indicator-Based Mating Selection, emphasizing the need for robust and scalable algorithms capable of handling the complexities of MaOPs.

Keywords: Indicator-Based Mating Selection · Many-Objective · Parent Selection Strategies

1 Introduction

Multi-Objective Optimization Problems (MOPs) are those that require the simultaneous optimization of two or more objective functions. These objectives are usually conflicting, resulting in a set of solutions rather than a single optimal solution. To find this set of solutions, Pareto optimality theory is generally used [1].

When a MOP has more than three objectives, it is called a Many-Objective Optimization Problem (MaOP). MaOPs present greater complexity compared to traditional MOPs due to the increase in the number of objectives [1]. This added

A. Paes and F. A. N. Verri (Eds.): BRACIS 2024, LNAI 15413, pp. 94–109, 2025.
https://doi.org/10.1007/978-3-031-79032-4_7

complexity makes it harder to accurately approximate the Pareto front, making it challenging to differentiate between high-quality and low-quality solutions. In addition, maintaining the convergence and diversity of solutions becomes more challenging, requiring advanced mechanisms to prevent premature convergence and ensure diversity.

In the context of Multi-Objective Evolutionary Algorithms (MOEAs), the Fuzzy Decomposition-Based Multi/Many-Objective Evolutionary Algorithm [11] stands out by combining problem decomposition with the principles of fuzzy logic. The FDEA employs a fuzzy decomposition approach to divide the MOP into a set of subproblems, each solved individually.

An essential component of the FDEA, as well as other evolutionary algorithms, is crossover selection. Crossover selection is a crucial step aimed at identifying pairs of individuals that can generate offspring with desirable characteristics [14]. This process is fundamental to ensure compatibility and complementarity of genetic traits, thereby enhancing the generation of superior solutions.

Related research indicates that only a few MOEAs utilize the crossover selection strategy, highlighting the need for more studies and methodological advances in this area. Indicator-based methods, such as R2-IBEA and MaOEA/IGD, demonstrate the effectiveness of crossover selection based on quality indicators to improve algorithm performance in various optimization scenarios.

This paper presents a study on using different crossover selection mechanisms in the FDEA algorithm. Four variants of the FDEA were compared, each with different selection strategies, to assess their impact on the algorithm's efficiency and effectiveness in solving complex MaOPs. The variants are: FDEA, the original variant that uses random selection; FDEA_TB, which employs binary tournament selection based on Pareto dominance; FDEA_CD, which uses binary tournament selection based on crowding distance; and FDEA_CDTB, which combines Pareto dominance and crowding distance to make the final selection.

The results indicate that incorporating quality indicators into the selection process enhances the efficiency and effectiveness of the FDEA, particularly in high-dimensional optimization scenarios. However, it is important to note that no single variant proved superior across all tested contexts, suggesting that the optimal selection strategy may vary depending on the specific characteristics of the problem at hand. This study underscores the potential for methodological advancements in Indicator-Based Mating Selection.

The remaining of this paper is organized as follows: some backgroud concepts are presented in Sect. 2. Section 3 presents the Related Work. In Sect. 4 an empirical study comparing the modification of the crossover selection method on the FDEA algorithm and Sect. 5 presents the conclusions.

2 Background

2.1 Many-Objective Optimization

A Multi-Objective Optimization Problem (MOP) is a type of problem that requires the simultaneous optimization of two or more objective functions. These objectives are usually in conflict, so these problems do not present only one optimal solution but a set of them. This set of solutions is usually found using Pareto optimality theory.

A solution $\mathbf{x} \in \Omega$ is optimal relative to Ω if and only if there is no $\mathbf{x'} \in \Omega$ that dominates $\mathbf{x}$ in all objectives [1]. In other words, a solution is Pareto optimal if there is no other solution that is better or equal in all objectives and better in at least one simultaneously.

In multi-objective optimization, it is inappropriate to use the term "optimal solution" since individual objectives often conflict. Instead, a set of sufficiently good solutions is sought, representing a balance between the different objectives.

When a multi-objective optimization problem (MOP) has more than three objectives, it is classified as a Many-Objective Optimization Problem (MaOP). MaOPs are more complex than MOPs due to the increased number of objectives, which makes approximating the Pareto front more difficult and reduces selective pressure, complicating the distinction between high and low-quality solutions. Additionally, maintaining convergence and solution diversity becomes more challenging, requiring advanced mechanisms to prevent premature convergence and preserve diversity [1].

Recently, there has been a growing interest in enhancing the performance of MOEAs in MaOPs [2]. The most popular methodologies include: (1) MOEAs that use relaxed Pareto dominance relations, applying alternative preferences to increase selection pressure [3]; (2) decomposition-based MOEAs, which transform an MOP into several single-objective optimization problems using scalarization functions; and (3) indicator-based MOEAs, which use quality indicators to evaluate solution sets and define selection mechanisms, aiming to optimize the population indicator value throughout the evolutionary process.

2.2 Quality Indicators

Evaluating the performance of a multi-objective optimizer is challenging, but there are three main requirements: convergence, which seeks results close to the true Pareto set; diversity, which looks for solutions widely distributed in the objective space to allow comprehensive choices; and pertinence, which directs convergence and diversity to relevant areas of the search space where the decision-maker has the most interest. These characteristics are desired to ensure the optimizer is efficient and provides meaningful options to the decision-maker in multi-objective problems [4].

Quality indicators are functions that evaluate approximation sets and can be used to define selection mechanisms in multi-objective evolutionary algorithms (MOEAs). The idea is to optimize the value of the population indicator throughout the evolutionary process [2].

Indicator-based selection involves identifying good parental solutions based on quality indicator values. This type of selection mechanism does not aim to solve or approximate the subset selection problem based on indicators; instead, it attempts to produce promising solutions to accelerate the evolutionary process.

Crowding Distance is a metric that estimates the density of solutions around a specific solution. It is calculated as the average of the normalized absolute distances between the objective function values of two adjacent solutions in each objective. The population is sorted according to the values of the objective function, and frontier solutions are given an infinite distance to ensure their selection. The distances of intermediate solutions are normalized between the minimum and maximum values of each objective, ensuring that all solutions have the same impact on the calculation of the metric. This metric is crucial for comparing solutions that belong to the same level of non-dominance, favoring those in less "crowded" regions [12].

IGD. Inverted Generational Distance (IGD) measures the average distance from each reference point in the objective space to the nearest solution in the approximation set. This metric considers both convergence and the diversity of solutions, ensuring that the solutions not only approach the true Pareto front but are also well distributed along it. IGD is useful to assess how well the solutions cover the entire Pareto front [5].

Hipervolume (HV) is a crucial metric in multi-objective optimization used to evaluate the quality of solution sets by measuring the volume of the objective space dominated by the solutions. It assesses both convergence to the Pareto front and the dispersion of solutions within the objective space, with higher HV values usually indicating better performance. Calculating HV can be computationally intensive and requires a carefully chosen reference point, as this significantly impacts the results. Overall, HV provides a comprehensive measure of an optimization algorithm's effectiveness by considering both the proximity of solutions to the Pareto front and their distribution across the objective space [2].

2.3 Benchmark Problems

Benchmark problems are widely used in the literature to evaluate the performance of multi-objective optimizers and compare algorithms. They are specifically designed for evaluation purposes and offer several advantages, such as known true Pareto fronts for precise assessment and control over difficulty levels, objectives, and decision variables, which facilitates comparisons between algorithms [9]. By isolating the optimizer's performance, these problems eliminate potential real-world interferences, providing an objective evaluation of the algorithms' capabilities, strengths, and limitations. However, it is important to note that benchmark problems are simplifications of real-world challenges, and the results obtained may not be directly transferable to real-world situations. Among the most commonly used benchmark problems in academia are the DTLZ and WFG problem sets, which are widely recognized and accepted for evaluating multi-objective optimization algorithms.

DTLZ. The DTLZ benchmark problem set, formulated by Deb et al. [10], was designed to provide a comprehensive and rigorous evaluation of multi-objective optimization algorithms, particularly those based on evolutionary principles. These problems are scalable, allowing researchers to adjust the number of objectives and decision variables to test an algorithm's ability to manage increasing complexities. This scalability is essential because challenges in multi-objective optimization typically increase with the problem's dimensionality, both in terms of search space and the complexity of the Pareto front.

Each function within the DTLZ set is constructed to evaluate distinct characteristics of evolutionary algorithms. For example, DTLZ1 tests the convergence of algorithms to a linear Pareto front, serving as an effective starting point for initial efficacy assessments. Conversely, DTLZ2 examines the ability of algorithms to converge to a concave Pareto front, necessitating a robust search for optimal solutions within a more complex solution space. Additionally, DTLZ3 increases complexity by incorporating multiple local optima into the DTLZ2 function. These problems, featuring non-convex and discontinuous Pareto front surfaces, challenge algorithms to find suitable solutions while navigating the complex topology of the solution space. Consequently, the DTLZ problems are invaluable to the scientific community, providing a robust foundation for systematic benchmarking and objective comparison of diverse multi-objective optimization strategies.

WFG. The WFG (Walking Fish Group) benchmark problem set is designed to evaluate the efficiency of multi-objective optimization algorithms. These problems are particularly valuable because they encompass a wide range of Pareto front shapes, allowing for the assessment of algorithm performance in various complex scenarios. WFG problems utilize a vector of underlying parameters, x, linked to a simple base problem that defines the fitness space. This vector results from sequential transformations from an initial parameter vector, z, adding layers of complexity such as multi-modality and non-separability.

All WFG problems follow a specific format that involves minimizing a set of functions with parameters, z and x, where M, represents the number of objectives. The set x defines underlying parameters related to distance and position, while z comprises working parameters divided into position-related elements and distance. The constants A_1 to A_{M-1} act as degeneracy factors, reducing the dimensionality of the Pareto front. The functions h_1 to h_M define the shape, and S_1 to S_M are scaling constants. These transformations and structures provide a robust framework for systematic benchmarking and comparing different multi-objective optimization strategies.

2.4 FDEA

In the domain of Multi-Objective Evolutionary Algorithms (MOEAs), the complexity of managing high-dimensional solution spaces presents unique challenges concerning algorithm complexity and performance. The Fuzzy Decomposition-Based Multi/Many-Objective Evolutionary Algorithm (FDEA) distinguishes itself by combining problem decomposition with the principles of fuzzy logic. This algorithm employs a fuzzy decomposition approach to divide a Multi-Objective Optimization Problem (MOP) into a set of subproblems, each solved individually to enhance the adaptability and precision of solutions in multi-objective problems [11].

Algorithm 1. FDEA(N, m, G_{max})

initialize P and set $G = 1$
while $G \leq G_{max}$ **do**
  generate the offspring population Q
  $U = P \bigcup Q$ and set $P = \varnothing$, $Q = \varnothing$
  normalize all solutions in U
  $(S_1, S_2, \ldots, S_N) =$ **Fuzzy-Decomposition**(U, N, m)
  $P =$ **Elite-Selection**($S_1, S_2, \ldots, S_N$)
  $G{+}{+}$
end while
return P

The FDEA begins by generating an initial random population and creates a new population by applying variation operators (crossover and mutation) to the parent population. The algorithm then performs the fuzzy decomposition process, which is divided into two main stages: fuzzy prediction and weight vector extraction.

Fuzzy Prediction: In this phase, the solution space is partitioned into subregions using fuzzy logic. Each solution is assigned to one or more subregions based on a membership function that determines the proximity of the solution to different reference vectors. This approach allows the FDEA to capture a variety of solutions representing different parts of the Pareto front.

Weight Vector Extraction: After fuzzy prediction, the algorithm extracts weight vectors that define the preferred directions in the objective space for each subproblem. These weight vectors are used to guide the search in each subregion, encouraging exploration in directions that promote a uniform coverage of the Pareto front.

Following the fuzzy decomposition, an elite selection procedure is implemented to form the next generation. This selection strategy ensures that high-quality solutions, which best represent the desired characteristics of the objective space (such as convergence and diversity), are preserved and propagated to subsequent generations. Elite selection in the FDEA uses a combination of Pareto dominance and crowding distance to select the most promising solutions from each subregion.

The FDEA adopts a normalization procedure derived from the NSGA-III algorithm to ensure uniform treatment of objectives during the optimization process. This normalization adjusts the solutions so that all objectives are treated equitably, preventing some objectives from unduly dominating others during the selection process. Evolutionary operators, such as Simulated Binary Crossover (SBX) and Polynomial Mutation (PM), are applied to pairs of parent solutions that are randomly selected or chosen based on indicator-based strategies. This normalization is crucial for the effective performance of the FDEA, especially in problems with highly conflicting objectives.

2.5 Crossover Selection

Crossover selection in evolutionary algorithms aims to identify pairs of individuals that can produce offspring with desirable characteristics. This process is crucial to ensure compatibility and complementary genetic traits, which enhance the generation of better solutions. Producing promising offspring is critical for efficient convergence to optimal or near-optimal solutions [14].

According to Falcón-Cardona and Coello [2], crossover selection involves using quality indicator values to identify good parental solutions. This selection mechanism focuses on producing high-quality offspring to accelerate the evolutionary process rather than solving the subset selection problem based on indicators. By prioritizing the generation of promising offspring, the mechanism aims to optimize the search in the solution space, increasing the probability of finding superior solutions in fewer generations [14].

Indicator-Based Mating Selection. Multi-objective evolutionary algorithms (MOEAs) traditionally use Pareto dominance to identify a set of optimal solutions. However, quality indicators can provide additional insight that reflects user preferences and enhances the conventional concept of optimality. Within this context, IB-Mating Selection leverages these indicators to select parental solutions, aiming to improve the generation of high-quality offspring [2].

IB-Selection, a broader category within indicator-based mechanisms, includes IB-Environmental Selection, IB-Density Estimation, and IB-Archiving. IB-Environmental Selection prioritizes solutions based on their quality and contribution to the population's environment, promoting diversity and overall set quality. IB-Density Estimation assesses the density of solutions in the search space, guiding algorithms to explore less dense regions or refine high-quality solutions in denser areas. IB-Archiving uses indicators to maintain a diverse and high-quality archive of solutions representing the Pareto front.

Specifically, IB-Mating Selection focuses on selecting μ parental solutions that, through genetic operations, produce λ offspring with high potential for quality and diversity. By prioritizing high-quality offspring, IB-Mating Selection drives the evolutionary process efficiently, ensuring a constant flow of superior solutions across generations.

3 Related Work

According to Falcón-Cardona and Coello [2], only three Multi-Objective Evolutionary Algorithms (MOEAs) use the crossover selection strategy: R2-IBEA,

MaOEA/IGD, and AR-MOEA. These algorithms represent distinct approaches to optimizing the selection of parent solutions, each with its own methodologies and specific focuses.

The R2-IBEA [15] is one of the algorithms that uses the binary R2 indicator to select solutions. This method ensures that the chosen solutions outperform others in terms of weak Pareto dominance, prioritizing high-quality solutions. The main advantage of this approach is the maintenance of a robust and competitive population of solutions, improving the overall performance of the algorithm in finding optimal solutions.

The MaOEA/IGD [16] implements a binary tournament where solutions are evaluated based on two main criteria: the dominance rank and the relative distance to the hyperplane, which serves as the reference set for the IGD indicator. Initially, the solutions are ranked by their dominance, with a preference for those with lower ranks. If there is a tie in the rank, the distance is used to determine the final selection. This method is effective in identifying solutions that not only have good convergence but also present an adequate distribution in the objective space.

The AR-MOEA [18] uses the IGD-NS metric to evaluate the solutions in the tournament, selecting those that offer the highest contribution in terms of fitness value. The solution with the highest contribution is selected for the parent set, ensuring that the generated solutions are optimal in terms of convergence. This focus on maximizing the contribution of each solution helps to enhance the efficiency of the evolutionary process, resulting in superior algorithm performance in multi-objective optimization problems.

Falcón-Cardona and Coello [2] observe a significant gap regarding the emphasis and in-depth study of Indicator-Based Crossover Selection, highlighting its underestimation and potential for new methodological advances. Indicator-Based Crossover Selection methods integrate Quality Indicators (QIs) as a central selection criterion, quantifying specific characteristics of solutions or the population for more informed and targeted evaluations.

4 Empirical Study

This section presents the results of the modification of the crossover selection operator in the FDEA algorithm. The study involved adapting the FDEA to use different selection strategies to select the parents, which were then subjected to evolutionary operators SBX and PM. An experimental study was carried out to compare the original FDEA selection (random) and the new selection strategies to assess their effectiveness. The following sections detail the standard and modified parameters, as well as the evaluation criteria used to assess the results.

4.1 Experimental Setup

In this study, the FDEA algorithm was executed with the same parameters as those outlined in the original article [11], as shown in Table 1, to maintain result

Table 1. Population and objective numbers

Objectives (m)	Population Size (N)	Maximum Generations (G_{max})
2	100	300
3	120	500
5	210	600
8	240	800
10	275	1000
15	240	1500

consistency and comparability. The number of objectives (m) ranged from 2 to 15, addressing various levels of MOP complexity. The population size (N) was modified based on the number of objectives. The maximum number of generations (Gmax) was set between 300 and 1500, depending on the objective numbers, to ensure adequate convergence. These parameters were kept to provide a thorough and comparable assessment of the FDEA, aiming to validate the proposed hypothesis and evaluate performance across different multi-objective optimization scenarios.

The study used four variants of the FDEA algorithm to assess the impact of different selection methods on its performance. The original FDEA variant uses random selection to choose pairs of individuals for crossover and mutation. This simple approach involves randomly selecting two solutions and ensuring that they are different if the set contains more than one solution.

The second variant, FDEA_TB, employs binary tournament selection based on Pareto dominance. In this method, two solutions are randomly selected and compared using a dominance comparator, if there is a domination relation between them, the non-dominated one is selected, otherwise, one of them is selected randomly.

The third variant, FDEA_CD, also uses binary tournament selection but compares solutions based on crowding distance, a metric that helps maintain diversity among non-dominated solutions by selecting the solution in the less crowded area.

The fourth variant, FDEA_CDTB, combines the approaches of FDEA_TB and FDEA_CD. The solutions are first compared using Pareto dominance and, if they are equally good, crowding distance is used to make the final selection.

The objective of exploring these variants is to determine how different selection strategies impact the algorithm's effectiveness, focusing on enhancing the evolutionary process by incorporating different criteria for parent selection.

The results of modifying the crossover selection method in the FDEA algorithm were obtained through a total of 30 independent runs for each variant of the algorithm, for each problem and number of objectives. After each run, the IGD of each obtained front was calculated. The results were subjected to a Kruskal-Wallis test with a significance level of 5%. In cases where there was no

Table 2. IGD_p results for DTLZ problems

Obj.	Algorithms	DTLZ1	DTLZ2	DTLZ3	DTLZ4
2	FDEA	**9.28E-05±2.17E-04**	**5.41E-03±3.59E-04**	**3.25E-03±3.52E-03**	**1.45E-01±3.11E-01**
	FDEA-TB	**5.11E-05±2.54E-06**	**5.31E-03±3.15E-04**	**2.93E-03±2.52E-03**	1.73E-01±3.34E-01
	FDEA-CD	5.44E-05±5.89E-06	**5.34E-03±2.56E-04**	**3.30E-03±2.66E-03**	**8.89E-02±2.51E-01**
	FDEA-CDTB	**1.77E-04±6.58E-04**	5.53E-03±3.14E-04	**4.29E-03±3.94E-03**	**8.91E-02±2.51E-01**
3	FDEA	**9.60E-03±7.01E-04**	**5.16E-02±2.06E-03**	**1.14E-03±1.80E-04**	**8.11E-02±1.66E-01**
	FDEA-TB	**9.44E-03±6.78E-04**	**5.14E-02±1.68E-03**	**1.19E-03±2.11E-04**	**4.99E-02±2.07E-03**
	FDEA-CD	**9.36E-03±5.03E-04**	**5.11E-02±1.68E-03**	**1.40E-03±9.11E-04**	**5.00E-02±1.95E-03**
	FDEA-CDTB	**9.26E-03±5.45E-04**	**5.20E-02±1.89E-03**	**1.19E-03±1.99E-04**	**4.95E-02±1.86E-03**
5	FDEA	**5.65E-04±1.41E-05**	**1.51E-01±2.97E-03**	**1.26E-03±1.74E-04**	**1.51E-01±2.78E-03**
	FDEA-TB	**5.66E-04±2.42E-05**	**1.51E-01±3.92E-03**	**1.27E-03±1.75E-04**	**1.50E-01±3.56E-03**
	FDEA-CD	**5.60E-04±1.54E-05**	**1.50E-01±2.71E-03**	1.46E-03±2.47E-04	**1.51E-01±2.29E-03**
	FDEA-CDTB	**5.78E-04±4.06E-05**	**1.51E-01±2.49E-03**	1.46E-03±2.83E-04	1.52E-01±2.65E-03
8	FDEA	**8.73E-04±6.12E-05**	2.58E-01±6.22E-03	**3.11E-03±2.35E-04**	**2.56E-01±5.39E-03**
	FDEA-TB	**8.57E-04±3.49E-05**	2.57E-01±7.76E-03	**3.12E-03±2.64E-04**	**2.55E-01±6.44E-03**
	FDEA-CD	**8.44E-04±3.04E-05**	**2.50E-01±7.19E-03**	**3.44E-03±1.18E-03**	**2.56E-01±5.35E-03**
	FDEA-CDTB	**8.69E-04±7.03E-05**	**2.52E-01±5.19E-03**	**3.14E-03±3.12E-04**	**2.58E-01±7.00E-03**
10	FDEA	**1.85E-03±1.14E-04**	**2.80E-01±8.34E-03**	**6.98E-03±1.46E-03**	**2.99E-01±5.56E-03**
	FDEA-TB	**1.86E-03±8.73E-05**	**2.77E-01±6.77E-03**	**7.31E-03±1.72E-03**	**3.00E-01±6.37E-03**
	FDEA-CD	**1.91E-03±1.48E-04**	**2.78E-01±6.11E-03**	**7.07E-03±1.15E-03**	**3.02E-01±5.68E-03**
	FDEA-CDTB	**1.90E-03±1.73E-04**	**2.77E-01±6.88E-03**	**7.34E-03±1.78E-03**	3.05E-01±6.43E-03
15	FDEA	**1.16E-03±1.07E-04**	**2.37E-01±1.96E-02**	**7.85E-04±1.54E-04**	**2.86E-01±1.60E-02**
	FDEA-TB	**1.16E-03±5.70E-05**	**2.35E-01±1.98E-02**	**8.40E-04±2.54E-04**	**2.89E-01±1.58E-02**
	FDEA-CD	**1.21E-03±7.85E-05**	**2.25E-01±2.30E-02**	1.03E-03±3.54E-04	2.99E-01±1.61E-02
	FDEA-CDTB	**1.22E-03±9.92E-05**	**2.36E-01±2.04E-02**	1.06E-03±3.46E-04	3.04E-01±1.54E-02

Obj.	Algorithms	DTLZ5	DTLZ6	DTLZ7
2	FDEA	**5.49E-03±2.27E-04**	**5.86E-03±4.55E-03**	**4.48E-03±2.61E-04**
	FDEA-TB	**5.46E-03±1.93E-04**	**5.71E-03±3.52E-03**	**4.64E-03±3.07E-04**
	FDEA-CD	**5.45E-03±2.10E-04**	**8.99E-03±4.74E-03**	**4.60E-03±2.76E-04**
	FDEA-CDTB	**5.42E-03±1.85E-04**	**7.58E-03±6.24E-03**	**4.59E-03±2.45E-04**
3	FDEA	**5.69E-03±3.34E-04**	3.88E-03±3.81E-03	**4.68E-02±6.67E-02**
	FDEA-TB	**5.50E-03±2.07E-04**	3.40E-03±2.58E-03	**4.64E-02±6.31E-02**
	FDEA-CD	**5.52E-03±1.65E-04**	**1.79E-03±6.53E-05**	**3.51E-02±1.50E-03**
	FDEA-CDTB	**5.57E-03±1.98E-04**	**1.79E-03±9.89E-05**	**4.73E-02±6.66E-02**
5	FDEA	**2.41E-02±3.12E-03**	3.25E-02±4.38E-03	**1.30E-01±2.87E-03**
	FDEA-TB	**2.48E-02±3.56E-03**	3.28E-02±3.74E-03	**1.30E-01±3.22E-03**
	FDEA-CD	**2.33E-02±2.75E-03**	**1.07E-02±2.68E-03**	1.33E-01±2.84E-03
	FDEA-CDTB	**2.30E-02±3.29E-03**	**9.90E-03±1.93E-03**	1.34E-01±2.48E-03
8	FDEA	**5.69E-02±1.29E-02**	6.46E-02±3.85E-03	**4.43E-01±1.31E-02**
	FDEA-TB	**5.53E-02±1.24E-02**	6.31E-02±3.71E-03	**4.42E-01±1.36E-02**
	FDEA-CD	**6.06E-02±1.06E-02**	**1.90E-02±3.58E-03**	**4.35E-01±1.06E-02**
	FDEA-CDTB	**5.79E-02±1.33E-02**	**2.00E-02±4.22E-03**	**4.40E-01±1.23E-02**
10	FDEA	**7.31E-02±1.70E-02**	7.82E-02±5.40E-03	6.91E-01±1.88E-02
	FDEA-TB	**7.33E-02±1.49E-02**	7.69E-02±3.07E-03	6.85E-01±1.59E-02
	FDEA-CD	**7.23E-02±1.95E-02**	**2.39E-02±6.33E-03**	**6.70E-01±1.93E-02**
	FDEA-CDTB	**7.69E-02±1.76E-02**	**2.50E-02±7.43E-03**	**6.74E-01±1.91E-02**
15	FDEA	**7.91E-02±1.99E-02**	1.24E-01±9.56E-03	1.25E+00±1.31E-02
	FDEA-TB	**8.09E-02±2.23E-02**	1.25E-01±9.07E-03	1.26E+00±1.25E-02
	FDEA-CD	1.01E-01±2.85E-02	**3.94E-02±9.69E-03**	**1.23E+00±1.70E-02**
	FDEA-CDTB	1.00E-01±2.58E-02	**4.14E-02±1.00E-02**	**1.23E+00±1.49E-02**

significant difference between any pairs, a post-hoc test was used to determine which pairs were equivalent.

The best results are displayed in Table 2 for the DTLZ problems and in Table 4 for the WFG problems, showing the indicator values for the analyzed algorithms. The tables are organized by the results of the statistical tests, highlighting the best results, and, in cases where there is no significant difference, those that are not significantly different from the best are also highlighted.

4.2 Results and Discussion for DTLZ Problems

For the DTLZ1 problem, known for its multi-modal decision space and linear Pareto front, all variants achieved no statistical difference, except for FDEA_CD which was outperformed by the other variants in two objectives.

In the DTLZ2 problem with a concave Pareto front, FDEA_CDTB had inferior performance in the 2-objective case. However, as objectives increased, FDEA_CD achieved good results in 3, 5, 8, and 15 objectives. FDEA and FDEA_TB were outperformed in the 8-objective setup, but all variants performed well and were statistically equivalent otherwise.

For the DTLZ3 problem, which includes multiple local Pareto fronts, FDEA and FDEA_TB had the best performance regardless of objective number. Conversely, FDEA_CD and FDEA_CDTB were outperformed for 5 and 15 objectives. For 2, 3, 8, and 10 objectives, all variants were statistically equivalent.

In the DTLZ4 problem, challenging for maintaining a diverse solution set, FDEA_TB was outperformed by the other variants for 2 objectives. FDEA_CDTB had the worst results in the 5 and 10-objective configurations, and both FDEA_CD and FDEA_CDTB performed poorly for 15 objectives. For 3 and 8 objectives, all variants were statistically equivalent. A possible explanation for this poor performance of CDTB over the others is that is may favor dominance resistant solutions, hence, lowering diversity.

For the DTLZ5 problem, which is aconcave and degenerate problem, FDEA_CD and FDEA_CDTB performed worst in the 15-objective setup. For other configurations, all variants were statistically equivalent.

In the DTLZ6 problem, which is degenerate and presents convergence challenges, all variants were statistically equivalent for two objectives. For 3, 5, 8, 10, and 15 objectives, FDEA_CD and FDEA_CDTB performed better than the others.

Finally, in the DTLZ7 problem, characterized by a disconnected Pareto front, all variants performed equivalently for 2, 3, and 8 objectives. For 5 objectives, FDEA and FDEA_TB are tied for the best performance. For 10 and 15 objectives, FDEA_CD and FDEA_CDTB tied for the best results.

Based on the results summarized in Table 3, a few patterns can be identified, first, in the non-degenerate problems that impose convergence challenge (DTLZ1 and DTLZ3) using the Crowding Distance that favors diversity does not seem to be a good choice. Second, in the easier problems (DTLZ1, DTLZ2 and DTLZ5) there are fewer differences between the variants, while in the DTLZ6 and DTLZ7 there are more differences. Third, original FDEA with random selection performs

Table 3. Lowest-performing of FDEA Variants for DTLZ Problems

Benchmark	Objectives	Variants	Observations
DTLZ1	2	FDEA_CD	Inefficient for multi-modal challenge
DTLZ2	2	FDEA_CDTB	Poor diversity in concave fronts
	8	FDEA; FDEA_TB	
DTLZ3	5, 15	FDEA_CD; FDEA_CDTB	Poor performance in multi-modal challenge
DTLZ4	2	FDEA_TB	Inconsistent performance in based problems
	5, 10	FDEA_CDTB	
	15	FDEA_CD; FDEA_CDTB	
DTLZ5	15	FDEA_CD; FDEA_CDTB	Difficulty with concave fronts
DTLZ6	3, 5, 8, 10, 15	FDEA; FDEA_TB	Poor performance in multi-modal challenge
DTLZ7	5	FDEA_CD; FDEA_CDTB	Very weak in disconnected fronts
	10,15	FDEA; FDEA_TB	

very well for two objectives, however its performance starts deteriorating as the number of objectives increase.

4.3 Results and Discussion for WFG Problems

For the WFG1 problem, which presents a convex geometry with a mixture of Pareto shapes, the FDEA_CD and FDEA_CDTB variants had the best performances for 2, 3, and 5 objectives. As the number of objectives increases, all variants show similar performances.

In the WFG2 problem, characterized by a convex and disconnected Pareto front, the results show that, regardless of the number of objectives, all variants had statistically equivalent performance.

In the WFG3 problem, characterized by a linear and degenerate geometry, the FDEA and FDEA_TB variants had the worst performance for 3 objectives. However, for 2, 5, 8, 10, and 15 objectives, all variants showed equivalent performances.

For the WFG4 problem, which has a concave Pareto front and is multi-modal, increasing the complexity of the problem, the FDEA and FDEA_TB variants had the worst performance for 10 and 15 objectives. However, for 2, 3, 5, and 8 objectives, all variants had equivalent performance.

In the WFG5 problem, which tests convergence ability, the FDEA and FDEA_TB variants had the worst performances for 2, 10, and 15 objectives. For 3, 5, and 8 objectives, all variants had equivalent performance.

In the WFG6 problem, a simple non-separable problem, all variants had statistically equivalent performance.

For the WFG7 problem, which is a biased problem, the FDEA variant had the worst performance for 10 objectives. For 2, 3, 5, 8, and 15 objectives, all variants had statistically similar performance.

For the WFG8 problem, characterized by distance parameters dependent on position parameters, resulting in a biased and highly complex problem, the

Table 4. IGD_p results for WFG problems

Obj.	Algorithms	WFG1	WFG2	WFG3	WFG4	WFG5
2	FDEA	3.0E-01±3.3E-02	**1.2E-01±7.7E-02**	**5.5E-03±4.7E-04**	**6.4E-03±4.7E-04**	2.6E-02±2.0E-04
	FDEA-TB	3.0E-01±3.8E-02	**1.2E-01±7.8E-02**	**5.5E-03±6.4E-04**	**6.3E-03±6.0E-04**	2.6E-02±1.9E-04
	FDEA-CD	**2.5E-01±4.6E-02**	**7.8E-02±8.1E-02**	**5.7E-03±4.6E-04**	**6.2E-03±4.9E-04**	**2.6E-02±2.0E-04**
	FDEA-CDTB	**2.4E-01±6.1E-02**	**7.3E-02±8.0E-02**	**5.7E-03±1.3E-03**	**6.2E-03±4.5E-04**	**2.6E-02±2.0E-04**
3	FDEA	2.8E-01±4.1E-02	**8.8E-02±5.0E-02**	3.2E-02±1.0E-02	**5.5E-02±3.3E-03**	**5.7E-02±1.1E-03**
	FDEA-TB	2.9E-01±3.5E-02	**7.8E-02±4.7E-02**	3.6E-02±1.1E-02	**5.4E-02±3.2E-03**	**5.8E-02±2.0E-03**
	FDEA-CD	**1.8E-01±5.0E-02**	**7.4E-02±4.5E-02**	**2.5E-02±4.2E-03**	**5.5E-02±3.1E-03**	**5.8E-02±1.8E-03**
	FDEA-CDTB	**1.7E-01±4.8E-02**	**8.1E-02±4.8E-02**	**2.7E-02±5.4E-03**	**5.5E-02±3.2E-03**	**5.8E-02±2.2E-03**
5	FDEA	3.6E-01±1.2E-02	**6.6E-02±1.9E-02**	**7.2E-02±2.3E-02**	**1.8E-01±1.7E-02**	**1.6E-01±5.4E-03**
	FDEA-TB	3.6E-01±9.3E-03	**6.2E-02±2.6E-03**	**7.6E-02±2.8E-02**	**1.8E-01±1.5E-02**	**1.6E-01±4.9E-03**
	FDEA-CD	**3.4E-01±3.1E-02**	**6.5E-02±1.9E-02**	**6.3E-02±2.8E-02**	**1.8E-01±1.7E-02**	**1.6E-01±4.1E-03**
	FDEA-CDTB	**3.3E-01±2.6E-02**	**6.4E-02±1.9E-02**	**6.0E-02±2.6E-02**	**1.9E-01±9.4E-03**	**1.6E-01±4.8E-03**
8	FDEA	**5.2E-01±2.0E-02**	**9.0E-02±3.4E-02**	**8.2E-02±1.7E-02**	**3.8E-01±3.6E-02**	**3.6E-01±1.2E-02**
	FDEA-TB	**5.0E-01±2.1E-02**	**8.0E-02±1.7E-02**	**7.7E-02±1.8E-02**	**3.7E-01±4.1E-02**	**3.5E-01±1.3E-02**
	FDEA-CD	**5.0E-01±2.3E-02**	**7.9E-02±1.7E-02**	**7.8E-02±2.3E-02**	**3.6E-01±3.4E-02**	**3.5E-01±1.2E-02**
	FDEA-CDTB	**5.1E-01±2.2E-02**	**7.9E-02±1.8E-02**	**8.2E-02±3.0E-02**	**3.7E-01±2.8E-02**	**3.5E-01±1.3E-02**
10	FDEA	**6.3E-01±2.1E-02**	**8.7E-02±2.0E-02**	**6.7E-02±1.7E-02**	4.3E-01±2.1E-02	4.4E-01±8.9E-03
	FDEA-TB	**6.3E-01±1.7E-02**	**9.3E-02±2.5E-02**	**5.9E-02±9.6E-03**	4.4E-01±2.1E-02	4.4E-01±9.9E-03
	FDEA-CD	**6.2E-01±2.1E-02**	**9.0E-02±1.9E-02**	**5.8E-02±1.6E-02**	**4.1E-01±2.3E-02**	**4.3E-01±9.6E-03**
	FDEA-CDTB	**6.3E-01±2.3E-02**	**9.2E-02±1.9E-02**	**5.9E-02±1.5E-02**	**4.2E-01±2.1E-02**	**4.3E-01±7.9E-03**
15	FDEA	**8.6E-01±3.7E-02**	**8.4E-02±2.3E-02**	**6.2E-02±1.8E-02**	5.4E-01±3.3E-02	6.2E-01±2.1E-02
	FDEA-TB	**8.5E-01±3.4E-02**	**7.9E-02±1.0E-02**	**6.4E-02±2.0E-02**	5.4E-01±3.5E-02	6.2E-01±2.2E-02
	FDEA-CD	**8.7E-01±2.4E-02**	**8.0E-02±1.1E-02**	**6.6E-02±1.9E-02**	**5.1E-01±3.0E-02**	**6.0E-01±2.7E-02**
	FDEA-CDTB	**8.7E-01±3.2E-02**	**8.2E-02±9.3E-03**	**5.9E-02±1.7E-02**	**5.3E-01±3.6E-02**	**6.0E-01±2.2E-02**

Obj.	Algorithms	WFG6	WFG7	WFG8	WFG9
2	FDEA	**2.3E-02±2.3E-03**	**5.9E-03±4.0E-04**	4.4E-02±1.4E-03	**1.5E-02±1.6E-02**
	FDEA-TB	**2.3E-02±2.9E-03**	**5.8E-03±3.3E-04**	4.4E-02±1.3E-03	**2.3E-02±2.0E-02**
	FDEA-CD	**2.3E-02±3.4E-03**	**5.7E-03±3.2E-04**	**4.2E-02±9.3E-04**	**3.0E-02±2.1E-02**
	FDEA-CDTB	**2.2E-02±2.3E-03**	**5.7E-03±3.7E-04**	**4.2E-02±9.7E-04**	**3.5E-02±2.0E-02**
3	FDEA	**5.7E-02±2.2E-03**	**5.4E-02±2.7E-03**	**9.3E-02±2.5E-03**	**5.6E-02±9.5E-03**
	FDEA-TB	**5.8E-02±2.4E-03**	**5.4E-02±2.3E-03**	**9.3E-02±2.2E-03**	**5.7E-02±1.0E-02**
	FDEA-CD	**5.7E-02±2.6E-03**	**5.5E-02±2.6E-03**	**9.3E-02±2.0E-03**	**6.4E-02±9.0E-03**
	FDEA-CDTB	**5.7E-02±2.5E-03**	**5.3E-02±2.0E-03**	9.4E-02±2.3E-03	**6.2E-02±9.1E-03**
5	FDEA	**1.6E-01±4.8E-03**	**1.8E-01±9.1E-03**	**1.8E-01±2.6E-03**	**1.5E-01±3.7E-03**
	FDEA-TB	**1.6E-01±3.9E-03**	**1.8E-01±7.0E-03**	**1.8E-01±2.6E-03**	**1.5E-01±4.0E-03**
	FDEA-CD	**1.6E-01±4.8E-03**	**1.7E-01±6.5E-03**	1.9E-01±1.9E-03	**1.5E-01±5.3E-03**
	FDEA-CDTB	**1.6E-01±5.0E-03**	**1.8E-01±5.7E-03**	1.9E-01±3.2E-03	**1.5E-01±3.6E-03**
8	FDEA	**3.5E-01±1.7E-02**	**3.7E-01±1.4E-02**	**3.5E-01±7.7E-03**	3.1E-01±2.0E-02
	FDEA-TB	**3.5E-01±1.7E-02**	**3.7E-01±1.9E-02**	**3.5E-01±9.7E-03**	**3.0E-01±1.9E-02**
	FDEA-CD	**3.6E-01±1.6E-02**	**3.6E-01±1.8E-02**	**3.5E-01±8.4E-03**	**3.0E-01±1.6E-02**
	FDEA-CDTB	**3.5E-01±1.7E-02**	**3.6E-01±1.9E-02**	**3.5E-01±8.3E-03**	**2.9E-01±1.9E-02**
10	FDEA	**4.8E-01±1.8E-02**	4.4E-01±1.3E-02	3.7E-01±1.8E-02	**4.3E-01±2.3E-02**
	FDEA-TB	**4.7E-01±1.7E-02**	**4.3E-01±1.3E-02**	3.7E-01±1.7E-02	**4.4E-01±1.9E-02**
	FDEA-CD	**4.8E-01±1.4E-02**	**4.3E-01±1.2E-02**	**3.6E-01±1.3E-02**	**4.2E-01±2.2E-02**
	FDEA-CDTB	**4.8E-01±1.5E-02**	**4.3E-01±1.5E-02**	**3.6E-01±1.5E-02**	**4.3E-01±2.3E-02**
15	FDEA	**6.6E-01±2.4E-02**	**5.8E-01±2.2E-02**	6.2E-01±9.4E-03	5.8E-01±2.6E-02
	FDEA-TB	**6.7E-01±2.2E-02**	**5.9E-01±1.8E-02**	6.2E-01±1.2E-02	**5.8E-01±2.5E-02**
	FDEA-CD	**6.6E-01±2.3E-02**	**5.8E-01±2.3E-02**	**6.0E-01±1.2E-02**	**5.7E-01±2.3E-02**
	FDEA-CDTB	**6.7E-01±2.0E-02**	**5.8E-01±2.3E-02**	**6.0E-01±1.5E-02**	**5.7E-01±2.4E-02**

FDEA and FDEA_TB variants had the worst performance for 2, 10, and 15 objectives. The FDEA_CDTB variant had the worst performance for 3 objectives, while for 5 objectives, the FDEA_CD and FDEA_CDTB variants showed the worst results. For 8 objectives, all variants had statistically equivalent performance.

Finally, for the WFG9 problem, which is a multi-modal, deceptive and biased problem, the FDEA variant had the worst results for 8 and 15 objectives. For the other numbers of objectives, all variants had statistically equivalent performances.

Table 5. Lowest-performing of FDEA Variants for WFG Problems

Benchmark	Objectives	Variants	Observations
WFG1	2,3,5	FDEA, FDEA_TB	Poor performance in biased challenges
WFG2, WFG6	All	All variants	Statistically equivalent performance
WFG3	3	FDEA, FDEA_TB	Poor performance in linear and degenerate geometry
WFG4	10, 15	FDEA, FDEA_TB	Inconsistent performance in multi-modal problems
WFG5	2, 10, 15	FDEA, FDEA_TB	Weak convergence ability on deceptive problems
WFG7	10	FDEA	Poor performance in biased challenges
WFG8	2, 10, 15	FDEA, FDEA_TB	Poor performance in biased, complex problems
	3	FDEA_CDTB	
	5	FDEA_CD; FDEA_CDTB	
WFG9	8, 15	FDEA	Poor performance in multi-modal, deceptive, and biased problems

In the WFG problems, where the results are summarized in Table 5, the introduction of the new operators had one of three effects: it made the results worse than the original algorithm, which only happened on WFG8 for 3 and 5 objectives, hence, two instances (problem/objective number); it does not make a significant difference, which happened for all instances of the WFG2 and WFG6 problems, and a total of 37 instances considering all problems; it improves the results of the original algorithm, which happened in 15 instances along the problems. These results indicate that the operators proposed are beneficial when solving problems with the characteristics presented in the WFG suite.

5 Conclusion

This paper investigated the impact of different crossover selection strategies on the performance of the FDEA algorithm for multi-objective optimization. Four variants of the FDEA were employed: FDEA (random selection), FDEA_TB (binary tournament with Pareto dominance), FDEA_TB (binary tournament with crowding distance), and FDEA_CDTB (combination of FDEA_TB and FDEA_CDTB). The performance was evaluated using the IGD metric on DTLZ and WFG problems with 2, 3, 5, 8, 10 and 15 numbers of objectives.

The results demonstrate that the choice of crossover selection strategy is capable of influencing the performance of the FDEA. FDEA_TB and FDEA_CDTB showed good performance across various problems, especially in configurations with a higher number of objectives. FDEA performed well in problems with multiple local Pareto fronts, and random selection (original FDEA) can be a simpler and computationally cheaper alternative, though with inferior results in some cases.

Despite the modifications, most variants maintained statistically equivalent performances in many of the tested scenarios. This suggests that while certain variants may excel in specific situations, there is no single variant that is superior in all configurations. This finding highlights the importance of adapting the choice of the FDEA variant based on the specific problem and its characteristics.

In conclusion, modifying the crossover selection method in the FDEA was effective in improving the algorithm's performance in different multi-objective optimization scenarios. These results encourage further studies on the hybridization of evolutionary algorithms, incorporating indicator-based selection methods to further explore the adaptive capabilities of algorithms in complex optimization environments. Future work includes exploring other variants of crossover selection, utilizing additional quality indicators, and conducting comprehensive tests in various multi-objective optimization scenarios. This will allow the identification of the most effective strategies for different problem characteristics and further refine the selection parameters to achieve consistent and robust performance across a wide range of situations.

References

1. Coello, C.A.C., et al.: Evolutionary Algorithms for solving multi-objective problems, 5th edn. Springer, New York (2007)
2. Falcón-Cardona, J.G., Coello, C.A.C.: Indicator-based multi-objective evolutionary algorithms: a comprehensive survey. ACM Comput. Surv. **53**(2), 1–35 (2020). https://doi.org/10.1145/3376916
3. Liefooghe, A., Derbel, B.: A correlation analysis of set quality indicator values in multiobjective optimization. Association for Computing Machinery, New York (2016). https://doi.org/10.1145/2908812.2908906
4. Adra, S.F.: Improving Convergence, Diversity and Pertinency in Multiobjective Optimization. UK (2007)
5. Ishibuchi, H., Masuda, H., Tanigaki, Y., Nojima, Y.: Modified distance calculation in generational distance and inverted generational distance. In: Gaspar-Cunha, A., Henggeler Antunes, C., Coello, C.C. (eds.) EMO 2015. LNCS, vol. 9019, pp. 110–125. Springer, Cham (2015). https://doi.org/10.1007/978-3-319-15892-1_8
6. Falcón-Cardona, J.G., Coello Coello, C.A.: Towards a more general many-objective evolutionary optimizer. In: Auger, A., Fonseca, C.M., Lourenço, N., Machado, P., Paquete, L., Whitley, D. (eds.) PPSN 2018. LNCS, vol. 11101, pp. 335–346. Springer, Cham (2018). https://doi.org/10.1007/978-3-319-99253-2_27
7. Luke, S.: Essentials of metaheuristics, 2nd edn (2012)
8. Derrac, J., García, S., Herrera, F.: A practical tutorial on the use of nonparametric statistical tests as a methodology for comparing evolutionary and swarm intelligence algorithms. Swarm Evol. Comput. 3–18 (2011)
9. Huband, S., et al.: A review of multiobjective test problems and a scalable test problem toolkit. IEEE Trans. Evol. Comput. 477–506 (2006)

10. Deb, K., et al.: Scalable multi-objective optimization test problems. IEEE Congr. Evol. Comput. 825–830 (2002)
11. Liu, S., et al.: A fuzzy decomposition-based multi/many-objective evolutionary algorithm. IEEE Trans. Cybern. 3495–3509 (2022)
12. Deb, K., et al.: A fast and elitist multiobjective genetic algorithm: NSGA-II. IEEE Trans. Evol. Comput. 182–197 (2002)
13. Deb, K., Jain, H.: An evolutionary many-objective optimization algorithm using reference-point-based nondominated sorting approach, part I: solving problems with box constraints. IEEE Trans. Evol. Comput. **18**, 577–601 (2014)
14. Esquivel, S.C., et al.: Selection mechanisms in evolutionary algorithms. Fundamenta Informaticae 17–33 (1998)
15. Phan, D.H., Suzuki, J.: R2-IBEA: R2 indicator based evolutionary algorithm for multiobjective optimization. IEEE Congr. Evol. Comput. 1836–1845 (2013)
16. Sun, Y., Yen, G.G., Yi, Z.: IGD indicator-based evolutionary algorithm for many-objective optimization problems. IEEE Trans. Evol. Comput. 173–187 (2019)
17. Hochstrate, N., Naujoks, B., Emmerich, M.: SMS-EMOA: multiobjective selection based on dominated hypervolume. Eur. J. Oper. Res. 1653–1669 (2007)
18. Tian, Y., et al.: An indicator-based multiobjective evolutionary algorithm with reference point adaptation for better versatilit. IEEE Trans. Evol. Comput. 609–622 (2018)

Improving LLMs' Reasoning and Planning with Finite-State Machines

Eduardo Faria Cabrera(✉), Marcel Rodrigues de Barros, and Anna Helena Reali Costa

Universidade de São Paulo, São Paulo, SP, Brazil
{eduardofcabrera,marcel.barros,anna.reali}@usp.br

Abstract. Large Language Models (LLMs) have shown remarkable capabilities across various applications but struggle with sequential decision-making tasks like planning. This work demonstrates that integrating LLMs with Finite-State Machines (FSMs) can enhance their reasoning and planning abilities, while also offering increased reliability. Several setup variants are compared, providing a better understanding of how validations, feedback loops, and restrictions enhance robustness and effectiveness. A comparison to the well known Chain-of-Thoughts approach is also provided. Our methods improve planning capabilities of all analysed LLMs, consistently increasing the success rate in solving tasks of varying complexities. A detailed analysis of the two best variants are provided, highlighting their respective strengths and weaknesses.

Keywords: Large Language Models · Machine Learning · Planning · Graph Representations

1 Introduction

Recently, Large Language Models (LLMs) have surged in development, revolutionizing AI capabilities. Scaling up these models has unlocked new feats, from crafting narratives and summarizing text to coding and enhancing reasoning skills [2,3,18]. Despite these advancements, LLMs still face challenges in basic logical reasoning, often producing implausible claims and struggling with rudimentary planning tasks [2,15]. Moreover, the opacity of their decision-making processes poses significant obstacles in interpreting and validating their outputs, especially in critical domains like healthcare and finance. Thus, there is an urgent need for methodologies to validate, guide, and interpret the decision pathways of LLMs.

One extensively studied form of reasoning task in the AI community is planning and making sequential decisions. Planning involves executing a series of

The authors thank the Centro de Ciência de Dados (C2D-Poli-USP) supported by Itaú Unibanco S.A., as well as the Center for Artificial Intelligence (C4AI-USP), supported by FAPESP (grant 2019/07665-4) and IBM Corporation. This work is partially supported by CNPq (grant number 312360/2023-1) and CAPES (Finance Code 001).

A. Paes and F. A. N. Verri (Eds.): BRACIS 2024, LNAI 15413, pp. 110–124, 2025.
https://doi.org/10.1007/978-3-031-79032-4_8

valid actions within a defined context to transition from an initial state to a goal state [11]. Even simple planning tasks pose challenges for LLMs, as they often struggle to achieve success rates and generate valid plans due to the requirement of validating each action's feasibility in every state, backtracking previous states and actions, and employing long-term reasoning [15]. Enhancing planning, reasoning, and introducing guarantees in multi-step decision making problems is crucial for improving LLMs' reliability, which would open up their usage in some risk averse applications.

The research community is actively developing new methods to address multi-step problems with LLMs by enhancing their reliability, interpretability, and robustness. One such technique is Chain-of-Thought (CoT) [19], a prompt-based approach that incorporates intermediate reasoning steps, or "thoughts," alongside task input/output pairs. By decomposing the original problem into these intermediate steps, CoT facilitates structured reasoning, thereby improving LLMs' ability to tackle procedural problems. Empirical evidence has demonstrated CoT's effectiveness in enhancing problem-solving capabilities [19]. Furthermore, advancements have extended CoT's capabilities, including techniques like Self-Consistency with CoT (CoT-SC) [17] and CausalCOT [9], as well as more complex and flexible approaches like Tree-of-Thoughts [20] and Graph-of-Thoughts [1]. However, these techniques primarily focus on refining prompting strategies without explicitly leveraging the logical structure of problems to validate and guide LLMs' reasoning processes.

Our approach associates LLMs to Finite-State Machines (FSMs), commonly used in systems that require reliability and predictability, as a guiding structure to the decision-making processes of LLMs. This association restricts the model's decision at each step, providing several additional guarantees enabling applications with tighter restrictions to benefit from modern AI capabilities.

The key contributions of our work are:

Novel Approach: We introduce a novel method that incorporates FSMs as a guiding structure for LLM decision-making processes.

Enhanced Problem-Solving: We provide evidence that validations, feedback loops, and restrictions enabled by FSMs, derived from standard planning language definitions, enhance LLMs' ability to solve planning problems.

Comprehensive Comparison: We present a detailed comparison of our most effective setups, thoroughly analyzing their strengths and weaknesses in the context of planning problems.

2 Background

This section covers essential concepts for this work. It briefly explains how auto-regressive LLMs function and their advantages in solving planning problems. It then provides a formal definition of a planning problem. Finally, it defines a Mealy Machine, a type of Finite-State Machine, which will be used to guide the LLM in addressing planning problems.

2.1 Large Language Models for Planning

Large language models are built upon a transformer architecture [16], which undergoes training on a vast corpus of text samples sourced from the web. Formally, we define a collection of examples $X = (x_1, x_2, \ldots, x_N)$, where each $x_i = (\alpha_1, \alpha_2, \ldots \alpha_{n_i})$ and every token α_i is drawn from a predefined vocabulary $\mathbf{V}$. Each LLM is equipped with a tokenizer responsible for converting raw input text into a sequence of tokens, each belonging to $\mathbf{V}$. LLMs typically undergo training in an auto-regressive manner, where the probability of each token is solely dependent on its preceding tokens in the sequence, expressed as $p(x_i) = \prod_{j=1}^{n_i} p(\alpha_j|\alpha_{0:j-1})$. During training, the model parameters are learned by maximizing the likelihood of the entire dataset, formulated as $p(X) = \prod_{i=1}^{N} p(x_i)$.

Addressing planning problems with pre-trained LLMs is highly advantageous as these models can leverage their extensive world knowledge, commonsense understanding, and logical reasoning abilities. The aim is to generate a sequence of actions $(a_1, a_2, \ldots, a_n)$ to reach a desired final state. Graph structures have been widely used to enhance the reasoning capabilities of LLMs, with empirical evidence showing significant improvements in solving complex problems [13,17, 19]. The CoT approach is fundamental, modeling the inference process as a linear chain and explicitly capturing intermediate reasoning steps. Building on CoT, methods like Tree-of-Thoughts and Graph-of-Thoughts introduce more complex topologies, including branches, backtracking, and aggregating reasoning steps.

Nonetheless, incorporating prompt topologies alone is insufficient to tackle planning problems. Studies such as [15] underscore the difficulties LLMs face in generating valid and optimal plans. Recent research, exemplified in [10,12], suggests that integrating feedback loops can significantly improve LLMs' accuracy in planning tasks. Consequently, implementing a mechanism for validating the model's outputs and providing feedback emerges as a promising strategy for enhancing its performance.

2.2 Planning Task Definition and Finite-State Machines

A planning problem can be defined as a tuple $\mathcal{P} = (\mathcal{S}, \mathcal{A}, s_0, \mathcal{G})$ [11], where:

- $\mathcal{S}$ is a finite set of states. Each state $s \in \mathcal{S}$ is a complete assignment of values to a set of state variables (or fluents).
- $s_0 \in \mathcal{S}$ is the initial state.
- $\mathcal{G} \subset \mathcal{S}$ is the set of goal states where the goal condition is satisfied. This condition is specified as a conjunction of literals.
- $\mathcal{A}$ is the set of actions, where each action $a \in \mathcal{A}$ is defined by its preconditions and effects:
  - Preconditions *pre*(a) are a set of literals that must be satisfied for the action to be applied.
  - Effects *eff*(a) are a set of literals that describe how the state changes after applying the action.

We define a transition model $\gamma : \mathcal{S} \times \mathcal{A} \rightarrow \mathcal{S}$ that describes deterministic state transitions in the planning problem. If a state $s \in \mathcal{S}$ satisfies $pre(a)$, then $\gamma(s, a) = s'$, where $s' \in \mathcal{S}$ is the new state resulting from applying $eff(a)$ to s. If s does not satisfy $pre(a)$, then the action cannot be applied.

The Planning Domain Definition Language (PDDL) is commonly used to represent planning problems, allowing for the encoding of a problem by defining its domain (types of objects, predicates, and actions) and its problem instance (initial state and goal state). There is extensive literature on planning algorithms for various tasks. An AI planner takes a PDDL domain and problem as input and produces a solution. Most state-of-the-art planners [6,7] are variants of heuristic-based forward search, where PDDL encodings are used to automatically derive a heuristic function.

Another common approach is the integration of Markov Decision Processes (MDPs) into automated planning. MDPs are best suited for modeling stochastic processes where maximizing rewards in uncertain contexts is crucial [4,11]. However, many mission-critical applications are modeled using deterministic finite-state machines due to reliability and predictability requirements.

Building on this foundation, our work investigates whether simply imposing an alphabet of interactively validated symbols can enhance LLMs' ability to solve planning tasks. This restriction of possible actions associated with deterministic transitions is best encapsulated by the FSMs formalism.

A Mealy Machine is an FSM defined by a 6-tuple $(\mathcal{S}, s_0, \Sigma, \Lambda, \mathcal{T}, \mathcal{O})$, where:

- $\mathcal{S}$ is a finite set of states.
- $s_0 \in \mathcal{S}$ is the initial state.
- Σ is a finite set of input alphabet.
- Λ is a finite set of output alphabet.
- $\mathcal{T} : \mathcal{S} \times \Sigma \rightarrow \mathcal{S}$ is a transition function.
- $\mathcal{O} : \mathcal{S} \times \Sigma \rightarrow \Lambda$ is the output function.

3 Associating LLMs to FSMs

In this setup, we denote the LLM as a function $L(p) = t$, where a prompt $p \in \mathbf{V}^*$ is provided, and the output $t \in \mathbf{V}^*$ is a text response. Note that both input and output can be any sequence of tokens $(\alpha_1, \ldots, \alpha_n) \in \mathbf{V}^*$, however, only a small fraction of this domain constitute valid actions.

The FSM can be derived from the problem's PDDL and is designed to receive an action $a \in \Sigma \subset \mathbf{V}^*$ and return an output $o \in \Lambda \subset \mathbf{V}^*$, which is appended to the next LLM prompt p. This establishes an interactive loop between the LLM and the FSM, where the language model proposes actions based on its context p, while the FSM models preconditions and effects. Note that in this setup there's no guarantee that the LLM output t is a valid action, which means the LLM can send invalid actions (i.e. symbols) to the FSM.

However, it is possible to project t onto Σ by providing a validation function f such that

$$f : (\mathbf{V}^*, \mathcal{S}) \rightarrow \Sigma \tag{1}$$

$$f(t_i, s_j) = \begin{cases} t_i, & t_i \text{ is a valid action in state } s_j \\ -1, & \text{otherwise} \end{cases} \tag{2}$$

This function checks whether t corresponds to a valid action for the current state, producing a common symbol -1 for invalid actions. Consequently, instead of the FSM's transition and output functions receiving t, they use the output from the validation function: $\mathcal{T}(s_j, f(t_i, s_j))$ and $\mathcal{O}(s_j, f(t_i, s_j))$.

To assess how FSM modeling can enhance the capabilities of LLMs, we evaluate seven distinct methods. These range from a straightforward preplanned input/output setup to more sophisticated approaches that utilize FSMs to receive information and provide feedback on their actions. Additionally, we examine the CoT approach, as outlined in [15], where the initial prompt induces the model to generate structured reasoning based on a similar example.

By analysing these variations we aim to measure three key aspects:

1. Which method proves more effective: generating the entire plan at once, or generating each action individually while interacting with the FSM?
2. Does integrating validation functions and feedback loops improve LLMs' effectiveness in solving planning tasks?
3. What methods can be employed to restrict the range of actions available to LLMs, thereby guiding them towards optimal outcomes?

Initially, all methods receive the same context prompt p_c sourced from [14]. This prompt includes a detailed description of the planning task, a rulebook outlining all predicates and actions, and information about the current and goal states, as shown in Fig. 1. Each setup has its own request prompt p_r, which solicits either the entire plan or individual actions, enabling customization based on each setup's specific approach.

When the inference begins, the epsilon-transition $\mathcal{T}(s_{\text{start}}, \epsilon) = s_0$ occurs and the respective output $\mathcal{O}(s_{\text{start}}, \epsilon) = p_c + p_r$ produces the initial prompt. Thus, the state space becomes $\mathcal{S} = \{s_{\text{start}}, s_{\text{end}}\} \cup \mathcal{S}_{\text{task}}$, where $\mathcal{S}_{\text{task}}$ is the set of all possible states of the task being planned.

The methods analyzed are combinations of three characteristics:

Preplanned vs Interactive: Preplanned methods involve requesting the whole sequence of actions to the LLM before performing any kind of validation. This setup was assessed in [15], revealing a low success rate and a high prevalence of invalid plans. In this setup, the LLM generates the entire plan $(t_1, t_2, \ldots, t_n)$ solely based on $p_c + p_r$. On the other hand, interactive methods require one action at a time, performing an interaction between FSM and LLM at each step, allowing for invalid actions to be recognized earlier, preventing a reset back to s_0.

No Validation vs Validated vs Restricted: Three different schemes of validation are analyzed, namely, *no validation*, *validated*, and *restricted*. The *no validation* scheme simply halts the process in the case of invalid plans or actions. The *validated* scheme makes use of f to guarantee LLM outputs are projected onto Σ and provides feedback in the case of invalid actions. Finally, *restricted* methods provide all valid actions (obtained from the PDDL) for a certain state to the LLM through $\mathcal{O}$. In this case the LLM is asked to choose the number t_i corresponding to the selected action, but can still output any $t \in \mathbf{V}^*$. To handle invalid choices and to convert the selected action into a valid symbol we make use of a modified f

$$f(t_i, s_j) = \begin{cases} a_{t_i}^{s_j} & t_i \in p_a(s_j), \\ -1 & \text{otherwise}. \end{cases} \tag{3}$$

where $a_{t_i}^{s_j}$ is the action that corresponds to option t_i at state s_j.

Standard vs Chain-of-Thoughts: The standard prompting setup simply describes the goal and requests a plan (or the next action). On the other hand, CoT prompting induces the LLM to perform a step-by-step reasoning process for each interaction with the FSM. Prompting strategy can be modified by changing the prompts produced by $\mathcal{O}$.

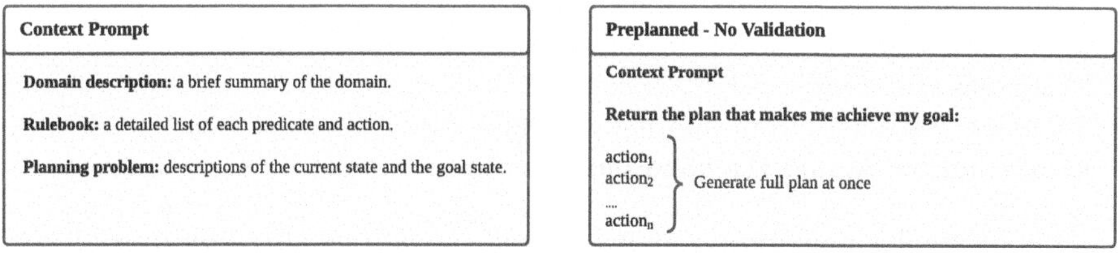

Fig. 1. (Left) Context prompt used in all proposed methods. (Right) Preplanned - No Validation: We prompt the LLM to generate the complete plan in one step.

4 Experimental Setup

The dataset is obtained from [14], which provides a platform to insert a valid PDDL domain and generate instances of it. For this study, we utilized the Blockworld domain, which captures common-sense block manipulations and consists of a set of blocks. Each block is uniquely identified by color and can be placed either on a table or on top of other blocks. In each instance, the initial state is a specific block arrangement, and the goal state is to arrange some of these blocks in a stack in a particular order (Fig. 2).

The Blockworld domain includes four different actions: *pick up*, *put down*, *stack*, and *unstack*. The preconditions for each action are as follows:

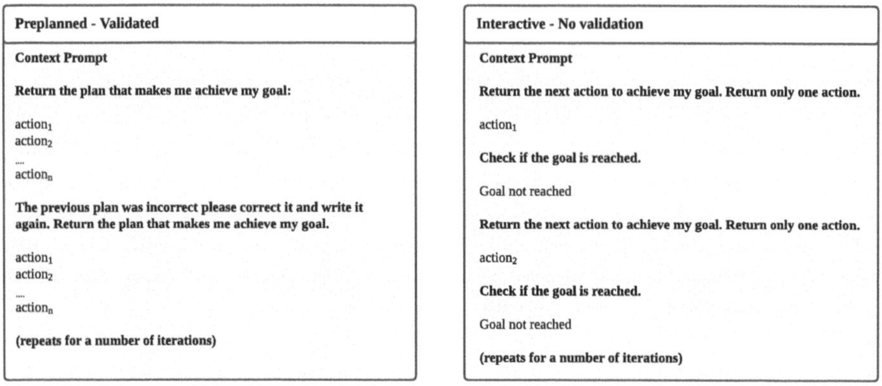

Fig. 2. (Left) Preplanned - Validated: We prompt the LLM to generate the entire plan in a single step. If the plan is invalid, we provide feedback and prompt the LLM to generate a new plan. (Right) Interactive - No Validation: We prompt the LLM to generate one action at a time and check if the goal is reached after each action. This process continues until the LLM indicates that the goal has been reached or the maximum number of iterations is exceeded.

- To *pick up* a block, it must be *clear* (no block stacked on it), *on the table*, and the hand must be *free.*
- To *put down* a block, it must be *on hand.*
- To *stack* a block on another, the first block must be *on hand* and the second block must be *clear.*
- To *unstack* a block from another, the first block must be *clear*, stacked on the second block, and the hand must be *free.*

This particular domain is widely adopted in the planning literature due to its simplicity and alignment with common-sense reasoning [5,8].

In this work, we generated a dataset with 768 instances, each containing 4 blocks. The dataset is balanced in terms of optimal plan length, which ranges from 2 to 12 actions. We use the optimal plan length as an indicator of the instance's complexity. For each instance, we use the same context prompt as in [14]. All prompts are zero-shot, except for methods using a CoT approach, where a one-shot approach is required to induce structured reasoning.

We evaluated three different LLMs in this work: Mixtral:8x22b, Llama3:70b, and GPT-4o. The temperature parameter is set to 0 for all models, ensuring that each experiment is deterministic. Our goal is to verify the effectiveness of the proposed methods independently of the specific LLM used. Additionally, a maximum of 8 tries for the Preplanned - Validated approach and 16 invalid actions for the Interactive approaches are enforced (Fig. 3).

The following metrics are calculated for each experiment:

- *Success Plan*: Indicates if the generated plan successfully reaches the goal without any errors.

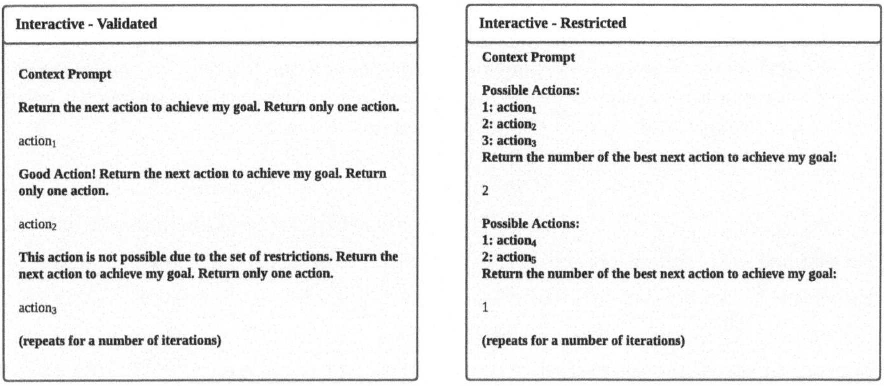

Fig. 3. (Left) Interactive - Validated: We prompt the LLM to generate one action at a time and provide feedback on whether the action is valid or not. This process continues until the goal is reached or the maximum number of iterations is exceeded. (Right) Interactive - Restricted: We prompt the LLM to choose one of all possible actions at the current state. This process continues until the goal is reached or the maximum number of iterations is exceeded.

- *Valid Plan*: Indicates if the generated plan is valid, meaning it does not contain any invalid actions.
- *Optimal Plan*: Indicates if the generated plan is the optimal plan.
- *Relative Plan Length*: The total number of actions in the generated plan divided by the optimal plan length.
- *Tokens*: The number of tokens processed during inference, measured in thousands.
- *Distance to Initial State*: The number of actions from the initial state to the final state of the generated plan.
- *Distance to Goal State*: The number of actions from the final state of the generated plan to the goal state.

5 Results

Numerical results are presented in Table 1, displaying average values of accuracy, optimal plan percentage, number of processed tokens and relative plan length metrics for all methods and models. The validated and restricted interactive methods consistently outperform other setups in most cases. They achieve higher accuracy than their preplanned and unvalidated counterparts regardless of the model.

An important finding is that the CoT approach achieved good results exclusively with the GPT-4o model, which is specifically trained to handle CoT reasoning. In contrast, the Mixtral and Llama models performed worse with the CoT scheme in both the preplanned and the interactive validated methods. This provides evidence that FSM-based validation provides increased accuracy in a more generic sense, being independent of how the particular LLM was trained.

Table 1. Numerical results for all methods and models. The metrics include accuracy percentage, number of tokens processed, and relative plan length, calculated based on all generated plans. The optimal plan percentage is calculated only for the successful plans. These results represent the average values across the entire dataset. Std refers to Standard prompt and CoT to Chain of Thought prompt.

	Accuracy(%) ↑			Optimal Plan (%) ↑		
	GPT-4o	Llama3	Mixtral	GPT-4o	Llama3	Mixtral
Preplanned - No validation - Std	7.2	2.9	0.78	100.0	63.3	83.3
Preplanned - No validation - CoT	53.4	7.9	4.2	81.5	32.8	53.1
Preplanned - Validated - Std	12.9	6.4	2.4	85.9	38.8	22.2
Interactive - No validation - Std	1.8	9.1	11.1	78.6	85.7	62.4
Interactive - Validated - Std	44.4	**33.5**	**21.9**	73.6	40.1	58.3
Interactive - Validated - CoT	**56.6**	2.5	1.1	34.6	14.3	9.1
Interactive - Restricted - Std	50.5	27.2	17.7	75.8	73.2	49.3
	Tokens (thousands) ↓			Relative Plan Length ↓		
	GPT-4o	Llama3	Mixtral	GPT-4o	Llama3	Mixtral
Preplanned - No validation - Std	0.55	0.56	0.56	1.1	1.1	1.1
Preplanned - No validation - CoT	1.5	1.5	1.5	1.2	1.4	1.3
Preplanned - Validated - Std	4.8	5.1	5.4	1.5	1.4	1.1
Interactive - No validation - Std	10.4	4.4	3.9	2.1	1.2	1.0
Interactive - Validated - Std	10.9	12.5	13.8	2.2	2.4	2.4
Interactive - Validated - CoT	46.5	57.2	60.2	1.7	3.2	3.1
Interactive - Restricted - Std	16.7	21.0	22.7	1.5	2.0	2.4

The optimal plan percentage, which is calculated for successful plans, is higher for preplanned setups for two main reasons: First, since preplanned methods generate fewer successful plans, often for simpler tasks, they tend to show a higher optimal plan percentage on average. Second, interactive validated setups often generate longer successful plans, because these methods have a deterministic function to hard stop inference when the goal state is reached, instead of relying on the LLM to recognize it. This allows for longer plans to eventually achieve the goal state in situations that other methods fail.

A downside of using interactive methods is their increased computational cost. As also shown in Table 1, the interactive setup processes an order of magnitude more tokens than the preplanned setups. This is because interactive schemes involve a sequence of prompts where all previous history is used as input, causing the size of each prompt to increase with each iteration. Special cases are those with the CoT scheme, which significantly increases the number of processed tokens for both preplanned and interactive setups. It's important to note that there is potential for prompt optimization in these schemes, which could make them faster and cheaper.

As displayed in Table 2 the preplanned and interactive methods without validation struggle with a low percentage of valid plans, as they are not robust against LLM hallucinations. In contrast, the interactive validated and restricted methods, by design, handle invalid inputs from the LLM, ensuring that all generated plans are valid. Once again, it is relevant to note that using CoT with Mixtral and Llama models frequently returned invalid actions, particularly in unfamiliar formats, demonstrating that such prompt engineering techniques can be sensitive to various factors and may work unpredictably across different models.

Table 2. Percentage of valid plans for methods that cannot handle invalid input. These results represent the average values across the entire dataset.

	Valid Plan (%) ↑		
	GPT-4o	Llama3	Mixtral
Preplanned - No validation - Std	13.7	15.5	12.9
Preplanned - Validated - Std	14.6	25.3	19.7
Preplanned - No validation - CoT	57.8	17.4	6.3
Interactive - No validation - Std	17.7	24.7	21.1

Considering the notably superior performance of the validated and restricted interactive methods compared to others, our analysis will primarily focus on these methods.

Figure 4 displays the average accuracy for each optimal plan length, serving as a complexity measure for each task in the dataset. As the optimal plan length increases, accuracy decreases, indicating that planning capabilities are sensitive to the complexity of the tasks. Furthermore, the two setups show different behaviors: the Interactive Validated setup demonstrates less sensitivity to optimal plan length, achieving a significant number of correct plans for more complex tasks, particularly with GPT-4o. In contrast, the Interactive Restricted setup shows more pronounced sensitivity, achieving higher accuracy for simpler tasks but experiencing a decrease in accuracy for more complex ones.

This behavior is further analysed in Figs. 5 and 6, which represent the distance from the final state of the generated plan to the goal and initial state, respectively. These metrics are only calculated for generated plans that didn't achieve the goal. Figure 5 shows that the two methods have a similar behavior: when they produce plans that don't achieve the goal state, these plans tend to finish far from the goal state. On average, the final state of the predicted plans is as far as the initial state of each task, indicating that the final state is not a good approximation of the goal state. However, in Fig. 6, one can see the distance from the final state of the produced plan to the initial state. This visualization shows that the distances for the Interactive Validated method are greater than the distances for the Interactive Restricted method. Thus, validated methods

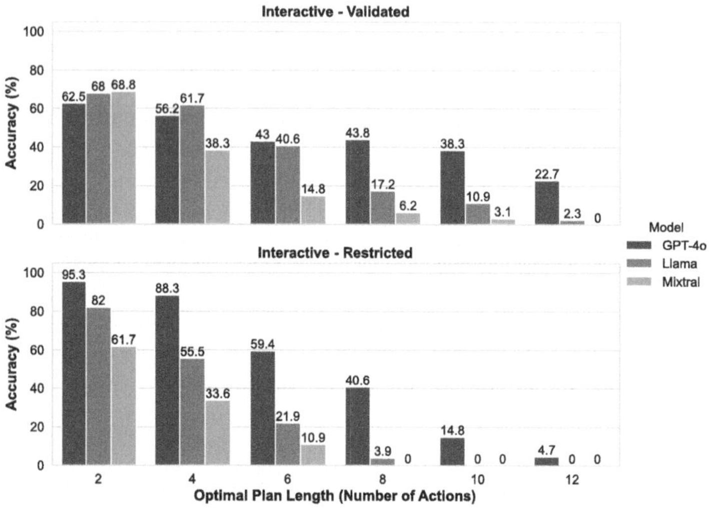

Fig. 4. The evaluation of average accuracy for the Interactive Validated and Restricted methods is divided by optimal plan length, which represents the minimal number of actions needed to achieve the goal state.

tends to explore more of the state space, resulting in final states farther away from the initial state, while restricted methods tend to perform local searches. This behavior is more pronounced for larger and more complex models, such as GPT-4o and Llama3.

Figure 7 presents further evidence for the difference in exploratory behavior between the methods. It shows the average number of times each model revisited states within a generated plan. We observe a significant difference: the Interactive Validated method exhibits a lower number of repeated states, indicating a tendency to explore diverse states. Conversely, the Interactive Restricted method frequently revisits states, suggesting a propensity to get stuck in loops. This looping behavior is likely inherent to the restricted setup. Restricted models receive previous actions as input, but often fail to recognize when they're trapped in a cycle. This potentially encourages short-term decision-making, hindering their ability to analyze past actions and states. We suspect this issue is exacerbated by the requirement to return action numbers instead of names in the restricted setting. This implicit task progress representation might make it challenging for the model to consider the entire historical trajectory when choosing the next action. Consequently, the restricted version tends towards local optimizations, leading to frequent loop formation.

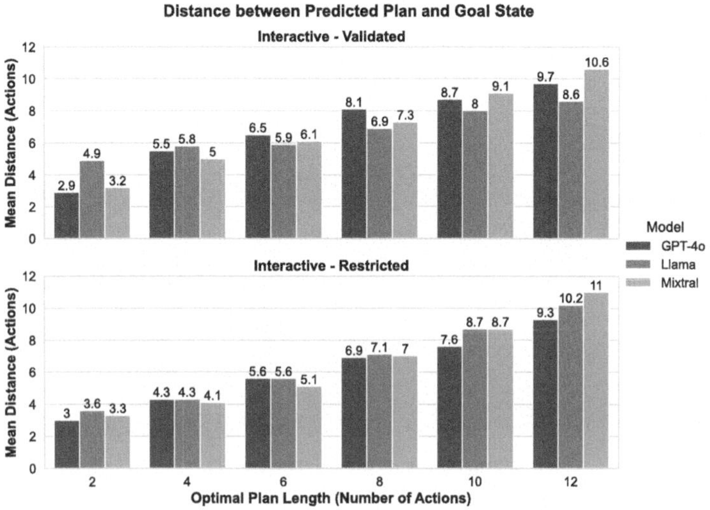

Fig. 5. The distance of the generated plan's final state from the goal state is measured in terms of the minimum number of actions required to transition from the final state of the generated plan to the goal state. It was only calculated for plans that didn't achieve the goal.

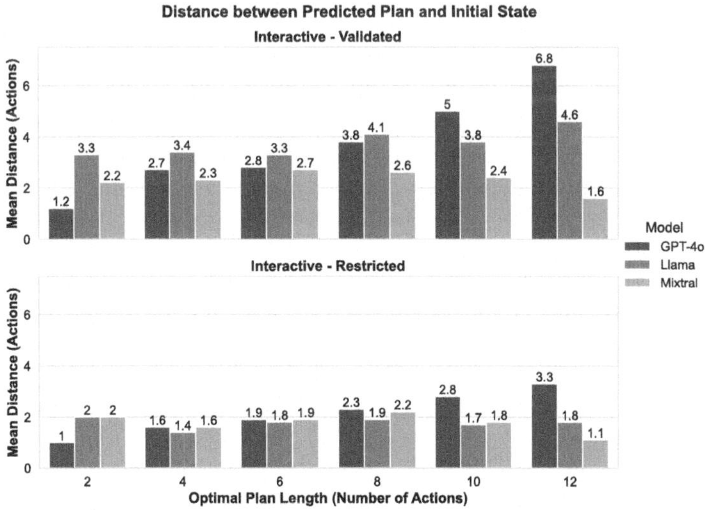

Fig. 6. The distance of the generated plan's final state from the initial state is measured in terms of the minimum number of actions required to transition from the final state of the generated plan to the initial state. It was only calculated for plans that didn't achieve the goal.

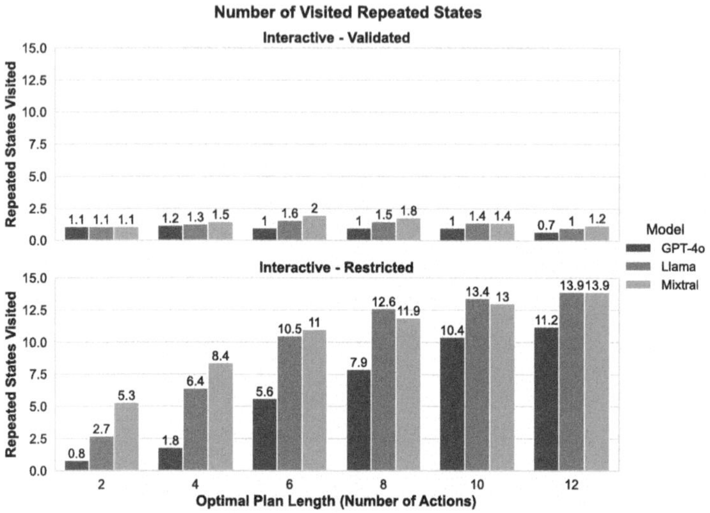

Fig. 7. We calculate the number of revisited states within a generated plan by summing up the occurrences of each state that appears more than once. This calculation is performed for all the plans generated.

6 Conclusion

Our results provide evidence that incorporating FSMs as a guiding structure enhances LLMs' ability to solve planning problems. By interacting with the FSM and leveraging integrated validations, feedback loops, and restrictions, LLMs demonstrably generate plans with increased accuracy across tasks of varying complexity. Although the accuracy achieved is not yet ideal, the proposed method demonstrates a valuable tool for improving model inference in reasoning tasks, offering benefits independent of other prompt engineering techniques.

FSM integration was also compared to the CoT prompting technique in both preplanned and interactive validated schemes. CoT showed superior performance exclusively for the GPT-4o model, while FSM integration greatly increased accuracy for all models. This provides evidence that FSM integration is a more consistent solution than prompt engineering techniques that can behave unpredictably across different scenarios.

The proposed Interactive Validated and Restricted setups offer a crucial advantage by being inherently designed to handle invalid inputs, thus guaranteeing that all generated plans are valid. This is a substantial improvement over traditional approaches, where LLMs frequently generate invalid plans. We observed that incorporating validation encourages exploratory behavior in the models, leading to better performance on complex tasks across all three compared models. Conversely, while incorporating restrictions can enhance performance on simpler tasks, it can also cause the models to enter loops, resulting in poorer performance on more complex tasks. A significant downside of interactive approaches is their increased computational cost. Results show that the number of tokens processed in these schemes is an order of magnitude higher than in preplanned setups, underscoring the necessity of prompt optimization in these scenarios.

Despite the simplicity of the experimental context, the findings open up new avenues for future research aimed at enhancing reasoning capabilities using logical structures that constrain the action space in problem-solving. The proposed method also introduces greater predictability and robustness into the system.

Our findings suggest promising avenues for future research. One direction involves developing approaches that can unify validation with feedback and restrictions within the LLM's decision-making process. This could lead to more reliable and robust models capable of both short-term and long-term reasoning.

References

1. Besta, M., et al.: Graph of thoughts: solving elaborate problems with large language models. In: Proceedings of the AAAI Conference on Artificial Intelligence, vol. 38, no. 16, pp. 17682–17690 (2024). https://doi.org/10.1609/aaai.v38i16.29720. https://ojs.aaai.org/index.php/AAAI/article/view/29720
2. Bubeck, S., et al.: Sparks of artificial general intelligence: early experiments with GPT-4 (2023)
3. Chen, M., et al.: Evaluating large language models trained on code (2021)
4. Ghallab, M., Nau, D., Traverso, P.: Automated Planning: Theory and Practice. The Morgan Kaufmann Series in Artificial Intelligence, Morgan Kaufmann, Amsterdam (2004). http://www.sciencedirect.com/science/book/9781558608566
5. Guan, L., Valmeekam, K., Sreedharan, S., Kambhampati, S.: Leveraging pretrained large language models to construct and utilize world models for model-based task planning. In: Oh, A., Naumann, T., Globerson, A., Saenko, K., Hardt, M., Levine, S. (eds.) Advances in Neural Information Processing Systems, vol. 36, pp. 79081–79094. Curran Associates, Inc. (2023)
6. Helmert, M.: The fast downward planning system. J. Artif. Intell. Res. **26**, 191–246 (2006). https://doi.org/10.1613/jair.1705
7. Hoffmann, J.: FF: the fast-forward planning system. AI Mag. **22**(3), 57 (2001). https://doi.org/10.1609/aimag.v22i3.1572. https://ojs.aaai.org/aimagazine/index.php/aimagazine/article/view/1572
8. Jiménez, S., De La Rosa, T., Fernández, S., Fernández, F., Borrajo, D.: A review of machine learning for automated planning. Knowl. Eng. Rev. **27**(4), 433–467 (2012). https://doi.org/10.1017/S026988891200001X
9. Jin, Z., et al.: Cladder: a benchmark to assess causal reasoning capabilities of language models (2023)
10. Raman, S.S., Cohen, V., Rosen, E., Idrees, I., Paulius, D., Tellex, S.: Planning with large language models via corrective re-prompting. In: NeurIPS 2022 Foundation Models for Decision Making Workshop (2022). https://openreview.net/forum?id=cMDMRBe1TKs
11. Russell, S., Norvig, P.: Artificial Intelligence: A Modern Approach, 3 edn. Prentice Hall (2010)

12. Stechly, K., Marquez, M., Kambhampati, S.: GPT-4 doesn't know it's wrong: an analysis of iterative prompting for reasoning problems. In: NeurIPS 2023 Foundation Models for Decision Making Workshop (2023). https://openreview.net/forum?id=PMtZjDYB68
13. Suzgun, M., et al.: Challenging big-bench tasks and whether chain-of-thought can solve them (2022)
14. Valmeekam, K., Marquez, M., Olmo, A., Sreedharan, S., Kambhampati, S.: Planbench: an extensible benchmark for evaluating large language models on planning and reasoning about change. In: Thirty-Seventh Conference on Neural Information Processing Systems Datasets and Benchmarks Track (2023). https://openreview.net/forum?id=YXogl4uQUO
15. Valmeekam, K., Marquez, M., Sreedharan, S., Kambhampati, S.: On the planning abilities of large language models - a critical investigation. In: Thirty-Seventh Conference on Neural Information Processing Systems (2023). https://openreview.net/forum?id=X6dEqXIsEW
16. Vaswani, A., et al.: Attention is all you need. In: Guyon, I., et al. (eds.) Advances in Neural Information Processing Systems, vol. 30. Curran Associates, Inc. (2017)
17. Wang, X., Wei, J., Schuurmans, D., Le, Q.V.: H. chi, sharan narang, aakanksha chowdhery, and denny zhou. self-consistency improves chain of thought reasoning in language models. In: The Eleventh International Conference on Learning Representations, vol. 1 (2023)
18. Wei, J., et al.: H. chi, tatsunori hashimoto, oriol vinyals, percy liang, jeff dean, and william fedus. emergent abilities of large language models. Trans. Mach. Learn. Res. **2022** (2022)
19. Wei, J., et al.: Chain-of-thought prompting elicits reasoning in large language models. In: Koyejo, S., Mohamed, S., Agarwal, A., Belgrave, D., Cho, K., Oh, A. (eds.) Advances in Neural Information Processing Systems, vol. 35, pp. 24824–24837. Curran Associates, Inc. (2022)
20. Yao, S., et al.: Tree of thoughts: deliberate problem solving with large language models. In: Oh, A., Naumann, T., Globerson, A., Saenko, K., Hardt, M., Levine, S. (eds.) Advances in Neural Information Processing Systems, vol. 36, pp. 11809–11822. Curran Associates, Inc. (2023)

Improving Short-Content Misinformation Detection Using Multiple Aspect Trajectories Classification Techniques

Juan Pablo Chavarro Sanchez[1](✉), Tarlis Tortelli Portela[2], and Jônata Tyska Carvalho[1]

[1] Universidade Federal de Santa Catarina, 476 88.040-900, Florianopolis, Brazil
{juan.pcs,jonata.tyska}@ufsc.br

[2] Instituto Federal do Paraná (IFPR), Curitiba, Brazil
tarlis.portela@ifpr.edu.br

Abstract. The proliferation of fake news and misinformation on social media poses a significant threat to social integrity. While there are approaches such as automatic content analysis and artificial intelligence detection to mitigate this problem, these methods face challenges in classifying misinformation in short content, like that posted on microblogging platforms or instant messaging services. In this study, we present an innovative approach to misinformation detection that combines content-based detection with multi-aspect trajectory classification. This approach models information propagation by considering each shared message as a moving object within the network, enriching the propagation trajectory with different semantic aspects. We implemented this approach on a misinformation dataset collected from WhatsApp and compared it with traditional content-based detection methods, and we combined these two detection orientations into a hybrid approach. Our findings indicate that the proposed hybrid approach achieves an F1-score of 0.89, surpassing baseline models by 8% in misinformation detection, even in short content. These results suggest that combining content-based detection with multi-aspect trajectory classification is a novel and promising strategy for addressing misinformation on social media.

Keywords: Multi-aspect trajectory · Misinformation detection · Online social networks

1 Introduction

In today's digital era, social media has taken a pivotal role in people's daily lives. These platforms provide users from around the world with the ability to communicate, share information, and express opinions quickly, affordably, and accessibly. According to Statistica's statistics [30], over 50% of the world's population is active on social media, emphasizing its significance as a tool for communication, information, and opinion sharing.

A. Paes and F. A. N. Verri (Eds.): BRACIS 2024, LNAI 15413, pp. 125–139, 2025.
https://doi.org/10.1007/978-3-031-79032-4_9

The growing reliance on social media for real-time news has increased the risk of misinformation, as malicious actors exploit these platforms [23,34]. Technological advancements, such as large language models, further enable the rapid spread of false information, reinforcing echo chambers and filter bubbles [3,21]. Additionally, the structure of social networks promotes the dissemination of misinformation [2], creating significant challenges for governments and society.

The detection of fake news and online misinformation has been approached in various ways, including two main approaches: *manual fact-checking* and *automatic detection*. Manual verification by news agencies faces limitations due to the vast volume of data, potential human bias, and impracticality for early detection [24]. Automatic detection can overcome most of these limitations. These methods can be based on the context, source of the information or content. Source-based detection assesses the credibility of the sources creating or sharing the information, while context-based detection analyzes user interaction patterns and information propagation. Content-based detection seeks patterns in the text or images to identify the news article's intent. Despite its effectiveness and the fact that this is the most studied analysis in the state of the art [11], this approach is vulnerable to the manipulation of news style, language, and adversarial attacks [37]. A significant and noteworthy challenge in the detection of misinformation is the recognition of short texts, typical on microblogging platforms such as Weibo and Twitter; and instant messages services such as WhatsApp or Telegram. This feature is particularly relevant in the context of fake news, as the concise messages on these platforms, due to their brevity, can be more complex to evaluate and verify accurately.

Given the complexity and heterogeneity of data related to information on online social networks, along with the limitations of content-based approaches, we propose two novel approaches: one based on the context of misinformation propagation and user interaction on social networks, and another hybrid approach that combines content-based and trajectory-based detection. In this paper, we introduce a model that conceptualizes the propagation path of a message on an online social network, specifically an instant messaging service, as a Multiple Aspects Trajectory (MAT) [18]. This type of data provides an effective and expressive representation of the diversity of data dimensions, namely the contexts associated with online information and its propagation. To validate the feasibility of these approaches, we turned to a dataset of messages spread in public WhatsApp groups focused on Brazilian politics [1]. Using the analogous definition of information as multiple aspects trajectories, we modeled the propagation paths of these messages based on forwarding chains as trajectories.

This study proposes a novel trajectory-based approach and a hybrid approach for misinformation detection in social networks, combining content-based and trajectory-based classification. The results show that the hybrid approach outperforms the baseline method by 8%, achieving an F1 score of 0.89, especially in the classification of short texts. The main contributions of this work are: a new dataset of information trajectories, a new application domain for trajectory-based classification methods, and a new classification method with great potential for detecting misinformation in social networks.

2 Related Work

In this section, we will analyze the related works on detecting misinformation and fake news, with a primary focus on those employing propagation-based approaches. These approaches take into careful consideration the social context in which such misleading information spreads, accounting for the ways data disseminates across online networks. Researchers have approached this challenge from various angles and have employed different input representations in their models [13,36].

The models that utilize cascade, tree, or "tree-like" structures directly represent the path of information propagation. In [4], Del et al. investigate how users in Facebook groups tend to share homogeneous content, leading to the formation of echo chambers. Within these echo chambers, homogeneity drives the spread of misinformation. Additionally, relationships between the trajectory size and the lifecycle of information are established. Similarly, [32] shows that false news spreads faster, farther, and to a broader audience on Twitter compared to true news, analyzing factors such as cascade depth and breadth. Additionally, [5] Ducci et al. introduces the "cascade-LSTM" model, which detects misinformation and analyzes retweet behavior.

Some other propagation-based detection methods utilize structural network information, such as text data and the propagation process, to portray the behavior of users involved in the dissemination of misinformation [13]. Often, these methods adopt a graph-based geometric orientation, and depending on the interaction between nodes, they can be classified as homogeneous, heterogeneous, or hierarchical. Homogeneous networks involve a single type of node and a single type of edge. In [16], Ma et al. propose a model based on recursive neural networks for rumor detection based on propagation trees that use user characteristics in the Twitter15 and Twitter16 datasets. In [10], Han et al. use a homogeneous network representation to classify news using Graph Neural Networks (GNNs), highlighting the advantages of propagation-based approaches in terms of news text independence. In [29], a model is proposed to capture temporal evolution patterns of news from Twitter15, Twitter16, FakeNewsSetGen, and Weibo using Homogeneous Dynamic Graph Neural Networks for fake news detection. In [19], a homogeneous graph is created without considering the textual content of each tweet. Instead, a tree structure is generated, and Graph Convolutional Networks (GCNs) are employed for fake news detection.

Heterogeneous networks involve multiple types of nodes and edges. For instance, in [27], relationships between authors, users, and news in a network called TriFN are proposed, and a classifier is employed to detect fake news. In the work Monti et al. [20], which employs similar dimensions to the previous studies and CNNs for classification, the model is found to be potentially independent of the text and geographic location of the users involved in the network. Additionally, the advantages of propagation-based approaches in terms of adversarial attacks, as these would require manipulating the social network.

Hierarchical networks involve various types of nodes and edges forming hierarchical relationships. This is exemplified in the work developed by Shu et al.

[26], where a hierarchical propagation network is constructed by evaluating different granularities to extract temporal, structural, or linguistic features.

An approach for representing propagation can be based on time series, more specifically, multivariate time series. In the work of Preveti et al. [22], rumor propagation is represented as the time series of a cascade, incorporating user-related and temporal aspects for classification. In the work of Liu et al. [15], the propagation path is represented as a multivariate time series, with points based on user and time-related aspects. These aspects are primarily numeric or binary, and convolutional and recurrent networks are applied to detect time series associated with fake news in the Twitter15, Twitter16, and Weibo datasets.

3 Dissemination of Misinformation as MAT

This section, divided into two parts, explores addressing misinformation in online social networks through multi-aspect trajectories. The first part covers the theoretical foundations, while the second focuses on applying these principles by converting a dataset into multi-aspect trajectories.

3.1 Misinformation and MAT

As discussed in Sect. 2, existing representations of news or information propagation paths generally use cascade structures, graphs, or time series, incorporating mainly numeric or textual features related to users, news, networks, or time. However, these representations lack semantic data. In this section, we introduce the essential definitions and concepts for our approach. We begin by presenting the definition of "Multiple-Aspect Trajectory" as suggested in [17], and then define what a "Multiple-Aspect Trajectory" is in the context of information in Online Social Networks (OSN), capable of capturing semantic data.

Definition 1 (Multiple - Aspect Trajectory). *A multiple - aspect trajectory is an ordered sequence of points* $T = \langle d_1, d_2, \ldots, d_n \rangle$, *with* $d_i = (x, y, t, \mathcal{A})$ *being the* $i-$*th point of the trajectory at location* (x, y) *and timestamp* t, *described by the set* $\mathcal{A} = \{a_1, a_2, \ldots, a_r\}$ *of* r *attributes.*

This type of sequential data allows for the inclusion of heterogeneous data of diverse nature and types. These trajectories can be multi-labeled, and for the classification task. According to the work by Petry et al. [17], the concept of a "trajectory classification" is defined as follows:

Definition 2 (Trajectory classification). *Given a set of labels* $\mathcal{L}$ *and a trajectory set defined by a set of pairs* $\mathcal{T} = \{(T_1, label(T_1)), \ldots, (T_{|\mathcal{T}|}, label(T_{|\mathcal{T}|}))\}$, *where each pair contains a trajectory* T_i *and its class label* $label(T_i) \in \mathcal{L}$. *Trajectory classification involves learning a prediction function (i.e., a model)* f *that assigns each trajectory* T_i *in the set* $\mathcal{T}$ *to one of the class labels in* $\mathcal{L}$.

Kumar et al. [14] differentiate between misinformation, which is false information without intent to deceive, and disinformation, which involves deceptive intent. This distinction is supported by other researchers [12,33], though some studies use a broader definition of misinformation to encompass all false or inaccurate information spread on social networks. This includes rumors, disinformation (intentional or unintentional), urban legends, fake news, spam, and hate speech. These phenomena, given their temporal evolution in a social network, can be represented as a sequence of multiple aspects.

In line with our objective to analyze the behavior of information in a social network, it is essential to have a general concept that represents the dissemination of information. Following Guo et al. [9], we define information dissemination in social networks by considering the content, social, and temporal dimensions, as outlined by Shu et al. [25].

- Let s be information disseminated in a social network N, which consists of two main sets: Publisher and Content. Publisher P_s includes a set of profile features to describe the original author, such as name, domain, age, among other attributes. Content C_s consists of a set of attributes that represent the news article and includes headline, text, image, etc.
- Let $E = \{e_{it}\}$ be the set of tuples to represent the process of how information spread over the time among n users $U = \{u_1, u_2, \dots u_n\}$ and their corresponding posts $P = \{p_1, p_2, \dots p_n\}$ on social media regarding information s. Each element e_{it} is described as $\{p_i, u_i, t\}$, indicating that a user u_i spreads a post s using p_i at time t.

Therefore, to consider information or news s as a trajectory, we must treat it as a sequence that passes through points of interest (POIs). Due to privacy concerns in social networks, it is not always possible to obtain a user's location. Thus, we initially need to define a temporal sequence of points along the trajectory it follows. In this case, each point corresponds to an interaction e_i at a time t, where a user u_i interacts with the post p_i corresponding to the POI at a specific time t. This i-th point in the temporal sequence is associated with content and context aspects provided by u_i and p_i. In this regard, and considering the definition of misinformation propagation in a social network adapted to the definition of multi-aspect trajectories given in the Definition 1, we present the following definition.

Definition 3 (Multiple-Aspect Trajectory of information). *A multi-aspect trajectory of information propagated in an Online Social Network (OSN) is an ordered sequence of points* $E = \langle e_1, e_2, \dots, e_n \rangle$*, where* $e_i = (u_i, t, \mathcal{A})$ *represents the* $i-th$ *point of the trajectory located at* p_i *at a timestamp* t*, described by a set of attributes* $\mathcal{A} = C_{s_i} \cup P_{s_i}$*, where* C_{s_i} *is the set of attributes related to the content of* s*, and* P_{s_i} *is the set of context attributes related to* s*.*

This study addresses the problem of binary classification of trajectories, using $\mathcal{L} = \{0, 1\}$, as specified in Definition 2.

3.2 Dataset

In this section, we start from the original dataset *"FAKEWHATSAPP.BR"*, which contains WhatsApp messages labeled as misinformation or not. From these messages, we generate the dataset *"MAT FAKEWHATSAPP.BR"*, which contains the multi-aspect trajectories of the messages, according to Definition 3.

FAKEWHATSAPP.BR. The original dataset, proposed in [1], consists of news collected from 59 public WhatsApp groups that were used during the political campaigns for the 2018 presidential elections in Brazil. This data collection took place from July 2, 2018, to October 29 of the same year, resulting in a total of 282,601 messages labeled in two classes: the positive class, represented as 1, corresponds to messages with misinformation, while the negative class, represented as 0, corresponds to messages without misinformation.

To effectively label the dataset, the findings of Vosoughi et al. [32] were considered, demonstrating that false information spreads faster, wider, and deeper. A subset of 5,284 unique messages was selected based on their virality and manually labeled following a classification protocol. The labeled subset comprises mostly brief messages, averaging 20 tokens for non-misinformative messages and 34 for misinformative messages.

To automatically expand the categories assigned to the subset of messages to the entire dataset, an algorithm generated the TF-IDF matrix from both the labeled subset and the complete corpus. It calculated the similarity between unlabeled and labeled messages, assigning labels to unlabeled messages when the cosine similarity exceeded 0.9. This process labeled 21,289 messages, with 6,926 being unique. Cabral [1] evaluated this dataset using classic natural language processing methods, including Bag of Words (BoW) and TF-IDF features with n-gram variations. The best results, with an F1 score of 0.73, were achieved using MLP, LSVM, and SGD algorithms. Excluding short messages and focusing on those with over 50 words increased the F1 score to 0.87.

MAT FAKEWHATSAPP.BR. Taking into account the Definition 3, we proceed to generate the set of multi-aspect trajectories using the aforementioned dataset. This set can be seen as a collection of trajectories passing through points of interest (POIs). In the context of WhatsApp, each POI represents a user who shares a message, and the moving object is the message being shared from one user to another. The semantic aspects are derived from the data obtained from the social network.

If we were to construct the trajectories as temporal sequences of message repetitions (forwards), we would obtain the 6,926 trajectories corresponding to the unique labeled messages in the dataset, composed of 21,289 points. However, a more detailed examination of the dataset reveals that the vector representation and similarity calculation fail to fully capture similar records that exhibit variations. To address this limitation, we will use a representation capable of capturing the context of the text and generating a better representation through word embeddings generated by BERT.

The process of creating these trajectories unfolds following the detailed steps below and can be followed in the Fig. 1.

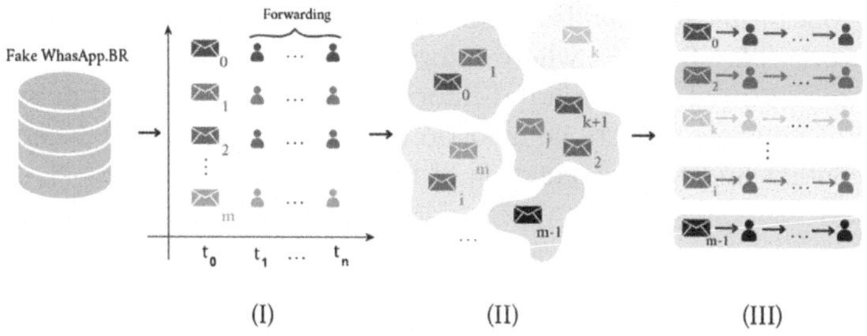

Fig. 1. Trajectory creation process.

I *Extraction and Grouping of Trajectories.*
Identification of Forwards: All forwarded messages in the conversation are identified, thus establishing the beginning of a trajectory.
Temporal Grouping: Forwarded messages are grouped into trajectories ordered by time, allowing us to capture the sequence of forwards and the spread of information in the chat.

II *Representation of Trajectories with Sentence Embeddings.*
Generation of Embeddings Each message in a trajectory is represented by a multilingual BERT sentence embedding. This captures the semantic and lexical context of each message in a vector space.
Similarity Calculation: We calculate cosine similarity between the embeddings of all trajectories. Those that exceed a similarity threshold of 0.9 are considered candidates for merging.
Merging Similar Trajectories: Trajectories or records with high similarity are candidates for merging and are combined into a single trajectory. This reflects the relationship between messages that share similar content.

III *Assignment of Unique Identifiers (TID).*
Temporal Sorting: The forwards of each message representing a trajectory are sorted chronologically to establish their temporal sequence.
TID Assignment: A unique TID is assigned to each message or trajectory, ensuring that each trajectory in the dataset corresponds to a unique representative message.

The process of creating the dataset involved selecting messages that were shared at least 10 times, thus representing trajectories with 10 or more points. This criterion reduced the dataset to 469 trajectories, composed of 177 labeled as non-misinformation and 292 as misinformation, reflecting the higher frequency of misinformation messages in the original dataset. These 469 trajectories span 10,251 points and are associated with 2,408 unique messages. The selected aspects for analysis are presented in Table 1. The selection of these aspects was guided by the attributes commonly used in location check-in datasets such as FourSquare [17].

The process of construction can be replicated using datasets from different messaging platforms such as Telegram, or it can be modified for use on other social networks like Twitter, where information is disseminated in brief texts. This representation of the spread of information on online social networks as a trajectory of multiple aspects is agnostic to the text itself, as it leverages data from the context and the propagation of information.

Table 1. Description of Dataset Attributes.

Attribute	Type	Range/Example	Unique values
User Id	Nominal	$\{15034, 31560, 11025 \ldots\}$	2161
Group	Nominal	$\{1, 2, 3, \ldots\}$	56
States of Brazil	Nominal	$\{AM, BA, \ldots, TO\}$	29
Month	Nominal	$\{Aug, Sep, Oct\}$	3
Days of the week	Nominal	$\{Mon, Tue, \ldots, Sun\}$	7
Days of the month	Numeric	$[1, 30]$	30
Hour	Numeric	$[0, 23]$	24
Text modified	Nominal	$\{True, False\}$	2

4 Trajectory Classification Models

In the field of multi-aspect trajectory analysis, various techniques have been developed to discover patterns and knowledge in this data, ranging from similarity analysis to clustering and classification, among others. Initially, the focus was on the raw trajectories of moving objects over time, limited to extractable attributes from space-time dimensions [28].

With the advent of semantic trajectories, defined as sequences of points of interest visited by a user, the problem of "Trajectory-user-linking" emerged, which aims to predict which user generated a trajectory. Despite advances such as TULVAE [35] and TULER [8], which have addressed these semantic trajectories, they still face limitations in other dimensions.

Finally, with multi-aspect trajectories, models capable of encompassing all dimensions have emerged. The Movelets method [6], a pioneer in this field, extracts representative subtrajectories of classes, continually improving the quality of the extracted Movelets and reducing computational costs. Additionally, the MARC [17] model, a recurrent neural network, can effectively handle all these dimensions. These models have found applications in a variety of datasets, including human trajectories, animal trajectories, transportation trajectories, hurricane trajectories, vessel trajectories, and storm trajectories. Below, we explore in detail how these models are capable of addressing multi-aspect trajectories [28].

4.1 MARC Model

One of the standout models in this context is the MARC Model [17], which relies on attribute embeddings and recurrent neural networks (RNNs). This model excels in processing trajectories defined by categorical or textual attributes. Its architecture comprises several key components:

Encoding of Trajectory Points: Initially, trajectory attributes are encoded using one-hot encoding. This encoded representation of our attributes is then multiplied by an embedding matrix, which has as many rows as possible values of the attribute. The rows of this embedding matrix represent the embedding of each of the attributes. Finally, an aggregation function is applied to these embedded attribute points. In the MARC model, three different aggregation functions can be used, which can be sum, average, or concatenation.

Recurrent Component: The encoded trajectory points move on to a recurrent component, utilizing Long Short-Term Memory (LSTM) cells. This neural network structure with LSTM units is fundamental for handling sequential data like trajectory points. LSTM units can capture relationships and patterns among different trajectory points and their attributes.

Fully Connected Layer and Sigmoid Function: The final component is a fully connected layer that receives information from the recurrent component and subsequently passes through a sigmoid function to calculate probabilities.

4.2 Movelet-Based Models

Another category of models aims to identify the best trajectory or subtrajectory features for input into a classifier. Various methods have been proposed for this task. These methods extract movelets [6], that are subsequences of the trajectory of any size and any combination of attributes. The movelets are evaluated by a quality score based on its discriminative power for each class, find the optimal alignment and division, and then assess its F-score, which determines its importance. The internal architecture and the way the candidate movelet is calculated differ for each model:

- **MasterMovelets** [7]: Calculates the distance from each candidate to all trajectories, making it the most computationally expensive method.
- **HiperMovelets** [31]: Extracts limited-size subtrajectories and generates movelet candidates based on their in-class frequency, currently being the fastest model.

5 Experimental Evaluation

In this section, we describe the methodology used to evaluate our approach for misinformation classification using multi-aspect trajectory classification. We compare it with a content-based detection approach similar to Cabral et al. [1], which serves as the baseline. All experiments will emphasize hyperparameter optimization through 5-fold stratified cross-validation, with the F1-score as the primary evaluation metric.

5.1 Content-Based Detection

In order to provide a comparative analysis, a classic content-based detection approach was established employing conventional natural language processing and machine learning methods. In this experiment, different textual content associated with each trajectory was utilized, meaning that if a trajectory represents the dissemination of certain information, all textual variants of that information are taken into account.

For each of the 2408 different messages linked to the 469 trajectories, the same preprocessing employed by Cabral et al. [1] was employed, included removing stopwords, normalization, lemmatization, retaining emojis as tokens, and normalizing URLs. Given their success with text vector representation using TF-IDF and various n-grams, the same vectorizer was used, incorporating unigrams, bigrams, and trigrams. Finally, the classification algorithms that yielded the best results with the aforementioned vector representation were selected, including Logistic Regression (LR), Multi-Layer Perceptron (MLP), and Linear Support Vector Machine (LSVM).

5.2 Trajectory-Based Detection

In evaluating the proposed trajectory propagation-based approach, the trajectory classification models discussed in the previous section were used, with preprocessing and transformation detailed in Sect. 3.

For the MARC model, experiments varied the aggregation method of the embeddings, testing different methods (add, average, concatenate).

For the model based on Movelets, HiperMovelets was chosen for its efficiency and quality in extracting the candidate Movelets matrix representing each trajectory. This matrix was used in three classification models: Random Forest (RF), Support Vector Machine (SVM), and Multilayer Perceptron (MLP).

5.3 Hybrid Approach

As pointed out by Cabral et al. [1], detecting misinformation based solely on content has limitations, especially with short texts. In addition to this limitation, it is important to consider the quantity of texts associated with an information, as content-based models employ frequency-based methods, and this can affect their performance if the frequency is low.

That's why the proposed hybrid classifier generates a classification threshold regarding the amount of words and texts associated with an information. If the text has fewer than a certain number of words and fewer than a certain number of different associated messages, the classification of this message will be determined by the classification assigned by the trajectory-based method to the trajectory to which that message belongs. Otherwise, the classification will be made by a content-based classification model.

For this approach, we evaluate the combination of all trajectory-based and content-based models with their optimized hyperparameters. Furthermore, for the estimation of this threshold, we consider the quantity of words and different texts as hyperparameters. The models are trained using cross-validation, evaluating the hybrid model on the validation fold. After finding the best hyperparameters, the hybrid model is trained on the complete training set, and its performance is evaluated on the test set.

6 Results

A total of 69 experiments were conducted to evaluate different approaches to misinformation detection. Nine combinations of TF-IDF with unigrams, bigrams, or trigrams were explored, using Multilayer Perceptron (MLP), Logistic Regression (LR), and Linear Support Vector Machines (LSVM) classifiers for content-based detection. Additionally, six trajectory-based models were evaluated: three variants of the MARC model and three models based on the creation of Movelets with HyperMovelets, classified with Multilayer Perceptron (MLP), Support Vector Machines (SVM), and Random Forest (RF). Finally, 54 experiments were implemented with the hybrid approach, combining TF-IDF with unigrams, bigrams, and trigrams along with the classifier for content-based detection, with the six trajectory classification models mentioned above.

Table 2. Results of misinformation classification using different approaches

		Accuracy	F1 score
Content-based	Trigram+MLP	0.746479	**0.815068**
	Bigram+MLP	0.732394	<u>0.811881</u>
	Unigram+MLP	0.715962	0.782765
Trajectory-Based	MARC-avg	0.786384	**0.819085**
	MARC-cct	0.791079	<u>0.811040</u>
	MARC-add	0.776995	0.809619
Hybrid approach	Bigram+MLP+Hiper+RF	0.850241	**0.894915**
	Unigram+MLP+Hiper+RF	0.830917	<u>0.882943</u>
	Unigram+MLP+MARC-cct	0.826086	0.880794

Given the number of experiments, Table 2 presents the top three results per approach, indicating their F1 score and accuracy. It is noteworthy that the best result is achieved when employing the hybrid approach that combines Bigram+MLP+Hyper+RF, which achieves the highest accuracy (0.85) and F1-score (0.89) among all evaluated models. Content-based models exhibit acceptable performance, primarily when considering the use of n-grams of any order and the MLP classifier. However, these models achieve the lowest accuracy. On the other hand, trajectory-based models are led by the MARC-based model, which in its three variants achieves the best results for this approach. Furthermore, the best trajectory-based model (MARC-avg) slightly outperforms the best content-based model in terms of F1 score, but the difference is more significant when observing the precisions achieved by these three models.

Table 2 presents the best results of the hybrid approach that combines trajectory-based and content-based models when determining a classification threshold. The hyperparameters of this threshold are, on average across all experiments, fewer than 40 words and fewer than 5 associated messages for an information. In such cases, classification is done by a trajectory-based method; otherwise, it is done by a content-based method.

Comparing the three best trajectory-based classifiers presented in Table 2, we observe that, on average, they correctly classify 77.5% of short messages with few different texts associated, while the three best content-based classifiers classify, on average, 62.1%, which is 15% less.

This demonstrates an advantage of trajectory-based classification: its independence from the text of messages associated with an information, as this approach only utilizes propagation data to represent a set of messages. When classifying a trajectory, all messages associated with it are classified, whereas in a content-based approach, only the unique text associated with an information is classified. This characteristic makes trajectory-based classification particularly suitable for short texts, where the information contained in the text may be limited. With the thresholds determined for each experiment, the best results are achieved when using unigrams and bigrams with the MLP classifier if the threshold condition is not satisfied, and Hyper+RF or MARC-cct otherwise. The best of these hybrid models achieves an F1-score of 0.89, which surpasses the best content-based method by 8%. These models also achieve superior accuracy. To better observe this, Fig. 2 presents the average accuracy and F1-score per classification orientation.

As shown in Fig. 2, trajectory-based and content-based methods deliver similar performance in terms of F1-score, with trajectory-based models achieving slightly higher accuracy but with greater variability. In contrast, the hybrid approach significantly outperforms both methods, demonstrating higher mean F1-score and accuracy with much lower variance, highlighting the advantage of combining both strategies for misinformation detection. We conducted a statistical significance analysis, starting with the Shapiro-Wilk test to assess normality ($W = 0.95$, $p = 0.75$), followed by a paired t-test ($t = 3.51$, p ¡ 0.02). The results indicated that the hybrid model significantly outperformed content- and trajectory-based models.

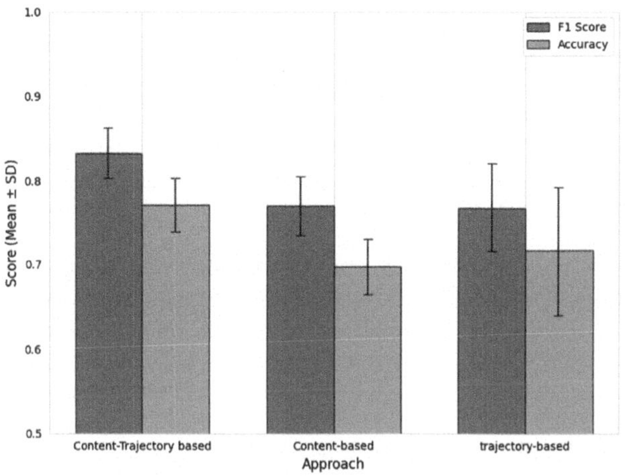

Fig. 2. Performance Comparison of Different Approaches

7 Conclusion

In this article, we have introduced an innovative definition of multi-aspect trajectories to address the spread of misinformation in social networks. Our approach views each shared message as a moving object within the social network and applies trajectory classification techniques, commonly used for classifying moving entities such as humans, animals, weather phenomena, among others.

Multi-aspect trajectory-based models have achieved comparable or even superior results to conventional misinformation detection methods, both in their trajectory-based variant and in combination with content-based models. The hybridization of both approaches has shown a significant improvement in classification capability.

This research opens a new horizon in the fight against misinformation. The application of trajectory-based models, particularly those based on Movelets, promises greater interpretability and accuracy in misinformation detection.

Acknowledgements. This work has been partially supported by CAPES (Finance code 001). We also thank the reviewers who contributed to the improvement of this work. The views and opinions expressed in this paper are the sole responsibility of the author.

References

1. Cabral, L., Monteiro, J.M., da Silva, J.W.F., Mattos, C.L.C., Mourao, P.J.C.: Fakewhastapp. br: NLP and machine learning techniques for misinformation detection in brazilian portuguese whatsapp messages. In: ICEIS (1), pp. 63–74 (2021)
2. Ceylan, G., Anderson, I.A., Wood, W.: Sharing of misinformation is habitual, not just lazy or biased. Proc. Natl. Acad. Sci. **120**(4), e2216614120 (2023)

3. Choi, D., Chun, S., Oh, H., Han, J., Kwon, T.T.: Rumor propagation is amplified by echo chambers in social media. Sci. Rep. **10**(1), 310 (2020)
4. Del Vicario, M., et al.: The spreading of misinformation online. Proc. Natl. Acad. Sci. **113**(3), 554–559 (2016)
5. Ducci, F., Kraus, M., Feuerriegel, S.: Cascade-LSTM: a tree-structured neural classifier for detecting misinformation cascades. In: Proceedings of the 26th ACM SIGKDD International Conference on Knowledge Discovery & Data Mining, pp. 2666–2676 (2020)
6. Ferrero, C.A., Alvares, L.O., Zalewski, W., Bogorny, V.: Movelets: exploring relevant subtrajectories for robust trajectory classification. In: Proceedings of the 33rd Annual ACM Symposium on Applied Computing, pp. 849–856 (2018)
7. Ferrero, C.A., Petry, L.M., Alvares, L.O., da Silva, C.L., Zalewski, W., Bogorny, V.: Mastermovelets: discovering heterogeneous movelets for multiple aspect trajectory classification. Data Min. Knowl. Disc. **34**(3), 652–680 (2020)
8. Gao, Q., Zhou, F., Zhang, K., Trajcevski, G., Luo, X., Zhang, F.: Identifying human mobility via trajectory embeddings. In: IJCAI, vol. 17, pp. 1689–1695 (2017)
9. Guo, B., Ding, Y., Yao, L., Liang, Y., Yu, Z.: The future of false information detection on social media: new perspectives and trends. ACM Comput. Surv. (CSUR) **53**(4), 1–36 (2020)
10. Han, Y., Karunasekera, S., Leckie, C.: Graph neural networks with continual learning for fake news detection from social media. arXiv preprint arXiv:2007.03316 (2020)
11. Hoy, N., Koulouri, T.: A systematic review on the detection of fake news articles. arXiv preprint arXiv:2110.11240 (2021)
12. Islam, M.R., Liu, S., Wang, X., Xu, G.: Deep learning for misinformation detection on online social networks: a survey and new perspectives. Soc. Netw. Anal. Min. **10**(1), 1–20 (2020). https://doi.org/10.1007/s13278-020-00696-x
13. Kondamudi, M.R., Sahoo, S.R., Chouhan, L., Yadav, N.: A comprehensive survey of fake news in social networks: attributes, features, and detection approaches. J. King Saud Univ.-Comput. Inf. Sci. **35**(6), 101571 (2023)
14. Kumar, S., Shah, N.: False information on web and social media: a survey. arXiv preprint arXiv:1804.08559 (2018)
15. Liu, Y., Wu, Y.F.: Early detection of fake news on social media through propagation path classification with recurrent and convolutional networks. In: Proceedings of the AAAI Conference on Artificial Intelligence, vol. 32 (2018)
16. Ma, J., Gao, W., Wong, K.F.: Rumor detection on twitter with tree-structured recursive neural networks. In: Proceedings of the 56th Annual Meeting of the Association for Computational Linguistics (Volume 1: Long Papers), pp. 1980–1989 (2018)
17. May Petry, L., Leite Da Silva, C., Esuli, A., Renso, C., Bogorny, V.: Marc: a robust method for multiple-aspect trajectory classification via space, time, and semantic embeddings. Int. J. Geogr. Inf. Sci. **34**(7), 1428–1450 (2020)
18. Mello, R.D.S., et al.: Master: a multiple aspect view on trajectories. Trans. GIS **23**(4), 805–822 (2019)
19. Michail, D., Kanakaris, N., Varlamis, I.: Detection of fake news campaigns using graph convolutional networks. Int. J. Inf. Manag. Data Insights **2**(2), 100104 (2022)
20. Monti, F., Frasca, F., Eynard, D., Mannion, D., Bronstein, M.M.: Fake news detection on social media using geometric deep learning. arXiv preprint arXiv:1902.06673 (2019)

21. NewsGuard: The next great misinformation superspreader: how chatgpt could spread toxic misinformation at unprecedented scale (2023). https://www.newsguardtech.com/misinformation-monitor/jan-2023/
22. Previti, M., Rodriguez-Fernandez, V., Camacho, D., Carchiolo, V., Malgeri, M.: Fake news detection using time series and user features classification. In: Castillo, P.A., Jiménez Laredo, J.L., Fernández de Vega, F. (eds.) EvoApplications 2020. LNCS, vol. 12104, pp. 339–353. Springer, Cham (2020). https://doi.org/10.1007/978-3-030-43722-0_22
23. Reuters Institute for the Study of Journalism: Digital news report 2023 (2023). https://reutersinstitute.politics.ox.ac.uk/digital-news-report/2023
24. Saikh, T., De, A., Ekbal, A., Bhattacharyya, P.: A deep learning approach for automatic detection of fake news. arXiv preprint arXiv:2005.04938 (2020)
25. Shu, K., Bernard, H.R., Liu, H.: Studying fake news via network analysis: detection and mitigation. In: Emerging Research Challenges and Opportunities in Computational Social Network Analysis and Mining, pp. 43–65 (2019)
26. Shu, K., Mahudeswaran, D., Wang, S., Liu, H.: Hierarchical propagation networks for fake news detection: investigation and exploitation. In: Proceedings of the International AAAI Conference on Web and Social Media, vol. 14, pp. 626–637 (2020)
27. Shu, K., Wang, S., Liu, H.: Beyond news contents: the role of social context for fake news detection. In: Proceedings of the Twelfth ACM International Conference on Web Search and Data Mining, pp. 312–320 (2019)
28. da Silva, C.L., Petry, L.M., Bogorny, V.: A survey and comparison of trajectory classification methods. In: 2019 8th Brazilian Conference on Intelligent Systems (BRACIS), pp. 788–793. IEEE (2019)
29. Song, C., Shu, K., Wu, B.: Temporally evolving graph neural network for fake news detection. Inf. Process. Manag. **58**(6), 102712 (2021)
30. Statista: Number of social network users worldwide from 2017 to 2025 (2021). https://www.statista.com/statistics/278414/number-of-worldwide-social-network-users/
31. Tortelli Portela, T., Tyska Carvalho, J., Bogorny, V.: Hipermovelets: high-performance movelet extraction for trajectory classification. Int. J. Geogr. Inf. Sci. **36**(5), 1012–1036 (2022)
32. Vosoughi, S., Roy, D., Aral, S.: The spread of true and false news online. Science **359**(6380), 1146–1151 (2018)
33. Wu, L., Morstatter, F., Carley, K.M., Liu, H.: Misinformation in social media: definition, manipulation, and detection. ACM SIGKDD Explor. Newsl. **21**(2), 80–90 (2019)
34. Zhang, X., Ghorbani, A.A.: An overview of online fake news: characterization, detection, and discussion. Inf. Process. Manag. **57**(2), 102025 (2020)
35. Zhou, F., Gao, Q., Trajcevski, G., Zhang, K., Zhong, T., Zhang, F.: Trajectory-user linking via variational autoencoder. In: IJCAI, pp. 3212–3218 (2018)
36. Zhou, X., Zafarani, R.: A survey of fake news: fundamental theories, detection methods, and opportunities. ACM Comput. Surv. (CSUR) **53**(5), 1–40 (2020)
37. Zhou, Z., Guan, H., Bhat, M.M., Hsu, J.: Fake news detection via NLP is vulnerable to adversarial attacks. arXiv preprint arXiv:1901.09657 (2019)

InRanker: Distilled Rankers for Zero-Shot Information Retrieval

Thiago Soares Laitz[1,2](✉), Konstantinos Papakostas[3], Roberto Lotufo[1,4], and Rodrigo Nogueira[1,2,3]

[1] School of Electrical and Computing Engineering, State University of Campinas (UNICAMP), Campinas, Brazil
[2] Maritaca AI, Campinas, Brazil
thiagolaitz@gmail.com
[3] Zeta Alpha, Amsterdam, Netherlands
[4] NeuralMind, Campinas, Brazil

Abstract. Despite multi-billion parameter neural rankers being common components of state-of-the-art information retrieval pipelines, they are rarely used in production due to the enormous amount of compute required for inference. In this work, we propose a method for distilling large rankers into their smaller versions focusing on out-of-domain effectiveness. We introduce InRanker, a version of monoT5 [25] distilled from monoT5-3B with increased effectiveness on out-of-domain scenarios. Our key insight is to use language models and rerankers to generate as much as possible synthetic "in-domain" training data, i.e., data that closely resembles the data that will be seen at retrieval time. The pipeline consists of two distillation phases that do not require additional user queries or manual annotations: (1) training on existing supervised soft teacher labels, and (2) training on teacher soft labels for synthetic queries generated using a large language model. Consequently, models like monoT5-60M and monoT5-220M improved their effectiveness by using the teacher's knowledge, despite being 50x and 13x smaller, respectively. Furthermore, we show that it is possible to transfer knowledge from English models to Portuguese fine-tuned models. Models and code are available at https://github.com/unicamp-dl/inranker.

Keywords: Deep Learning · Language Models · Information Retrieval

1 Introduction

It is well known that the effectiveness of IR pipelines increases with larger models [2,22,24,25,28]. For instance, multi-billion parameter rankers and dense models achieve top positions on leaderboards of IR benchmarks and competitions [10–12]. These large models leverage increased representation capacity, enabling them to encode features that might elude smaller models. However, deploying these large models is not without its challenges. The computational costs are substantial, often requiring specialized hardware such as GPUs or TPUs to operate in

A. Paes and F. A. N. Verri (Eds.): BRACIS 2024, LNAI 15413, pp. 140–154, 2025.
https://doi.org/10.1007/978-3-031-79032-4_10

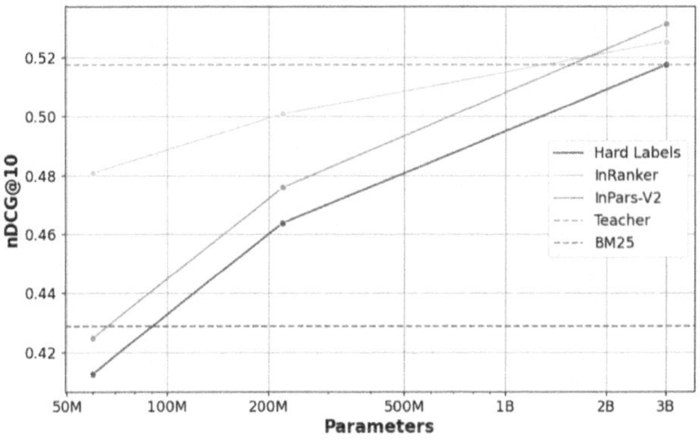

Fig. 1. Effectiveness on the BEIR benchmark [31]. All models are based on monoT5 [25], applying different fine-tuning methods.

latency-critical applications. The high cost is directly related to the large number of parameters that these models contain, as they require hardware with high memory and compute capacity and because the latency scales almost linearly with the number of parameters. In a production environment, this means higher operating costs and reduced scalability.

To address these challenges, there have been efforts to create more efficient models without significantly reducing effectiveness. One such approach is model distillation [17]. Distilled models, such as MiniLM [32], use a teacher or an ensemble of larger models to transfer knowledge to a smaller student model. Rosa et al. [30] show that MiniLM surpassed the zero-shot effectiveness of monoT5-base, which is a seq2seq model trained for binary classification, in IR tasks despite being an order of magnitude smaller in size. This has shown that knowledge transfer via model distillation is not only feasible but also effective. However, most distillation techniques have been geared towards optimizing effectiveness on specific benchmark tasks and do not focus on out-of-domain effectiveness. Rosa et al. also show that while smaller models are capable of achieving high in-domain results, similar to their larger counterparts, the disparity in effectiveness becomes evident in out-of-domain scenarios. As the concept of out-of-domain is subjective, we define it as a test distribution that is significantly different from the training distribution. A straighforward example of a out-of-domain scenario is when a model is trained on chemistry-related data and tested on legal data. However, we recognize that this distinction blurs in many scenarios.

Usually, training a retrieval model requires human-annotated hard labels informing which passage is relevant for each query. However, with the advance of Large Language Models (LLMs), it has become possible to generate synthetic queries for passages, providing a feasible approach for data augmentation [1,2,4,19,26]. Our work introduces a method for the generation of synthetic data specifically designed for distilling rankers that increases their out-of-domain effectiveness. We present InRanker, a distilled model derived from

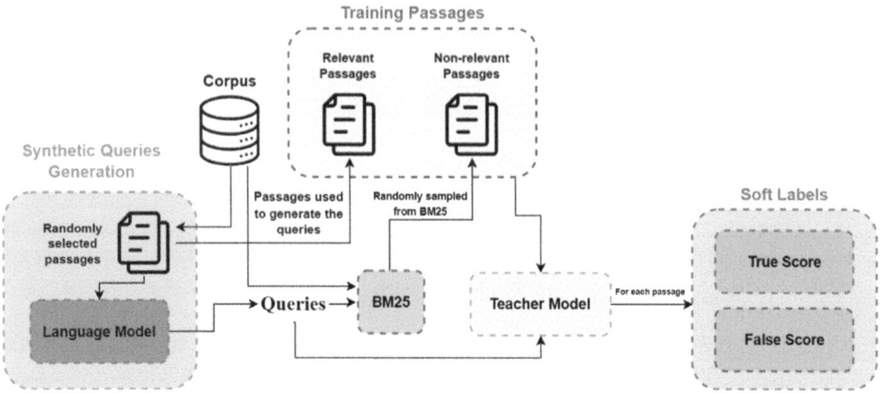

Fig. 2. Pipeline for generating the synthetic triples <query, passage, soft label> for the InRanker model.

monoT5-3B [25], that uses the predictions of the teacher directly with both synthetic, generated from an out-of-domain corpus, and real query-document pairs. Effectively, this approach converts any corpus to be in-domain, since the model will be trained using queries from the target domain. As a result, this approach leads to reduced model sizes while maintaining improved out-of-domain effectiveness as presented in Fig. 1. The methodology, results, and ablation experiments are presented in detail in the following sections.

2 Related Work

The research community has been using LLMs in a variety of tasks aimed at increasing the availability of data and improving the effectiveness of existing systems. Magister et al. [20] employed synthetic text generated by PaLM 540B [8] and GPT-3 175B [5] to transfer knowledge to smaller models such as T5. Fu et al. [15] successfully specialized student models in multi-step reasoning using FlanT5 [9] and code-davinci-002 as teachers. However, all these works rely on training the student models using synthetic text rather than directly using the soft labels. Furthermore, Muhamed et al. [21] distilled cross-attention scores of a language model for click-through-rate prediction, achieving better results when exposed to contextual features such as tabular data. Wang et al. [32] distilled the self-attention module, which is a crucial part of transformers, and successfully transferred knowledge to a variety of tasks.

Previous studies have also explored training a student from soft labels produced by a teacher: Hofstätter et al. [18] proposed a cross-architecture knowledge distillation approach using the MarginMSE loss. Similarly, Formal et al. [14] used the MarginMSE loss to distill knowledge to sparse neural models. Finally, Hashemi et al. [16] proposed a method for generating synthetic data for domain adaptation of dense passage retrievers. This approach involves creating new queries and a target collection, along with pseudo-labels extracted using a BERT cross-encoder. However, they did not evaluate the model's effectiveness on datasets to which it was not domain-adapted. The existing research has mainly focused on in-domain evaluation, where the goal has been to increase the effectiveness of the student model on test datasets whose domain is similar to the datasets it was trained on. Our study also focuses on the robustness of the student and its ability to perform well even in out-of-domain scenarios, similar to the abilities of the larger teacher model.

3 Methodology

Our proposed method consists of two key phases of distillation, each designed with specific objectives to maximize the model's zero-shot effectiveness. The first phase uses real-world data to familiarize the student model with the ranking task, while the second phase uses synthetic data designed to improve zero-shot generalization and improve the model's effectiveness on a specific dataset. The dataset used to distill InRanker consists of {query, passage, logits} triplets, where the logits (soft labels) originate from a teacher model that has been trained for the relevance task. For the first stage, we chose to use query-document pairs from the MS MARCO [23] dataset, given their variety, the large number of annotated pairs, and its demonstrated effectiveness in enhancing retrieval effectiveness [29]. Next, we source the synthetic queries from InPars [2], which used an LLM to generate queries for the datasets in BEIR in a few-shot manner.

Distilling rerankers involves using the Mean Squared Error (MSE) loss to match the logits of the teacher and the student, as part of a two-phase pipeline illustrated in Fig. 2. The first phase consists of two steps: (1) generating the teacher logits given a query and either a positive (relevant) or a negative (non-relevant) passage, where the negatives are randomly sampled using BM25 on the top-k = 1000 candidates, and the positives are sampled from the human-annotated pairs; and (2) training InRanker given the queries and passages as input using the MSE loss to match the student logits to those of the teacher, who remains frozen during training. This approach can be beneficial as it removes the need for making hard decisions about a passage's relevance, i.e. determining a threshold to obtain binary relevance labels, and instead focuses on a soft target objective aimed at aligning the student's perception of relevance with that of the teacher.

The second phase, with a focus on zero-shot effectiveness, uses the same two steps. However, instead of employing real queries sourced from a costly human-annotation process, it uses synthetic queries generated by an LLM based on randomly sampled documents from the corpus. In this scenario, the positive document is the one used to create the query, and the negatives are collected using the same top-k sampling approach as before.

We also perform zero-mean normalization on the teacher logits for each query-document pair, independent of the overall dataset distribution. This approach intends to make the data distribution symmetric for each query-document pair, thereby minimizing the bias that InRanker is required to learn. Formally:

$$\begin{aligned} L'_{\text{true}} &= L_{\text{true}} - \frac{L_{\text{true}} + L_{\text{false}}}{2} \\ L'_{\text{false}} &= L_{\text{false}} - \frac{L_{\text{true}} + L_{\text{false}}}{2} \end{aligned} \tag{1}$$

with L_{true} and L_{false} denoting the teacher's logits for the relevant and non-relevant classes, respectively, and L' being the normalized values. This results in the following loss for each training example:

$$\mathcal{L}_{\text{MSE}} = ([Y_{\text{true}} - L'_{\text{true}}]^2 + [Y_{\text{false}} - L'_{\text{false}}]^2) \tag{2}$$

with Y_{true} and Y_{false} representing the logits of the student.

Due to the training objective described in Eq. (2), the model no longer determines the relevance of passages and instead focuses on replicating the teacher's output, thus eliminating the need for tunning a relevance threshold that would be needed to produce a binary label. With this approach, we can easily expand the out-of-domain knowledge of distilled models by generating new queries for documents using an LLM and fine-tuning the distilled model using the teacher's logits. In the experiments section, we demonstrate the effectiveness of this approach in enhancing the student model's effectiveness across 16 datasets of BEIR simultaneously. We present the hyperparameters used for training and the dataset curation in Appendix A, and we discuss variations of the training loss in Appendix C.

4 Experiments

4.1 English Knowledge Distillation Results

We distilled monoT5-3B to models with parameters ranging from 60M to 3B, using combinations of the following configurations:

Human Hard: representing the common approach for training rankers with human-annotated hard (i.e., binary) labels from the MS MARCO passage ranking dataset. In this case, a vanilla cross-entropy loss is used:

$$\mathcal{L}_{\text{CE}} = -\log P_{\text{relevant}} - \log P_{\text{non-relevant}} \tag{3}$$

where P_{relevant} and $P_{\text{non-relevant}}$ are the probabilities assigned by the model to the relevant and non-relevant query-document pair, respectively. Non-relevant pairs are sampled from the top-1000 retrieved by BM25.

Human Soft: representing a distillation step for matching the logits of a teacher and a student model, using real (human-generated) queries from the ranking dataset as inputs, but without the binary relevance judgments for targets.

Synthetic Soft: representing a distillation step for matching the logits of the two models, similar to the previous configuration, but using exclusively synthetic queries generated from the corresponding BEIR corpora with InPars [2,19].

Table 1. Distillation results (nDCG@10) on 16 BEIR datasets. The model marked with * represents the teacher model. We did not train InRanker-3B on human soft labels due to computational constraints.

Model	Training Configurations			Avg. Score
	Human Hard	Human Soft	Synthetic Soft	
(1) monoT5-60M	✓			0.4125
(2) ↪ w/ soft human		✓		0.4356
(3) InRanker-60M		✓	✓	**0.4807**
(4) monoT5-220M	✓			0.4638
(5) ↪ w/ soft human		✓		0.4870
(6) InRanker-220M		✓	✓	**0.5008**
(7) monoT5-3B*	✓			0.5174
(8) InRanker-3B	✓		✓	**0.5253**

From Table 1, we see that both distillation steps were essential for improving the average nDCG@10 score compared to the model trained solely using human hard labels from MS MARCO.[1] As a result, InRanker-60M (row 3) and InRanker-220M (row 6), despite being 50x and 13x smaller than the teacher model, were able to improve their effectiveness on the BEIR benchmark significantly. Moreover, models trained exclusively on MS MARCO soft labels (rows 2 & 5) saw an increase in effectiveness in comparison to training on solely hard labels (rows 1 & 4), corroborating findings from previous studies regarding the effectiveness of soft labels [14,16–18]. Furthermore, we observed an increase in the effectiveness even in self-distillation training (row 8), where the student learns soft labels generated by itself. We hypothesize that the improvement stems from the extra knowledge provided by the language model used to generate the synthetic queries. We did not provide results for the 3B model trained on both human soft and synthetic soft due to computational costs.

Furthermore, in Table 2, we present a effectiveness comparison between InRanker, Promptagator [13], and RankT5 [33]. Although we used monoT5-3B as a teacher for our experiments, which has a lower effectiveness on average when compared to Promptagator or RankT5-3B, our method is model-agnostic and thus one could use a stronger teacher model and anticipate even stronger results. Nonetheless, InRanker remains competitive in both model groups of 220M and 3B parameters, outperforming the other two baselines in 6 out of the 10 evaluated datasets, despite the average score not reflecting this due to Promptagator and RankT5 attaining a significantly higher score in two datasets: ArguAna and Touché.

[1] Results per dataset are shown in Appendix D.

Table 2. Comparison of the effectiveness for various reranking models, measured by nDCG@10 on the BEIR benchmark. The model marked with * represents the teacher model used for training InRanker. Bolded scores correspond to the best effectiveness on a specific dataset for a given model size, while underlined scores indicate the best effectiveness overall.

Dataset	InRanker 60M	InRanker 220M	Promptagator 110M + 110M	RankT5-Enc 220M	InRanker 3B	monoT5 3B*	RankT5-Enc 3B
TREC-COVID	0.7775	**0.7984**	0.7620	0.7896	0.8175	0.7936	**0.8237**
NFCorpus	0.3547	0.3658	0.3700	**0.3731**	0.3825	0.3801	**0.3990**
HotpotQA	0.7563	**0.7742**	0.7360	0.7269	**0.7800**	0.7595	0.7536
Climate-FEVER	0.2729	**0.2914**	0.2030	0.2462	**0.2931**	0.2835	0.2753
DBPedia	0.4451	**0.4650**	0.4340	0.4373	**0.4762**	0.4719	0.4598
ArguAna	0.2466	0.2873	**0.6300**	0.3094	**0.4243**	0.3824	0.4069
Touché-2020	0.2883	0.2897	0.3810	**0.4449**	0.2924	0.3026	**0.4869**
SCIDOCS	0.1788	0.1911	**0.2010**	0.1760	**0.1990**	0.1978	0.1918
SciFact	0.7490	**0.7618**	0.7310	0.7493	**0.7831**	0.7773	0.7600
FiQA-2018	0.4043	0.4431	**0.4940**	0.4132	0.5027	**0.5068**	0.4932
Average	0.4474	0.4668	**0.4942**	0.4666	0.4951	0.4856	**0.5050**

4.2 Portuguese Knowledge Distillation Results

To further assess the efficacy of the technique in different languages, we evaluated InRanker on a Portuguese dataset for information retrieval: QUATI [6]. Instead of using the same T5 model, we started from PTT5, a Portuguese fine-tuned version of T5 [7,27]. We used the same two-step training approach as before, but with a strategy that allowed us to distill a Portuguese model using an English teacher (monoT5-3B). Given the availability of a translated version of MS MARCO in Portuguese [3], the first step (human soft) involved training the model using the Portuguese text, while matching the soft labels generated by the teacher using the original English text. This approach enabled us to leverage a stronger model that is not available in Portuguese for the distillation process. In the second step, involving synthetic soft labels from BEIR, we trained using the English text, as there is no translated version of BEIR available in Portuguese.

The results of this evaluation are presented in Table 3. We conclude that, similarly to English, the distillation process was able to improve the effectiveness of models in a zero-shot manner, as the models were trained using only real data from MS MARCO and synthetic data from BEIR. Remarkably, InRanker-740M surpassed the effectiveness of the mT5-3.7B on QUATI. Note that for the QUATI evaluation we used the same prompt presented in the paper to annotate all unjudged documents using gpt-4-turbo. Therefore all results are presented with a jugded@10 of 100%. Further results using synthetic data from QUATI and mixed training data are presented in Appendix E.

4.3 Ablations

In this section, we present our ablation experiments aimed at validating the best configuration for distilling monoT5-3B into smaller T5-based models, as well

Table 3. InRanker results on QUATI, a Portuguese evaluation dataset for information retrieval using PTT5-v2 [27]. All synthetic soft labels were generated using the BEIR datasets.

Model	Training Configurations			Avg. Score
	Human Hard	Human Soft	Synthetic Soft	
(1) PTT5-v2-60M	✓			0.4225
(2) ↪ w/ soft human		✓		0.4372
(3) InRanker-60M		✓	✓	**0.5121**
(4) PTT5-v2-220M	✓			0.5662
(5) ↪ w/ soft human		✓		0.5693
(6) InRanker-220M		✓	✓	**0.6108**
(7) PTT5-v2-740M	✓			0.5917
(8) ↪ w/ soft human		✓		0.6362
(9) InRanker-740M		✓	✓	**0.6624**
(10) monoT5-3B	✓			0.4864
(11) mT5-3.7B	✓			**0.6593**

as assessing their zero-shot capabilities. The initial experiments we conducted focused on evaluating how distillation would affect the model's effectiveness on novel dataset distributions that were not seen during training, i.e., we did not generate synthetic queries for them. To achieve this, we created two subsets, each containing 8 randomly selected datasets from 16 datasets of BEIR[2], which we named sample sets 1 and 2 and used only one set for training per experiment. The datasets that were used for training are designated as the "in-domain" category, while the remaining datasets, i.e. the other 8 datasets that are not part of the training set, represent the "out-of-domain" (O.O.D.) category.

Impact of Soft Knowledge Distillation on O.O.D. Effectiveness. Our first ablation experiment focused on evaluating the initial distillation process using the MS MARCO dataset with soft labels. To accomplish this, we generated logits with monoT5-3B and trained both T5-base and T5-small models for 10 epochs. As shown in Table 4, rows 1–2 & 5–6, both models demonstrated an improvement in their nDCG@10 scores compared to the baseline, which was trained using the hard labels from MS MARCO. Remarkably, the overall score increased in both scenarios, even though the models were not exposed to any BEIR passages during this phase.[3]

Adding Soft Synthetic Targets as a Second Distillation Phase. For the next experiment, we applied a second distillation step with synthetic soft labels on top of the model that we acquired from the last phase (monoT5 w/ soft human

[2] We list their exact composition in Appendix B.

[3] The individual results for each dataset are presented in Appendix D.

Table 4. Comparison of the in-domain vs out-of-domain effectiveness of our method, measured by nDCG@10. The model marked with * represents the teacher model used for the knowledge distillation process.

T5 Model	Training Configurations			Sample Set 1		Sample Set 2	
	Human Hard	Human Soft	Synthetic Soft	In-domain	O.O.D.	In-domain	O.O.D.
(1) 60M (monoT5)	✓			0.4141	0.4109	0.4817	0.3434
(2) 60M		✓		0.4422	0.4290	0.5124	0.3587
(3) 60M (InRanker)		✓	✓	**0.4768**	**0.4716**	**0.5558**	**0.3852**
(4) 60M	✓		✓	0.4475	0.4587	0.5355	0.3617
(5) 220M (monoT5)	✓			0.4647	0.4629	0.5475	0.3801
(6) 220M		✓		0.4867	0.4873	0.5692	0.4048
(7) 220M (InRanker)		✓	✓	**0.4945**	**0.5028**	**0.5874**	**0.4083**
(8) 220M	✓		✓	0.4905	0.4942	0.5832	0.3941
(9) 3B* (monoT5)	✓			0.5095	0.5253	0.6053	0.4295

labels). For that, we used the 100K synthetic queries generated by InPars for each dataset indicated as "in-domain" and trained for 10 epochs. As shown in Table 4, rows 3 & 7, while it was expected that the in-domain datasets would have an increase in their nDCG@10 scores, we observe that the out-of-domain datasets also had improvements, suggesting that the model's generalization capabilities were enhanced.

Using Hard Human Targets for the First Distillation Phase. Finally, we investigated the impact of skipping the first phase of distillation on MS MARCO logits, and instead starting from a model that was trained on hard human labels (monoT5-small and monoT5-base) and directly training using the synthetic soft BEIR targets. As we can see in Table 4, rows 3–4 & 7–8, when comparing with the model that was trained using the soft human targets, the overall effectiveness was reduced. From this, we conclude that the distillation step that includes the soft human targets on MS MARCO is beneficial, as it improves the model's effectiveness in both in-domain and out-of-domain scenarios.

Upper Bound for Soft Distillation. To estimate the upper bound of the effectiveness that these models could attain through distillation, we repeated the process using *real queries* from BEIR, (i.e., the validation queries) instead of the synthetic ones. Results presented in Table 5 show that for both model sizes, there was an increase in effectiveness for the in-domain datasets, as the model was exposed to the evaluation queries during training. However, we also observed an increase in effectiveness for out-of-domain datasets, indicating that the synthetic queries used for training could be improved.

Table 5. Upper bound effectiveness (nDCG@10) using real queries from BEIR for the distillation datasets. Bold indicates the best between using synthetic and real queries.

Model	Sample Set 1		Sample Set 2	
	In-domain	O.O.D.	In-domain	O.O.D.
InRanker-60M	0.4768	0.4716	0.5558	**0.3852**
↪ w/ real queries	**0.4975**	**0.4719**	**0.5860**	0.3813
InRanker-220M	0.4945	0.5028	0.5874	0.4083
↪ w/ real queries	**0.5242**	**0.5175**	**0.6159**	**0.4202**

5 Conclusion

This paper introduces a method for distilling the knowledge of information retrieval models and improve upon previous work how to better use synthetic data, aimed at improving the out-of-domain effectiveness of students. The study reveals that, through this knowledge distillation process, smaller models can achieve results comparable to the teacher, even in the context of multilingual transfer knowledge. This approach is particularly significant for applications where computational resources are limited, in production environments, or for languages with a lack of available models that can serve as a teacher. The methodology involves two steps of distillation: (1) using a human-curated corpus, and (2) using synthetic data generated by an LLM. Consequently, our work shows that it is possible to improve a reranker's capabilities in specific domains without requiring additional human-annotated labels. Finally, we observe that synthetic query generation could be improved since the real queries achieved a better out-of-domain effectiveness compared to the model trained solely on synthetic ones. However, the presented method has limitations. Specifically, it is not clear how to adapt the proposed loss to train dense retrievers, which tipically use a contrastive loss.

Disclosure of Interests. The authors have no competing interests to declare that are relevant to the content of this article.

A Training Details

This appendix presents the parameters used for training the models using an A100 GPU with 80GB of VRAM. All experiments were conducted using the same learning rate of 7e-5 and the AdamW optimizer with its default hyperparameters in HuggingFace. The batch size was set to 32. For the 3B model, we used gradient checkpointing and gradient accumulation (to achieve an effective batch size of 2×16) due to memory constraints. During the generation of soft labels using the teacher model, we sampled 9 non-relevant passages passage for each relevant passage, leading to 10 pairs of logits per query. Differently from InPars and

Promptagator, which train a separate model for each dataset, InRanker is a single model trained on all 16 datasets from BEIR.

B Datasets Used in Ablations

This section shows the datasets that were *randomly* chosen for inclusion in each sample set, resulting in the use of 12 out of the 16 BEIR datasets (as some were not used for training at all). Sample Set 1 includes the following datasets from the BEIR benchmark: NFCorpus, NQ, HotpotQA, DBPedia, Quora, SCIDOCS, FiQA-2018, and Signal-1M. Sample Set 2 comprises TREC-COVID, BioASQ, NQ, HotpotQA, Robust04, SCIDOCS, SciFact, and FiQA-2018. Climate-FEVER, TREC-NEWS, ArguAna, and Touché-2020 are not included in either sample set.

C Loss Function Ablation

We tested different loss functions, including the KL divergence and MSE, to match the logits of the two models. The results indicates that KL divergence was slightly worse for T5-small and that using only the true label in MSE as opposed to using both true and false labels also reduced the effectiveness. For a model with 60M parameters, the MSE with normalized logits yielded a result of 0.4807, while using the true logits only resulted in 0.4748. The KL divergence for the same model size was 0.4712. For a larger model with 220M parameters, the MSE with normalized logits produced a result of 0.5008, and the KL divergence was 0.5012. These results reflect the average nDCG@10 on 16 datasets of the BEIR benchmark with the varying loss functions.

D Complete Results on BEIR

Table 6 presents the results obtained after distilling the models using soft labels from MS MARCO and BEIR. We can observe the impact of both proposed distillation steps, namely using soft human labels and soft synthetic labels, which bring significant effectiveness improvements over the base models. In particular, using logits from MS MARCO leads to an average of a 2-point nDCG@10 improvement for each model, while the subsequent fine-tuning phase with the synthetic BEIR queries further enhances their effectiveness by 4.5 points for T5-small and approximately 1.4 points for T5-base.

E Complete Results on QUATI

Table 7 shows the results obtained by fine-tuning ptt5-v1 and ptt5-v2 [27] using different training sets. In particular, ptt5-v2 has a better overall nDGG@10, showing that even though both versions have the same number of parameters, a better pre-training process can improve downstream tasks such as information retrieval. For the QUATI soft labels generation we used the MT5-3.7B as a teacher instead of T5-3B, since this model has a better performance on Portuguese text.

Table 6. nDCG@10 values for each dataset after two steps of distillation.

Dataset	T5-small (60M)			T5-base (220M)			T5-3B
	Baseline	1st Step	+ 2nd Step	Baseline	1st Step	+ 2nd Step	Teacher
TREC-COVID	0.6928	0.7247	**0.7775**	0.7775	0.7643	**0.7984**	0.7936
NFCorpus	0.3180	0.3475	**0.3547**	0.3570	0.3639	**0.3658**	0.3801
BioASQ	0.4880	0.4648	**0.5516**	0.5240	0.5281	**0.5652**	0.5652
NQ	0.4733	0.5214	**0.5469**	0.5674	0.5855	**0.5971**	0.6251
HotpotQA	0.5996	0.6842	**0.7563**	0.6950	0.7546	**0.7742**	0.7595
Climate-FEVER	0.2116	0.2488	**0.2729**	0.2451	0.2739	**0.2914**	0.2835
DBPedia	0.3437	0.3745	**0.4451**	0.4195	0.4446	**0.4650**	0.4719
TREC-NEWS	0.3848	0.4478	**0.4646**	0.4475	**0.4808**	0.4695	0.4806
Robust04	0.4222	0.4782	**0.5386**	0.5016	0.5588	**0.5774**	0.6171
ArguAna	0.1274	0.1098	**0.2466**	0.1946	0.2431	**0.2873**	0.3824
Touché-2020	0.2643	0.2557	**0.2883**	0.2773	**0.2991**	0.2897	0.3026
Quora	0.8259	0.8246	**0.8335**	0.8230	0.8418	**0.8427**	0.8347
SCIDOCS	0.1436	0.1526	**0.1788**	0.1649	0.1746	**0.1911**	0.1978
SciFact	0.6963	0.7022	**0.7490**	0.7356	0.7505	**0.7618**	0.7773
FiQA-2018	0.3377	0.3712	**0.4043**	0.4136	0.4374	**0.4431**	0.5068
Signal-1M	0.2711	0.2612	**0.2820**	0.2771	0.2910	**0.2926**	0.3004
Average	1|0.4125	0.4356	**0.4807**	0.4638	0.4870	**0.5008**	**0.5174**

Table 7. Complete results (nDCG@10) on QUATI. All training datasets were used to distill knowledge from a teacher model.

Parameters	Training Datasets			QUATI Evaluation	
	MSMARCO	BEIR	QUATI	nDCG@10 ptt5-v1	nDCG@10 ptt5-v2
60M	✓			0.3838	0.4372
	✓	✓		0.4706	0.5121
	✓		✓	-	0.4993
	✓	✓	✓	0.4994	0.4999
220M	✓			0.5545	0.5693
	✓	✓		0.6129	0.6108
	✓		✓	-	0.5998
	✓	✓	✓	0.6270	0.6232
740M	✓			0.5818	0.6362
	✓	✓		0.6129	0.6624
	✓		✓	-	0.6468
	✓	✓	✓	0.6270	0.6570

References

1. Alaofi, M., Gallagher, L., Sanderson, M., Scholer, F., Thomas, P.: Can generative LLMs create query variants for test collections? An exploratory study. In: Proceedings of the 46th International ACM SIGIR Conference on Research and Development in Information Retrieval, pp. 1869–1873. SIGIR 2023. Association for Computing Machinery, New York (2023). https://doi.org/10.1145/3539618.3591960
2. Bonifacio, L., Abonizio, H., Fadaee, M., Nogueira, R.: InPars: unsupervised dataset generation for information retrieval. In: Proceedings of the 45th International ACM SIGIR Conference on Research and Development in Information Retrieval, pp. 2387–2392. SIGIR 2022. Association for Computing Machinery, New York (2022). https://doi.org/10.1145/3477495.3531863
3. Bonifacio, L., Campiotti, I., de Alencar Lotufo, R., Nogueira, R.F.: mmarco: a multilingual version of MS MARCO passage ranking dataset. CoRR abs/2108.13897 (2021). https://arxiv.org/abs/2108.13897
4. Boytsov, L., et al.: Inpars-light: cost-effective unsupervised training of efficient rankers (2023)
5. Brown, T., et al.: Language models are few-shot learners. In: Larochelle, H., Ranzato, M., Hadsell, R., Balcan, M., Lin, H. (eds.) Advances in Neural Information Processing Systems, vol. 33, pp. 1877–1901. Curran Associates, Inc. (2020). https://proceedings.neurips.cc/paper/2020/file/1457c0d6bfcb4967418bfb8ac142f64a-Paper.pdf
6. Bueno, M., de Oliveira, E.S., Nogueira, R., Lotufo, R.A., Pereira, J.A.: Quati: a Brazilian Portuguese information retrieval dataset from native speakers (2024)
7. Carmo, D., Piau, M., Campiotti, I., Nogueira, R., Lotufo, R.: PTT5: pretraining and validating the T5 model on Brazilian Portuguese data (2020)
8. Chowdhery, A., Narang, S., Devlin, J., et. al: Palm: scaling language modeling with pathways (2022)
9. Chung, H.W., et al.: Scaling instruction-finetuned language models (2022)
10. Craswell, N., Mitra, B., Yilmaz, E., Campos, D.: Overview of the TREC 2020 deep learning track (2021)
11. Craswell, N., Mitra, B., Yilmaz, E., Campos, D., Lin, J.: Overview of the TREC 2021 deep learning track. In: Soboroff, I., Ellis, A. (eds.) Proceedings of the Thirtieth Text REtrieval Conference, TREC 2021, online, 15–19 November 2021. NIST Special Publication, vol. 500-335. National Institute of Standards and Technology (NIST) (2021). https://trec.nist.gov/pubs/trec30/papers/Overview-DL.pdf
12. Craswell, N., et al.: Overview of the TREC 2022 deep learning track. In: Soboroff, I., Ellis, A. (eds.) Proceedings of the Thirty-First Text REtrieval Conference, TREC 2022, online, 15–19 November 2022. NIST Special Publication, vol. 500-338. National Institute of Standards and Technology (NIST) (2022). https://trec.nist.gov/pubs/trec31/papers/Overview_deep.pdf
13. Dai, Z., et al.: Promptagator: few-shot dense retrieval from 8 examples. In: The Eleventh International Conference on Learning Representations (2023). https://openreview.net/forum?id=gmL46YMpu2J
14. Formal, T., Lassance, C., Piwowarski, B., Clinchant, S.: From distillation to hard negative sampling: making sparse neural IR models more effective. In: Proceedings of the 45th International ACM SIGIR Conference on Research and Development in Information Retrieval, pp. 2353–2359. SIGIR 2022. Association for Computing Machinery, New York (2022). https://doi.org/10.1145/3477495.3531857

15. Fu, Y., Peng, H., Ou, L., Sabharwal, A., Khot, T.: Specializing smaller language models towards multi-step reasoning. In: Krause, A., Brunskill, E., Cho, K., Engelhardt, B., Sabato, S., Scarlett, J. (eds.) Proceedings of the 40th International Conference on Machine Learning. Proceedings of Machine Learning Research, vol. 202, pp. 10421–10430. PMLR (2023). https://proceedings.mlr.press/v202/fu23d.html
16. Hashemi, H., Zhuang, Y., Kothur, S.S.R., Prasad, S., Meij, E., Croft, W.B.: Dense retrieval adaptation using target domain description. In: Proceedings of the 2023 ACM SIGIR International Conference on Theory of Information Retrieval, pp. 95–104. ICTIR 2023. Association for Computing Machinery, New York (2023). https://doi.org/10.1145/3578337.3605127
17. Hinton, G., Vinyals, O., Dean, J.: Distilling the knowledge in a neural network (2015)
18. Hofstätter, S., Althammer, S., Schröder, M., Sertkan, M., Hanbury, A.: Improving Efficient Neural Ranking Models with Cross-Architecture Knowledge Distillation (2020)
19. Jeronymo, V., et al.: Inpars-v2: large language models as efficient dataset generators for information retrieval (2023)
20. Magister, L.C., Mallinson, J., Adamek, J., Malmi, E., Severyn, A.: Teaching small language models to reason. In: Rogers, A., Boyd-Graber, J., Okazaki, N. (eds.) Proceedings of the 61st Annual Meeting of the Association for Computational Linguistics (Volume 2: Short Papers), pp. 1773–1781. Association for Computational Linguistics, Toronto, Canada (2023). https://doi.org/10.18653/v1/2023.acl-short.151. https://aclanthology.org/2023.acl-short.151
21. Muhamed, A., et al.: CTR-BERT: cost-effective knowledge distillation for billion-parameter teacher models. In: NeurIPS Efficient Natural Language and Speech Processing Workshop (2021)
22. Neelakantan, A., et al.: Text and code embeddings by contrastive pre-training (2022)
23. Nguyen, T., et al.: MS MARCO: A Human Generated MAchine Reading COmprehension Dataset (2016)
24. Ni, J., et al.: Large dual encoders are generalizable retrievers. In: Goldberg, Y., Kozareva, Z., Zhang, Y. (eds.) Proceedings of the 2022 Conference on Empirical Methods in Natural Language Processing, pp. 9844–9855. Association for Computational Linguistics, Abu Dhabi, United Arab Emirates (2022). https://doi.org/10.18653/v1/2022.emnlp-main.669. https://aclanthology.org/2022.emnlp-main.669
25. Nogueira, R., Jiang, Z., Pradeep, R., Lin, J.: Document ranking with a pretrained sequence-to-sequence model. In: Cohn, T., He, Y., Liu, Y. (eds.) Findings of the Association for Computational Linguistics: EMNLP 2020, pp. 708–718. Association for Computational Linguistics, Online (2020). https://doi.org/10.18653/v1/2020.findings-emnlp.63. https://aclanthology.org/2020.findings-emnlp.63
26. Penha, G., Palumbo, E., Aziz, M., Wang, A., Bouchard, H.: Improving content retrievability in search with controllable query generation. In: Proceedings of the ACM Web Conference 2023, pp. 3182–3192. WWW 2023. Association for Computing Machinery, New York (2023). https://doi.org/10.1145/3543507.3583261
27. Piau, M., Lotufo, R., Nogueira, R.: PTT5-V2: a closer look at continued pretraining of T5 models for the Portuguese language (2024)
28. Pradeep, R., Nogueira, R., Lin, J.: The expando-mono-duo design pattern for text ranking with pretrained sequence-to-sequence models (2021)
29. Ren, R., et al.: A thorough examination on zero-shot dense retrieval. In: Bouamor, H., Pino, J., Bali, K. (eds.) Findings of the Association for Computational Linguistics: EMNLP 2023, pp. 15783–15796. Association for Computational Linguistics,

Singapore (2023). https://doi.org/10.18653/v1/2023.findings-emnlp.1057. https://aclanthology.org/2023.findings-emnlp.1057

30. Rosa, G.M., et al.: No parameter left behind: how distillation and model size affect zero-shot retrieval (2022)
31. Thakur, N., Reimers, N., Rücklé, A., Srivastava, A., Gurevych, I.: BEIR: a heterogeneous benchmark for zero-shot evaluation of information retrieval models. In: Thirty-Fifth Conference on Neural Information Processing Systems Datasets and Benchmarks Track (Round 2) (2021). https://openreview.net/forum?id=wCu6T5xFjeJ
32. Wang, W., Wei, F., Dong, L., Bao, H., Yang, N., Zhou, M.: Minilm: deep self-attention distillation for task-agnostic compression of pre-trained transformers. In: Larochelle, H., Ranzato, M., Hadsell, R., Balcan, M., Lin, H. (eds.) Advances in Neural Information Processing Systems, vol. 33, pp. 5776–5788. Curran Associates, Inc. (2020)
33. Zhuang, H., et al.: Rankt5: fine-tuning T5 for text ranking with ranking losses. In: Proceedings of the 46th International ACM SIGIR Conference on Research and Development in Information Retrieval, pp. 2308–2313. SIGIR 2023. Association for Computing Machinery, New York (2023). https://doi.org/10.1145/3539618.3592047

Investigating Behavior Cloning from Few Demonstrations for Autonomous Driving Based on Bird's-Eye View in Simulated Cities

Eric Aislan Antonelo[1(✉)], Gustavo Claudio Karl Couto[1], Christian Möller[2], and Pedro Henrique Fernandes[1]

[1] Automation and Systems Engineering Department, Federal University of Santa Catarina, Florianópolis, Brazil
eric.antonelo@ufsc.br, gustavo.karl.couto@posgrad.ufsc.br
[2] Faculty of Science and Engineering, Information Technology, Åbo Akademi University, Turku, Finland
christian.moller@abo.fi

Abstract. This paper investigates the use of Behavior Cloning (BC) for autonomous driving from a bird's-eye view (BEV) perspective in simulated urban environments. BC uses supervised learning to mimic expert driving behaviors. Previous works have applied BC in the CARLA simulator but did not fully address the challenges of traffic light compliance. Our approach enhances BC by integrating a kernel density estimator to adjust training sample weights based on action density, thereby improving the learning of rare but critical actions such as stopping at red lights and accelerating at green lights, specially in scenarios of scarce number of expert demonstrations. Using BEV inputs, which provide an abstract top-down view of the driving environment, our method simplifies the policy learning process. The trained convolutional neural network (CNN) outputs steering and acceleration actions based on these BEV inputs and additional state variables. Experimental results in the CARLA simulator demonstrate that our weighted BC method significantly improves driving performance, achieving higher route completion compared to standard BC. This weighted approach proved to be crucial in learning correct driving behaviors, particularly in test environments not encountered during training, highlighting its potential for enhancing autonomous vehicle navigation.

Keywords: Autonomous vehicles · Expert demonstrations · Imitation learning · kernel density estimation · CARLA Simulator

1 Introduction

Autonomous driving has traditionally been implemented using separate modules for perception, planning, and control, organized in a sequential pipeline [13,15,

A. Paes and F. A. N. Verri (Eds.): BRACIS 2024, LNAI 15413, pp. 155–168, 2025.
https://doi.org/10.1007/978-3-031-79032-4_11

16,21]. However, learning-based approaches have been gaining popularity as they attempt to address the complexities of autonomous driving in various scenarios, whether in simulation or real-world environments.

Most of these learning-based approaches rely on Behavior Cloning (BC), which employs supervised learning on a collection of expert demonstrations gathered offline [2,3,6,7,19]. For example, a human driver generates a set of input (observations) and corresponding desired output pairs.

Codevilla [6,7] employed BC for autonomous driving in the CARLA simulator. He collected a large dataset of human driving and augmented it using image processing techniques to train end-to-end policies conditioned on the desired route. Further improvements in BC were introduced in [4], where a large deep ResNet network was used for feature extraction. This approach combined data from cameras, LiDAR, and radar to create comprehensive feature maps. The work also incorporated a probabilistic motion planner to account for uncertainties in the trajectory.

Couto [11] has shown results on behavior cloning on the CARLA simulator, but the agent did not have to respect traffic lights. In this work, we investigate an approach where the agent is able to navigate the city while respecting traffic lights, i.e., breaking when the light is red, and crossing the traffic light when it is green. Only relying on regular BC to train an agent is not enough in a scenario with limited amount of training routes and where the agent needs to break at the red light and accelerate from stillness when light turns to green. To solve that, we employ a kernel density estimator [18] to estimate the actions density from the expert dataset. Then, the training samples are weighted in the BC loss function individually by setting the weight equal to the inverse of the actions' density. In this way, we can account for the underrepresented actions in the expert set, i.e., those happening at traffic light transitions. This weighting is needed because we are considering only eight routes as training data.

Instead of learning directly from raw camera pixels as in [11,12], the agent in this work has structured image inputs that are more abstract and meaningful to the navigation task. To decouple feature representation from policy learning [8,14] has its advantages as potentially easing up the learning task. In our context, this decoupling is accomplished by using a mid-level representation of the vehicle's raw camera images. For example, Bird's-Eye View (BEV) representations of the road ahead have been used as mid-level inputs for trajectory generation and motor control networks in [2] and [20], respectively. In BEV representation, the camera scene is projected onto a top-down view, segmenting areas of interest, like vehicle lanes and sidewalks, into different colors [21]. By using BEV as input for the control policy, the learning process can focus on the navigation problem, leveraging a more abstract and simplified view of the road ahead. The policy here is represented by convolutional neural networks, mapping the BEV image to control actuators such as steering and acceleration.

Experimental results in the CARLA simulator demonstrate that our weighted BC method significantly improves driving performance, achieving higher route completion compared to standard BC. This weighted approach proved to be essential in learning correct driving behaviors, particularly in test environments not encountered during training.

In Sect. 2, we present the bird's-eye view representation employed as input to the convolutional neural network, the behavior cloning loss for the agent, and the kernel density estimation used to generate the weights per sample for the modified loss function. The agent architecture is presented in Sect. 3. Finally, the experiments and conclusions are shown in Sects. 4 and 5.

2 Methods

2.1 Bird's-Eye View - BEV Representation

The bird's-eye view of a vehicle represents its position and movement in a top-down coordinate system [2]. The vehicle's location, heading, and speed are represented by p_t, θ_t, and s_t respectively. The top-down view is defined so that the agent's starting position is always at a fixed point within an image (the center of it). Furthermore, it is represented by a set of images of size $W \times H$ pixels, at a ground sampling resolution of ϕ meters/pixel. The BEV of the environment moves as the vehicle moves, allowing the agent to see a fixed range of meters in front of it. For instance, the BEV representation for the vehicle whose three frontal cameras are shown in Fig. 1(a) is given in Fig. 1(b), where the desired route, drivable area and lane boundaries are shown as three BEV channels. Besides, other three BEV channels for the traffic lights representing the state of the traffic lights (green or red) at different timesteps are shown in Fig. 1(c). Figure 1(d) shows the integrated six-channel BEV image used as input to the agent.

2.2 Behavior Cloning (BC)

Behavior cloning is a supervised learning approach used in imitation learning, where the goal is to learn a policy that mimics the behavior of an expert. In this work, we want to learn a policy for an autonomous vehicle that outputs two actions (acceleration, steering) based on a Bird-Eye-View (BEV) representation of the environment and additional scalar state variables as input. For this reason we need to train a CNN to output these two continuous actions based on input from the BEV and the state variables.

The loss function for behavior cloning is typically based on the difference between the actions predicted by the learned policy and the actions taken by the expert in the training data. Let: $\pi_\theta(s)$ be the policy parameterized by θ, which maps a state s to an action a; and $\mathcal{D} = \{(s_i, a_i)\}_{i=1}^N$ be the dataset of N state-action pairs collected from the expert, where s_i is a state and a_i is the corresponding action taken by the expert.

Although the behavior cloning loss function can be defined as the mean squared error (MSE) between the predicted actions and the expert actions (or the cross-entropy loss if the actions are discrete), in this work, we model a stochastic

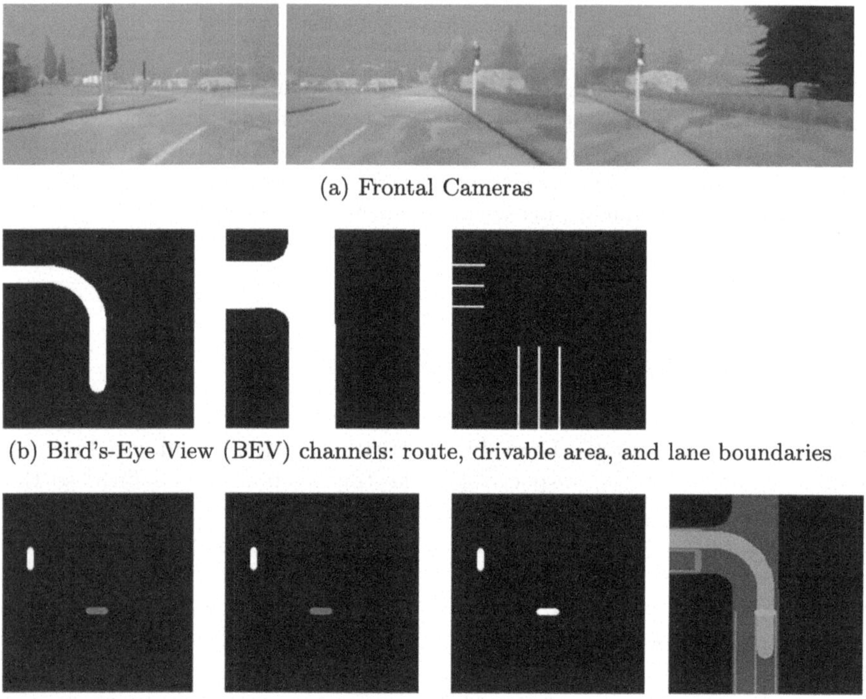

(a) Frontal Cameras

(b) Bird's-Eye View (BEV) channels: route, drivable area, and lane boundaries

(c) BEV's traffic lights channels at timesteps (-1, -9, -16) (d) final BEV image

Fig. 1. a) Images from three frontal cameras positioned on the left, center, and right side of the vehicle. These images were taken after the initial interactions of the agent in the CARLA simulation environment. Each camera captures a 256×144 RGB image. The corresponding bird's-eye view channels that our agent uses are shown in (b), (c) and (d), computed at the same moment as shown in (a). b) Three BEV 192×192 channels that represent, from left to right, the desired route, drivable area, and lane boundaries. c) BEV's traffic lights channels at timesteps (−1, −9, −16), i.e., representing the past states of the traffic light channel (grey color is red light off; white color means red ligh on). d) The final image displays all three channels combined in different colors. (Color figure online)

policy and, for that reason, the policy is represented by a Beta distribution. The Beta distribution is suitable for modeling variables that are bounded within an interval, such as acceleration and steering actions within $[-1, 1]$.

Given the actions $a_i = (a_i^{\text{acc}}, a_i^{\text{steer}})$ and the policy $\pi_\theta(s_i) = (\alpha_i^{\text{acc}}, \beta_i^{\text{acc}}, \alpha_i^{\text{steer}}, \beta_i^{\text{steer}})$, where α and β are the shape parameters of the Beta distribution, the loss function will be based on the negative log-likelihood of the Beta distribution.

The probability density function of the Beta distribution [10] is:

$$f(x; \alpha, \beta) = \frac{x^{\alpha-1}(1-x)^{\beta-1}}{B(\alpha, \beta)},$$

where $B(\alpha, \beta)$ is the Beta function, a normalization constant to ensure that the total probability is 1. To transform actions in $[-1, 1]$ to $[0, 1]$, we use:

$$a_i' = \frac{a_i + 1}{2}$$

The negative log-likelihood loss for the Beta distribution can then be written as:

$$\mathcal{L}(\theta) = -\frac{1}{N}\sum_{i=1}^{N}\left[\log f\left(\frac{a_i^{\text{acc}}+1}{2}; \alpha_i^{\text{acc}}, \beta_i^{\text{acc}}\right) + \log f\left(\frac{a_i^{\text{steer}}+1}{2}; \alpha_i^{\text{steer}}, \beta_i^{\text{steer}}\right)\right] \tag{1}$$

This loss function accounts for the probabilistic nature of the Beta distribution, ensuring that the model learns to predict parameters (α, β) for the distribution of actions rather than the actions directly. This approach allows for modeling uncertainty and variability in the predicted actions.

2.3 Kernel Density Estimation

A Kernel Density Estimator (KDE) is a non-parametric way to estimate the probability density function (PDF) of a random variable. For each data point in the sample, the kernel function is centered at that point. The KDE is then computed by summing up all these kernel functions and normalizing by the number of data points. Mathematically, the KDE for a point x is given by:

$$\hat{f}(x) = \frac{1}{nh}\sum_{i=1}^{n} K\left(\frac{x - x_i}{h}\right), \tag{2}$$

where $\hat{f}(x)$ is the estimated density at point x, n is the number of data points, h is the bandwidth, x_i are the data points, and K is the kernel function.

The Gaussian kernel, often used in KDE, is a type of kernel function that assigns weights to data points based on their distance from the point where the density is being estimated. The Gaussian kernel is defined by the standard normal distribution, and it ensures that closer points have a higher influence on the density estimate than points further away. The mathematical form of the Gaussian kernel K is:

$$K(u) = \frac{1}{\sqrt{2\pi}} e^{-\frac{u^2}{2}},$$

where u is the standardized distance between the point where the density is being estimated and the data point, scaled by the bandwidth h.

In this work, the KDE is used to weigh the training samples in the BC loss in (1). By using the inverse of the kernel density estimation of the actions as weights, we give more importance to less frequent actions. This approach can help the model learn better on underrepresented parts of the action space.

The resulting weighted BC loss function can be written as:

$$\mathcal{L}(\theta) = -\frac{1}{N}\sum_{i=1}^{N} w_i \left[\log f\left(\frac{a_i^{\text{ac}}+1}{2}; \alpha_i^{\text{ac}}, \beta_i^{\text{ac}}\right) + \log f\left(\frac{a_i^{\text{steer}}+1}{2}; \alpha_i^{\text{steer}}, \beta_i^{\text{steer}}\right)\right] \tag{3}$$

where: $w_i = \frac{1}{\text{KDE}(a_i)}$; and $\text{KDE}(a_i)$ is the kernel density estimation evaluated at the action a_i. The KDE is estimated for the actions in the training set. Then the weights w_i are calculated as the inverse of the KDE for each action.

This approach ensures that the loss function accounts for the density of actions in the training set, giving higher importance to less frequent actions.

3 Agent

3.1 Input Representation

The input s to the agent's policy is a six-channel 192×192 image produced by a module in CARLA simulator that has access to the city map. This image represents the mid-level bird's-eye view of the vehicle's current position. In this work, the BEV contains three channels, the road, the route and the lane, and also three more channels representing the history of the traffic lights at time steps -16, -9 and -1.

Additionally, the agent receives the current speed of the vehicle and the last values of the policy actuators (previous acceleration and steering). These additional inputs are incorporated further down in the network, specifically in the first fully connected layer.

3.2 Output Representation

The vehicle in CARLA has three actuators: steering $[-1, 1]$, throttle $[0, 1]$, and brake $[0, 1]$. Our agent's action space is $\mathbf{a} \in [-1, 1]^2$, where the two components correspond to steering and acceleration. Braking is implied when acceleration is negative, ensuring the agent cannot brake and accelerate simultaneously [17].

Instead of using the Gaussian distribution for the policy's actions, we employ the Beta distribution $\mathcal{B}(\alpha, \beta)$. The Beta distribution's bounded support is suitable for real-world applications like autonomous driving, where actions are naturally constrained [17]. This approach avoids the need for clipping or squashing inputs to enforce constraints, allowing explicit computation of the policy loss $\mathcal{L}_P$.

The Beta distribution also handles extreme driving scenarios, such as sharp turns and sudden braking, effectively. The parameters α and β, which are outputs of the policy neural network π_θ, control the shape of the distribution, enabling the policy to adapt to these critical behaviors.

3.3 Network Architecture

In this work, we train a network architecture, shown in Fig. 2, to learn to control an autonomous vehicle, using steering and acceleration actuators, based on the current BEV and some state variables as inputs. The Convolutionary Neural Network (CNN) maps the BEV into an embedding vector of size 256. Similarly, the scalar state variables are embedded into a vector of size 256 by a standard MLP (multi-layer perceptron layer). Together they form the current representation of the environment the agent faces. This concatenated embedding vector is processed by another MLP to map to the desired action.

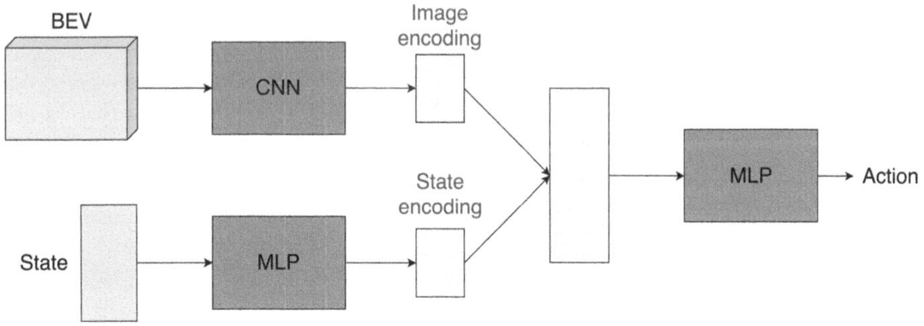

Fig. 2. Network architecture. The BEV (state) input is processed by a CNN (MLP) to create an image (state) encoding. Both image and state encodings are concatenated and fed into a final MLP that outputs the actions.

Figure 3 shows the detailed network architecture.

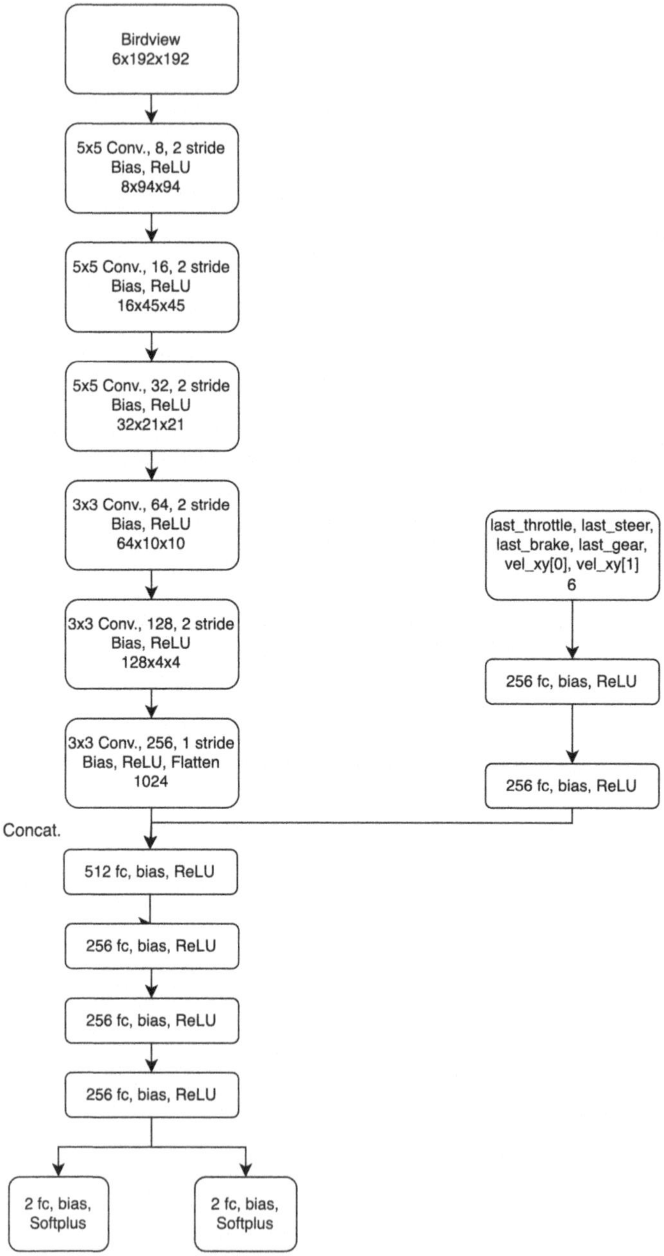

Fig. 3. Detailed network architecture. The BEV input consists of 6 channels of 192×192 resolution, while the State input correponds to the last executed actions, and the current speed measurements. The final policy output corresponds to the two parameters α and β of the Beta distribution for both acceleration and steering (thus, the two heads).

4 Experiments

In this section, the proposed BC with weigthed loss function is compared with the standard BC approach for autonomous vehicle navigation tasks in urban scenarios using the CARLA simulator.

4.1 Dataset Generation

The environment and trajectories are sourced from the CARLA Leaderboard evaluation platform [1]. Specifically, the *town01* environment from this platform, along with ten predefined trajectories, is used to create the expert dataset.

The expert dataset is generated using a deterministic agent that navigates with a dense point trajectory and a classic PID controller [5]. The dense point trajectory offers numerous points at a high resolution, while a sparse point trajectory has significantly fewer points, giving the agent only a general direction. Therefore, the dense point trajectory is employed to generate the expert's training data, whereas the sparse point trajectory is used by the agent for broader guidance. Thus, the agent has no access to the dense point trajectory. Besides, the expert breaks at the detection of red traffic lights and accelerates when the light turns back to green.

In Fig. 4, one of the 10 routes executed by the expert to create the labeled training set of demonstrations is illustrated. The line that starts in yellow and ends in red represents the desired trajectory (not visible to the agent as it is). The sparse trajectory appears as yellow dots, generated every 50 m or when the vehicle is about to change its movement (e.g., from *straight* to *turn* and vice versa).

The ten trajectories in the expert set were recorded at a rate of 10 Hz, producing 10 observation-action pairs per second. For the shortest route of 1480 samples (average route of 2129 samples), this equates to 2.5 min (3.5 min) of simulated driving. Altogether, the ten trajectories resulted in 21,287 labeled samples (30 GB of uncompressed data). The entire set corresponds to approximately 36 min or 8 km of driving.

4.2 Training and Inference

The network shown in Fig. 3 is trained with 8 routes from the expert dataset of *town01* environment, where the agent must also respect traffic lights. The first agent (Weighted BC) weighs the samples according to the density of the actions using (3) as loss function, whereas the second agent (BC) considers all actions with same weight, i.e., it removes the higher importance of less frequent actions, using (1) as the conventional BC loss function.

4.3 Evaluation

Although training is offline in *town01* environment, we still evaluate both BC and weighted BC agents as their model's parameters are updated. Every eight

(a) *town01*

(b) *town02*

Fig. 4. (a) The *town01* environment of the agent includes one of the routes used by the expert to collect data. This highlighted path measures 740 m, featuring 20 points in the sparse trajectory (indicated by yellow dots) and 762 points in the dense point trajectory (not shown). (b) The *town02* is used to evaluate the trained agent in a new environment. (Color figure online)

training epochs, both agents are evaluated using all ten routes from the same training town (Fig. 5). As four random runs are executed for each agent, the mean and standard deviation of the number of incomplete routes is plot in Fig. 5. One incomplete route means that the vehicle committed an infraction before completing the route. As the total number of routes is ten for *town01*, both BC and Weighted BC start at 10, the minimum performance. As training evolves, both agents improve the performance by committing less infractions, such as going out of the correct lane or passing a red light, but the Weighted BC agent attains the minimum error faster, i.e., it can complete all routes without any infractions and respecting traffic lights.

To understand what the action's density given by KDE and, consequently, the weights in the BC loss function, look like for the first route in *town01* (shown in Fig. 4), we record both actuators's values, acceleration and steering, from our expert as it follows the route as well as the corresponding logarithm of the action's density for each timestep of the route. These values are plot in Fig. 6, where we can see that different actions will produce different weights in the loss function since some of them are less frequent than others in the expert dataset. The highest action density corresponds to going straight with constant acceleration and zero steering, for instance, between timesteps 470 and 900. The least frequent action happens after a traffic light turns from red to green, which corresponds to a higher, but shorter, acceleration period.

The driving performance for both agents as well as the expert is presented in Table 1. The understanding of the ↑ or ↓ in the table is as follows. For

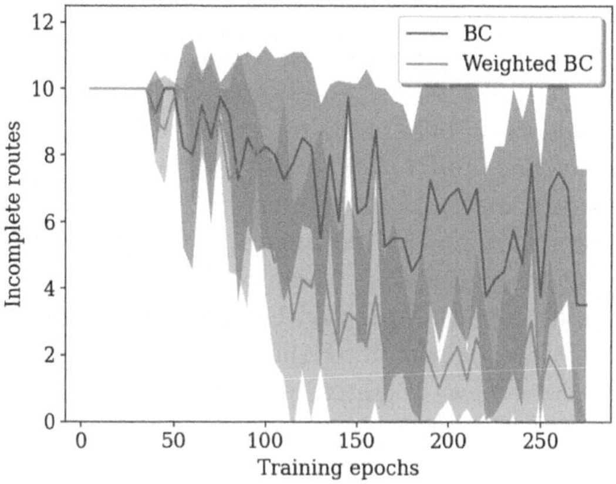

Fig. 5. Evaluation of both BC policies (weighted and regular) throughout its learning epochs. At every eight training epochs, both BC and Weighted BC agents are evaluated in all ten routes of *town01*. The vertical axis presents the number of incomplete routes. Four runs were executed for each agent, with mean and standard deviation shown by the solid line and the shaded area, respectively. In the beginning, both agents commit infractions in all routes and, thus, the number of incomplete routes is ten since one infraction ends the route uncompleted. The Weighted BC agent is able to achieve zero infractions by the end of training, unlike the BC agent.

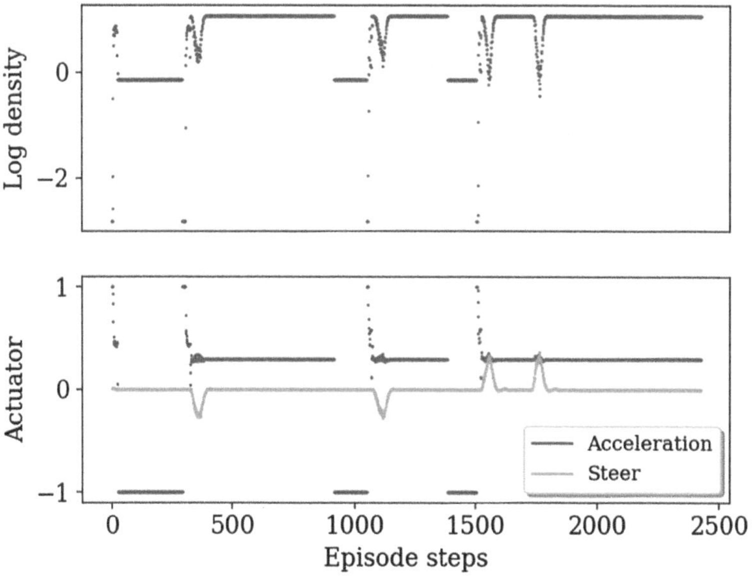

Fig. 6. Logarithm of the density of actions, given by KDE, during navigation of a route in *town1*, shown in Fig. 4. The expert encounters three red traffic lights, during which it breaks (acceleration is −1), and subsequently accelerates as soon as it turns green. Two left turns and two right turns can be observed by examining the steer actuator of the expert. (Color figure online)

instance, the lower (↓) the Red Light Infraction, which is #/Km or the number of infractions per kilometer, the better. In the same vein, the higher the Infraction Penalty (or the Driving score), the better is the performance. The expert has maximum driving score, being able to complete all routes with no red light infraction in both *town01* and *town02*. The number of crossed traffic lights is just a non-performance metric, just counting the times that the vehicle cross them. An equivalent result is achieved by Weighted BC in *town01*, the training environment, but not by regular BC, which presents a mean driving score of 81. In *town02*, a test environment, Weighted BC performs significantly better than BC. Videos with the agent controlling the vehicle in CARLA are available at https://sites.google.com/view/weighted-bc, comparing both BC and Weighted BC agents.

Table 1. Driving performance and infraction analysis for different agents on Town01 and Town02.

Agent	Town	Metrics				
		Driving Score (%,↑)[a]	Route Completion (%,↑)	Infraction Penalty (%,↑)	Red Light Infraction (#/Km,↓)	Crossed traffic lights (#)
Expert	*town01*	100 ± 0	100 ± 0	100 ± 0	0 ± 0	4.7 ± 1.9
	town02	100 ± 0	100 ± 0	100 ± 0	0 ± 0	9.0 ± 2.16
BC	*town01*	81 ± 20	83 ± 18	97 ± 5	0.39 ± 0.68	3.78 ± 1.03
	town02	34 ± 18	53 ± 18	66 ± 20	3.08 ± 2.45	2.79 ± 2.27
Weighted BC	*town01*	100 ± 0	100 ± 0	100 ± 0	0 ± 0	4.7 ± 1.9
	town02	98 ± 0.03	100 ± 0	98 ± 0.03	0 ± 0	9.0 ± 2.16

[a]The *Driving Score*, a metric from CARLA Leaderboard, is calculated by the product of *Route Completion* (the percentage of the route distance completed) and an *Infraction Penalty* which accounts for all violations committed. For instance, if an agent ran two red lights during a route and the penalty for running one red light is 0.7, then the final penalty would be $0.7 \times 0.7 = 0.49$.

5 Conclusion

In this work, we proposed a weighted behavior cloning approach that improves agent learning by giving higher weights to underrepresented actions in the expert training set of driving behaviors. The application consisted of route navigation in two simulated cities in the CARLA realistic simulator for autonomous vehicles, while respecting traffic lights. As the action space of the expert is not uniformly visited, and the expert training dataset is somewhat limited (only 10 routes in *town01*), a Kernel Density Estimator was employed to infer the action density which, in turn, was converted to a weight per observation-action pair. In our investigated scenario, the weighted loss function for BC proved to be crucial to learn the correct behavior, showing significantly better driving performance than regular BC, specially for the test city *town02* not used during training. Future work include extensions to cities with pedestrians and other vehicles and more advanced behavior cloning approaches such as implicit behavioral cloning [9] to account for multimodal action distributions.

Disclosure of Interests. The authors have no competing interests to declare that are relevant to the content of this article.

References

1. (2020). https://leaderboard.carla.org/
2. Bansal, M., Krizhevsky, A., Ogale, A.: Chauffeurnet: learning to drive by imitating the best and synthesizing the worst (2018). https://doi.org/10.48550/ARXIV.1812.03079. https://arxiv.org/abs/1812.03079
3. Bojarski, M., et al.: End to end learning for self-driving cars (2016). https://doi.org/10.48550/ARXIV.1604.07316. https://arxiv.org/abs/1604.07316
4. Cai, P., Wang, S., Sun, Y., Liu, M.: Probabilistic end-to-end vehicle navigation in complex dynamic environments with multimodal sensor fusion. IEEE Rob. Autom. Lett. **5**(3), 4218–4224 (2020). https://doi.org/10.1109/LRA.2020.2994027
5. Chen, D., Zhou, B., Koltun, V., Krähenbühl, P.: Learning by cheating (2019)
6. Codevilla, F., Müller, M., Dosovitskiy, A., López, A.M., Koltun, V.: End-to-end driving via conditional imitation learning. In: 2018 IEEE International Conference on Robotics and Automation (ICRA), pp. 1–9 (2017). https://api.semanticscholar.org/CorpusID:3672301
7. Codevilla, F., Santana, E., Lopez, A., Gaidon, A.: Exploring the limitations of behavior cloning for autonomous driving. In: 2019 IEEE/CVF International Conference on Computer Vision (ICCV), pp. 9328–9337 (2019). https://doi.org/10.1109/ICCV.2019.00942
8. Couto, G.C.K., Antonelo, E.A.: Hierarchical generative adversarial imitation learning with mid-level input generation for autonomous driving on urban environments. arXiv preprint arXiv:2302.04823 (2023)
9. Florence, P., et al.: Implicit behavioral cloning. In: Conference on Robot Learning, pp. 158–168. PMLR (2022)
10. Gelman, A., Carlin, J.B., Stern, H.S., Rubin, D.B.: Bayesian Data Analysis. Chapman and Hall/CRC, Boca Raton (1995)
11. Karl Couto, G.C., Antonelo, E.A.: Generative adversarial imitation learning for end-to-end autonomous driving on urban environments. In: 2021 IEEE Symposium Series on Computational Intelligence (SSCI), pp. 1–7 (2021). https://doi.org/10.1109/SSCI50451.2021.9660156
12. Liang, X., Wang, T., Yang, L., Xing, E.: CIRL: controllable imitative reinforcement learning for vision-based self-driving. In: Ferrari, V., Hebert, M., Sminchisescu, C., Weiss, Y. (eds.) ECCV 2018. LNCS, vol. 11211, pp. 604–620. Springer, Cham (2018). https://doi.org/10.1007/978-3-030-01234-2_36
13. Montemerlo, M., et al.: Junior: the stanford entry in the urban challenge. J. Field Rob. **25**, 569–597 (2008). https://doi.org/10.1002/rob.20258
14. Ota, K., Jha, D.K., Kanezaki, A.: Training larger networks for deep reinforcement learning. arXiv preprint arXiv:2102.07920 (2021)

15. Paden, B., Cáp, M., Yong, S.Z., Yershov, D.S., Frazzoli, E.: A survey of motion planning and control techniques for self-driving urban vehicles. IEEE Trans. Intell. Veh. **1**, 33–55 (2016)
16. Parizotto, L.M.S., Antonelo, E.A.: Cone detection with convolutional neural networks for an autonomous formula student race car. In: 26th ABCM International Congress of Mechanical Engineering (COBEM 2021) (2021)
17. Petrazzini, I.G., Antonelo, E.A.: Proximal policy optimization with continuous bounded action space via the beta distribution. In: 2021 IEEE Symposium Series on Computational Intelligence (SSCI), pp. 1–8. IEEE (2021)
18. Silverman, B.W.: Density Estimation for Statistics and Data Analysis. Routledge, Abingdon (2018)
19. Xu, H., Gao, Y., Yu, F., Darrell, T.: End-to-end learning of driving models from large-scale video datasets (2016). https://doi.org/10.48550/ARXIV.1612.01079. https://arxiv.org/abs/1612.01079
20. Zhang, Z., Liniger, A., Dai, D., Yu, F., Van Gool, L.: End-to-end urban driving by imitating a reinforcement learning coach. In: 2021 IEEE/CVF International Conference on Computer Vision (ICCV), pp. 15202–15212 (2021). https://doi.org/10.1109/ICCV48922.2021.01494
21. Zhu, H., Yuen, K.V., Mihaylova, L., Leung, H.: Overview of environment perception for intelligent vehicles. IEEE Trans. Intell. Transp. Syst. **18**(10), 2584–2601 (2017). https://doi.org/10.1109/TITS.2017.2658662

Investigating Universal Adversarial Attacks Against Transformers-Based Automatic Essay Scoring Systems

Igor Cataneo Silveira(✉), André Barbosa, Daniel Silva Lopes da Costa, and Denis Deratani Mauá

Universidade de São Paulo, São Paulo, Brazil
{igorcs,abarbosa,ddm}@ime.usp.br, dslcosta2016@usp.br

Abstract. Automatic Essay Scoring promises to scale up student feedback on written input, addressing the excessive cost and time demand associated with human grading. State-of-the-art automatic scorers are based on Transformers-based neural networks. While such models have shown impressive results in reasoning tasks, learned models often produce answers that arise from statistical clues in datasets and are misaligned with human objectives. Such systems are thus potentially fragile for scenarios where users are incentivized to deceive the system, as in a classroom setting. In this work, we evaluate the susceptibility of state-of-the-art automatic scorers to attacks made by non-expert users, such as students interacting with an automatic grader. We develop a methodology to simulate such student attacks and test them against scorers based on BERT, Phi-3 and Gemini models. Our findings suggest that (i) a BERT-based grader can be deceived using simple feature-based attacks; (ii) although Google's Gemini has a solid agreement with graders, it can assign undeservedly high grades for small sentences; (iii) a Phi-3-based grader was less susceptible than BERT, but it still assigned relatively high grades to some of our attacks.

Keywords: Automatic Essay Scoring · Adversarial Attack · Natural Language Processing

1 Introduction

Automatic Essay Scoring (AES) systems ease the burden on teachers while enhancing student learning by providing meaningful, timely and personalized feedback [32]. The technology is currently mature enough to be employed in high-stakes standardized exams such as the GRE and TOEFL [3,5].

Like many Machine Learning based solutions, AES systems are prone to malicious usage and are sensitive to spurious correlations in the training data [22]. This is particularly relevant in educational and scoring settings, where users might be motivated to exploit the system's vulnerability, and where a correct prediction for the wrong reason might be as harmful as an incorrect prediction.

A. Paes and F. A. N. Verri (Eds.): BRACIS 2024, LNAI 15413, pp. 169–183, 2025.
https://doi.org/10.1007/978-3-031-79032-4_12

Unintended uses of predictive systems are known in the Machine Learning literature as adversarial attacks [21]. The more recent literature cares specifically about attacks that are carried out by changing an existing input in a human imperceptible way and such that the prediction shifts towards or away from some desired target value [13,44]. For example, in AES that might amount to determining words or expressions in an essay that, if changed, maintain the text's overall characteristics as perceived by a human while leading the predictive model to raise the score significantly. Such attacks usually require knowledge and access to the system's characteristics. Thus, they are usually difficult to be performed by regular users of AES systems.

A more straightforward approach is the universal attack, which consists of finding input-unrelated rules that cause the predictive model to increase scores. For example, it has been noticed that essay length is a good predictor of essay quality [26]. A malicious user can exploit this fact by appending unrelated text to an essay to increase its score. Such an attack does not require deep knowledge of the system and can be formulated by repeated use.

Designers of AES systems have long been aware of the pitfalls of malicious usage. For instance, authors have discussed whether to remove certain predictive features such as text length in order to improve the system's robustness against attacks [42]. The matter is complicated by the fact that textual features are often correlated, so that removing a feature might hurt performance without actually making the system less susceptible to attacks.

Following the trend in Natural Language Processing, state-of-the-art AES systems are based on Transformer deep neural networks [27]. While in principle such models can learn rich representations and enable logical and commonsense reasoning, there is evidence that learning algorithms and spurious correlations in data often lead to predictions being based on simple statistical cues [43]; hence such systems might be as exposed to attacks as traditional feature-based systems.

In this work, we investigate whether **Transformers-based AES systems are susceptible to universal adversarial attacks that might occur in a classroom setting**. In particular, we consider a fictitious (but realistic) case where users (i.e., students) interact with the system by submitting an essay (input) and receiving a score (the output), while having no knowledge of the predictive model being used (including parameters, derivatives, etc.). The hypothetical system is non-adaptive (i.e., no learning after deployment) and students use the system repeatedly (e.g. grading different essays along the academic term or being able to submit an improved version) and share information about usage among themselves. Given that cheating in the classroom is prevalent [20,41], some students are expected to exploit any vulnerability to improve their grade without necessarily improving the quality of their texts.

To simulate how such non-expert users might learn vulnerabilities, we train an interpretable predictive model (a linear regressor) based on handcrafted features and select the most predictive features to compose attacks, such as inserting lists of adjectives of adverbs, employing more adjectives and adverbs than usual, and repeating the same sentence multiple times. The rationale is that by repeated interaction and information sharing, users might develop a simplified model of the system's inner work and use such a model to come up with such simple attacking strategies.

The effectiveness of the attacks is evaluated with respect to BERT, Phi-3 and Gemini-based AES systems on a benchmark of human annotated (Brazilian) Portuguese essays [34]. We show that all models are vulnerable to simple attacks.[1] For example, on a scale of 0 to 1000, where higher numbers are better, the BERT and Phi-3 based models assigned, respectively, 800 and 560 points to a universal attack (i.e., a fabricated essay) based on repeating a generic conclusion employing many adverbs and adjectives. The Gemini-based model showed resilience against attacks based on repetition, but for instance assigned a relatively high score of 640 points to an essay consisting of only a small list of adjectives.

In the rest of the document, we present related work in Sect. 2, discuss our methodology in Sect. 3, and show the experimental results in Sect. 4. We conclude with a summary and final remarks in Sect. 5.

2 Related Work

The evolution of AES systems mimics that of other NLP tasks. The initial systems were based on handcrafted features, which had limited predictive power and require extensive expertise to be designed [32]. Those systems were incrementally superseded by systems that learned representations directly from data. First, by approaches that extracted shallow representations such as word embeddings [37]. More recently, by approaches that learn representation by deep models such as Transformer neural networks [27].

Several AES systems for (Brazilian) Portuguese have been proposed, mostly focusing on grading essays in the format required for the Brazilian national entrance exam (ENEM) [2,4,12,15,28,34]. An early work used words as features of a Naïve Bayes [4]. Subsequent approaches ranged from defining handcrafted features to combining word-embedding with Recurrent Neural Networks [2,15,29]. Our study intersects with the existing literature by evaluating the performance of previously unused models, specifically the Phi-3 and Gemini-based graders. Our goal is not pursuing high agreement, but presenting how susceptible these novel models are to adversarial attacks. A recent work compared several (Brazilian) BERT-based graders [34]. We incorporate these models into our study, alongside new state-of-the-art models, for comparative analysis.

Different AES systems were assessed based not on their QWK performance, but rather on their oversensibility and overstability [21]. The authors investigated how perturbations in the input affected the models' output. The ASAP challenge dataset [18] was used to test five models, ranging from a feature-based to a BERT-based grader. They found that, except for the feature-based and one of

[1] The models are available at: https://github.com/kamel-usp/aes_enem/.

the word-embedding-based, all models were overstable. Moreover, increasing the disturbance percentage lead to increased standard deviation in the grades. Their work investigated whether adding, removing or changing content of an essay caused it to be graded differently. This can be seen as students rewriting their essays after grading. Our scope is different, for they do not consider universal attacks, and we consider ill-intended attacks that could be done by students with minimal effort.

Without any scenario restrictions, there are possibly infinite adversarial attack setups. We rule out possible attacks by restricting ourselves to a very specific scenario. Firstly, students do not have access to any parts of the model, so they are not allowed to attack the embedding space [16], the gradient [11], nor the calculated features, to name a few. Secondly, they do not know the underlying model, which prevents them from using the pre-trained version of the model to generate attacks, such as [25]. Lastly, the students are not too invested in breaching the system, so attacks such as [14,40] are not applicable. The first requires (possibly) testing combinations of the whole vocabulary used by the tokenizer of the underlying model. This is ruled out of our scenario as they do not have access to this information, some models do not have a vocabulary, and Gemini's vocabulary has 256k tokens, so it would require at least this many attacks. Similarly, the second attack requires changing the whole essay to have the same word, which results in a massive number of needed attacks. Thus, the attack we devise here is much closer to what one could expect to happen at a classroom level and has not been used yet in the literature.

The task of Automatic Short Answer Grading (ASAG) is very similar to AES, the difference is that, instead of (long) essays, its input is a relatively small text. Adjectives and adverbs have already shown to be useful to cheat ASAG systems based on Transformers, such as BERT and T5 [13]. The scenario of their study is similar to ours. The differences are: they are concerned with generating answers that cannot be identified by humans, and that their labels are: correct, incorrect or contradictory. By inserting adjectives and adverbs in the sentences, they could fool the models to change the answer from incorrect to correct between 8 and 22% of the cases. Showing that they are strong candidates to fool the model. Additionally, their investigation found that humans did not find the adversarial examples suspicious, but they were not natural either.

To evaluate the robustness of available models against such adversarial attacks, we rely on the LMSys benchmark [9] to choose candidates. LMSys is a comprehensive benchmarking platform designed to evaluate the performance of large language models across a wide array of tasks, including natural language understanding, generation, and interaction. By providing standardized evaluation metrics and diverse benchmarks, it enables a consistent comparison of model capabilities, thus serving as an invaluable resource for researchers and developers. As of the current date, the 10 best-ranked models available there are proprietary and cannot be easily fine-tuned due to their restricted access. Among the available ones, Nemotron [31] and the top-performing models from the Llama family [39] are notably expensive to run, limiting their practicality for widespread use.

In contrast to this trend, there is a growing interest in "Small" Language Models, such as Llama3-8B and the Phi-family [17]. The Phi-family has recently released the Phi-3 models [1], which include three distinct model sizes: 3.8B (Phi-3-mini), 7B (Phi-3-small), and 14B (Phi-3-medium). These models, despite their smaller size, are claimed by the authors to outperform some proprietary models, such as GPT-3.5, in various tasks.

We chose the Phi-3 model for our research due to its open-source nature, which allows for easier access and adaptability. Additionally, the Phi-3 model was trained exclusively on English data and has not been extensively tested in other languages. Our study aims to evaluate the performance of the Phi-3 model using Portuguese-only data, contributing to the understanding of its capabilities and limitations in multilingual contexts.

The GPT family (Davinci, 3.5 turbo and 4.0) have been tested in AES and ASAG. The authors of [30] showed that the Davinci model could achieve 0.38 QWK in the TOEFL dataset, while linguistic features could achieve almost 0.6 QWK. GPT 3.5 and 4.0 were used to grade short answers in Finnish [7]. The conclusion was that the models could not be directly employed and that they are more lenient than human graders, assigning fewer failing grades. While OpenAI models are the most famous, their cost is prohibitive, thus we explore here the usage of Gemini's free API and how resilient it is against adversarial attacks.

3 Methodology

We start by selecting the ENEM-AES dataset and pre-trained models [34], which are easily accessible through Hugging Face API. Essays in this dataset are evaluated according to five difference criteria, or competencies, and separated predictors have been trained for each specific criterion. We use the BERT-base Ordinal model available for each competence, as they are relatively small and achieve high performances. Additionally, we used the splits available using the "propor2024" parameter.

We also used the scikit-learn package [33] to train, for each individual competence, a Linear Regressor that uses 72 features calculated by the open API of NILC-Metrix [24]. These features are interesting because they calculate various linguistic levels of written language, such as the ratio of a certain part of speech, how ambiguous the words are, text length, Flesch score and others. Although they might be far from what the ENEM guiding book states that are being evaluated, they might relate closer to what the neural models are evaluating. For this grader, as we did not optimize any hyperparameters, we joined the training and validation split to increase the number of training samples. As this model outputs a real number and the ENEM grades are in steps of 40, we round it to the closest allowed grade.

For fine-tuning a Large Language Model, we chose to fine-tune a Phi-3-medium model, balancing model size with our available hardware. Although smaller than some other models, it still demands significant computational power for training. We had access to an NVIDIA GPU RTX A6000, and we applied several optimizations for fine-tuning [35], including Gradient Checkpointing, Flash Attention V2 [10], and LoRA [19].

The available model is instruction fine-tuned.[2] During our tests, we found it necessary to pass an instruction to the system role before prompting a given essay. The prompt structure is shown in Fig. 1.

```
<|system|>CONCEPT_SYSTEM<|end|>
<|user|>Grade the following essay: {essay_example}<|end>
<|assistant|>
```

Fig. 1. Prompt-structure used for each Concept fine-tuning through Phi-3.

In this structure, `CONCEPT_SYSTEM` is a placeholder for the competence being fine-tuned (e.g., ENEM Competences that vary from 1 to 5), and `essay_example` represents the essay we are interested in fine-tuning.

To define the concepts, we copied the definitions directly from the ENEM 2023 Student's Reference manual[3], which provides clear instructions on how each grade is applied.

Following these optimizations, we added a sequence classification head to the existing model and fine-tuned it according to the hyperparameters outlined in Table 1. It is noteworthy that these models were instruction fine-tuned through supervised fine-tuning and direct preference optimization for English conversations. Fine-tuning capabilities for languages other than English were not investigated prior to the completion of this study [1]. We found it interesting to notice an ability of the final Phi-3 to generalize and outperform models that were trained solely on Portuguese data—the prior model based on BERTimbau [36].

Finally, we employ Google's Gemini chatbot. We chose Gemini over ChatGPT because Gemini offers a free API. The current model available is the Gemini 1.5 Flash [38]. Its size and training set are not publicly disclosed, so we can only speculate about its dimensions. According to the LMSys Leaderboard [9], it is ranked near the LLama3 70B model, suggesting it has at least 70 billion parameters. We do not train this model; we use it exclusively through its API. Additionally, we did not employ any prompt engineering techniques, instead applying a zero-shot learning process [6,8,23], as we only asked it to grade the essays using the ENEM criteria. However, the model produces grades that do not follow the 40-point step, so we round the numbers according to the same criteria as the Linear Regressor.

[2] Specifically, microsoft/Phi-3-medium-128k-instruct on the Hugging Face Hub: https://huggingface.co/microsoft/Phi-3-medium-128k-instruct.

[3] https://download.inep.gov.br/publicacoes/institucionais/avaliacoes_e_exames_da_educacao_basica/a_redacao_no_enem_2023_cartilha_do_participante.pdf.

Table 1. Phi-3 Fine Tuning Hyperparameters

Hyperparameter	Value
Warmup steps	10
Gradient Checkpoint	True
Num train epochs	12
Learning rate	5e–5
Weight decay	0.01
BFloat16 use	True
LoRA Attention Dimension	256
LoRA Alpha Scaling	32
LoRA Dropout	0.15
LoRA Task Type	Text Classification
LoRA Target Modules	All Linear

Using these graders, we test them—except for the BERT one, as its performance has already been published in [34]—and compare them against the baselines. Then, we use the weights of the Linear Regression to identify the relevant features. Finally, we use them to create possible universal adversarial attacks that students could create.

There are no limits to the number of adversarial attacks that can be created. In our investigation, we restricted ourselves to a scenario where (1) the students do not know the underlying AES system; (2) they do not have access to any part of the system except for an interface where they can type the essay and see their final score; (3) students can share their intuitions with each other; (4) they do not want to properly write the essay, but still they want the best possible grade. Having these restrictions, we devise some attacks and test them against all graders.

4 Experiments

We now discuss the empirical findings of our investigation.

4.1 Testing the New Models

The first step before devising our adversarial attacks is comparing the performance of an interpretable model against the neural ones. We train the models presented in the previous section, namely a Linear Regressor (LR), a Phi-3 and a Gemini-based grader. Their performances, measured in Quadratic Weighted Kappa (QWK), are presented in Table 2 and compared against the BERTs available and presented in [34].

The first thing to notice is that their sizes have different orders of magnitude, but the performances are still not that distant. The Linear Regressor was

Table 2. Performance comparison of different automatic graders, as measured by QWK.

Model	Size	C1	C2	C3	C4	C5
LR	72	0.23	0.40	0.47	0.34	0.22
Phi-3	14B ($\approx$ 892M Trainable)	0.46	0.35	0.52	0.29	0.61
Gemini	$\geq$70B?	0.41	0.40	0.40	0.36	0.35
BERTs	110M–330M	0.29–0.37	0.23–0.37	0.42–0.50	0.28–0.42	0.26–0.53

expected to have the worst performance across all competences, but it happened only in two out of the five. When it had the lowest performance, it was only by a margin of 0.07 and 0.04 compared to the BERT baseline. The similar performance to the others may point out that the neural models are exploring simple features of the essays.

The Phi-3 model performs best in three competences, with a surprising 0.61 in the fifth competence. Notably, this model loses twice to the Linear Regression model, but it is always at least competitive with the BERT baseline.

Finally, Gemini is the largest model, with an unknown size and training set. It drew with the Linear Regressor for the best performance in Competence 2 and had the worst performance in Competence 3. In general, this model is the most stable one, with performances between 0.41 and 0.35. Its relatively poor performance must be seen with a grain of salt, as it is the only model that was not fine-tuned to actually grade the essays.

Having shown that the performances between the different Transformers-based graders are not that different from the Linear Regressor, we check which features receive the largest weights. The names of the five more important (positive) features are presented in Table 3. Although named with a different pattern, *adverbs* and *verbs* are features that measure the ratio of such parts of speech. *Content words* and *Function words* also account for the ratio of words. The former aggregates nouns, verbs, adjectives, adverbs, and discourse markers. The latter accounts for articles, conjunctions, interjections, numerals, pronouns, prepositions, conjunctive and subordinating adverbs. Related to the previous two, content density is the ratio of *content words* to *functional words*. Finally, *cau neg conn ratio* is the ratio of negative causal connectives.

Competences 1 to 3 have the same set in the same order, which does not match what is expected from the official grading, as each competence is supposed to evaluate a different aspect of the text. Competence 5 has the same set as the previous three but in a different order. Competence 4 is supposed to evaluate the proper usage of connectives, so having *function words* and *cau neg conn ratio* between the top 5 makes sense.

With this information, we can assume that students will come up with intuitions like (1) it is important to have a lot of adverbs, as it is the most important feature in 3 competencies, it is part of the top feature in competence 4 and is the third most important in competence 5; (2) it is important to have a lot

Table 3. The four most important features for each competence according to the Linear Regressor.

Competence 1	Competence 2	Competence 3	Competence 4	Competence 5
adverbs	adverbs	adverbs	content words	adjective ratio
adjective ratio	adjective ratio	adjective ratio	function words	verbs
noun ratio	noun ratio	noun ratio	cau neg conn ratio	adverbs
verbs	verbs	verbs	content density	noun ratio

of adjectives, as it is the most important in one competence, the second most important in three, and part of the top feature in the other.

4.2 Defining the (Universal) Adversarial Attacks

Bearing the previous intuitions in mind, we can devise three goals: increase the number of adverbs, increase the number of adjectives, and increase both. We consider that although the feature is the ratio of such classes, it is not trivial to identify that the importance is the ratio and not the total number.

To increase the number of one of the classes, we can consider some strategies, such as (a) writing a list of words belonging to this class, (b) copying and pasting the previous list so that it becomes similar to the four paragraphs structure; (c) similar to the previous, but with more copying and pasting and using a memorized sentence that uses many of the desired part of speech.

We have 3 goals, and each goal can be reached by 3 strategies, so we can define 9 adversarial attacks:

Attack 1a: We devise a list of adverbs, such as: "Well, badly, enormously, smally, certainly, wrongly, rapidly, slowly, fairly, unfairly".

Attack 1b: We repeat the previous list four times, one in each paragraph.

Attack 1c: We create a sentence that uses more adverbs than usual, such as: "Undeniably, progressing slowly, leisurely, carefully, silently while deeply breathing and thinking intensively about the given problem", copy it 10 times in each paragraph, in 4 paragraphs.

Attack 2a: We devise a list of adjectives, such as: "Good, bad, big, small, best, worst, right, wrong, last, first, fair, unfair".

Attack 2b: We repeat the previous list four times, one in each paragraph.

Attack 2c: We create a sentence that uses more adjectives than usual, such as: "The constant and innovative development of modern technology has brought significant and deep changes to this issue, creating diverse and thrilling opportunities for a more promising and sustainable future." then we copy it 10 times in each paragraph, in 4 paragraphs.

Attack 3a: We concatenate the sentences of 1a and 2a.

Attack 3b: We repeat the previous list four times, one in each paragraph.

Attack 3c: We create a sentence that uses more adjectives and adverbs than usual, such as: "Consequently and undeniably, the constant innovative development slowly and progressively brought some significant and deep changes to constantly needed problems", then we copy it 10 times in each paragraph, in 4 paragraphs.

4.3 Testing the Attacks

We now use the sentences created for each attack and submit them to the models. The grades assigned by each model are displayed in Table 4. Schools in Brazil usually require students to get at least 6 or 7 to be approved, so we can assume that an adversarial student would be happy with getting a 600 or higher with the lowest possible effort, and this will count as fooling the model.

Table 4. Grades assigned by all models, in all competences, to each attack.

Att.	Model	C1	C2	C3	C4	C5	Total	Att.	C1	C2	C3	C4	C5	Total	Att.	C1	C2	C3	C4	C5	Total
1a	LR	200	200	200	200	0	800	2a	200	200	200	200	0	800	3a	200	200	200	200	0	800
	BERT	120	80	40	120	0	360		120	80	40	120	0	360		120	120	40	120	0	400
	Gemini	120	120	80	120	120	560		120	120	120	120	160	640		120	80	80	120	120	520
	Phi-3	0	40	40	0	0	80		80	120	40	0	0	240		80	40	40	0	0	160
1b	LR	200	200	200	200	0	800	2b	200	200	200	200	0	800	3b	200	200	200	200	0	800
	BERT	120	120	80	120	0	440		120	120	80	120	0	440		120	120	120	120	0	480
	Gemini	80	80	80	80	80	400		80	80	80	80	80	400		80	40	40	40	80	280
	Phi-3	80	120	80	120	0	400		80	120	120	120	0	440		80	120	80	120	0	400
1c	LR	200	200	200	200	200	1000	2c	120	200	200	200	200	920	3c	200	200	200	200	200	1000
	BERT	160	160	160	160	0	640		160	160	160	160	0	640		160	160	160	160	40	680
	Gemini	40	40	0	40	40	160		0	0	0	0	0	0		40	0	0	0	0	40
	Phi-3	80	120	40	80	0	320		80	40	40	80	0	240		80	40	40	80	0	240

We can see that our simplest attacks, the ones with suffix *a*, are sufficient to fool a simple Linear Regressor on all but the Competence 5. This is not surprising, as the attacks were defined according to most correlated predictor variables. Changing from one sentence to four paragraphs (i.e., changing an attack from type *a* to type *b*) did not increase the score, while changing from type *b* to type *c* did improve the score. We note that either changing from type *a* to type *b* or from type *b* to *c* leads to a significant increase of the text length. The number of words and paragraphs are features available to the model and have a positive coefficient, so increasing them naturally increases the grade. Additionally, as this Linear Regressor has a limit of 2k words, it is easy to get a full grade by increasing certain word frequencies.

More interestingly, the BERT grader was not completely fooled by the simplest attacks; it assigned attacks *1a* and *2a* with 360 points, and attack *3a* with 400 points. This suggests that BERT-based predictors are either sensitive to diversity in parts of speech or that their inner representation considers word frequencies. Changing from attack type *a* to *b* increased scores, which could point to the fact that BERT is sensitive to the four paragraph structure—this could be further emphasized by Competence 3's scores going up. But since BERT does not have a token for line break, it is not clear whether that effect is due to the presence of longer sentences or due to the increased word frequency. Finally, we see that going from attacks of type *b* to *c* resulted in an increase of 200 points. This was the only attack that received a score above zero in Competence 5 with this model.

Gemini, on the other hand, seems to recognize attacks that increase essay length. The simplest attacks *1a* and *2a* were assigned scores of 560 and 640 points, respectively, by this model, which are relatively high and way higher than expected. However, the relatively longer essays produced by attacks of type *b* received lower scores, showing that the Gemini is sensitive to repetition. This is further evidenced by the fact that is assigns only 40 points to attack *3c*, which is the one with most repetitions. This model was the only one to identify one attack (*2c*) and it almost identified attack *3c*.

Finally, the Phi-3-based model was the most resilient to the simplest attacks od type *a*. Even though attacks *1a* and *2a* are very similar, the model was more sensitive to the latter one. Attack *3a*, which somehow combines attack *1a* and *2a*) received an intermediary score, suggesting that the model is not composing the representation to produce scores. Phi-3's tokenizer has a token for the line break, which can make it sensitive to the four paragraphs structure. This appears to be the case, since attacks of type *a* received smaller scores than attacks of type *b*. Finally, the attacks of type *c*, which are based on repetition of sentences were not as effective as the attacks of type *b* which are based of repetition of lists. We highlight that Phi-3 achieved the highest QWK score for Competence 5, across all models and competences, and was not fooled by any attack—that is, it scored all attacks with a zero.

Seeing that Competence 5 is the one that most often receives a zero, it is possible to imagine that students would try to target it. The fifth competence is responsible for evaluating the conclusion. Accordingly, we devised the **Attack 4** based on the previous results: a sentence that resembles a conclusion using adjectives and adverbs, namely: "Consequently, it is up to the fair and democratic Federal Government to rapidly approve laws that rapidly reduce the occurrence of these horrendous problems. Following, the dear Brazilian population must abide by the undeniable laws, and the fast police must arrest those that committed any inhuman crime", copied in 7 paragraphs. The grades assigned by each model for this attack are presented in Table 5.

Surprisingly, the Linear Regressor was the only grader not deceived in the fifth competence. Nonetheless, the model as a whole assigned 800 points, a high score. The grade can be further boosted by adding more repetitions and thus achieving a perfect score.

As intended, the BERT grader assigned a high grade in Competence 5, demonstrating that it is not only using word frequency. Still, this attack uses a

Table 5. Grades assigned by each model when grading the Attack 4.

Attack	Model	C1	C2	C3	C4	C5	Total
4	LR	200	200	200	200	0	800
4	BERT	160	160	160	160	160	800
4	Gemini	40	40	0	40	40	160
4	Phi-3	160	120	120	120	40	560

lot of words, which could cause lead to other competences receiving high scores. It is noteworthy that when trained, this grader did not have access to many training instances of perfect grade, which causes it to rarely (if ever) assign 200 points in a competence. Thus, 800 points is the highest possible grade in this method, which means that it was completely deceived by this attack.

If Gemini is sensitive to repetition and we repeat the same sentence 7 times, then we would expect a low grade assigned by this model. That is exactly what happens. The zero assigned in Competence 3 is rounded from 20. In our rounding we always round down, but maybe systems used in classrooms would round up to motivate students. If we had rounded it up, this attack would have been the attack based on repetition that fooled Gemini the most. As an additional test made only for this grader, we tested asking it to grade only the first paragraph of Attack 4; it assigned a total of 440 points.

Finally, this attack was the one in which Phi-3 assigned the highest grades, for the first time assigning points in the fifth competence. Overall, Phi-3 was more resilient than BERT and the Linear Regressor but less than Gemini. This might suggest that being an order of magnitude pays off by having higher-level features. Nonetheless, the grade assigned by this model was surprisingly high.

5 Conclusion

In this work we investigated whether Transformers-based Automatic Essay Scorers are susceptible to universal adversarial attacks that might occur in a classroom setting.

To simulate realistic student behavior, we trained a Linear Regressor based on standard NLP features to predict scores, then used the most important features to craft simple essays that should receive low scores according to grading guidelines yet might fool statistical AES systems. Such essays varied from simple list of adjectives and adverbs to repetitions of template texts unrelated to the required prompt.

We found that the BERT-based grader is susceptible to both long sentences and to the presence of adjectives and adverbs, similarly to reported in the literature [13].

The Gemini-based grader was the only model to completely identify an attack, meaning that it assigned 0 to such essays (as a human grader would). The more repetition an essay had, the lower was the score assigned by that model.

This contrasts with the BERT-based model, which assigned higher scores to longer (and more repetitive) texts. Despite that relative improvement, this type of model still assigned undeserving high grades to most of the attacks. This is aligned with the findings of the ChatGPT-based grader [7], which found that the model was more lenient than a human-grader. It is however interesting to note that this model assigned the highest score to essays consisting only of a list of adjectives or adverbs.

The Phi-3-based model, which is the best performing model for most competences, performed similarly to the other models. In particular, it was more susceptible than Gemini to an attack that included a unrelated but sensible template conclusion to essay. This is in line with findings that fine-tuned models improve performance at the expense of being more susceptible to attacks [40]. Interestingly, although the model was originally trained only texts in English and fine-tuned with texts in Portuguese, it achieved the best performances, even when compared to models trained exclusively with Portuguese texts.

This work shows that even sophisticated AES systems can be fooled by simple attacks, which should be of great concern to real-world systems deployed. Future work should better investigate whether standard techniques such as data augmentation and loss function shaping can mitigate such an effect.

Acknowledgments. The authors of this work would like to thank the Center for Artificial Intelligence (C4AI-USP) and the support from the São Paulo Research Foundation (FAPESP grants #2019/07665-4 and 2022/02937-9) and from the IBM Corporation. This work was also supported by CNPq grant no. 305136/2022-4 and CAPES Finance Code 001.

References

1. Abdin, M., Jacobs, S.A., Awan, A.A., Aneja, J., Awadallah, A., et al.: Phi-3 technical report: a highly capable language model locally on your phone (2024)
2. Amorim, E., Veloso, A.: A multi-aspect analysis of automatic essay scoring for Brazilian Portuguese. In: Proceedings of the Student Research Workshop at the 15th Conference of the European Chapter of the Association for Computational Linguistics, pp. 94–102 (2017)
3. Attali, Y., Burstein, J.: Automated essay scoring with e-rater v.2. J. Technol. Learn. Assess. **4**(3) (2006)
4. Bazelato, B.S., Amorim, E.: A bayesian classifier to automatic correction of Portuguese essays. In: Conferência Internacional sobre Informática na Educação (TISE), vol. 18, pp. 779–782 (2013)
5. Beigman Klebanov, B., Madnani, N.: Automated evaluation of writing – 50 years and counting. In: Proceedings of the 58th Annual Meeting of the Association for Computational Linguistics. pp. 7796–7810 (2020)
6. Brown, T., Mann, B., Ryder, N., Subbiah, M., Kaplan, J.D., et al.: Language models are few-shot learners. In: Advances in Neural Information Processing Systems, vol. 33, pp. 1877–1901. Curran Associates, Inc. (2020)
7. Chang, L.H., Ginter, F.: Automatic short answer grading for finnish with chatgpt. In: Proceedings of the AAAI Conference on Artificial Intelligence, vol. 38, pp. 23173–23181 (2024)

8. Chang, M.W., Ratinov, L., Roth, D., Srikumar, V.: Importance of semantic representation: dataless classification. In: Proceedings of the 23rd National Conference on Artificial Intelligence, vol. 2, pp. 830–835 (2008)
9. Chiang, W.L., et al.: Chatbot arena: an open platform for evaluating llms by human preference (2024)
10. Dao, T.: Flashattention-2: faster attention with better parallelism and work partitioning (2023)
11. Ebrahimi, J., Rao, A., Lowd, D., Dou, D.: HotFlip: white-box adversarial examples for text classification. In: Proceedings of the 56th Annual Meeting of the Association for Computational Linguistics, vol. 2: Short Papers, pp. 31–36 (2018)
12. Filho, A.H., Concatto, F., do Prado, H.A., Ferneda, E.: Comparing feature engineering and deep learning methods for automated essay scoring of Brazilian national high school examination. In: International Conference on Enterprise Information Systems (2021)
13. Filighera, A., Ochs, S., Steuer, T., Tregel, T.: Cheating automatic short answer grading with the adversarial usage of adjectives and adverbs. Int. J. Artif. Intell. Educ. **34**(2), 616–646 (2024)
14. Filighera, A., Steuer, T., Rensing, C.: Fooling automatic short answer grading systems. In: Bittencourt, I.I., Cukurova, M., Muldner, K., Luckin, R., Millán, E. (eds.) AIED 2020. LNCS (LNAI), vol. 12163, pp. 177–190. Springer, Cham (2020). https://doi.org/10.1007/978-3-030-52237-7_15
15. Fonseca, E.R., Medeiros, I., Kamikawachi, D., Bokan, A.: Automatically grading Brazilian student essays. In: Proceedings of International Conference on Computational Processing of the Portuguese Language, pp. 170–179 (2018)
16. Gao, H., Oates, T.: Universal adversarial perturbation for text classification. arXiv preprint arXiv:1910.04618 (2019)
17. Gunasekar, S., et al.: Textbooks are all you need (2023)
18. Hamner, B., Morgan, J., Lynnvandev, M., Ark, T.: The hewlett foundation: automated essay scoring. Kaggle (2012)
19. Hu, E.J., et al.: Lora: low-rank adaptation of large language models (2021)
20. Klein, H., Levenburg, N., McKendall, M., Mothersell, W.: Cheating during the college years: how do business students compare? J. Bus. Ethics **72**, 197–206 (2007)
21. Kumar, Y., Bhatia, M., Kabra, A., Li, J.J., Jin, D., Shah, R.R.: Calling out bluff: attacking the robustness of automatic scoring systems with simple adversarial testing. CoRR arxiv:2007.06796 (2020)
22. Kumar, Y., Parekh, S., Singh, S., Li, J.J., Shah, R.R., Chen, C.: Automatic essay scoring systems are both overstable and oversensitive: explaining why and proposing defenses. Dial. Discour. **14**(1), 1–33 (2023)
23. Larochelle, H., Erhan, D., Bengio, Y.: Zero-data learning of new tasks. In: Proceedings of the 23rd National Conference on Artificial Intelligence, vol. 2, pp. 646–651 (2008)
24. Leal, S.E., Duran, M.S., Scarton, C.E., Hartmann, N.S., Aluísio, S.M.: Nilc-metrix: assessing the complexity of written and spoken language in Brazilian Portuguese. Lang. Res. Eval. **58**(1), 73–110 (2024)
25. Li, L., Ma, R., Guo, Q., Xue, X., Qiu, X.: BERT-ATTACK: adversarial attack against BERT using BERT. In: Proceedings of the 2020 Conference on Empirical Methods in Natural Language Processing, pp. 6193–6202 (2020)
26. Lim, C.T., Bong, C.H., Wong, W.S., Lee, N.K.: A comprehensive review of automated essay scoring (aes) research and development. Pertanika J. Sci. Technol. **29**(3), 1875–1899 (2021)

27. Liu, J., Xu, Y., Zhu, Y.: Automated essay scoring based on two-stage learning (2019)
28. Marinho, J., Anchiêta, R., Moura, R.: Essay-br: a Brazilian corpus of essays. In: Anais do III Dataset Showcase Workshop, pp. 53–64 (2021)
29. Marinho, J., Cordeiro, F., Anchiêta, R., Moura, R.: Automated essay scoring: an approach based on enem competencies. In: Anais do XIX Encontro Nacional de Inteligência Artificial e Computacional, pp. 49–60 (2022)
30. Mizumoto, A., Eguchi, M.: Exploring the potential of using an AI language model for automated essay scoring. Res. Methods Appl. Linguist. **2**(2), 100050 (2023)
31. Nvidia, Adler, B., Agarwal, N., Aithal, A., Anh, D.H., et al.: Nemotron-4 340b technical report (2024)
32. Page, E.B.: The imminence of... grading essays by computer. In: The Phi Delta Kappan, pp. 238–243 (1966)
33. Pedregosa, F., et al.: Scikit-learn: machine learning in python. J. Mach. Learn. Res. **12**, 2825–2830 (2011)
34. Silveira, I.C., Barbosa, A., Mauá, D.D.: A new benchmark for automatic essay scoring in Portuguese. In: Proceedings of the 16th International Conference on Computational Processing of Portuguese, vol. 1. pp. 228–237 (2024)
35. Singh, A., Pandey, N., Shirgaonkar, A., Manoj, P., Aski, V.: A study of optimizations for fine-tuning large language models (2024)
36. Souza, F., Nogueira, R., Lotufo, R.: BERTimbau: pretrained BERT models for Brazilian Portuguese. In: Cerri, R., Prati, R.C. (eds.) BRACIS 2020. LNCS (LNAI), vol. 12319, pp. 403–417. Springer, Cham (2020). https://doi.org/10.1007/978-3-030-61377-8_28
37. Taghipour, K., Ng, H.T.: A neural approach to automated essay scoring. In: Proceedings of the 2016 Conference on Empirical Methods in Natural Language Processing, pp. 1882–1891. Association for Computational Linguistics (2016)
38. Team, G., Georgiev, P., Lei, V.I., Burnell, R., Bai, L., et al.: Gemini 1.5: unlocking multimodal understanding across millions of tokens of context (2024)
39. Touvron, H., et al.: Llama: open and efficient foundation language models (2023)
40. Wangkriangkri, P., Viboonlarp, C., Rutherford, A.T., Chuangsuwanich, E.: A comparative study of pretrained language models for automated essay scoring with adversarial inputs. In: IEEE Region 10 International Conference TENCON, pp. 875–880 (2020)
41. Whitley, B.E.: Factors associated with cheating among college students: a review. Res. High. Educ. **39**(3), 235–274 (1998)
42. Woods, B., Adamson, D., Miel, S., Mayfield, E.: Formative essay feedback using predictive scoring models. In: Proceedings of the 23rd ACM SIGKDD International Conference on Knowledge Discovery and Data Mining, pp. 2071–2080 (2017)
43. Zhang, H., Li, L.H., Meng, T., Chang, K.W., Van den Broeck, G.: On the paradox of learning to reason from data. In: Proceedings of the Thirty-Second International Joint Conference on Artificial Intelligence, pp. 3365–3373 (2023)
44. Zhang, J., Wang, J., Luo, X., Ma, B., Xiong, N.: Imperceptible and reliable adversarial attack. In: Cao, C., Zhang, Y., Hong, Y., Wang, D. (eds.) FCS 2021. CCIS, vol. 1558, pp. 49–62. Springer, Singapore (2022). https://doi.org/10.1007/978-981-19-0523-0_4

Likelihood Estimator for Multi Model-Based Reinforcement Learning

Guilherme Albarrans(✉) and Valdinei Freire

Universidade de São Paulo, São Paulo, SP, Brazil
guilherme.albarrans@usp.br

Abstract. Efficiently learning and generalizing from interactions with the environment is a concern in reinforcement learning. Model-based approaches offer promising potential to address these challenges, leveraging the underlying structure of the world to speed learning and improve generalization. By assuming a structured environment, such approaches aim to capture key patterns from limited interactions and synthesize experiences for enhanced exploitation and generalization to unexplored territories. In this paper, we test a model-based reinforcement learning framework aimed at further advancing the frontiers of efficiency and generalization. Motivated by the insight that environments often exhibit distinct regions with varying dynamics, we introduce additional assumptions about the structure of the world to facilitate faster generalization. Specifically, we construct a model comprising a classifier that, given a state input, selects parameters characterizing the operating regime of the environment. Both the model and the regional parameters are learned from past experiences using a likelihood function. Leveraging this model, we employ a rapid exploring random tree based planner to generate new real experiences, capitalizing on the identified structural nuances within the environment. To evaluate the efficacy learning a segmented model, we conduct a comparative analysis against a traditional method that employs a single general model to learn the dynamics of the entire environment. Our results demonstrate the superiority of the segmented approach in terms of both efficiency and generalization, underscoring the benefits of incorporating additional assumptions about the structure of the world into model-based reinforcement learning paradigms.

Keywords: Model Based Reinforcement Learning · Planning · Likelihood Estimator

1 Introduction

Model-based reinforcement learning (RL) has a focus on efficiency and generalization through model construction. By leveraging these models, the agent can simulate possible outcomes of actions without directly interacting with the environment, or explore a state [11]. This can potentially reduce the number of interactions needed to learn an effective policy. However, model-based methods may

A. Paes and F. A. N. Verri (Eds.): BRACIS 2024, LNAI 15413, pp. 184–198, 2025.
https://doi.org/10.1007/978-3-031-79032-4_13

suffer from errors or inaccuracies in the learned model, leading to sub-optimal performance [2].

The concerns of model-based RL lies in the balance between the accuracy of the learned model and the performance achieved by utilizing it to derive a policy. By incorporating learned models into the decision-making process, these methods can adapt more readily to new situations and exploit previously acquired knowledge to achieve better performance. As demonstrated by Young et al. [14] the benefits of the model-based approach could go even further at the point of presenting better solutions for some problems, as in cases where the model's error account for an unintended exploration that can have a high return payoff.

We intend to delve into the nuanced advantages of model-based solutions, particularly in environments characterized by distinct regions with varying dynamics. Consider a scenario where a robot operates in an environment comprising two areas with markedly different frictional properties on their floors. In one area, the effects of actions accounting for movement differ significantly from those in the other.

This paper aim to explore how assumptions like the existence of distinct regions in model-based solutions can enhance its capabilities of planning. Constructing multiple models as done by Doya et al. [3] tailored to individual regions exhibiting stable dynamics, rather than attempting to capture the complex dynamics of both regions within a single model could yield better results. We propose an approach wherein distinct models are built for each identifiable region, and selected for use based on agent past experience.

The learned models are leveraged for plan generation through the utilization of a rapidly exploring random tree (RRT) based algorithm. Our planning strategy draws inspiration from the work of Lin and Zhang [9], which introduced the integration of RRT with reinforcement learning for planning tasks.

By adopting this multi-model approach, we aim to investigate its efficacy in accurately representing the environment's dynamics and its impact on the agent's decision-making process. We hypothesize that by aligning models with specific regions, the agent can better understand and adapt to the unique characteristics of each area, ultimately leading to improved performance and efficiency in task execution.

Through empirical evaluation and analysis, we aim to provide insights into the comparative advantages and limitations of both traditional single-model and proposed segmented model approaches in addressing challenges posed by environments with heterogeneous dynamics. Our findings aim to contribute to a deeper understanding of the applicability of model-based solutions in real-world scenarios characterized by spatially varying dynamics.

In Sect. 2 of our paper, we delve into related works that have explored the concept of leveraging multiple models to enhance generalization in RL. Section 3 formalizes the segmented environment problem in terms of a Markov decision problem. Section 4 introduces our approach based on segmented models problem that forms the basis of our comparative analysis. Moving on to Sect. 5, we provide a concise description of the simple environment utilized for collecting empirical results. Finally, in Sect. 6, we present the results of our empirical evaluations, detailing the performance metrics and outcomes observed across different experimental conditions.

2 Related Works

Several existing works have explored the concept of constructing a segmented model of the environment's dynamics, aiming to create a comprehensive understanding of the environment. One notable example is the work by Doya et al. [3], which introduces a concept akin to a mixture of specialists. In this approach, the environment's dynamics are modeled by a collection of smaller models, each specialized for specific states or moments where distinct behaviors are observed. The fundamental difference between observed approach [3] and our methodology lies in the mechanism for combining the outputs of these specialized models. In Multiple model based reinforcemnt learning [3], the specialists contribute to the overall model through a continuous responsibility signal, allowing for a potentially blended output that may not be directly attributable to any individual specialist. In contrast, our approach adopts a structured mechanism where the output is determined by the specialist that maximizes its signal. This ensures that the output is directly linked to the most relevant specialist, facilitating clearer interpretation and utilization of the model's predictions.

An approach was proposed [5], which introduces a novel representation of the problem by augmenting the traditional Markov decision process (MDP) with contextual information. This augmentation enables the inference of different underlying models based on the context provided, offering a flexible framework for adapting to diverse environmental dynamics. One notable aspect of this work is the similarity in notation used for representing context, which bears resemblance to the representation of our models. However, a significant distinction lies in the treatment of context over the course of a trajectory. Unlike our approach, which allows for the possibility of encountering multiple contexts within the same episode, the contextual MDP framework [5] assumes a fixed context for the entire trajectory.

The idea of dynamically identifying the context have been addressed [12] with a solution capable of learning a new local model whenever it detects a context not yet experienced. Also using as a basis the prediction error of models already learned. This approach resulted in solutions that proved to be very suitable for dealing with non-stationary environments, as detailed by Alegre, Bazzan e Silva (2021) [1] who explored the power of rapid adaptation by identifying change points with high confidence.

Li et al. [7,8], have delved into hierarchical approaches, where segments of information are compounded in a hierarchical manner. This hierarchical structuring enables more efficient learning and decision-making processes, particularly in complex environments with multiple levels of abstraction.

There are works [10] that have focused on incremental learning techniques, which have shown promising results in adapting to changing environments and evolving tasks. By incrementally updating models based on new data and experiences, these approaches offer flexibility and scalability in learning complex tasks over time.

Offline learning settings, [6] have demonstrated the effectiveness of model-based approaches, showcasing their utility in offline learning scenarios where access to real-time data may be limited or impractical. Leveraging offline data, these approaches enable efficient learning and decision-making based on previously collected experiences.

The utilization of a likelihood estimator in the context of building options [4], inspired our exploration into employing a similar approach for constructing models tailored to different regions within the world. The likelihood estimator framework proved instrumental in our efforts to construct region-specific models, enhancing the efficiency and effectiveness of our reinforcement learning approach in complex and heterogeneous environments as it was used in the options context.

3 *k*-Models MDP

The Markov decision process (MDP) [13] framework serves as the foundational formalism for representing the core aspects of reinforcement learning. An MDP is typically characterized by the tuple $< \mathcal{S}, \mathcal{A}, \mathcal{P}, \mathcal{R} >$, where $\mathcal{S}$ denotes the set of states, $\mathcal{A}$ represents the set of actions, $\mathcal{P}$ defines the transition probabilities between states given actions, and $\mathcal{R}$ denotes the reward function associated with each state.

If we assume a know $\mathcal{P}$ parameterized function a natural implementation for a model based approach would be to find the parameters that fit a observed experience, and if we assume that the dynamics of the world vary spatially, meaning that different regions exhibit distinct behaviors, we can define a function $\beta : \mathcal{S} \mapsto \theta$ that give us the parameter for the model based on a given state. We are going to extend this concept now based in a assumption that the world has k regions where the parameters can be different

In the context of modeling segmented dynamics of the world, we extend the MDP framework to $< \mathcal{S}, \mathcal{A}, \mathcal{P}, \mathcal{R}, \Theta, \beta, \mathcal{G}, \mathcal{S}_0 >$ where the transition function $\mathcal{P}$ is parameterized by θ, so a segment of the world is represented by a θ that dictates the dynamics applied by $\mathcal{P}$ on that region of the state space, and to define that region we introduce a new function $\beta : \mathcal{S} \times \theta \mapsto [0, 1]$, which assigns a probability of the parameters θ based on the current state. So given a state s and a parameter θ the $\beta(s, \theta)$ function yields the probabilities for parameter θ to be the one that parameterize $\mathcal{P}$ on s.

The set Θ represents the union of all possible parameters evolved in functions $\mathcal{P}$ that maps a region in the environment, in a way that for a world with k distinct regions $\Theta = \{\theta_1, \theta_2 ..., \theta_k\}$. Each θ_i represents a distinct parameter for $\mathcal{P}$ that captures the dynamics of the environment within a specific region.

We define $\mathcal{G} \subset \mathcal{S}$ as the set of goal states used by the planning module to construct plans to reach them, and $\mathcal{S}_0 \subset \mathcal{S}$ as the set of initial states used to randomly define the initial state of each episode.

4 Likelihood Solution for a k-Models MDP

The model-based approach to solving MDPs involves learning a model that describes the transitions between states in the environment. This model is acquired by observing real transitions that occur as the agent interacts with the environment. Over time, as more experience is gathered, the model can be refined and adjusted to more accurately reflect the dynamics of the environment.

The process of learning the model typically involves estimating transition probabilities or dynamics from observed state-action-state transitions. We bring a likelihood approach to find the best parameters of the model based on the observed experience. As the agent interacts with the environment and accumulates more experience, we can update the parameters so the learned model becomes increasingly accurate and reliable. By learning and utilizing this model, the agent can plan and make decisions more effectively, for solving the reinforcement learning problem at hand.

To facilitate performance comparison, we have implemented a separate planner module capable of generating plan sequences to reach a goal state given a model of the world. This planner module can be utilized with various implementations, including the k-models approach, a standard single-model approach, or even with real-world models.

By integrating these components, we create a comprehensive framework for modeling segmented dynamics, learning multiple models, and conducting performance evaluations across different approaches. This approach enables us to explore the effectiveness of our proposed method in improving sample efficiency and generalization in reinforcement learning tasks.

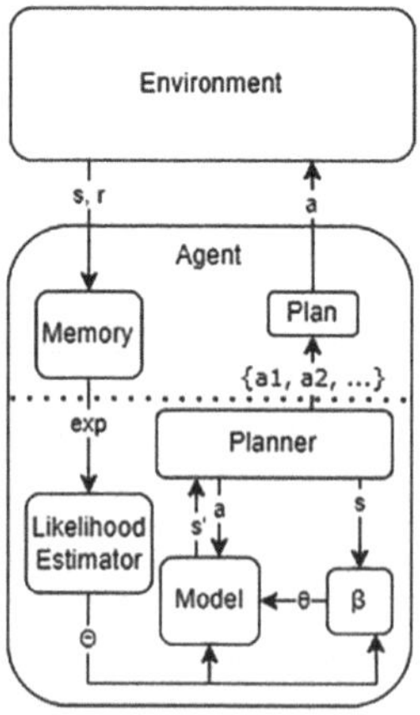

Fig. 1. Modules that compose a k-model's agent that interacts with the environment

Figure 1 shows a x-ray of the proposed framework that brings k parameters that are updated with likelihood estimation based on experienced transitions.

We can see above the dashed line the components that are activated every time step, like the memory that every interaction receives new data to be stored and the current executing plan that is driving the decisions the agent is taking every step. Below the dashed line are the components that are not activated every time step like the planner, that uses the models to generate a new plan, and the estimator that relies on the memory containing new observed transitions together with past experiences, to update the parameters of the models.

4.1 Likelihood Estimator

Drawing inspiration from the likelihood methodology used to find deep options [4], we devise an estimation approach for all parameters Θ within our framework. for the options it was utilized the likelihood in a gradient-based expectation-maximization algorithm to identify the optimal parameterized options from observed trajectories. Leveraging this insight, we develop a likelihood function tailored to our parameters Θ.

Our likelihood function captures the probability of observing the transitions in the environment given the current parameters Θ. By maximizing this likelihood function, we aim to iteratively update and refine our parameter estimates to better align with observed data. To achieve this, we employ a stochastic gradient ascent algorithm, which iteratively adjusts the parameters Θ based on online experience. By maximizing the log likelihood of the observed data, we can effectively update our parameter estimates in a principled manner, leveraging the information gleaned from interactions with the environment.

This approach allows us to adaptively learn and refine the parameters Θ over time, thereby improving the accuracy and effectiveness of our multi model-based reinforcement learning system. By continuously updating our parameter estimates based on observed data, we can enhance the performance and efficiency of our algorithm in learning the dynamics of segmented environments and generating optimal plans for achieving desired goals.

The likelihood function for a trajectory $\xi = s_1, a_1, s_2, a_2, ..., s_T$ in a standard MDP with a model parameterized by Θ is calculated using the transition probabilities as follows:

$$\mathcal{L}[\xi|\Theta] = \sum_{t=1}^{T-1} \log \mathcal{P}_\Theta(s_{t+1}|s_t, a_t) \tag{1}$$

To adapt this likelihood to the context of k models, we incorporate the β function and account for all $\mathcal{P}_\theta$ as follows:

$$\mathcal{P}_\Theta(s_{t+1}|s_t, a_t) = \sum_{\theta \in \Theta} \mathcal{P}_\theta(s_{t+1}|s_t, a_t)\beta_\theta(s_t) \tag{2}$$

We can use this equation in the traditional likelihood equation. By iteratively applying stochastic gradient descent on the negative log likelihood, we aim to optimize the parameters θ:

$$\Theta = \Theta + \eta \nabla \mathcal{L}[\xi|\Theta] - \alpha \Delta\Theta \tag{3}$$

where η represents the learning rate and α is the decay factor for momentum. This iterative process allows us to update the parameters Θ in a principled manner, leveraging the observed trajectory ξ to refine our model and improve its accuracy in capturing the dynamics of the environment.

4.2 Planner

With the established models we have chosen a planning approach for obtaining value of the representation of the world, inspired by the implementation RRT based algorithms with RL [9] for optimizing policies to find a path to the goal state, we decides to implement our modified version of the RRT algorithm that makes use of a model to come with a plan for reaching the goal. The implementation builds a tree of state transitions based on the execution of actions on a model passed as input, so the model can generate synthetic transitions that are incorporated to the tree plan, when completed the tree starts from a start state, and have a depth of a certain max number of steps that we wish the plan to have. The Algorithm 1 is responsible for building this plan tree.

Algorithm 1. RRT Based Planner

procedure PLAN(*tree*, *model*, *s*, *max*)
    if $max > 0$ **then**
        $tree.extend(null, s)$
    for all $a \in A$ **do**
        $s' \sim model(s, a)$
        $tree.extend(a, s')$
        PLAN(*tree*, *model*, s', $max - 1$)

Once the plan tree is constructed, we can extract a path to the goal state or to the state within the tree that is closest to the goal. To accomplish this, we only require a function to calculate the relative distance between states. Algorithm 2 provides a systematic method for extracting a path from the input tree to either the goal state or the closest state found using the $tree.NEAREST$ function.

The $tree.NEAREST$ function plays a crucial role in this process by identifying the closest state to the goal within the plan tree. In this case it is used the euclidean distance to compare the states distances. Once this closest state is determined, the algorithm proceeds to construct a path from the initial state to either the goal state or the identified closest state. By leveraging Algorithm 2, the agent can effectively obtain a plan of actions to take.

The quality of plans generated by the algorithm is directly influenced by the accuracy and fidelity of the underlying models used to construct them. Initially, when the models are far from reflecting the real transitions in the environment,

Algorithm 2. Get Path

function PATH(*tree*, *goal*)
    node $\leftarrow$*tree*.NEAREST(*goal*)
    path $\leftarrow$ [*node*]
    while $\exists$*node.parent* **do**
        node $\leftarrow$ *node.parent*
        path.insert(*node*)
      return *path*

the generated plans may be misguided or sub optimal. However, as the agent gathers more experience and refines its models to better reflect the true dynamics of the environment, the quality of the generated plans improves significantly.

This highlights the iterative nature of model-based reinforcement learning, and works [14] have even shown that this initial misguidance of poor models could bring valuable exploration experiences in some specific problems.

5 Applied Environment

Our challenge lies in simultaneously learning the k parameters θ and the probability distribution given by β across states. To address this challenge, we assume that the function $\mathcal{P}$ is based in a Gaussian parameterized, with $\theta_i = \{\tau_i, \sigma_i\}$ representing the parameters that will be used in the transitions.

To create a simulation environment reflecting practical problems characterized by segmented dynamics, we developed a two-dimensional space where states are represented by x and y coordinates specifying $\mathcal{S} \in \mathbb{R}^2$. In this environment, agents have four possible actions corresponding to movement in the cardinal directions then giving us $\mathcal{A} \in \{\uparrow, \leftarrow, \downarrow, \rightarrow\} = \{(0,-1), (-1,0), (0,1), (1,0)\}$.

The uncertainty of the agents' movements, are formulated in transitions as linear transformations of the state with respect to the chosen action with a random factor drawn from a Gaussian distribution. Notably, the parameters of this distribution vary across distinct regions of the state space. So if we have k regions resulting state s' after executing action a in state s would be $s' \sim \mathcal{N}(s + a\tau_i, \sigma_i)$ with τ_i and σ_i being parameters θ_i of region M_i where $M_i \subset \mathcal{S}; s \in M_i$ and $s' \in \mathcal{S}$; $a \in \mathcal{A}$;

This give us a transition function that is derived from the probability density function dependent of the region i:

$$\mathcal{P}_\theta(s'|s,a) = \frac{1}{\sigma\sqrt{2\pi}} e^{-\frac{1}{2}\left(\frac{s'-(s+a\tau)}{\sigma}\right)^2} \tag{4}$$

The probability defined by Eq. 4 will be used to determine the likelihood function,as elaborated in Sect. 4.

An illustration of such an environment can be observed in Fig. 2, providing a visual representation of the two-dimensional space with segmented dynamics. This figure effectively encapsulates the concept of distinct regions within the state space, each characterized by unique parameters of the Gaussian distribution governing uncertainty in transitions.

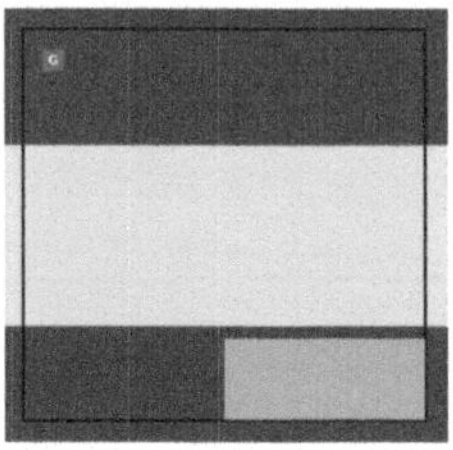

Fig. 2. A representation of a instance of environment,with the goal states represented by the square with G, and a set of initial states represented by the white rectangle, and the two regions of different dynamics represented with the colors yellow and blue (Color figure online)

By implementing this approach, we effectively model environments where dynamics vary between different regions. Adjusting the parameters of the Gaussian distribution enables us to replicate diverse levels of uncertainty and variability within each region. Consequently, agents must adapt their strategies accordingly as they navigate through the environment.

This simulation framework offers a versatile platform for investigating and evaluating model-based reinforcement learning algorithms in scenarios where the state space exhibits heterogeneous dynamics. Through experimentation and analysis within this environment, researchers gain valuable insights into the effectiveness and adaptability of different learning approaches in complex real-world settings.

5.1 Functions Aproximators

In our approach, we leverage the likelihood estimators to optimize both the parameters of the β function and the k-models, which can also be represented by parameterized functions. For simplicity and flexibility, in our experiments, we employed a feedforward neural network with an softmax activation function in the output layer to represent the β functions. Additionally, we assumed a known distribution function for the k-models, specifically a Gaussian distribution.

The parameters of the neural network, along with the means and variances of the Gaussian distributions representing the k-models, collectively form the parameter set Θ that we aim to estimate using the likelihood-based optimization method described previously. Replacing the structural assumptions for the k-model's functions with a universal function approximation, similar to what we've done for the β function, is a natural step towards generalizing this solution for more practical problems, and for that, no changes in the likelihood estimator is necessary, since it's agnostic to the functions structure, optimizing the parameters directly.

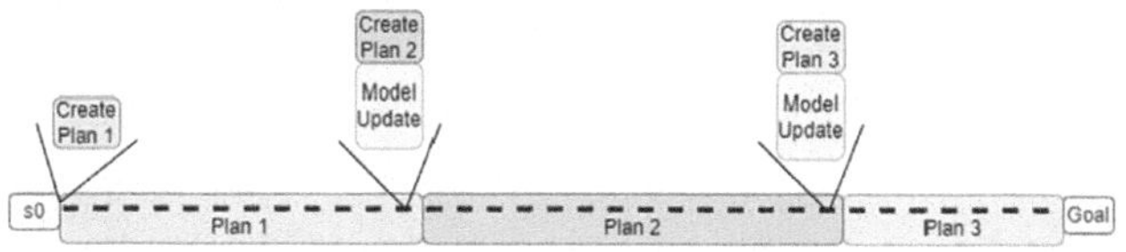

Fig. 3. A representation of an episode, showing the steps where the model is updated and a new plan is generated

The dynamics for updating the models based on new experience and the generation of a new plan base on the new model can be visualized on Fig. 3 that show how an episode starts on s_0 and a plan is generated to start guiding the agent's initial actions, and after a certain number of steps, the model can be updated based on the new gained information, which enable the agent to come with a new plan, and this goes on, until the agent reaches the goal.

There are two aspects that can be adjusted in our approach that can change the impact of the learning of the models. The first one is the memory size, since we use real past experience to update the parameters via likelihood, we have decided to specify the size of the memory, in other words, how many experiences the agent will remember, and use to recalculate the likelihood. As the size increases, the better the updates reflects the reality, but given a infinity horizon problem, this could became a problem.

The second relevant aspect is the frequency in which the models are updated. We control that frequency by specifying the number of steps that the agent will be forced to exploit the models after each update. So every fixed number of steps, the experience collected are added to the memory and the parameters can be recalculated afterwards with the extended trajectory.

With all that put together we have a solution that start roaming for some steps and then uses its short experience to fit a model, and then uses the model to plan the next actions generating new experiences. Following this loop for some time, eventually the models are able to get precise and the plans actually leads to the goal.

The Fig. 4 show us two examples of tree plan generated with the models learned trough a episode, on the left is the tree generated in the beginning and in the right at the end of the episode. We can see how in the end of the episode, the quality of the models were improved resulting in a more consistent plan.

6 Results

To compare the advantages of using a segmented representation of the world against the standard approach of a single unified model, we designed a series of experiments aimed at highlighting the differences in performance.

First we design the experiment in which we conducted ten runs of two episodes with at most three hundred steps each, where both models interacted with the same environment. By comparing their performances in this controlled

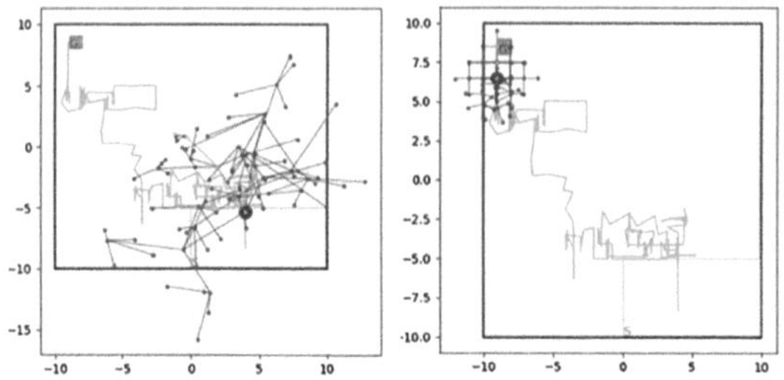

Fig. 4. In blue is the three step tree plan generated with the models learned trough a episode, on the left is the tree generated in the beginning and in the right at the end of the episode. The red line shows the entire path taken in the episode. (Color figure online)

setting, we expect to have both solution being very similar in the first episode. We limited the horizon to three hundred steps.

So with this experiment setup we develop two scenarios, the first one just as we described but the second one we change the environment between first and second episode. As the agent maintain the knowledge of the previous episode, we expected to see a better adaption by the k-model solution in a different environment that brings familiar dynamics.

So, for the first experiment, we run the process into two episodes. In this setup, each solution retained knowledge from previous episodes, allowing us to observe how their performance evolved over time. We expected both models to improve across episodes, leveraging their accumulated experience to make increasingly informed decisions.

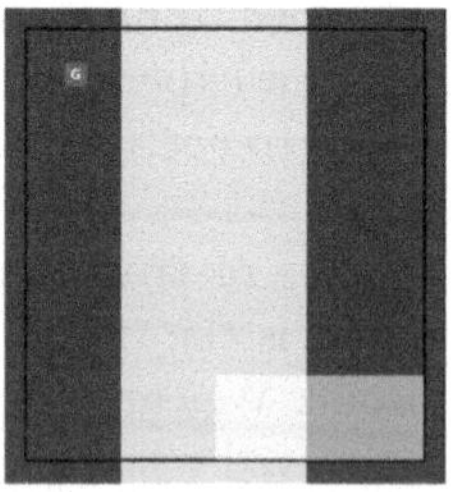

Fig. 5. The different environment presented in scenario 2 where the environment changes between episodes.

In the second experiment, we introduced a novel twist by presenting the agent with a different environment in the second episode. While the dynamics of this environment remained very similar to the one previously encountered, the regions within it were arranged in different shapes and positions, as depicted in Fig. 5.

Despite this change, both models retained persistent knowledge from the previous episodes. However, we anticipated that the segmented model would exhibit superior adaptability in this scenario. This expectation stemmed from the segmented model's ability to identify previously learned models and adjust its routing network accordingly to accommodate the new configuration of regions.

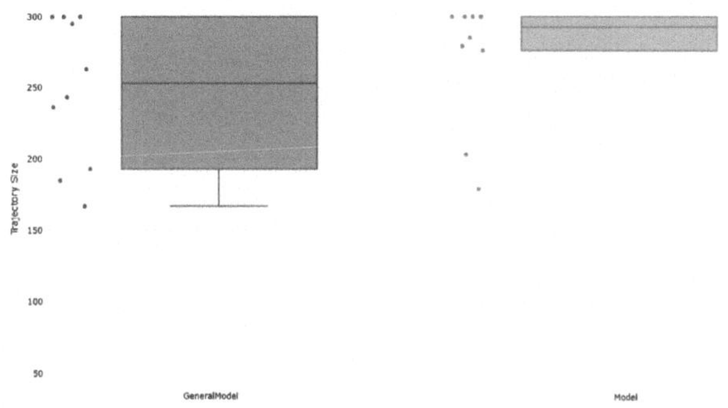

Fig. 6. Box plot showing distribution of steps that each model spent to get to the goal across first episode of the first scenario with blue being the standard single model approach and orange being the k-model segmented approach. (Color figure online)

The analysis presented in Fig. 6 reveals interesting insights into the performance of both approaches across episodes. Initially, in the first episode, both the single-model and the k-models exhibit comparable performances, as evidenced by the similar duration of the episodes with the k-model being slightly worse. This suggests that both models are capable of effectively learning to navigate through the environment in a similar fashion.

However, in the subsequent episode shown in Fig. 6 a notable disparity emerges between the two approaches. The k-models demonstrate a significant reduction in episode length in most of the runs, compared to the single model. This observation suggests that the k-models have a better reset power to retrieve the knowledge from the beginning of the previous episode, while the changes presented through the episode had more impact in the single model approach, resulting in a better performance compared to that observed in the first episode but not as good as the improvement observed by the k-models.

The improvement became even more evident when analysing the performances in the second episode for the scenario when the environment changes. The Fig. 7 show us the difference between the two approaches on this scenario. The k-models demonstrate a significant reduction in episode length in most of the runs, compared to the single model as it did in the first scenario, but this time we can't see the same improvement in the single model approach. This observation suggests that the single model struggles to transfer knowledge from the previous episode to the next when the environment changes, resulting in performance similar to that observed in the first episode, where it had to learn the environment from scratch. In contrast, the k-models leverage the dynamics learned in the first episode to adapt and improve their performance in the second episode. By dynamically adjusting their routing network, the k-models demonstrate a remarkable ability to rapidly adapt to the new environment, resulting in shorter episode duration.

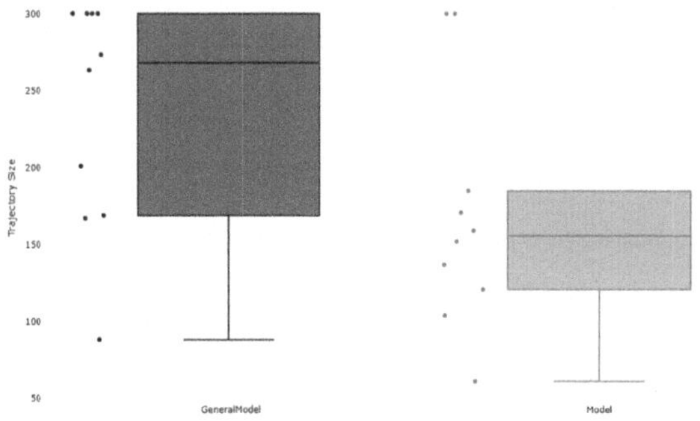

Fig. 7. Box plot showing distribution of steps that each model spent to get to the goal across the second episode of the second scenario with blue being the standard single model approach and orange being the k-model segmented approach (Color figure online)

Overall, these findings underscore the effectiveness of the segmented k-models approach in leveraging accumulated knowledge from previous episodes to facilitate faster adaptation and improved performance in subsequent episodes. This highlights the advantages of employing a segmented representation of the world's dynamics, particularly in scenarios where the environment exhibits dynamic changes between episodes.

7 Conclusion

This work aims to demonstrate the benefits of leveraging segmented dynamics in reinforcement learning problems. By recognizing that the environment's dynamics can be effectively segmented into distinct regions, we introduce a novel approach that enhances the efficiency of reinforcement learning solutions.

Through the utilization of a structured framework comprised of smaller models, the agent gains the ability to discern situations where it can switch between models. This segmentation allows for improved generalization across diverse contexts and facilitates faster adaptation to familiar dynamics within different environments.

By adopting this approach, our research highlights the importance of understanding and leveraging the underlying structure of the environment's dynamics. By incorporating segmented models into the learning process, we empower agents to navigate complex environments more effectively, ultimately leading to enhanced performance and adaptability in a variety of reinforcement learning tasks.

One limitation of our approach is the requirement to specify the number of models as input, which may be unknown in practical applications. This presents a challenge, as determining the optimal number of models can be non-trivial and may require prior knowledge or experimentation.

To address this limitation, future research could explore methods for automatically determining the appropriate number of models based on the characteristics of the environment or through data-driven approaches. Techniques such as model selection criteria, Bayesian optimization, or clustering algorithms could be employed to dynamically adjust the number of models based on observed data and performance metrics.

Additionally, the flexibility of our approach allows for the incorporation of adaptive mechanisms that can dynamically adjust the number of models based on feedback from the learning process. By incorporating such adaptive mechanisms, our approach can become more robust and applicable to a wider range of real-world scenarios where the number of models may vary or change over time.

Overall, addressing the limitation of requiring a predefined number of models is an important direction for future research in order to enhance the scalability and applicability of our approach in practical reinforcement learning applications.

Acknowledgments. This study was supported in part by the *Coordenação de Aperfeiçoamento de Pessoal de Nível Superior* (CAPES) – Finance Code 001, and by the Center for Artificial Intelligence (C4AI-USP), with support by FAPESP (grant #2019/07665-4) and by the IBM Corporation.

References

1. Alegre, L.N., Bazzan, A.L.C., da Silva, B.C.: Minimum-delay adaptation in non-stationary reinforcement learning via online high-confidence change-point detection. In: Proceedings of the 20th International Conference on Autonomous Agents and MultiAgent Systems, AAMAS '21, pp. 97–105. International Foundation for Autonomous Agents and Multiagent Systems, Richland (2021)
2. Chitnis, R., Silver, T., Tenenbaum, J., Kaelbling, L.P., Lozano-Pérez, T.: Glib: efficient exploration for relational model-based reinforcement learning via goal-literal babbling, vol. 13B, pp. 11782–11791 (2021). https://www.scopus.com/inward/record.uri?eid=2-s2.0-85100285730&partnerID=40&md5=4e2e2fca957818475950cdcd634527fe
3. Doya, K., Samejima, K., Katagiri, K.i., Kawato, M.: Multiple model-based reinforcement learning. Neural Comput. **14**(6), 1347–1369 (2002). https://doi.org/10.1162/089976602753712972
4. Fox, R., Krishnan, S., Stoica, I., Goldberg, K.: Multi-level discovery of deep options (2017)
5. Hallak, A., Castro, D., Mannor, S.: Contextual Markov decision processes (2015)
6. Kidambi, R., Rajeswaran, A., Netrapalli, P., Joachims, T.: Morel: Model-based offline reinforcement learning. In: NIPS 2020. Curran Associates Inc., Red Hook (2020)

7. Li, J., Tang, C., Tomizuka, M., Zhan, W.: Hierarchical planning through goal-conditioned offline reinforcement learning. IEEE Rob. Autom. Lett. **7**(4), 10216–10223 (2022). https://doi.org/10.1109/LRA.2022.3190100
8. Li, T., Lambert, N., Calandra, R., Meier, F., Rai, A.: Learning generalizable locomotion skills with hierarchical reinforcement learning. In: 2020 IEEE International Conference on Robotics and Automation (ICRA), pp. 413–419 (2020). https://doi.org/10.1109/ICRA40945.2020.9196642
9. Lin, D., Zhang, J.: A reinforcement learning based rrt algorithm with value estimation. In: 2022 3rd International Conference on Big Data, Artificial Intelligence and Internet of Things Engineering (ICBAIE), pp. 619–624 (2022). https://doi.org/10.1109/ICBAIE56435.2022.9985840
10. Ng, J.H.A., Petrick, R.P.A.: Incremental learning of planning actions in model-based reinforcement learning. In: Proceedings of the 28th International Joint Conference on Artificial Intelligence, IJCAI 2019, pp. 3195–3201. AAAI Press (2019)
11. Rybkin, O., Zhu, C., Nagabandi, A., Daniilidis, K., Mordatch, I., Levine, S.: Model-based reinforcement learning via latent-space collocation, vol. 139, pp. 9190–9201 (2021). https://www.scopus.com/inward/record.uri?eid=2-s2.0-85161314975&partnerID=40&md5=0f9769abe528c97cbf983f602ecff397
12. da Silva, B., Basso, E., Bazzan, A., Engel, P.: Dealing with non-stationary environments using context detection, vol. 2006, pp. 217–224 (2006). https://doi.org/10.1145/1143844.1143872
13. Sutton, R., Barto, A.: Reinforcement Learning: An Introduction. MIT Press, Cambridge (2018)
14. Young, K., Ramesh, A., Kirsch, L., Schmidhuber, J.: The benefits of model-based generalization in reinforcement learning. In: Proceedings of the 40th International Conference on Machine Learning. ICML 2023. JMLR.org (2023)

LLM-Driven Chest X-Ray Report Generation With a Modular, Reduced-Size Architecture

Talles Viana Vargas, Helio Pedrini(✉), and André Santanchè

Institute of Computing (IC), University of Campinas, Campinas, SP, Brazil
{helio,santanche}@ic.unicamp.br

Abstract. Large Language Models (LLMs) have been widely employed in various text processing tasks. In computer vision, these models have found application in generating captions and text from natural images, as well as in Visual Question Answering (VQA) systems. In the field of medical imaging, there are studies based on text generation proposing automated diagnoses of X-rays, magnetic resonance imaging scans, computed tomography scans, and other modalities. Few initiatives seek to apply and harness the potential of LLMs in medical text generation; they use models with tens of billions of parameters and are thus computationally expensive. This work addresses this gap by evaluating the use of frozen pre-trained models (CXAS U-Net and BioGPT) for chest X-ray report generation. We adapt the BLIP-2 modular architecture where only a cross-modal alignment module must be trained in order to generate text from images. We were able to achieve competitive scores over Clinical Efficacy (CE) metrics compared to some state-of-the-art (SOTA) methods, while obtaining lower scores for Natural Language Generation (NLG) metrics. Our findings suggest that NLG metrics may not serve as suitable proxies for evaluating models in the chest X-ray generation task.

Keywords: Radiology Report Generation · Language Models · Lightweight Training · Chest X-Rays

1 Introduction

Chest X-rays are a fundamental diagnostic tool in healthcare, containing valuable information that can assist healthcare professionals in diagnosing various pulmonary, cardiac or traumatic conditions [1]. They provide a snapshot of a patient's internal chest structure, revealing critical details that can guide medical practitioners in their decision-making processes.

However, interpreting and extracting the findings from chest X-rays is a complex, repetitive and often time-consuming task. The analysis of these images traditionally relies on the expertise of radiologists, who interpret the visual information contained within the X-rays and then transcribe their findings into textual

A. Paes and F. A. N. Verri (Eds.): BRACIS 2024, LNAI 15413, pp. 199–211, 2025.
https://doi.org/10.1007/978-3-031-79032-4_14

reports. This manual process is not only labor-intensive but is also prone to variations in interpretation and reporting. It is in this context that innovative solutions are needed to enhance the efficiency, accuracy, and accessibility of chest X-ray analysis [12].

In recent years, the advancements in artificial intelligence (AI) and natural language processing (NLP) are revolutionizing the way humans interact with the machines [20,31]. Some Large Language Models (LLMs) applications, such as ChatGPT, are showing off their exceptional text generation capabilities, being able to output texts with impressive syntax and semantic quality. They also show a good performance while performing tasks they were not previously trained for, for example, one can prompt them to generate a clinical diagnostic, or write a code to solve a given problem.

Several works are harnessing the developments in AI and NLP to automate and streamline the process of transforming visual information from chest X-rays into informative textual descriptions, however, only a few have capitalized on the potential of LLMs. These existing methods predominantly rely on resource-intensive models that present challenges in accessibility for those with limited computational resources.

Therefore, to leverage the capabilities of LLMs while minimizing the resources required for training and deploying a chest X-ray report generation system, this study proposes adapting the BLIP-2 modular architecture [16] to utilize frozen in-domain pre-trained models for both vision encoding and text generation. We employ BioGPT [18] as our text generation model, which is based on GPT-2 with 354M parameters, offering significantly lower resource requirements compared to models used in other studies [6,30,36] and widely-used LLMs currently used deep learning area [20,31].

By adopting this modular architecture, where the image encoder and text decoder weights are frozen, only a vision-text alignment module needs to be trained, reducing training complexity and computational costs compared to having to train a large LLM.

By harnessing the power of LLMs while seeking computational efficiency, this research contributes to the advancement of AI-powered medical image analysis. This advancement has the potential to reduce radiologists' workload by serving as a decision support, increasing throughput, minimizing diagnosis delays, and allowing them to focus on complex analyses where their expertise is most valuable [11].

2 Background

The literature review focuses on three areas: Image Captioning, Medical Captioning, and Language Models. In the first two areas, we make an comprehensive exploration focusing on different approaches, architectures, benchmarks and applicable metrics for our task. Lastly, we investigate some Language Models, aiming to identify their sizes and number of parameters.

2.1 Image Captioning

The image captioning task targets the automatic generation of description of natural images. In the literature, we find many methods designed to achieve this goal [16,17,23,33,37,39]. These works vary based on: their selection of vision encoder, language model, and their approach to cross-domain alignment, which involves aligning features between the image and text domains.

Vinyals et al. [33] and Rennie et al. [23] used CNN models to extract the features from images and LSTM network to regressively generate the desired text. The CNN features provide a simple and compact representation of an image, but they can hinder further fine-grained description due to this compression. The LSTM network, due to its sequential nature, can be very slow to output the text.

Zhou et al. [37] and Li et al. [17] leveraged the potential of the Transformer [32] along region image features extracted by Fast-RCNN [9], an widely used CNN for feature extraction. Zhou et al. [37], for example, used a shared multi-layer Transformer responsible for both vision encoding and text decoding steps, hence this unique module is responsible for aligning the image features and generating the caption.

Very recent works [16,39] take advantage of the few shot potential of LLMs [3]. Zhu et al. [39] introduce MiniGPT-4, which combines a Vision Transformer image encoder with an open-source LLM based on Llama [31], aligning both image and text domain using solely a single projection layer. The authors named it MiniGPT-4 due to its similar capability on description generation compared to GPT-4.

Built on pre-trained models, Li et al. [16] introduced the BLIP-2 framework, which requires training only an cross-modal alignment module called Q-Former to bridge the gap between image and text representations. This approach leverages pre-trained image encoders and large language models (LLMs) by freezing their weights (i.e., not training them).

The core component, Q-Former, is a BERT-based module responsible for aligning the features extracted from images and text. By training only the Q-Former, BLIP-2 reduces training complexity and computational costs.

2.2 Medical Captioning

Medical captioning involves the automatic generation of descriptive and informative captions for medical images, such as X-rays, MRI scans, CT scans, and more. The goal is to provide accurate textual descriptions that convey relevant medical information, possibly helping medical professionals make faster and more accurate diagnoses, facilitating research and training, and ultimately improving patient care. Considerable progress has been made in this field, with numerous proposed architectures and approaches [4–6,26,30,35,36].

Some works, particularly those focused on chest X-ray captioning, employ CNNs as image encoders and Transformers as text decoders [4,5]. Chen et al.

[4] introduced the use of a novel relational memory module alongside the Transformer to enhance the caption generation, while Chen et al. [5] focused on the alignment of image and text domain, proposing the use of a cross-modal memory network to facilitate the interactions across modalities (image and text). On the other hand, a recent study [35] adopts a vision transformer as its image encoder and introduces Expert Tokens, which are designed to interact with the extracted images patches and with one another. As a result, each token can focus on a different part of the image.

Recently, other papers have started to explore the potential of LLMs for medical caption generation [6,30,36]. Yang et al. [36] adopted the BLIP-2 architecture, although they do not follow the original training process, instead, they directly use a large vision encoder and a LLM decoder, and fine-tune the Q-Former and the LLM to the ImageClef task, which involves caption prediction for general X-ray images. They compared their performance solely based on the task competition ranking.

Thawkar et al. [30] used the Vision Transformer from MedCLIP [34] as the vision encoder, applied a simple linear transformation layer to align features, and employed an LLM based on LLaMA [31] as the text decoder. This study does not compare their approach with others. Danu et al. [6] took a different approach by first multi-classifying diseases on chest X-ray images with bounding boxes, and then using these abnormality features as input for the LLM.

Med-PaLM [26,27] is a large-scale generalist biomedical AI system, capable of interpreting various biomedical data modalities, including tasks like chest X-ray report generation and medical visual question answering. It employs a ViT-based vision encoder and PaLM as the LLM decoder.

Building on prior work, Nicolson et al. [19] investigated the impact of combining different pre-trained models for chest X-ray report generation. They experimented with various combinations of vision encoders and text decoders, both from the medical domain (in-domain) and general domains. Interestingly, their findings revealed that a combination of an out-of-domain vision encoder, CvT-21, and a general-domain text decoder, DistilGPT-2, achieved superior performance on chest X-ray report generation benchmarks after fine-tuning. This finding suggests that leveraging out-of-domain models, when carefully chosen and fine-tuned, can lead to state-of-the-art (SOTA) results.

2.3 Language Models

In recent years, significant advancements have been made in the development of language models. A significant milestone was the introduction of BERT [8] (Bidirectional Encoder Representations from Transformers), which has 110 million parameters in its base version and 340 million in its large version. BERT's bidirectional training approach on masked language modeling tasks enabled it to achieve state-of-the-art results on a variety of NLP benchmarks.

The transformer-based architectures have been further scaled to develop models with even larger parameter sizes. For instance, OpenAI's GPT-2 [21], with 1.5 billion parameters on its larger model, demonstrated the potential of unsupervised language models to generate coherent and contextually relevant text. GPT-2's architecture paved the way for its successor, GPT-3 [3], which boasts an unprecedented 175 billion parameters, showcasing remarkable capabilities in few-shot learning and diverse NLP tasks without requiring task-specific fine-tuning.

Despite their impressive performance, these large-scale models necessitate substantial computational resources for training and inference, limiting their accessibility. This challenge has driven the development of more resource-efficient models. DistilBERT [24], with 66 million parameters, and ALBERT [14], with 12 million parameters for its base version, offer more compact alternatives that maintain competitive performance levels.

In the biomedical domain, specialized language models have been developed to address the unique challenges posed by medical texts. BioBERT [15], based on BERT with approximately 110 million parameters, and ClinicalBERT [2], also based on BERT, have been fine-tuned on large corpora of biomedical literature and clinical notes, resulting in improved performance on domain-specific tasks.

Following the LLMs line, BioGPT [18], a model derived from GPT-2 with 354 million parameters, specifically tailored for biomedical text generation, provides a balanced approach, offering the extensive capabilities of larger models while maintaining lower resource requirements. This efficiency makes BioGPT a practical choice for potentially generating high-quality biomedical text, ensuring accessibility for a broader range of researchers.

Wrapping up, we observe that many works in medical captioning, specifically chest X-ray captioning, are focused on enhancing performance metrics. Furthermore, those concentrating on harnessing the power of LLMs do not appear to prioritize the search for reduced-size architectures that would enable simple computers with limited computational resources to run these models.

3 Materials

This work adopts the MIMIC-CXR dataset [13], a widely-used dataset for generating captions for chest X-rays. It is currently the largest public[1] chest x-ray dataset, containing 377,100 radiograms and 227,835 free-text reports. The dataset has an official data split allowing fair comparison of our models against literature methods and different techniques.

Most related works also utilize the IU Xray dataset [7]. However, in this study, we have chosen to use only the MIMIC-CXR due to its larger size. In the future, we plan to include additional datasets to assess generalization.

Since MIMIC-CXR includes sample reports with multiple views, such as frontal and lateral X-rays, we employed the following strategy for inference:

- For reports with two images: We randomly selected two images for analysis.
- For reports with only one image: We duplicated the single image to ensure the pipeline could process it consistently.

[1] MIMIC-CXR webpage: https://physionet.org/.

Following this selection process, all images were resized to a standard dimension of 512×512 pixels. Additionally, normalization was applied using the mean and variance values specific to the chosen vision encoder.

4 Method

This section presents the proposed model architecture, the training stages, and the metrics used during the evaluation process.

Model Architecture. This works leverages the architecture and training pipeline presented by BLIP-2 [16], a captioning model for natural images that utilizes frozen pre-trained models.

The key advantage of this architecture is its modularity. This allows for the integration of different vision encoders and text decoders, with only the alignment module (Q-Former) requiring specific training. The source code was based and adapted from BLIP-2 implementation on LAVIS repository[2].

Image Encoder Stage. The architecture, as illustrated in Fig. 1, starts with the image encoder, responsible for extracting pertinent features from chest X-ray images.

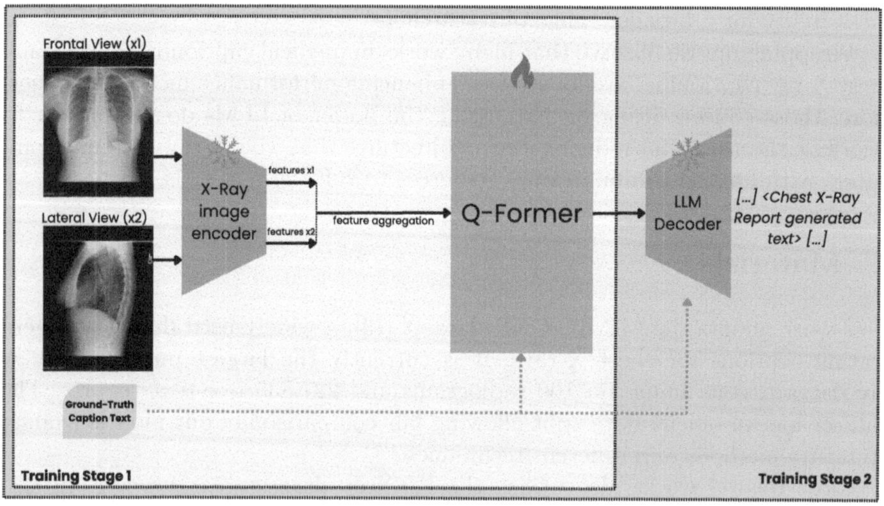

Fig. 1. Proposed architecture for chest X-ray report generation.

This work adopted a U-Net [25] formerly pre-trained on a chest x-ray segmentation dataset, therefore we can leverage its prior in-domain knowledge. This choice offers a significant advantage, as U-Net has nearly 15 times fewer parameters compared to the Vision Transformer (ViT-g/14) employed in the original

[2] LAVIS repository: https://github.com/salesforce/LAVIS.

BLIP-2 work. At this point, we utilized the feature map output of the first upsampling layer in the network, flattened the height and width dimensions, and then switched channels dimension with this flattened spatial dimension. This allows having features that are ready to be processed by the next neural network.

Regarding the image feature extraction, the study utilizes two images for every sample as aforementioned, hence we followed other studies [4,35], sequentially extracting features from both images using the selected image encoder, and averaging those features.

Q-Former Alignment Module. The cornerstone of the architecture, The Q-Former is responsible for aligning extracted images features with the text decoder input.

As the Q-Former is a BERT-based model, we leveraged available BioClinical-BERT [2] pre-trained weights, which was trained on PubMed abstracts and also on MIMIC III notes. Initializing language models with in-domain pre-trained weights, instead of using the standard BERT initialization, potentially boosts the capacity of feature alignment of the model; behavior seen in other medical tasks [15]. We used the same configuration from the base source code.

Text Decoder Module. The text decoder is the module responsible for generating the chest X-ray descriptions given the features aligned by the Q-Former.

Similar to BLIP-2 architecture, we adopted a Large Language Model as our text decoder, due to its text generation capability. Generally, an LLM receives text prompts on its input which work as an start point, from where the model generates the text output. However, in this work, as the LLM inputs come from the Q-Former aligned features, these inputs can be considered as soft prompts, i.e., they are not explicit text embeddings, but a feature vector that must help and guide the LLM decoder on its text generation task.

This work utilizes BioGPT [18], an LLM based on GPT-2 [21] architecture, since it was specifically pre-trained on biomedical domain using PubMed abstracts. It is a lightweight model compared to the LLM from the original BLIP-2, containing 345 million parameters instead of 2.7 billion.

4.1 Training Stages

The training process is divided into two distinct stages, each serving a specific purpose in enhancing the capabilities of a portion of our architecture, as illustrated in Fig. 1.

Representation Learning. In this initial training stage, the Q-Former is trained in conjunction with a frozen pre-trained image encoder. The primary objective is to enable the Q-Former to serve as a bridge between image and text domains. This step prepares the module to align, understand, and also, generate meaningful captions, enhancing the performance of the architecture as a whole. During this phase, the Q-Former outputs are optimized using three distinct methods (cost functions), as outlined in the BLIP-2 work [16].

We adopted AdamW optimizer with weight decay of 0.05. The learning rate followed a cosine decay schedule during 10 epochs, starting at 1e-4 and ending at 1e-5. A warm-up period of one epoch was employed with learning rate set to 1e-6. Also, the best model checkpoint was selected based on the lowest loss overall.

Generative Learning. In the second training stage, our whole architecture is assembled. The previously trained Q-Former module is integrated with the frozen pre-trained BioGPT. Subsequently, the Q-Former is fine-tuned by optimizing the outputs generated by the LLM decoder. During this fine-tuning process, the Q-Former acts as a guide, influencing BioGPT's text generation based on the input chest X-ray images.

We adopted AdamW optimizer with weight decay of 0.05. The learning rate followed a cosine decay schedule during 10 epochs, starting at 1e-5 and ending at 1e-6. A warm-up period of two epochs was employed with learning rate set to 1e-8. At this point, we observed better results by unfreezing the BioGPT and freezing all the other modules. Same as before, the final model checkpoint was selected based on the lowest loss value. All training stages were performed on AWS Cloud using a *ml.g5.xlarge* instance containing a 24Gb GPU and 16Gb memory.

4.2 Metrics and Evaluation

Following Chen et al. [4,35,38], our evaluation aimed to assess both the quality of the generated text descriptions and their potential clinical usefulness.

We adopted two approaches, leveraging established Natural Language Generation (NLG) metrics (BLEU, METEOR and ROUGE-L) and additionally, we incorporated clinical efficacy (CE), where we employed the CheXbert system [28] to automatically label the generated reports within 14 categories.

By comparing these labels with the ground truth diagnoses, we calculated precision, recall, and F1-score, providing insights into the model's ability to generate clinically accurate reports. This work presents both micro and macro averages of precision, recall and F1-score [29].

5 Experimental Results

This section presents the findings of our evaluation on the effectiveness of the proposed method for generating clinical text reports. We assessed performance using a combination of Natural Language Generation (NLG) metrics and clinical efficacy (CE) metrics over the MIMIC-CXR dataset.

Our architecture consists of approximately 450 million parameters. In comparison with other techniques that also adopt LLM in the decoder stage, XRayGPT [30] has around 7 billion parameters solely in its large language model (LLM). Similarly, Yang et al. [36] employs the ChatGLM-6B LLM model, which comprises approximately 6 billion parameters. Med-PaLM [26], on the other hand, incorporates 540 billion parameters in its PaLM LLM. Unfortunately, these studies do not evaluate their models using commonly used datasets, thus our comparison is limited to the number of parameters.

In order to quantitatively compare our approach with other studies in the area, we evaluated the clinical efficacy of the reports using CheXbert labels. Precision, recall, and F1-score were calculated for each CheXbert finding. Table 1 summarizes the overall CE performance compared to existing methods.

Our method achieved the highest precision (0.504) among the evaluated methods, indicating a low rate of false positive findings. However, the recall score (0.346) suggests it may overlook some relevant findings compared to CvT21-2DistilGPT2 [19].

Table 1. Comparison of the performance of different methods using precision, recall, and F1 metrics.

Method	Precision	Recall	F1
R2Gen [4]	0.333	0.273	0.276
CMN [5]	0.334	0.275	0.278
METransformer [35]	0.364	0.309	0.311
COMG [10]	0.424	0.291	0.345
CvT21-2DistilGPT2 [19]	0.398	**0.497**	**0.442**
Ours	**0.504**	0.346	0.410

Table 2 presents a comparison of NLG metrics between other techniques and our model. Our model has low scores if compared to other techniques. This could be explained due to the fact the those NLG metrics were developed to mainly evaluate machine translation, which is clearly not the task here. Since our core model is a pre-trained GPT-2 model, our predictions may not align as closely with the ground truth. Consequently, this misalignment could account for the observed low scores in our proposed model's metrics.

Table 2. Comparison of the NLG metrics between our model and other techniques.

Method	BLEU-1	BLEU-2	BLEU-3	BLEU-4	METEOR	ROUGE-L
R2Gen [4]	0.353	0.218	0.145	0.103	0.142	0.277
CMN [5]	0.353	0.218	0.148	0.106	0.142	0.278
METransformer [35]	0.386	0.250	0.169	0.124	0.152	0.291
COMG [10]	0.346	0.216	0.145	0.104	0.137	0.279
CvT21-2DistilGPT2 [19]	0.462	0.295	0.214	0.165	0.192	0.370
Ours	0.284	0.165	0.106	0.074	0.120	0.201

A further analysis revealed an interesting nuance between clinical efficacy (CE) metrics and Natural Language Generation (NLG) metrics in the context of chest X-ray report generation. Specifically, within a subset of high-performing predictions based on CE metrics—where predictions effectively conveyed observations from the X-rays—the corresponding NLG metrics displayed lower scores (see Table 3). This observation suggests that NLG metrics alone may not be the most reliable indicator of model quality for this task.

Table 3. NLG metric scores (e.g., BLEU, ROUGE-L) for a subset of model predictions that achieved the highest precision, recall, and F1-score in the clinical efficacy (CE) evaluation. Low NLG scores on these high-performing predictions suggest that NLG metrics alone may not be the most reliable indicator of model quality for chest X-ray report generation.

BLEU-1	BLEU-2	BLEU-3	BLEU-4	METEOR	ROUGE-L
0.275	0.171	0.117	0.086	0.138	0.222

Even predictions which are clinically accurate by CE metrics might not receive high NLG scores. This highlights the limitations of NLG metrics in capturing the full spectrum of factors that contribute to good clinical reports, such as factual accuracy, nuanced phrasing, and adherence to reporting conventions.

Table 4. Clinical Efficacy (CE) metrics for each observation within the MIMIC-CXR test set. The Positive Cases and Negative Cases columns indicate the distribution of positive and negative labels in the ground truth data, allowing for verification of class imbalance.

Observation	Positive Cases	Negative Cases	Precision	Recall	F1
Support Devices	1126	1642	0.688	0.678	0.683
Pleural Effusion	986	1782	0.695	0.445	0.543
Cardiomegaly	1018	1750	0.595	0.498	0.542
Atelectasis	750	2018	0.385	0.288	0.330
Lung Opacity	963	1805	0.469	0.152	0.229
No Finding	193	2575	0.135	0.534	0.216
Edema	564	2204	0.490	0.124	0.198
Enlarged Cardiomediastinum	200	2568	0.170	0.040	0.065
Pneumonia	155	2613	0.154	0.039	0.062
Consolidation	150	2618	0.087	0.013	0.023
Lung Lesion	151	2617	0.000	0.000	0.000
Pneumothorax	74	2694	0.000	0.000	0.000
Fracture	117	2651	0.000	0.000	0.000
Pleural Other	92	2676	0.000	0.000	0.000
Macro-Average	–	–	0.276	0.201	0.206
Micro-Average	–	–	0.504	0.346	0.410

Finally, as shown in Table 4, the model performs well on identifying findings such as *Support Devices* (0.683 F1) and *Cardiomegaly* (0.521 F1). However, it struggles with less prevalent findings like *Atelectasis* (0.330 F1) and *Lung Opacity* (0.229 F1). This may be due to limitations in the training data or the model's capacity to capture subtle variations in these specific findings.

These results demonstrate the potential of our proposed method for generating clinically relevant and accurate text reports. The high precision ensures a low false positive rate, while further investigation is needed to improve recall, particularly for less frequent findings.

6 Conclusions

In conclusion, our work presented an approach for generating clinical text reports from chest X-ray images. The selection of relative small-sized models, and the modular characteristic of BLIP-2 has made the training process less-hardware intensive, and we hope it foster future investigations of this type of architecture.

Our method achieved the highest precision among the evaluated techniques. However, our evaluations also reveals challenges to achieve a high recall, particularly for less prevalent observations. Hence, the model needs further investigations to improve model's capacity.

Furthermore, our analysis of the NLG metrics highlights the limitations of using machine translation metrics for our specific task, which do not capture the actual quality of the report. Future efforts should focus on developing custom evaluation methods that align with our specific goals.

Overall, our results demonstrate the potential of using in-domain, pre-trained, reduced-size models to generate clinically relevant and accurate text reports. However, ongoing research and development efforts are still needed to improve model's recall, to refine to overall's capability of the model. With this improvements, this approach could become a valuable tool, contributing to more efficient and accurate diagnosis.

As future directions, we aim to assess this framework with other datasets, such as BRAX [22] and IU XRay [7], to validate model's generalization.

References

1. Al-Dasuqi, K., Johnson, M.H., Cavallo, J.J.: Use of artificial intelligence in emergency radiology: an overview of current applications, challenges, and opportunities. Clin. Imaging **89**, 61–67 (2022)
2. Alsentzer, E., Murphy, J.R., Boag, W., Weng, W.-H., Jin, D., Naumann, T., McDermott, M.: Publicly available clinical BERT embeddings. arXiv preprint arXiv:1904.03323 (2019)
3. Brown, T., et al.: Language models are few-shot learners. Adv. Neural. Inf. Process. Syst. **33**, 1877–1901 (2020)
4. Chen, Z., Song, Y., Chang, T.-H., Wan, X.: Generating radiology reports via memory-driven transformer. arXiv preprint arXiv:2010.16056, 2020

5. Chen, Z., Shen, Y., Song, Y., Wan, X.: Cross-modal memory networks for radiology report generation. arXiv preprint arXiv:2204.13258, 2022
6. Danu, M.D., et al.: Generation of Radiology Findings in Chest X-Ray by Leveraging Collaborative Knowledge. arXiv preprint arXiv:2306.10448 (2023)
7. Demner-Fushman, D., et al.: Preparing a collection of radiology examinations for distribution and retrieval. J. Am. Med. Inf. Assoc. **23**(2), 304–310 (2016)
8. Devlin, J., Chang, M.-W., Lee, K., Toutanova, K.: BERT: pre-training of deep bidirectional transformers for language understanding. arXiv preprint arXiv:1810.04805 (2018)
9. Girshick, R.: Fast R-CNN. In: IEEE International Conference on Computer Vision, pp. 1440–1448 (2015)
10. Gu, T., Liu, D., Li, Z., Cai, W.: Complex organ mask guided radiology report generation. In: IEEE/CVF Winter Conference on Applications of Computer Vision, pp. 7995–8004 (2024)
11. Irmici, G., et al.: Chest x-ray in emergency radiology: what artificial intelligence applications are available? Diagnostics **13**(2), 216 (2023)
12. Jalal, S., Parker, W., Ferguson, D., Nicolaou, S.: Exploring the role of artificial intelligence in an emergency and trauma radiology department. Can. Assoc. Radiol. J. **72**(1), 167–174 (2021)
13. Johnson, A.E., et al.: MIMIC-CXR-JPG, a large publicly available database of labeled chest radiographs. arXiv preprint arXiv:1901.07042 (2019)
14. Lan, Z., et al.: Albert: A lite bert for self-supervised learning of language representations. *arXiv preprint*arXiv:1909.11942 (2019)
15. Lee, J., et al.: BioBERT: a pre-trained biomedical language representation model for biomedical text mining. Bioinformatics **36**(4), 1234–1240 (2020)
16. Li, J., Li, D., Savarese, S., Hoi, S.: BLIP-2: bootstrapping language-image pre-training with frozen image encoders and large language models. *arXiv preprint*arXiv:2301.12597 (2023)
17. Li, X., et al.: Oscar: object-semantics aligned pre-training for vision-language tasks. In: Vedaldi, A., Bischof, H., Brox, T., Frahm, J.-M. (eds.) ECCV 2020. LNCS, vol. 12375, pp. 121–137. Springer, Cham (2020). https://doi.org/10.1007/978-3-030-58577-8_8
18. Luo, R., et al.: BioGPT: generative pre-trained transformer for biomedical text generation and mining. Brief. Bioinform. **23**(6), 409–421 (2022)
19. Nicolson, A., Dowling, J., Koopman, B.: Improving chest X-ray report generation by leveraging warm starting. Artif. Intell. Med. **144**, 102633 (2023)
20. OpenAI. GPT-4 Technical Report (2023)
21. Radford, A., Wu, J., Child, R., Luan, D., Amodei, D., Sutskever, I., et al.: Language models are unsupervised multitask learners. OpenAI Blog **1**(8), 9 (2019)
22. Reis, E.P., et al.: Brax, brazilian labeled chest x-ray dataset. Sci. Data **9**(1), 487 (2022)
23. Rennie, S.J., Marcheret, E., Mroueh, Y., Ross, J., Goel, V.: Self-critical sequence training for image captioning. In: IEEE Conference on Computer Vision and Pattern Recognition (2017)
24. Sanh, V., Debut, L., Chaumond, J., Wolf, T.: DistilBERT, a distilled version of BERT: smaller, faster, cheaper and lighter. arXiv preprint arXiv:1910.01108 (2019)
25. Seibold, C., et al.: Accurate fine-grained segmentation of human anatomy in radiographs via volumetric pseudo-labeling. arXiv preprint arXiv:2306.03934 (2023)
26. Singhal, K., et al.: Large language models encode clinical knowledge. *arXiv preprint*arXiv:2212.13138 (2022)

27. Singhal, K., et al.: Towards expert-level medical question answering with large language models. arXiv preprint arXiv:2305.09617 (2023)
28. Smit, A., et al.: Chexbert: combining automatic labelers and expert annotations for accurate radiology report labeling using bert. arXiv preprint arXiv:2004.09167 (2020)
29. Sorower, M.S.: A literature survey on algorithms for multi-label learning. Oregon State University, Corvallis **18**(1), 25 (2010)
30. Thawkar, O., et al.: XRayGPT: chest radiographs summarization using medical vision-language models. arXiv preprint arXiv:2306.07971 (2023)
31. Touvron, H., et al.: Llama 2: open foundation and fine-tuned chat models. arXiv preprint arXiv:2307.09288 (2023)
32. Vaswani, A., et al.: Attention is all you need. Adv. Neural Inf. Process. Syst. **30** (2017)
33. Vinyals, O., Toshev, A., Bengio, S., Erhan, D.: Show and tell: a neural image caption generator. In: IEEE Conference on Computer Vision and Pattern Recognition, pp. 3156–3164 (2015)
34. Wang, Z., Wu, Z., Agarwal, D., Sun, J.: MedCLIP: contrastive learning from unpaired medical images and text. arXiv preprint arXiv:2210.10163 (2022)
35. Wang, Z., Liu, L., Wang, L., Zhou, L.: METransformer: radiology report generation by transformer with multiple learnable expert tokens. In: IEEE/CVF Conference on Computer Vision and Pattern Recognition, pp. 11558–11567 (2023)
36. Yang, B., Raza, A., Zou, Y., Zhang, T.: Customizing general-purpose foundation models for medical report generation. arXiv preprint arXiv:2306.05642 (2023)
37. Zhou, L., Palangi, H., Zhang, L., Hu, H., Corso, J., Gao, J.: Unified vision-language pre-training for image captioning and VQA. In: AAAI Conference on Artificial Intelligence, vol. 34, pp. 13041–13049 (2020)
38. Zhou, Z., Shi, M., Wei, M., Alabi, O., Yue, Z., Vercauteren, T.: Large model driven radiology report generation with clinical quality reinforcement learning. arXiv preprint arXiv:2403.06728 (2024)
39. Zhu, D., Chen, J., Shen, X., Li, X., Elhoseiny, M.: MiniGPT-4: enhancing vision-language understanding with advanced large language models. arXiv preprint arXiv:2304.10592 (2023)

Multilingual Extractive Summarization: Investigating State-of-the-Art Methods for English and Brazilian Portuguese

Germano Antonio Zani Jorge[1(✉)], Davi Alves Bezerra[1], Clarissa Castellã Xavier[2], and Thiago Alexandre Salgueiro Pardo[1]

[1] Núcleo Interinstitucional de Linguística Computacional (NILC), Instituto de Ciências Matemáticas e de Computação, Universidade de São Paulo, São Carlos, SP, Brazil
germano.jorge@alumni.usp.br
[2] SiDi, Campinas, SP, Brazil

Abstract. Automatic Text Summarization (ATS) is a Natural Language Processing (NLP) task essential for handling large volumes of information. ATS can be classified into two main types: extractive and abstractive. Extractive summarization selects sentences or phrases directly from the source text(s), while abstractive summarization generates new sentences that try to capture the original meaning of the source text(s). This paper describes our efforts to perform extractive single-document summarization in multilingual contexts. Although various summarization methods, such as PreSumm and HiStruct+, have shown promising results on English corpora like CNN/DM, there is a significant gap in applying these methods to other languages, especially Brazilian Portuguese. Additionally, these summarizers were evaluated with traditional metrics like ROUGE, which has limitations as it primarily measures superficial text overlap. To fill these gaps, we evaluate the effectiveness of these state-of-the-art methods on the CSTNews corpus (with news texts in Brazilian Portuguese) employing ROUGE and the recent BLANC metric, which measures how much the generated summary aids a pre-trained language model (like BERT) in understanding the document. Our contributions include the results and comparison of adapted models, the discussion of the BLANC metric in contrast to ROUGE, and the expansion of resources available to the Portuguese and multilingual NLP community.

Keywords: Extractive Summarization · Multilinguality · Natural Language Processing

1 Introduction

Automatic Text Summarization is a fundamental task in the field of Natural Language Processing (NLP), essential for handling large amounts of information. This task aims to condense information from one or more documents, producing summaries that preserve the main points of the original content [13].

A. Paes and F. A. N. Verri (Eds.): BRACIS 2024, LNAI 15413, pp. 212–223, 2025.
https://doi.org/10.1007/978-3-031-79032-4_15

Summarization can be classified into two main types: extractive and abstractive. Extractive summarization selects sentences or phrases directly from the original text(s), while abstractive summarization generates new sentences that capture the original meaning of the text(s) [17]. Additionally, summarization can be categorized as single-document, when summarizing an unique text, or multi-document, integrating information from multiple texts [7].

More recently, Large Language Models (LLMs) have been employed for several NLP applications, including generation tasks as text summarization [1]. Despite their popularity, LLMs present an explainability deficit and have some limitations, such as the expensive necessary computational infrastructure and the occurrence of hallucinations, which, in the case of summarization, may hinder the proposal of preserving the original meaning of the texts. On the other hand, extractive summarization methods generally do not suffer from these disadvantages. Interestingly, there has been a resurgence of interest in these methods, with the proposal of novel methods that incorporate classical ideas with new perspectives and also bring new approaches to the task.

Various extractive methods have been proposed and tested mainly on English corpora like CNN/DM [9]. Some notable methods are PreSumm [12], HiStruct+ [20], RankSum [10], MemSum [8] and MatchSum [22], which have demonstrated promising performances in generating summaries. However, there is a significant gap in the application of these methods to other languages and, therefore, limited evaluation on their multilingual capacity. With the web, multilingual summarization is a very relevant field, and there is a growing need for resources and techniques that support multiple languages, expanding the applicability of NLP advances to diverse linguistic communities.

Evaluation in summarization (whether it is multilingual or not) is crucial for advancing research in the field. Traditionally, the ROUGE (Recall-Oriented Understudy for Gisting Evaluation) metric [11] has been widely used to measure the overlap of n-grams between automatic summaries and human reference summaries. However, ROUGE is limited since it only computes superficial text similarity. To address these limitations, other metrics have arisen, as BLANC [21]. BLANC evaluates summaries by using them to help pre-trained language models (such as BERT [5]) to perform the token unmasking task, with the supposition that a "more helpful" summary may improve the model success on the task.

In this context, this work aims at evaluating the effectiveness of some more recent extractive methods for Brazilian Portuguese, trying to assess their multilingual performance. We run some selected state-of-the-art and more classical methods on news corpora for English and Brazilian Portuguese, using well-known reference summarization corpora. We compare the results using both ROUGE and BLANC metrics, in order to provide a more comprehensive evaluation.

Next section briefly describes the main related work. The datasets that we use are detailed in Sect. 3. Section 4 brings our experiment setup, while discussion of results and final remarks are made in Sects. 5 and 6, respectively.

2 Related Work

Several automatic extractive summarization methods have been developed in the area, especially using deep learning models and word embeddings. Here, we briefly overview classical and more recent methods, including those that are considered state-of-the-art in the area.

TextRank [14] is a graph-based ranking algorithm used for both sentence and keyword extraction. Inspired by Google's PageRank algorithm, TextRank applies the idea of "voting" to determine the importance of textual units. It constructs a graph where nodes represent sentences or words, and edges indicate co-occurrence or similarity relations. The algorithm iterates over the graph until the importance scores of the nodes converge. TextRank is an unsupervised method that has proven competitive in established benchmarks like DUC-2002, demonstrating robustness and portability across different domains and languages.

Centroid-embedding [19] summarization leverages the compositional capabilities of word embeddings to capture semantic relationships between words and sentences. This method constructs a centroid vector, which represents the central theme of the document, by summing the embeddings of the most significant words, determined by their TF-IDF scores. Each sentence is then represented as a sum of its word embeddings, and sentences closest to the centroid vector are selected for the summary. This approach addresses the limitations of traditional bag-of-words models, which often fail to capture semantic similarities between sentences with different words but similar meanings. The centroid-based method has proven effective in both multi-document and multilingual summarization tasks, offering competitive performance compared to more complex deep learning models. Its simplicity and robustness make it an attractive option for extractive summarization across various domains and languages.

PreSumm [12], also known as BERTSUM, has extractive and abstractive summarization strategies. It uses the BERT architecture to understand the context of sentences and select the most relevant ones for the summary. The technique includes pretraining on general language tasks and fine-tuning specific to the summarization task. PreSumm performs summarization in two phases: (1) an extraction phase where the most important sentences are selected, and (2) an abstractive generation phase where these sentences can be refined to improve the fluency and cohesion of the final summary. For this work, we used only the extractive strategy of PreSumm.

HiStruct+ [20] incorporates hierarchical structure information into pre-trained language models like BERT and RoBERTa to improve extractive summarization. This model is designed to handle the intrinsic structure of documents, considering the hierarchy of sections and subtitles when selecting sentences for the summary. HiStruct+ is particularly effective for long and complex documents with clear hierarchical structures, such as scientific articles. The approach improves summarization accuracy by preserving the logical and semantic structure of the documents.

Other relevant methods are RankSum [10], MemSum [8] and MatchSum [22], but we do not address them in this paper because they have been outperformed by some of the above methods in the literature and also in some of our initial experiments (that we do not report in this paper).

3 Datasets

For our experiments for Brazilian Portuguese, we adopt the **CSTNews** corpus[1] [2]. It is a reference and widely known corpus developed to support research in automatic summarization of journalistic texts in Brazilian Portuguese. The corpus comprises 50 clusters of news articles, totaling 150 texts and their respective summaries. Each cluster contains 2 to 3 news articles about the same event, collected from various Brazilian media sources such as Folha de São Paulo, Estadão, O Globo, Jornal do Brasil, and Gazeta do Povo, according to their impact at the time of publication. The corpus contains 2,088 sentences and 47,240 words, with an average of 2.8 texts, 41.76 sentences, and 944.8 words per cluster. In addition to the original texts, each cluster includes single-document manual summaries, multiple-document manual summaries, and automatic summaries. The corpus is also manually annotated in various ways for syntax, semantics and discourse information.

CNN/Daily Mail [9] is a corpus widely used in automatic summarization research for English. It comprises approximately 286,817 article-summary pairs, each containing a news article and an associated summary. The articles are sourced from the CNN and Daily Mail news websites, and the summaries are human-generated bullet points summarizing the main points of the news articles. The corpus was created initially for the passage-based question-answering task but was adapted for summarization research by restoring the bullet points to form multi-sentence summaries. In the training set, the source documents have an average of 766 words distributed over 29.74 sentences, while the summaries consist of 53 words distributed over 3.72 sentences. The corpus is available in two versions: one with real entity names and another with anonymized entities that were replaced by document-specific IDs, facilitating vocabulary reduction and experimentation with deep learning models.

4 Experiment Setup

The corpora data was divided into 70%, 15%, and 15% for training, validation, and testing for the experiments (as some methods needed training). The original configurations of each summarizer were maintained, adapting them to the CSTNews corpus. The experimental process is summarized in the following subsections.

[1] https://sites.icmc.usp.br/taspardo/sucinto/cstnews.html.

4.1 Pre-processing

The NLTK library was used for tokenization, stopword removal, and normalization of Portuguese texts. This step is crucial to ensure that the models, originally trained in English, can operate effectively in the new language. Tokenization segments the text into smaller units, such as words or subwords. Stopword removal eliminates common words that do not significantly contribute to the task. Normalization adjusts the text to a more consistent form, facilitating subsequent processing. For HiStruct+ and PreSumm, we performed similar pre-processing using the Stanford CoreNLP tool.

4.2 Model Training and Adjustment

We applied the methods' original codes with minimal modifications. We also fine-tuned the model parameters to optimize their performance on CSTNews. For this, we adapted the tokenization processes and changed some hyperparameters specific to the language and style of the texts in the CSTNews corpus.

We aimed to replicate the original training conditions of the models as closely as possible despite our limitations. The model used in PreSumm was BERT-base-uncased, while HiStruct+ employed RoBERTa-base, as these were the models used in the original works.

Since we only had access to a CPU, the number of epochs and batch sizes had to be adjusted. PreSumm and HiStruct+ were trained for 7 epochs with a batch size of 14.

The learning rates were the same as in the original works: 2e-3 for PreSumm and HiStruct+.

4.3 Evaluation

The generated summaries were evaluated using ROUGE and BLANC. ROUGE measures the overlap of n-grams between the automatic and human reference summaries. The most commonly used variants are ROUGE-1, which measures unigram (individual words) overlap, ROUGE-2, which measures bigram (word pairs) overlap, and ROUGE-L, which measures the longest common subsequence overlap. Each variant can be evaluated in terms of F1 score, precision, and recall. The F1-score is the harmonic mean of precision and recall, balancing the two. Precision measures the proportion of n-grams in the generated summary that is present in the reference summary. In contrast, recall measures the proportion of n-grams in the reference summary that is present in the generated summary. Although ROUGE is widely used for its simplicity and effectiveness, it is criticized for not adequately capturing summaries' semantic quality and coherence, focusing mainly on superficial text similarity. Conversely, BLANC measures how much the generated summary aids a pre-trained language model (like BERT) in understanding the document. It uses the masked token task (Cloze task) to evaluate the functional utility of a summary. In BLANC-help mode (which is the one we used), the summary is concatenated to each document sentence during inference, and the model's ability to predict masked words with the summary is measured. The difference in prediction accuracy with and without the summary indicates the summary's quality. A higher BLANC score means the summary provides more contextual information that helps the model to understand the document better. Unlike ROUGE, BLANC does not require human-written reference summaries.

5 Results and Discussion

Tables 1 and 2 show the ROUGE results for the summarization methods on the CSTNews and CNN/DM corpora. For CSTNews, PreSumm demonstrates the highest performance in most ROUGE metrics, particularly in ROUGE-1 F1 (55.55), ROUGE-1 Precision (67.91), and ROUGE-2 Precision (49.64). This indicates its high effectiveness in selecting relevant sentences that closely match the reference summaries at the unigram and bigram levels. However, its lower recall scores suggest that, while maintaining high precision, it may miss some relevant content. For the CNN/DM dataset, PreSumm also shows the highest performance with ROUGE-1 score of 53.37. TextRank and Centroid-embedding have lower scores (30.14 and 18.12), consistent with their lower performance on the CSTNews corpus too and demonstrating the need for advanced models to handle complex text summarization tasks.

Table 1. Results on ROUGE metrics for CSTNews dataset

Metric	PreSumm	HiStruct+	TextRank	Centroid-embedding
ROUGE-1 F1	**55.55**	54.58	46.47	44.79
ROUGE-1 Precision	67.91	66.05	53.14	31.09
ROUGE-1 Recall	49.03	48.56	42.77	80.63
ROUGE-2 F1	39.89	37.85	26.42	29.36
ROUGE-2 Precision	49.64	46.58	30.23	19.42
ROUGE-2 Recall	34.70	33.21	24.31	60.49
ROUGE-L F1	46.20	44.81	33.34	43.30
ROUGE-L Precision	56.77	54.48	38.30	30.06
ROUGE-L Recall	40.60	39.70	30.55	77.95

Table 3 presents the BLANC results for the same models. Centroid-embedding achieves the highest BLANC score (57.74) for CSTNews. Despite its high ROUGE scores, PreSumm produces the lowest BLANC score, which shows the importance of using complementary metrics for a more holistic evaluation. For the CNN/DM dataset, TextRank achieves the highest BLANC score (26.20). Centroid-embedding also performs well, with score of 24.35. PreSumm again shows a relatively low BLANC score.

Table 2. Results on ROUGE metrics for CNN/DM dataset

Metric	PreSumm	HiStruct+	TextRank	Centroid-embedding
ROUGE-1 F1	**53.37**	52.15	30.14	18.12
ROUGE-1 Precision	38.18	39.58	22.51	10.56
ROUGE-1 Recall	43.05	43.40	60.38	81.48
ROUGE-2 F1	24.06	24.53	12.70	08.63
ROUGE-2 Precision	17.93	18.69	09.57	05.03
ROUGE-2 Recall	20.15	20.42	25.22	38.88
ROUGE-L F1	48.85	47.81	19.58	12.05
ROUGE-L Precision	35.00	36.36	14.69	07.01
ROUGE-L Recall	39.44	39.84	39.29	54.89

Table 3. Results on BLANC metric for CSTNews and CNN/DM

Model	CSTNews	CNN/DM
PreSumm	25.66	15.87
HiStruct+	26.01	15.37
TextRank	26.57	26.20
Centroid-embedding	57.74	24.35

As illustration, Table 4 shows a summary generated by PreSumm, allowing comparing it with the human-written summary. The summary generated by PreSumm accurately captured Fabiana Murer's victory and the matching of the medal record, but omitted details about the competitors and the achieved marks.

Table 5 shows three CSTNews summaries generated by PreSumm, the summarizer which achieved the best ROUGE results. These summaries were randomly chosen for a manual human evaluation. We used some of the known TAC (Text Analysis Conference) criteria [4], which include **overall responsiveness** and **readability**. The assessment of overall responsiveness examines how effectively a summary addresses the information need outlined in the topic statement, considering both the content and linguistic quality of the summary. The readability score evaluates the summary's fluency and structure, independent of content, and is based on factors such as grammatical correctness, lack of redundancy, referential clarity, focus, structure, and coherence. These criteria are evaluated on a five-point scale: from very poor to very good.

We evaluated these three random summaries as "very good" based on the above criteria once they show no redundancy, no referential problems, and very good structure and coherence. Their high ROUGE scores can reflect the responsiveness of the summaries. Once the summaries generated are close to the golden standard, that is, the summary generated by humans, they are expected to cover the document's main content, thus, having a very good responsiveness.

Table 4. Comparison between summary generated by PreSumm and the human-written summary

Human-written summary
como esperado, a atleta fabiana murer conquistou a medalha de ouro - a 29.ł brasileira - no salto com vara nos jogos pan-americanos do rio, nesta segunda-feira, quebrando o recorde pan-americano. esta conquista iguala o número de medalhas de ouro faturadas em santo domingo (2003). ela atingiu a marca de 4m60, seguida pela norte-americana april steiner com a marca de 4m40. o bronze pertence à cubana yarisley silva, com a marca de 4m30. a disputa entre a brasileira e a norte-americana foi acirrada
Presumm summary
como esperado, a atleta fabiana murer conquistou a medalha de ouro - a 29. łbrasileira - no salto com vara nos jogos pan-americanos do rio, nesta segunda - feira, no estádio joão havelange. de quebra, esta conquista iguala o número de medalhas de ouro faturadas em santo domingo (2003), quando o brasil também somou 29

Table 5. Three CSTNews summaries generated by PreSumm

Summary 1
Se a eleição fosse hoje, o presidente Luiz Inácio Lula da Silva, candidato à reeleição, teria 44 % das intenções de voto, contra 25 % do tucano Geraldo Alckmin, de acordo com a pesquisa cni/ibope divulgada nesta sexta - feira. Considerando apenas os votos válidos, Lula venceria as eleições de outubro ainda no primeiro turno. Em junho, Lula tinha 57 % e Heloísa 21 %
Summary 2
O presidente Luiz Inácio Lula da Silva abriu nesta terça - feira a assembléia geral da organização das nações unidas (ONU) reforçando o discurso do secretário - geral Ban Ki - Moon sobre a importância dos países desenvolvidos liderarem os esforços pela preservação do meio ambiente. Diante de representantes de mais de 150 países que participam da assembléia - centrada este ano nas questões ambientais - , Lula defendeu os esforços brasileiros na redução do desmatamento, e propôs a realização, em 2012, de um novo encontro internacional nos moldes da ec0 - 92, sediada pelo Rio há 20 anos. Não nos iludamos: se o modelo de desenvolvimento global não for repensado, crescem os riscos de uma catástrofe ambiental e humana sem precedentes
Summary 3
A seleção brasileira de vôlei mais uma vez mostrou o porquê de ser o melhor time do mundo ao vencer a rússia por 3 sets a 1, com parciais de 18 a 25, 25 a 23, 28 a 26 e 25 a 22, na tarde deste domingo, em Katowice, Polônia, na final da liga mundial de vôlei, conquistando o heptacampeonato da competição. Agora, a meta é a medalha de ouro dos jogos pan-americanos, que estão sendo disputados no rio de janeiro. Apresentando erros em seus ataques, o Brasil deixou a Rússia dominar o primeiro set, que foi fechado em 25 a 18, em 22 minutos.

Table 6 shows three CSTNews summaries generated by Centroid-embedding, the summarizer which achieved the best BLANC results. These summaries were also randomly chosen for a manual human evaluation and evaluated under the TAC criteria. Similarly to the previous summaries, we also evaluate these three as "very good" regarding the overall responsiveness and readability.

These results raise an important question: why did the Centroid-embedding achieve such high BLANC scores? Since BLANC measures how much the generated summary aids a pre-trained language model in understanding the original document, its high BLANC score suggests a high amount of informational

Table 6. Three CSTNews summaries generated by Centroid-embedding

Summary 1
Um dos destaques desta temporada do esporte brasileiro, a ginasta Jade Barbosa foi escolhida, na noite desta terça-feira, para ser a representante do Brasil no revezamento da tocha dos Jogos Olímpicos de Pequim. Em votação pela internet, a ginasta recebeu mais de 100 mil votos e superou o nadador Thiago Pereira, que ganhou seis ouros nos Jogos Pan-Americanos. O Brasil não faz parte do trajeto da tocha olímpica. Na América do Sul, a chama passará por Buenos Aires, onde Jade participará do revezamento, no dia 11 de abril. Aos 16 anos, Jade conquistou três medalhas no Pan: ouro na disputa dos saltos, prata na apresentação por equipes e bronze no solo. Ao todo, a chama olímpica percorrerá 20 países antes de chegar a Pequim para a abertura da competição, no dia 8 de agosto.
Summary 2
Um terremoto de 6,8 graus na escala Richter atingiu a costa noroeste do Japão nesta segunda-feira, 16, matando pelo menos sete pessoas na cidade de Kashiwazaki e deixando outros 700 feridos. Equipes de resgate continuam procurando por pessoas em meio aos escombros. Chamas e rolos de fumaça preta foram vistos na usina nuclear de Kashiwazaki, que foi automaticamente fechada durante o terremoto. Segundo a TV NHK, o fogo atingiu um transformador de eletricidade e não houve vazamento de radiação. A força do abalo danificou estradas e pontes na costa. Cerca de 2 mil moradores de Kashiwazaki, a cidade mais afetada pelo tremor, tiveram de deixar suas casas. Cerca de 300 edifícios da cidade ficaram destruídos. Rachaduras de um metro de largura foram vistas no solo ao longo do litoral. A Agência de Meteorologia divulgou alertas de tsunami para o estado de Niigata e a região costeira das redondezas, mas por fim o aviso foi cancelado.
Summary 3
A seleção brasileira masculina de vôlei, que é treinada por Bernardinho, venceu a Finlândia por 3 sets a 0, parciais de 25/17, 25/22 e 25/21, nesta sexta-feira, em Tampere (FIN), e manteve sua invencibilidade na Liga Mundial-06. Este foi o sétimo triunfo consecutivo dos brasileiros na competição –antes, o país conquistou quatro vitórias contra a seleção argentina e duas diante de Portugal. Os dois times voltam a se enfrentar às 12h30 deste sábado, no mesmo ginásio, que normalmente é utilizado para competições de hóquei no gelo. O resultado de hoje deixou o Brasil perto de conquistar a única vaga do Grupo B da Liga Mundial, que classifica o melhor de cada uma das quatro chaves, a Rússia (país-sede) e mais um time convidado pela Federação Internacional de Vôlei, para a fase final, de 23 a 27 de agosto, em Moscou (Rússia). A equipe brasileira masculina já conquistou cinco vezes a Liga Mundial –1993, 2001, 2003, 2004 e 2005. Com oito títulos, a Itália é a maior vencedora da competição.

content in the summaries. However, based on these three randomly chosen summaries, we could not hypothesize why they had higher BLANC scores than PreSumm, for example. Thus, we encourage further research and new experiments with both ROUGE and BLANC metrics.

While providing valuable insights into the performance of various summarization models on the CSTNews corpus, the present study is subject to several methodological limitations. One significant limitation is the size of the dataset. Although a reference in the area, the CSTNews corpus contains only 150 texts, and more than this limited number of texts is needed for training robust summarization models. So, we emphasize that we need more resources in languages other than English, to enhance the reliability and generalizability of summarization models. We are aware of a recent corpus, RecognaSumm [15], which contains over 130.000 journalistic texts in Brazilian Portuguese, and we plan to perform experiments with it in the future.

Another limitation of this study is the computational resources used. Due to the unavailability of GPU resources, the experiments were conducted using CPUs. The lack of GPU access limited our ability to fine-tune model parameters optimally, potentially impacting the performance of the summarizers. With the availability of GPUs, future studies could perform more extensive parameter tuning, leading to better results.

More efforts are also needed to refine and develop evaluation metrics beyond traditional measures like ROUGE. Metrics such as BLANC, which supposedly consider semantic coherence and fluency, should be further explored and validated. Developing new metrics that can more accurately reflect the quality of summaries in different languages and contexts will be crucial for advancing the field.

6 Final Remarks

We contributed to the investigation of multilingual summarization by evaluating state-of-the-art extractive summarization models on the Brazilian Portuguese CSTNews corpus and the English CNN/DM corpus. Our study utilized ROUGE and the more recently developed BLANC metric, providing insights into the models' performance. The findings revealed that PreSumm performed well according to ROUGE metrics, while Centroid-embedding had good BLANC scores for both languages.

Future work includes testing other methods for the previous languages, as well as other languages, including languages of different linguistic typologies. We also aim at using corpora of diverse genres and domains. Such evaluation is necessary for determining the potential of current and future summarization methods for multilingual summarization.

Another relevant issue consists in exploring multilingual multi-document summarization, leveraging some previous researches and datasets that exist for Brazilian Portuguese, as the ones of [3] and [6]. Exploring the multilingual performance of some classic single and multi-document methods that were created for Portuguese may be other interesting endeavor, as it is the case of GistSumm [16] – the first summarization system for Portuguese – and RSumm [18].

The interested reader may find the datasets, the source codes of the methods and other details about this work at the POeTiSA project website (at https://sites.google.com/icmc.usp.br/poetisa).

Acknowledgments. To SiDi and CAPES (*Coordenação de Aperfeiçoamento de Pessoal de Nível Superior*), for supporting this work. The author Thiago Alexandre Salgueiro Pardo also thanks the Center for Artificial Intelligence of the University of São Paulo (C4AI - http://c4ai.inova.usp.br/), with support by the São Paulo Research Foundation (FAPESP grant #2019/07665-4) and by the IBM Corporation, and the Ministry of Science, Technology and Innovation, with resources of Law N. 8,248, of October 23, 1991, within the scope of PPI-SOFTEX, coordinated by Softex and published as Residence in TIC 13, DOU 01245.010222/2022-44.

Disclosure of Interests. The authors have no competing interests to declare that are relevant to the content of this article.

References

1. Bommasani, R., et al.: On the opportunities and risks of foundation models (2022)
2. Cardoso, P.C., et al.: Cstnews-a discourse-annotated corpus for single and multi-document summarization of news texts in Brazilian Portuguese. In: Proceedings of the 3rd RST Brazilian Meeting, pp. 88–105 (2011)
3. Chaud, M., Felippo, A.D.: Exploring content selection strategies for multilingual multi-document summarization based on the universal network language (unl). REVISTA DE ESTUDOS DA LINGUAGEM **26**(1), 45–71 (2017)
4. Dang, H.T., Owczarzak, K., et al.: Overview of the tac 2008 update summarization task. In: TAC (2008)
5. Devlin, J., Chang, M.W., Lee, K., Toutanova, K.: BERT: pre-training of deep bidirectional transformers for language understanding (2018). https://doi.org/10.48550/arXiv.1810.04805. http://arxiv.org/abs/1810.04805v2
6. Di Felippo, A., Tosta, F.E.S., Pardo, T.A.S.: Applying lexical-conceptual knowledge for multilingual multi-document summarization. In: Silva, J., Ribeiro, R., Quaresma, P., Adami, A., Branco, A. (eds.) PROPOR 2016. LNCS (LNAI), vol. 9727, pp. 38–49. Springer, Cham (2016). https://doi.org/10.1007/978-3-319-41552-9_4
7. El-Kassas, W.S., Salama, C.R., Rafea, A.A., Mohamed, H.K.: Automatic text summarization: a comprehensive survey. Expert Syst. Appl. **165**, 113679 (2021). https://doi.org/10.1016/j.eswa.2020.113679. https://linkinghub.elsevier.com/retrieve/pii/S0957417420305030
8. Gu, N., Ash, E., Hahnloser, R.: MemSum: extractive summarization of long documents using multi-step episodic Markov decision processes. In: Muresan, S., Nakov, P., Villavicencio, A. (eds.) Proceedings of the 60th Annual Meeting of the Association for Computational Linguistics, vol. 1: Long Papers, pp. 6507–6522. Association for Computational Linguistics, Dublin (2022). https://doi.org/10.18653/v1/2022.acl-long.450. https://aclanthology.org/2022.acl-long.450
9. Hermann, K.M., et al.: Teaching machines to read and comprehend. Adv. Neural Inf. Process. Syst. **28** (2015)
10. Joshi, A., Fidalgo, E., Alegre, E., Alaiz-Rodriguez, R.: Ranksum–an unsupervised extractive text summarization based on rank fusion. Expert Syst. Appl. **200**, 116846 (2022)
11. Lin, C.Y.: ROUGE: a package for automatic evaluation of summaries (2004)
12. Liu, Y., Lapata, M.: Text summarization with pretrained encoders. In: Inui, K., Jiang, J., Ng, V., Wan, X. (eds.) Proceedings of the 2019 Conference on Empirical Methods in Natural Language Processing and the 9th International Joint Conference on Natural Language Processing (EMNLP-IJCNLP), pp. 3730–3740. Association for Computational Linguistics, Hong Kong (2019). https://doi.org/10.18653/v1/D19-1387. https://aclanthology.org/D19-1387
13. Mani, I.: Automatic summarization. Natural language processing (2001). https://doi.org/10.1075/nlp.3
14. Mihalcea, R., Tarau, P.: Textrank: bringing order into text. In: Proceedings of the 2004 Conference on Empirical Methods in Natural Language Processing, pp. 404–411 (2004)

15. Paiola, P.H., Garcia, G.L., Jodas, D.S., Correia, J.V.M., Sugi, L.A., Papa, J.P.: Recognasumm: a novel brazilian summarization dataset. In: Proceedings of the 16th International Conference on Computational Processing of Portuguese, pp. 575–579 (2024)
16. Pardo, T.A.S., Rino, L.H.M., Nunes, M.G.V.: GistSumm: a summarization tool based on a new extractive method. In: Mamede, N.J., Trancoso, I., Baptista, J., das Graças Volpe Nunes, M. (eds.) PROPOR 2003. LNCS (LNAI), vol. 2721, pp. 210–218. Springer, Heidelberg (2003). https://doi.org/10.1007/3-540-45011-4_34
17. Radev, D.R., McKeown, K., Hovy, E.: Introduction to the Special Issue on Summarization (2002)
18. Ribaldo, R., Christina Figueira Cardoso, P., Alexandre Salgueiro Pardo, T.: Exploring the subtopic-based relationship map strategy for multi-document summarization. Revista de Informática Teórica e Aplicada **23**(1), 183–211 (2016). https://doi.org/10.22456/2175-2745.59104. https://seer.ufrgs.br/index.php/rita/article/view/RITA-VOL23-NR1-183
19. Rossiello, G., Basile, P., Semeraro, G.: Centroid-based text summarization through compositionality of word embeddings. In: Giannakopoulos, G., Lloret, E., Conroy, J.M., Steinberger, J., Litvak, M., Rankel, P., Favre, B. (eds.) Proceedings of the MultiLing 2017 Workshop on Summarization and Summary Evaluation Across Source Types and Genres, pp. 12–21. Association for Computational Linguistics, Valencia (2017). https://doi.org/10.18653/v1/W17-1003. https://aclanthology.org/W17-1003
20. Ruan, Q., Ostendorff, M., Rehm, G.: HiStruct+: improving extractive text summarization with hierarchical structure information. In: Muresan, S., Nakov, P., Villavicencio, A. (eds.) Findings of the Association for Computational Linguistics: ACL 2022, pp. 1292–1308. Association for Computational Linguistics, Dublin (2022). https://doi.org/10.18653/v1/2022.findings-acl.102. https://aclanthology.org/2022.findings-acl.102
21. Vasilyev, O., Dharnidharka, V., Bohannon, J.: Fill in the blanc: human-free quality estimation of document summaries. arXiv preprint arXiv:2002.09836 (2020)
22. Zhong, M., Liu, P., Chen, Y., Wang, D., Qiu, X., Huang, X.: Extractive summarization as text matching. In: Jurafsky, D., Chai, J., Schluter, N., Tetreault, J. (eds.) Proceedings of the 58th Annual Meeting of the Association for Computational Linguistics, pp. 6197–6208. Association for Computational Linguistics, Online (2020). https://doi.org/10.18653/v1/2020.acl-main.552. https://aclanthology.org/2020.acl-main.552

Multimodal and Hybrid Models for Predicting SCD Risk in Chagas Cardiomyopathy

Isabelly P. da Costa[1], Bruno M. P. Takazono[1], Carlos H. L. Cavalcante[1](✉), João P. V. Madeiro[3], and Roberto C. Pedrosa[2]

[1] Federal Institute of Education and Technology of Ceara, Maracanau, Ceara, Brazil
{isabelly.pinheiro.costa08,bruno.matsuyama.pereira42}@aluno.ifce.edu.br, henriqueleitao@ifce.edu.br

[2] Edson Saad Heart Institute – Federal University of Rio de Janeiro, Rio de Janeiro, Brazil
coury@hucff.ufrj.br

[3] Universidade Feferal do Ceará (UFC), Fortaleza, CE, Brazil
jpaulo.vale@dc.ufc.br

Abstract. Predicting sudden cardiac death (SCD) in Chagas cardiomyopathy (CC) remains a challenge. An estimated 6 to 7 million people worldwide are affected by this condition. To address this challenge, we propose a novel intelligent system that utilizes multimodal and hybrid machine learning (ML) models to predict SCD risk in CC patients. We evaluated different approaches to combine clinical features (tabular data) and temporal features extracted from ECG-Holter exams. Our results indicate that the hybrid model (RNN with MLP) achieved the highest recall (sensitivity) of 91.63% and 81.08% of AUC. This study aims to develop a predictive model to support clinical decision-making in the context of SCD in CC. The model can potentially aid in risk stratification, enabling early intervention and potentially saving lives.

Keywords: Chagas disease · Sudden Cardiac Death · Machine Learning · Recurrent Neural Network

1 Introduction

The World Health Organization (WHO) estimates that between 6 and 7 million people are infected with Chagas disease (CD). CD is endemic in Brazil and Latin America, and its incidence is increasing steadily across North America, Europe, Japan and Australia.

CD, caused by the parasite Trypanosoma cruzi, is a persistent and potentially lethal illness. It affects up to 1.5 million (30%) people worldwide who have Chagas cardiomyopathy (CC). This condition significantly compromises the health

A. Paes and F. A. N. Verri (Eds.): BRACIS 2024, LNAI 15413, pp. 224–237, 2025.
https://doi.org/10.1007/978-3-031-79032-4_16

of infected individuals, leading to a shortened lifespan or death due to sudden cardiac events (SCD) [27].

SCD is one of the leading causes of death worldwide, often presenting as an abrupt collapse with a documented loss of vital signs, which may lead to attempts to restore circulation through cardiopulmonary resuscitation (CPR) [8]. In the context of CD, SCD accounts for approximately 45% of deaths among patients diagnosed with CC [19], with an occurrence rate (2.4% per year) higher than that of the general population [20].

The search for early and advanced treatments for patients with CC who are susceptible to SCD has led to measures such as implantable cardioverter-defibrillators (ICDs), which can extend patients' lifespans [20]. However, these interventions are limited by the unpredictability of SCD, a challenge that is well-documented in the literature [12,14,19]. In this context, providing systems that aid in the diagnosis of SCD risk in CC patients becomes a valuable tool, especially given the high prevalence of this disease in some countries and the unpredictable nature of SCD.

Therefore, this study proposes a model architecture that allows for the incorporation of various data types (tabular and serial data) and hybrid ML models, eventually supporting clinical decision-making in the context of SCD in CC. Additionally, this work developed a first prototype of a user-friendly computational system to assist doctors in patient care. The system utilizes a predictive model built with a minimal feature set, focusing on patients not considered at high risk of SCD.

By combining the output of a recurrent neural network (RNN), which effectively captures temporal features, with tabular data, the approach leverages both sequential and non-sequential information. This combined approach ensures the model's scalability to handle diverse input data types and tasks. Consequently, it enables the exploration of potential impacts and the relevance of new features for predicting SCD risk in CC patients.

This study is organized into five sections. The first section provides an introduction to the research problem. The second section discusses related works that align with the themes of the present study. The third section explores the methodological aspects of the research, outlining the proposed model and system architecture. The fourth section presents the experimental results, detailing the system's development and the model's performance. Finally, the last section offers concluding considerations of the developed model and suggestions for future work.

2 Related Work

Intelligent Systems (IS) with machine learning (ML) have a significant impact on assisting diagnosis in various areas of healthcare. Their use can lead to safer and more precise medical practices [13]. In the specific context of CC, some studies have shown that ML algorithms can classify patients with CD with the same or higher precision than physicians [3].

Despite the global significance of SCD [15], its occurrence in the specific context of CC has been scarcely explored, especially regarding the use of ML computational tools by end-users. Furthermore, a comprehensive review of the literature reveals a current lack of studies dedicated to addressing this knowledge gap.

One study has a limited approach, as it only performed classic linear analysis and did not fully explore the potential of combining different parameters [23]. Another research study utilized eight variables from 78 patients, applying heart rate variability (HRV) and heart rate turbulence (HRT) techniques to ECG-Holter recordings to investigate SCD in Chagas heart disease (ChHD). The attribute set was reduced using forward and reverse stepwise approaches. This work employed the k-nearest neighbors (KNN) classifier and leave-one-out (LOO) cross-validation (CV) [1]. However, the lack of sensitivity and methodology for handling imbalanced data makes reproducing this study unfeasible.

A third study employed several classic ML methods in a dataset with 218 patients, using seven cardiac restitution metrics (CRM) extracted from a 4-hour segment of a 24-hour ECG-Holter along with clinical data. This study reported sensitivity results of 90% and identified the Gaussian Naive Bayes algorithm as the best-performing model [4]. However, there was no exclusion of high-risk patients in this research.

[18] investigated the use of action potential duration restitution (APDR) dynamics, using a cross-sectional study of 221 patients, the authors extracted APDR dynamics metrics from 4-hour ECG segments and analyzed them alongside clinical variables using a classification tree. The study identified the %QTend TendQ $>$ 1, as the most relevant predictor of SCD in the model with a 0.96 AUC. However, it is limited by its inclusion criteria that included patients who met the following: (1) had regular appointments at the Clementino Fraga Filho University Hospital (HUCFF-UFRJ), Rio de Janeiro, Brazil, (2) had resided outside an area with endemic Chagas disease transmission for at least 20 years, (3) were classified as having a low-intermediate Rassi risk score, (4) had their vital status checked until the end of February 2017, and (5) had an established diagnosis of Chagas cardiomyopathy (CC), confirmed by at least two positive serologic tests for antibodies against Trypanosoma cruzi and characteristic 12-lead electrocardiogram changes.

3 Methodology

About the scientific method, this work adheres to the action research methodology, aiming to propose and test solutions for problems within a specific context [17]. As for its approach, this study can be categorized as quantitative, centered on problem-solving through theory testing and the numerical quantification of variables [10]. The project was organized into developing prediction models and creating a prototype web system, aspects that are detailed below.

3.1 Data Collection

The study utilized data from a clinical follow-up program conducted between 1992 and 2023 at Clementino Fraga Filho University Hospital of the Federal University of Rio de Janeiro. Patients with a high degree of cardiac compromise were excluded, as they already had indications for potential interventions to ensure survival, such as CRT implantation or other treatments. Samples meeting the following criteria were removed from the dataset: NYHA class 4, severe dysfunction classification, high Rassi score, 2005 Classification Guideline of C or D, and all patients with a Teicholz EF below 0.40. After this exclusion, 120 patients remained for analysis, including 19 (16%) who experienced SCD and 101 (84%) who did not. The gender distribution among these patients was 42 (35%) males and 78 (65%) females.

3.2 Standardization Process

The dataset used consists of two subsets: the first derived from tabular data and the second from time series extracted from ECG-Holter signals using the sames methods used by [16,21]. Regarding the number of attributes, the dataset consists of 49 tabular features and 27 time series features.

All features are grouped into 7 categories: clinical data, echocardiogram (ECHO), electrocardiogram (ECG), electrocardiogram Holter (ECG-Holter), cardiology guideline classification (CGC), treatments, and cardiac restitution metrics (CRM). Notably, the CRM group exclusively contains the 27 time series features, while the remaining 49 tabular features are distributed among the other four categories. The groups and features used in this work are detailed in Table 1.

Initially, 14 features were excluded due to confidentiality concerns or because they contained irrelevant information for this study, which resulted in 49 final tabular attributes. Among these attributes, 30 were binary, 12 were scalar, and 7 were categorical. The seven categorical attributes underwent one-hot encoding, transforming each category into a binary attribute, thereby bringing the dataset back to 63 attributes. Subsequently, the data was normalized using the MinMax technique.

Regarding the time series extracted from the ECG-Holter, each series initially had 30 attributes, but two were excluded—one constant and the other the target variable indicating SCD. This left 28 final attributes: 27 corresponding to CRM and 1 indicating the time of each sample in the series; all of these attributes were scalar. These 27 time series features represent an initial result of an ongoing study that is processing ECG-Holters from Chagas disease patients. In this work, we propose an architecture that utilizes both sequential and non-sequential data for an initial analysis of the results. This architecture concatenates the output of a recurrent neural network (RNN) with tabular data as input to a multilayer perceptron (MLP).

Table 1. Description of the sequential and non-sequential features used in this work, grouped into 7 groups. The 49 tabular features are classified into the groups: clinical data, ECHO, ECG, ECG-Holter, CGC, and treatments. The 27 time series features extracted from ECG-Holter are exclusively in the CRM.

Group	Qnt	Attributes
Clinical data	19	CVA, Cancer, Coronary heart disease, PVD, T2DM, SAH, Age at ECG-Holter, Dyslipidemia, Atrial Fibrillation/Flutter, BMI, Heart failure, Chronic renal insufficiency, Pacemaker, SCD, TSH, Sedentary lifestyle, Gender, Syncope, Smoking
ECG	8	Eletrical Inactive Area, Primary Alteration, $\geq$ 3 s pause, Atrioventricular Conduction Disorder, SVES, VE, NSVT, Heart rate
ECG-Holter	8	Sinus node dysfunction, Atrioventricular Conduction Disorder, VE, Total VE, FEVE, Atrial Fibrillation/Flutter, NSVT, Average heartrate
ECO	7	Classification, Segmental dysfunction, Diastolic dysfunction, Left Atrial Diameter, EF Teicholz, LVAD, LVSD
CGC	4	2005 Classification Guideline, NYHA, Rassi score, Rassi point
Treatments	3	Ablations, Amiodarone, ICD
CRM	27	SampEn - Scale1, SampEn - Scale2, SampEn - Scale3, SampEn - - Scale4, SampEn - Scale5, QTend - Median, QTpeak - Median, TendQ - 5ž Percentile, TpeakQ - 5ž Percentile, Percentage - - QTend/TendQ (m1) > 1, QT/TQ ratio (m1) - 98ž Percentile Percentage - QTpeak/TpeakQ(m1) > 1, QTpeak/TpeakQ ratio (m1) - 98ž Percentile, QT/TQ ratio (m2) - 98ž Percentile Percentage - QTend/TendQ (m2) > 1, QTpeak/TpeakQ ratio (m2) - 98ž Percentile, Percentage - QTpeak/TpeakQ (m2) > 1 Tp-Te - 5ž Percentile, QRS duration - Median, QRS duration - - 5ž Percentile, QRS duration - 98ž Percentile, QRS Amplitude - - Median, QRS Amplitude - 5ž Percentile, QRS Amplitude - 98ž Percentile, T Wave Amplitude- Median, T Wave Amplitude - 5ř Percentile, T Wave Amplitude - 98ž Percentile

3.3 Tools

The Python programming language was chosen for this study due to its widespread adoption in the field of ML. The Google Colab platform was utilized, taking advantage of its cloud-hosted Jupyter notebooks and the CPU resources provided by Google, which facilitated the execution of experiments for the prediction model.

For model development, two primary libraries were used: scikit-learn and PyTorch. These libraries were selected for their wide range of algorithms and functionalities available for developing and evaluating ML models. The predictive model with the best performance was deployed through an Application Programming Interface (API) using the pickle5 library.

Regarding the system prototype, a REST architecture with Docker was defined for both the backend and frontend applications. The Flask framework was employed for backend construction, integrated with a PostgreSQL database. Frontend development was implemented using Next.js.

3.4 Classification Experiments

The experiments to create the predictive model were conducted under two scenarios. In each scenario, the data were split into a 67% training set and a 33% test set for robust evaluation and all experiments were repeated 10 times to calculate the average of the metrics. Synthetic data generation was not considered for either scenario due to the specific nature of the data, scope limitations, and time constraints.

In the first scenario, exclusively tabular data from 120 patients was used. Three models were employed: Multilayer Perceptron (MLP), XGBoost (XGB), and Random Forest (RF). The selection of tree-based models was motivated by their established performance in handling tabular data, as documented in the literature [7,25].

For the experiments in this scenario, scikit-learn was utilized. The hyperparameters for the classification algorithms were chosen from the following values:

- **Multilayer Perceptron (MLP):** *hidden_layer_size*: (200, 50, 30), (100, 50, 10), (100, 50), (200, 100), (500, 250), (20,), (50,), (100,), (10,), (200,); *activation*: Tangencial e ReLU.; *solver*: SGD e Adam; *alpha*: 0.0001, 0.005, 0.05; *learning_rate*: constant e adaptive.
- **Random Forest (RF):** *n_estimators*: 100, 300, 500, 800, 1200; *max_depth*: 5, 8, 15, 25, 30; *min_samples_split*: 2, 5, 10, 15, 100; *min_samples_leaf*: 1, 2, 5, 10; *max_features*: 1, 2, 3, 4, 5.
- **XGBoost (XGB):** *min_child_weight*: 1, 5, 10.; *gamma*: 0.5, 1.0, 1.5, 2.0, 5.0; *subsample*: 0.6, 0.8, 1.0; *colsample_bytree*: 0.6, 0.8, 1.0; *max_depth*: 3, 4, 5, 8, 12; *eta*: 0.3, 0.2, 0.1, 0.05, 0.01, 0.005.

For the second scenario, tabular data and time series data from each patient's ECG-Holter exams were utilized. In this scenario, we present a formal description of a composite system employing an Elman RNN and a Multilayer Perceptron (MLP) for tasks involving time series and tabular data. The overall process can be decomposed into two stages: (1) the Elman RNN, responsible for handling the inherent sequential nature of the time series data [5], and (2) the Multilayer Perceptron, which integrates the outputs generated by the RNN with the provided tabular data.

The purpose of choosing a hybrid model using RNN + MLP is to observe, through the use of the RNN, whether the time series extracted from ECG exams add value to the SCD problem in CC when concatenated with tabular data, in order to obtain a more efficient multimodal hybrid model. Another point is that backpropagation ensures that both parts of the model (RNN + MLP) are adjusted in a coordinated and efficient manner, maximizing the model's ability to capture and integrate temporal and tabular information.

The output of the RNN at time step t can be mathematically described by equation below (i), where x_t represents the input vector at time t, h_t is the hidden state at time t, y_t is the output at time t, b_y is the bias vector for output, and h_t is the hidden state.

$$y_t = W_{hy}h_t + b_y \qquad (i)$$

The h_t is given by the following equation (ii), where W_{xh} represents the weight matrix from input to hidden state, W_{hh} is the weight matrix from previous hidden state to current hidden state, W_{hy} is the weight matrix from hidden state to output, and b_h is the bias vector for hidden state.

$$h_t = \sigma(W_{xh}x_t + W_{hh}h_{t-1} + b_h) \qquad (ii)$$

Assume that the vector z_t is the concat of the output of an RNN (y_t) and tabular data (d_t). The mathematical expression for processing z_t as the input to an MLP is described as follows, where

$$z_t = [y_t || d_t] \qquad (iii)$$

The input to the MLP is $a^{(0)} = z_t$. For an MLP with L hidden layers and an output layer *(L + 1)*, the final output $\hat{y}$ is showed in followed equitation (iv), where $W^{(l)}$ represents the weight matrix of layer l, $b^{(l)}$ is the bias vector of layer l, $a^{(l)}$ is the activation vector of layer l, and σ is the an activation function, such as ReLU, sigmoid, or tanh.

$$\hat{y} = a^{(L+1)} = \sigma(W^{(L+1)}a^{(L)} + b^{(L+1)}) \qquad (iv)$$

The summary of the process used in this work can described in four pass: (i) pass the time series through the RNN to obtain y_t; (ii) concatenate y_t with d_t to get z_t, (iii) pass z_t through the MLP to obtain the final output $\hat{y}$ and (iv) the compute the error and propagate the gradients back through the MLP and then through the RNN to update the weights and biases.

In scenario 1, the experiments were conducted using a feature selection process, considering a range of 5, 10, 15, 20, and 25 features, as well as using all 63 features. The GridSearchCV method was employed to select the top features in each configuration. All experiments were evaluated using leave-one-out (LOO) and 5-fold CV.

In the second scenario, two approaches were executed. In the first approach, the output of the RNN was directly concatenated with all tabular data features. In the second approach, the RNN output was concatenated only with each of the top selected features from the MLP model in scenario 1. Both the RNN and MLP were implemented using PyTorch. To facilitate data manipulation, two tensors were created for the tabular data: one containing the target labels and the other containing the attributes.

During model training in scenario 2, the *CrossEntropyLoss* loss function was automatically calculated gradients. Backpropagation was carried out using the loss.backward() function to optimize the model by learning from errors and enhancing performance. The gradients were propagated through the entire network, encompassing both the RNN layer and the MLP.

To handle the time series extracted from the ECG-Holter exams of the 120 patients, two tensors were created as well. One tensor stored the values of the 27

CRM variables, while the other stored the length of each time series. Hyperparameter selection was carried out using the ParameterGrid function from scikit-learn. The RNN hyperparameters were chosen from the following values:

- **Recurrent Neural Network (RNN):** *hidden_layer_size*: 8, 10, 16, 32, 50, 64, 100, 128; *num_layers*: 1; *nonlinearity*: ReLu; *solven*: SGD e Adam.; *learning_rate*: 0.0001, 0.005, 0.001, 0.05 ou 0.01.

4 Results

4.1 Model Prediction Performance

Experiments were conducted using all features (63) and subsets (5, 10, 15, 20, and 25) of the most relevant features selected through a feature selection process. After the feature selection process was completed, the subset of features that yielded the best performance for each algorithm was selected. Interestingly, certain features were frequently selected across various combinations of top features that led to the models (MLP, RF, XG) best performance. Table 2 provides a detailed breakdown of the selected features for each model.

Table 2. Selected tabular features present in the best performance of the experiments conducted with the models (MLP, RF, XGBoost).

Model	Qnt	Features
MLP	5	NSVT, Amiodarone, Moderate Classification, B2 2005 Classification Guideline, Intermediate Rassi escore
RF	5	NSVT, Amiodarone, Moderate Classification, B2 2005 Classification Guideline, Total VE
XGB	10	NSVT, Amiodarone, Moderate Classification, B2 2005 Classification Guideline, Intermediate Rassi escore, Total VE, Low Rassi escore, Rassi points, NYHA class 1, NYHA class 2

The results demonstrate four features (NSVT, Amiodarone, "Moderate Classification", and the "B2 2005 Classification Guideline") were selected in all models. Additionally, the features Total VE and "Intermediate Rassi Score" appear in two of the models. This consistency indicates the potential relevance of these features to the prediction of SCD, which is crucial for comparing and validating these selected attributes with the state-of-the-art in medical literature.

The best results from the feature selection process and from using all the features are presented for each model. In the first scenario, the best performance of the feature selection process was achieved by the Multilayer Perceptron (MLP) with 5 features, the Random Forest (RF) with 5 features, and the XGBoost (XGB) with 10 features. These models were trained using only tabular data as input. In the second scenario, the RNN output was concatenated with the 5 best-performing features selected from the MLP model in scenario 1 and all 63 tabular features.

To select the best performance for each tested configuration, we prioritized recall (sensitivity) and AUC (Area Under the ROC Curve). This choice reflects the critical need to accurately identify patients at high risk for SCD, maximizing the detection of potential cases consistent with other studies [1,4,6,22,23]. The Table 3 displays the models used, the number of features employed, and the evaluation metrics including ACC (accuracy), AUC, Recall (sensitivity), precision, and F1 Score, along with their respective results and standard deviations. Important to note, the feature quantity of 63 in the table is regarding the 49 features after one-hot encoding.

Table 3. Prediction results of the XGB, MLP, and RF models in scenario 1 using only tabular data. Results are presented with all 63 features and the best result with feature selection using the two types of cross-validation.(S= scenario; NF = Number of features)

Model	S	NF	ACC	AUC	Recall	Precision	F1-Score
MLP	1	5	93.58 ± 0.01	90.51 ± 0.03	83.68 ± 0.03	77.74 ± 0.04	80.55 ± 0.03
MLP	1	63	94.88 ± 0.01	92.60 ± 0.02	82.31 ± 0.05	85.64 ± 0.05	83.91 ± 0.04
RF	1	5	93.00 ± 0.06	95.38 ± 0.08	89.33 ± 0.18	78.40 ± 0.21	81.03 ± 0.15
RF	1	63	94.50 ± 0.02	96.49 ± 0.01	77.69 ± 0.09	87.14 ± 0.04	81.89 ± 0.06
XGB	1	10	94.63 ± 0.02	91.44 ± 0.06	83.85 ± 0.08	83.55 ± 0.08	83.54 ± 0.07
XGB	1	63	94.19 ± 0.02	95.03 ± 0.03	76.85 ± 0.08	85.48 ± 0.06	80.82 ± 0.07
RNN+MLP	2	27+5	90.13 ± 0.06	81.08 ± 0.14	90.13 ± 0.06	90.94 ± 0.07	89.80 ± 0.07
RNN+MLP	2	27+63	91.63 ± 0.07	81.89 ± 0.17	91.63 ± 0.07	91.20 ± 0.09	90.88 ± 0.07

Regarding only the scenario 1 with tabular data, the RF model performed better with 5 features obtained 89.33% recall and 95.38% AUC. However, it presented a significant standard deviation in recall of almost 20%. In second place, the XGB model with 10 features acquired 83.85% recall and 91.44% AUC, performing slightly better than the MLP with 5 features, that reached 83.68% recall and 90.51% AUC.

When evaluating the three models (MLP, RF, XGB), we observed that feature selection did not result in a significant performance loss; in fact, it improved the recall metric. This finding suggests that some of the features might not be relevant to the context. The high dimensionality of data can be a challenge for developing intelligent systems. However, reducing the number of features offers several advantages. These include creating systems that are interoperable for domain experts, leading to faster, more accurate decision-making and improved patient care (or other relevant outcome).

In the second scenario, where the time series is added, the experiments obtained 91.63% recall with all features (27 temporal series + 63 tabular features) and 90.13% with selected features (27 temporal series + 5 tabular features). Interestingly, the approach that concatenates all features (63 tabular) yielded better results for all metrics compared with concatenating only the feature selected (5 tabular).

Analyzing all results of both scenarios, the RNN+MLP hybrid model applied in Scenario 2, using sequential and non-sequential information, obtained a recall improvement. To assess the significance of this improvement, we performed a hypothesis test with a significance level (alpha) of 0.05, using the techniques Wilcoxon signed-rank test [26] and the Student's t-test [9]. The null hypothesis (H0) was defined as follows: the recall of the combined sequential and non-sequential data is not significantly higher than the recall using only non-sequential data.

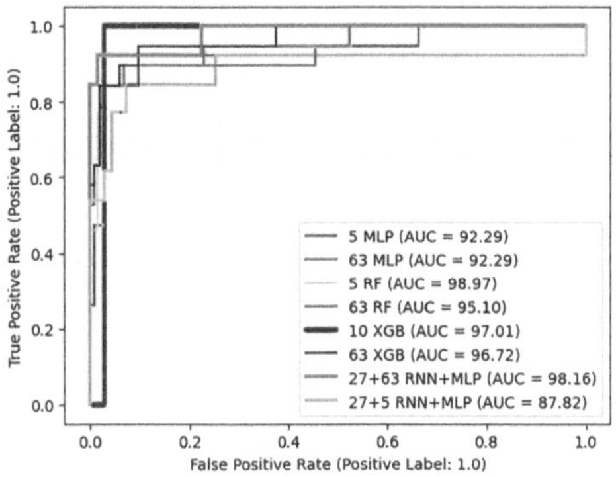

Fig. 1. ROC curves from all models (MLP, RF, XGB, RNN+MLP) in scenarios 1 and 2.

The p-values obtained for the RNN+MLP model with 27 temporal features and 63 tabular features were highly significant (1.95e-3 for Wilcoxon and 1.47e-5 for Student's t-test), leading to a strong rejection of H0. Similarly, for the model with 5 features, the p-values (1.95e-2 for Wilcoxon signed-rank and 7.21e-3 for Student's t-test) also resulted in H0 rejection. These results support the conclusion that combining time series and tabular data statistically improves recall. The Figure (Fig. 1) shows the ROC curves for all models (MLP, RF, XGB, RNN+MLP) in scenarios 1 and 2.

For comparative purposes, Table 4 shows the performance of SCD-related works in ChHD and non-ChHD contexts with the same parameter (recall) as this study. The results from both approaches were similar to the other works, the recall from RNN+MLP (27 temporal features + 63 tabular features) being the second highest one. Despite methodological differences, our results seem relevant to the SCD scenario in ChHD.

Table 4. Performance of the machine learning model of studies on predicting SCD risk, considering sensitivity as a reference parameter.

Study	SCD in CC	Recall
[24]	No	70.00%
[28]	No	96.00%
[11]	No	89.00%
[2]	No	88.40%
[4]	Yes	90.63%
RNN+MLP (27 temporal features + 5 tabular features)	Yes	90.13%
RNN+MLP (27 temporal features + 63 tabular features)	**Yes**	**91.63%**

4.2 User Interface

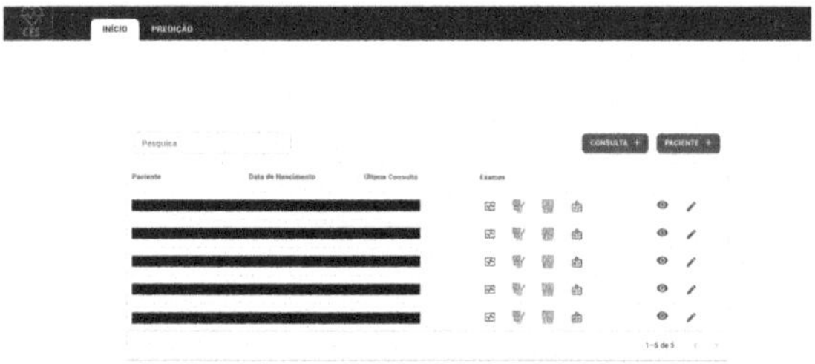

Fig. 2. Home Screen that provides access to all system functionalities.

To empower domain experts with the ability to predict SCD in CC, the MLP model with 5 features from Scenario 1 was exported using the Pickle5 library. This exported model served as the initial foundation for a initial Web prediction system. We proposed a prototype intelligent system with six key functionalities: Login, Patient Registration, Appointment Scheduling, Exam Registration, Prediction, and Model Visualization. The application's login was created using OAuth authentication with the Google platform, that offers secure and convenient access by leveraging existing Google credentials. Upon successful login, the user is redirected to the application's main screen, depicted in Fig. 2.

This screen serves as a central hub for managing patients. It lists all registered patients, displaying their names, birth dates, last appointment details, and quick access buttons for requesting relevant cardiac exams (e.g., ECG, ECO, ECG-Holter) to aid in SCD risk assessment. Additionally, users can register patient data, input clinical or exam-related information, allowing the system to function as basic medical record software.

Regarding SCD risk prediction functionality, users can access the dedicated screen by clicking on the 'Predict' in top menu of the main screen, as depicted

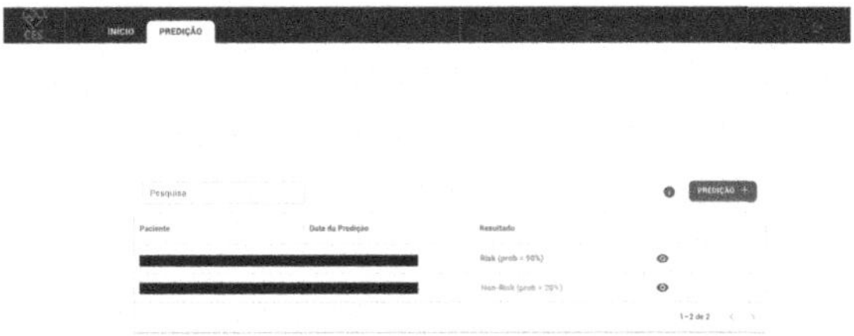

Fig. 3. Cardiology risk prediction listing screen

in Fig. 3. This screen displays a grid summarizing all SCD risk predictions made in the system. Each entry in the grid shows the patient's name, prediction date, predicted result (Risk or non-Risk, and Probability), and a button to view the features values used for the prediction.

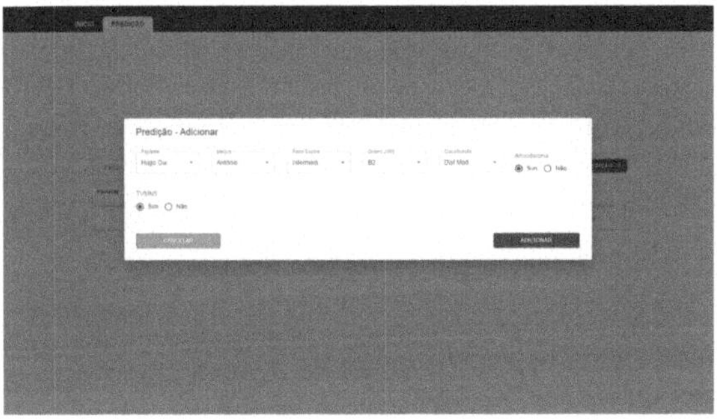

Fig. 4. Add prediction screen

After clicking the "Predict +" button, the system redirects the user to a screen for inputting new prediction data (Fig. 4). This screen includes all fields used as input for the predictive model, including patient demographics (name, physician) and five selected tabular features (NSVT, Amiodarone classification, Rassi score). The very simple and user-friendly prototype system was developed as a web platform to facilitate easy access and use by doctors, ultimately assisting them in patient care.

5 Conclusion

This work proposes the use of Multimodal and Hybrid Models to assess SCD risk in non-high-risk CC patients. Furthermore, a prototype IS has been developed to support healthcare professionals in identifying patients with potential SCD risk, aiding in early diagnosis and intervention.

Encouragingly, despite a limited sample size, the study achieved promising results. Combining time series data with tabular data using a hybrid RNN+MLP model improved recall, precision, and F1-score performance, regardless of feature selection. It's important to note that time series feature extraction is an ongoing area of research, and this analysis represents an initial exploration of the findings. This work serves as a foundation for future research aimed at developing even more robust models that can assist in the early diagnosis of SCD risk in CC patients who are not classified as high-risk. These models have the potential to address the current limitations in accurately predicting SCD in this patient population.

Future work includes reprocessing the models applying feature selection techniques to the temporal data, comparing the results of classical models with results of other models that have the capability to handle different types of data, applying XAI techniques to enhance the interpretability of predictions for specialists, and exploring other RNN combination approaches to potentially improve model performance and generalizability.

Acknowledgments. The authors thank the Federal Institute of Education, Science, and Technology of Ceará (IFCE) for its support through the PIBIC Nž7/2023 and PIBITI Nž9/2023 calls from the PRPI/REITORIA-IFCE.

References

1. Alberto, A.C., Pedrosa, R.C., Zarzoso, V., Nadal, J.: Association between circadian holter ecg changes and sudden cardiac death in patients with chagas heart disease. Physiol. Meas. **41**(2), 025006 (2020)
2. Atallah, J., Gonzalez Corcia, M.C., Walsh, E.P.: Ventricular arrhythmia and life-threatening events in patients with repaired tetralogy of fallot. Am. J. Cardiol. **132**, 126–132 (2020)
3. Barker, J., et al.: Machine learning in sudden cardiac death risk prediction: a systematic review. Europace **24**(11), 1777–1787 (2022)
4. Cavalcante, C.H., et al.: Sudden cardiac death multiparametric classification system for chagas heart disease's patients based on clinical data and 24-hours ecg monitoring. Math. Biosci. Eng. **20**(5), 9159–9178 (2023)
5. Elman, J.L.: Finding structure in time. Cogn. Sci. **14**(2), 179–211 (1990)
6. Goldstein, B.A., Chang, T.I., Mitani, A.A., Assimes, T.L., Winkelmayer, W.C.: Near-term prediction of sudden cardiac death in older hemodialysis patients using electronic health records. Clin. J. Am. Soc. Nephrol. **9**(1), 82–91 (2014)
7. Grinsztajn, L., Oyallon, E., Varoquaux, G.: Why do tree-based models still outperform deep learning on typical tabular data? Adv. Neural. Inf. Process. Syst. **35**, 507–520 (2022)
8. Hinkle, L.E., Jr., Thaler, H.T.: Clinical classification of cardiac deaths. Circulation **65**(3), 457–464 (1982)
9. Johnston, L.: Student's t-test. J. Qual. Technol. **2**(4), 243–245 (1970)
10. Knechtel, M.d.R.: Metodologia da pesquisa em educação: uma abordagem teórico-prática dialogada. Curitiba: Intersaberes, 106 (2014)

11. Lee, S., et al.: Territory-wide cohort study of brugada syndrome in hong kong: predictors of long-term outcomes using random survival forests and non-negative matrix factorisation. Open Heart **8**(1), e001505 (2021)
12. Leoni, L., et al.: Bridging implantable cardioverter-defibrillator patients undergoing radiotherapy with a wearable cardioverter-defibrillator. Cardiology: Open Access **8**(1), 24–27 (2023)
13. Lopez-Jimenez, F., et al.: Artificial intelligence in cardiology: present and future. In: Mayo Clinic Proceedings, vol. 95, pp. 1015–1039. Elsevier (2020)
14. Malik, L.H., Singh, G.D., Amsterdam, E.A.: Chagas heart disease: an update. Am. J. Med. **128**(11), 1251-e7 (2015)
15. Members, A.F., Priori, S.G., Blomström-Lundqvist, C., Mazzanti, A., Blom, N., Borggrefe, M., Camm, J., Elliott, P.M., Fitzsimons, D., Hatala, R., et al.: 2015 esc guidelines for the management of patients with ventricular arrhythmias and the prevention of sudden cardiac death: The task force for the management of patients with ventricular arrhythmias and the prevention of sudden cardiac death of the european society of cardiology (esc) endorsed by: Association for european paediatric and congenital cardiology (aepc). Ep Europace **17**(11), 1601–1687 (2015)
16. Nicolson, W.B., et al.: Prospective evaluation of two novel ecg-based restitution biomarkers for prediction of sudden cardiac death risk in ischaemic cardiomyopathy. Heart **100**(23), 1878–1885 (2014)
17. Nunes, J.M., Infante, M.: Pesquisa-ação: uma metodologia de consultoria. FIOCRUZ, Rio de Janeiro (1996)
18. Pedrosa, R.C., et al.: Risk stratifier for sudden cardiac death beyond the left ventricular ejection fraction in chagas cardiomyopathy. Pacing Clini. Electrophysiol. **47**(2), 312–320 (2024)
19. Pérez-Molina, J.A., Molina, I.: Chagas disease cardiomyopathy treatment remains a challenge-authors' reply. Lancet **391**(10136), 2209–2210 (2018)
20. Rassi, A., Jr., Rassi, A.: Another disappointing result with implantable cardioverter-defibrillator therapy in patients with chagas disease. Europace **15**(9), 1383 (2013)
21. Schwartz, P., La Rovere, M., Vanoli, E.: Autonomic nervous system and sudden cardiac death. experimental basis and clinical observations for post-myocardial infarction risk stratification. Circulation **85**(1 Suppl), I77–91 (1992)
22. Shakibfar, S., et al.: Predicting electrical storms by remote monitoring of implantable cardioverter-defibrillator patients using machine learning. Ep Europace **21**(2), 268–274 (2019)
23. de Souza, A.C.J., et al.: Development of a risk score to predict sudden death in patients with chaga's heart disease. Inter. J. Cardiol. **187**, 700–704 (2015)
24. Tse, G., et al.: Incorporating latent variables using nonnegative matrix factorization improves risk stratification in brugada syndrome. J. Am. Heart Assoc. **9**(22), e012714 (2020)
25. Uddin, S., Lu, H.: Confirming the statistically significant superiority of tree-based machine learning algorithms over their counterparts for tabular data. PLoS ONE **19**(4), e0301541 (2024)
26. Woolson, R.F.: Wilcoxon signed-rank test. Encyclopedia Biostatist. **8** (2005)
27. World Health Organization: Chagas disease (American trypanosomiasis) (2021). https://www.who.int/news-room/fact-sheets/detail/chagas-disease-(american-trypanosomiasis)
28. Zoni-Berisso, M., Molini, D., Viani, S., Mela, G.S., Delfino, L.: Noninvasive prediction of sudden death and sustained ventricular tachycardia after acute myocardial infarction using a neural network algorithm. Ital. Heart. J. **2**, 612–620 (2001)

On the Equivalence Between Logic Programs and Bipolar Argumentation Frameworks

Renan Cordeiro(✉) and João Alcântara

Department of Computer Science, Federal University of Ceará, Fortaleza, Brazil
renandcsc@alu.ufc.br, jnando@dc.ufc.br

Abstract. In this paper, we extend the connection between argumentation formalisms and logic programs, by showing the equivalence between bipolar argumentation β-semantics and various 3-valued logic programming semantics. In particular, we show that β-semi-stable semantics corresponds to L-stable semantics, which has been previously shown not to be expressible by traditional attack-only argumentation frameworks. Besides enhancing our understanding of how logic programs relate to argumentation with support, we provide direct translations between them and their semantics, allowing their interchangeable use.

Keywords: Equivalence · Bipolar argumentation · Logic programming

1 Introduction

Formal argumentation and logic programming are closely related paradigms [7] for representing possibly contradictory, uncertain or incomplete information. In formal argumentation, an argument represents a claim whose acceptability is determined by some criteria (referred to as semantics) that depend on how arguments interact with each other. Dung introduced Abstract Argumentation Frameworks (AAFs) in his seminal work [10], defining an AAF by a set of arguments and an attack relation between them. In AAFs, an argument is treated as an abstract entity without any internal structure, which means that semantics can only take into account the interactions between arguments. In Normal Logic Programs (NLPs), a claim is represented by an atom whose evaluation consists in determining whether it can be inferred from logical rules.

Despite expressing information in distinctive perspectives, formal argumentation and logic programming are closely connected. Their connection is a very relevant line of research, as it allows one to view logic programs as argumentation frameworks and vice versa, allowing the most suitable formalism to be employed given a specific application. Besides, algorithms, techniques and semantics developed for a formalism can be adapted to the other. Not surprisingly, equivalences were extensively investigated. There is correspondence between complete and

A. Paes and F. A. N. Verri (Eds.): BRACIS 2024, LNAI 15413, pp. 238–253, 2025.
https://doi.org/10.1007/978-3-031-79032-4_17

partial stable semantics [16], grounded and well-founded semantics [10], preferred and regular semantics [7], stable (argumentation) and stable (logic programming) semantics [10]. However, there exists a logic programming semantics (L-stable) that cannot be captured by any abstract argumentation semantics [7].

In this paper, we continue this line of research by considering instead Bipolar Argumentation Frameworks (*BAF*s [5]), in which additionally there is a support relation between arguments: a positive interaction independent from the attack relation. Although indirectly and not for all semantics, bipolar argumentation can be linked to logic programming by the current literature. In [9], *BAF*s are represented by Assumption-Based Argumentation (*ABA*) for admissible, preferred and stable semantics and four interpretations of support. As *ABA* is linked to logic programming [8], the transitive nature of these correspondences induces an indirect connection between *BAF*s and *NLP*s. However, the semi-stable semantics of *ABA* does not correspond in general to the L-stable semantics of *NLP*s [8]. A common aspect of these interpretations of support is their asymmetric treatment of attacks and supports. The former are favoured, as an argument is always rejected when it is both attacked and supported by accepted arguments. We instead resort to the interpretation of support from the β-semantics [1], in which an argument is always accepted if it is supported by an accepted argument.

Several works also relate more expressive argumentation formalisms to logic programming. For instance, abstract dialectical frameworks [2,6,14], *ABA* [8], claim-augmented argumentation [13,15], frameworks with collective attacks [3,15] and many others [4] are connected to logic programming. The work closest to ours is [13], where the authors relate Bipolar Claim-augmented Argumentation Frameworks (*BCAF*s) to *NLP*s. They could establish a one-to-one correspondence between *BCAF*s and *NLP*s in both directions, whilst preserving the main acceptability semantics, including the L-stable semantics. The main distinction between *BCAF*s and *BAF*s (and hence between their work and ours) is that *BCAF*s have additional structure, as each argument is associated to an explicit claim, whereas arguments in a *BAF* are abstract: the β-semantics depend exclusively on the attack and support relation between arguments. In contrast, the semi-stable semantics of *BCAF*s [13] take into account the conclusion of arguments, information not inferred solely from the attack and support relations.

This paper is divided as follows: in Sect. 2, we briefly recall the foundations of Bipolar Argumentation Frameworks (*BAF*s) and their β-semantics, and of Normal Logic Programs (*NLP*s) and their 3-valued semantics; in the sequence, we provide a translation from *NLP*s to *BAF*s and between their semantics; the inverse direction from *BAF*s to *NLP*s is discussed in Sect. 4; we collect our conclusions in Sect. 5.

2 Preliminaries

2.1 Bipolar Argumentation Frameworks (*BAFs*)

In [5], Dung's Abstract Argumentation Frameworks (*AAF*s) were extended to incorporate an explicit notion of support between arguments, independent from the attack relation. The resulting framework, called *BAF*, is defined next:

Definition 1 (BAF [5]). *A Bipolar Argumentation Framework (BAF) is a tuple* $(\mathcal{A}, Att, Sup)$, *in which* $\mathcal{A}$ *is a set of arguments,* $Att \subseteq \mathcal{A} \times \mathcal{A}$ *is the attack relation and* $Sup \subseteq \mathcal{A} \times \mathcal{A}$ *is the support relation. For an argument* $A \in \mathcal{A}$, *we define* $Att(A) = \{B \in \mathcal{A} \mid (B, A) \in Att\}$ *as the set of attackers of* A *and* $Sup(A) = \{B \in \mathcal{A} \mid (B, A) \in Sup\}$ *as the set of direct supporters of* A.

In *AAF*s, the interaction between arguments is specified only by the attack relation. In *BAF*s, the novelty is that arguments can also support other arguments. This means that *BAF*s $(\mathcal{A}, Att, Sup)$ with $Sup = \emptyset$ amount to (standard Dung) *AAF*s.

We are interested not only in the direct supporters of an argument, but also in the indirect ones. They are indistinctly called the supporters of an argument:

Definition 2 (Supporters [1]). *Let* $\mathcal{B} = (\mathcal{A}, Att, Sup)$ *be a BAF and* $A \in \mathcal{A}$ *an argument. We define the supporters of* A *recursively as follows:* A *is a supporter of* A *in* $\mathcal{B}$*; if* A' *is a supporter of* A *in* $\mathcal{B}$ *and* $(B, A') \in Sup$, *then* B *is a supporter of* A *in* $\mathcal{B}$. *We denote the set of all supporters of* A *in* $\mathcal{B}$ *by* $\mathfrak{Sup}(A)$.

An acceptance criteria, referred to as semantics, for arguments in a *BAF* is how we determine the accepted or rejected arguments in a framework. Semantics can be equivalently characterised in terms of extensions (sets of accepted arguments) or labellings (mapping arguments to their acceptability) [1]. We will focus on labelling-based semantics as it allows a more direct comparison to logic programming semantics.

Definition 3 (Labelling). *Let* $\mathcal{B} = (\mathcal{A}, Att, Sup)$ *be a BAF. A labelling (of* $\mathcal{B}$*) is a total function* $\mathcal{L} : \mathcal{A} \rightarrow \{\mathtt{in}, \mathtt{out}, \mathtt{undec}\}$.

When $\mathcal{L}$ is a labelling, we write $\mathtt{in}(\mathcal{L})$ to denote the set $\{A \mid \mathcal{L}(A) = \mathtt{in}\}$, $\mathtt{out}(\mathcal{L})$ for $\{A \mid \mathcal{L}(A) = \mathtt{out}\}$ and $\mathtt{undec}(\mathcal{L})$ for $\{A \mid \mathcal{L}(A) = \mathtt{undec}\}$. As a labelling defines a partition among arguments, when convenient we write $\mathcal{L}$ as the triple $(\mathtt{in}(\mathcal{L}), \mathtt{out}(\mathcal{L}), \mathtt{undec}(\mathcal{L}))$. Intuitively, the label `in` indicates acceptance, the label `out` indicates rejection and the label `undec` indicates that the argument is undecided, i.e., neither accepted nor rejected. We can now describe the *BAF* semantics studied in this paper:

Definition 4 (Labelling-based Semantics for *BAF*s *[1]***).** *Let* $\mathcal{B} = (\mathcal{A}, Att, Sup)$ *be a BAF. A labelling* $\mathcal{L}$ *is a* β*-complete labelling of* $\mathcal{B}$ *if for any* $A \in \mathcal{A}$,

- $\mathcal{L}(A) = \mathtt{in}$ *if and only if there exists* $A' \in \mathfrak{Sup}(A)$ *such that for every* $B \in Att(A')$, *it holds* $\mathcal{L}(B) = \mathtt{out}$.

- $\mathcal{L}(A) = \texttt{out}$ *if and only if for every* $A' \in \mathfrak{Sup}(A)$, *there exists* $B \in Att(A')$ *such that* $\mathcal{L}(B) = \texttt{in}$.

Additionally, we say a β*-complete labelling* $\mathcal{L}$ *of* $\mathcal{B}$ *is* β*-grounded (resp.* β*-preferred) if* $\texttt{in}(\mathcal{L})$ *is minimal (resp. maximal) w. r. t.* $\subseteq$ *among the* β*-complete labellings of* $\mathcal{B}$*;* β*-stable if* $\texttt{undec}(\mathcal{L}) = \emptyset$*;* β*-semi-stable if* $\texttt{undec}(\mathcal{L})$ *is minimal (w. r. t.* $\subseteq$*) among the* β*-complete labellings of* $\mathcal{B}$.

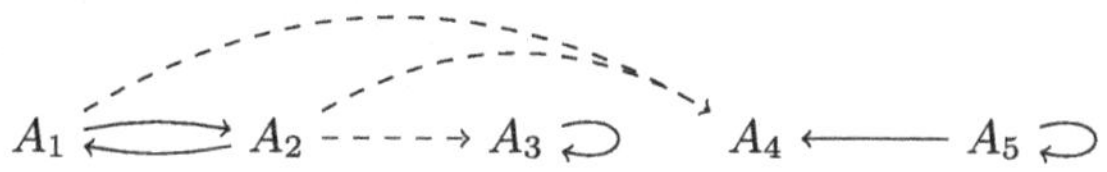

Fig. 1. *BAF*

Labelling-based semantics of *BAF*s preserve remarking properties of the corresponding labelling-based semantics of *AAF*s [1]: for every *BAF* there is always a uniquely defined β-complete labelling; every β-stable labelling is a β-preferred labelling; and if there exists β-stable labellings, they coincide with β-semi-stable and β-preferred labellings.

Example 1. Consider the *BAF* depicted by the directed graph in Fig. 1, where nodes represent arguments, solid arrows represent attacks, and dashed arrows represent the support relation. We obtain labellings as follows: β-complete labellings are $\mathcal{L}_1 = (\emptyset, \emptyset, \mathcal{A})$, $\mathcal{L}_2 = (\{A_1, A_4\}, \{A_2\}, \{A_3, A_5\})$ and $\mathcal{L}_3 = (\{A_2, A_3, A_4\}, \{A_1\}, \{A_5\})$; β-grounded labelling is $\mathcal{L}_1$; β-preferred labellings are $\mathcal{L}_2$ and $\mathcal{L}_3$; there are no β-stable labellings; and β-semi-stable labelling is $\mathcal{L}_3$.

2.2 Logic Programs and Semantics

Now, we take a look at Propositional Normal Logic Programs. To delve into their definition and semantics, we will follow the presentation outlined in [7], which draws from the foundation laid out in [12].

Definition 5. *A* rule r *is an expression*

$$r : c \leftarrow a_1, \ldots, a_m, \texttt{not } b_1, \ldots, \texttt{not } b_n \tag{1}$$

where $(m, n \geq 0)$*;* c*, each* a_i $(1 \leq i \leq m)$ *and each* b_j $(1 \leq j \leq n)$ *are atoms, and* $\texttt{not}$ *represents negation as failure. A literal is either an atom* a *(positive literal) or a negated atom* $\texttt{not } a$ *(negative literal). Given a rule* r *as above,* c *is called the* head *of* r*, which we denote as* $head(r)$*, and* $body(r) = \{a_1, \ldots, a_m, \texttt{not } b_1, \ldots, \texttt{not } b_n\}$ *is called the* body *of* r*. Further,* $body(r)$ *is divided in two sets* $body^+(r) = \{a_1, \ldots, a_m\}$ *and* $body^-(r) = \{\texttt{not } b_1, \ldots, \texttt{not } b_n\}$*. A* fact *is a rule where* $m = n = 0$*. A Normal Logic Program (NLP) or simply a* logic program P *is a finite set of rules. If every* $r \in P$ *has* $body^-(r) = \emptyset$*,* P *is a positive program. The* Herbrand Base *of* P *is the set* HB_P *of all atoms appearing in* P.

A wide range of *NLP* semantics are based on the 3-valued interpretations of programs [12]:

Definition 6 (3-Valued Herbrand Interpretation [12]**).** *A 3-valued Herbrand Interpretation $\mathcal{I}$ (or simply interpretation) of P is a total function $\mathcal{I} : HB_P \rightarrow \{\mathbf{t}, \mathbf{f}, \mathbf{u}\}$.*

Given an interpretation $\mathcal{I}$ of P, we write $\mathbf{t}(\mathcal{I})$ to denote the set $\{a \in HB_P \mid \mathcal{I}(a) = \mathbf{t}\}$, $\mathbf{f}(\mathcal{I})$ for $\{a \in HB_P \mid \mathcal{I}(a) = \mathbf{f}\}$ and $\mathbf{u}(\mathcal{I})$ for $\{a \in HB_P \mid \mathcal{I}(a) = \mathbf{u}\}$. As an interpretation defines a partition among atoms in HB_P, when convenient we write $\mathcal{I}$ as the triple $(\mathbf{t}(\mathcal{I}), \mathbf{f}(\mathcal{I}), \mathbf{u}(\mathcal{I}))$. Intuitively, the value $\mathbf{t}$ indicates the atom is true in $\mathcal{I}$, the value $\mathbf{f}$ indicates the atom is false in $\mathcal{I}$, and the value $\mathbf{u}$ indicates the atom is undefined in $\mathcal{I}$, i.e., neither true nor false.

Now we will consider the main semantics for *NLP*s. Let $\mathcal{I}$ be a 3-valued Herbrand interpretation of P; the reduct of P with respect to $\mathcal{I}$ (written as $P/\mathcal{I}$) is the *NLP* constructed using the following steps:

1. Remove any $a \leftarrow a_1, \ldots, a_m, \texttt{not } b_1, \ldots, \texttt{not } b_n \in P$ such that $\mathcal{I}(b_j) = \mathbf{t}$ for some j $(1 \leq j \leq n)$;
2. Afterwards, remove any occurrence of $\texttt{not } b_j$ from P such that $\mathcal{I}(b_j) = \mathbf{f}$.
3. Then, replace any occurrence of $\texttt{not } b_j$ left by a special atom $\mathbf{u}$ ($\mathbf{u} \notin HB_P$).

In the above procedure, $\mathbf{u}$ is assumed to be an atom not in HB_P which is undefined in all interpretations of P (a constant). Note that $P/\mathcal{I}$ is a positive program since all negative literals have been removed. As a consequence, $P/\mathcal{I}$ has a unique least 3-valued model [12] denoted as $\Psi_P(\mathcal{I})$[1] with minimal $\mathbf{t}(\Psi_P(\mathcal{I}))$ and maximal $\mathbf{f}(\Psi_P(\mathcal{I}))$ (w.r.t. set inclusion) such that, for every $a \in HB_P$, (i) $a \in \mathbf{t}(\Psi_P(\mathcal{I}))$ if there is a rule $r' \in P/\mathcal{I}$ with $head(r') = a$ and $body^+(r') \subseteq \mathbf{t}(\Psi_P(\mathcal{I}))$; and (ii) $a \in \mathbf{f}(\Psi_P(\mathcal{I}))$ if for every rule $r' \in P/\mathcal{I}$ with $head(r') = a$, it holds $body^+(r') \cap \mathbf{f}(\Psi_P(\mathcal{I})) \neq \emptyset$.

We can now describe the *NLP* semantics studied in this paper:

Definition 7 *Let P be an NLP and $\mathcal{I}$ be an interpretation: $\mathcal{I}$ is a partial stable model of P iff $\Psi_P(\mathcal{I}) = \mathcal{I}$ [12]; $\mathcal{I}$ is a well-founded (resp. regular) model of P iff $\Psi_P(\mathcal{I}) = \mathcal{I}$ and $\mathbf{t}(\mathcal{I})$ is minimal (resp. maximal) w.r.t. set inclusion among all partial stable models of P [11, 12]; $\mathcal{I}$ is a (2-valued) stable model of P iff $\Psi_P(\mathcal{I}) = \mathcal{I}$ and $\mathbf{u}(\mathcal{I}) = \emptyset$ [12]; $\mathcal{I}$ is an L-stable model of P iff $\Psi_P(\mathcal{I}) = \mathcal{I}$ and $\mathbf{u}(\mathcal{I})$ is minimal w.r.t. set inclusion among all partial stable models of P [11].*

Although some of these definitions are not standard in logic programming literature, their equivalence is proved in [7]. This format helps us to relate *NLP* and *BAF* semantics due to the structural similarities between Definition 7 and Definitions 3 and 4. We illustrate these semantics in the following example:

[1] The above definition consists of a least fix-point of the immediate consequence operator Ψ defined in [12], which is guaranteed to exist and be unique for positive programs.

Example 2. Consider the following *NLP* P:

$$r_0 : d \leftarrow \texttt{not } d \qquad r_1 : c \leftarrow \texttt{not } c, \texttt{not } d$$
$$r_2 : a \leftarrow \texttt{not } b \qquad r_3 : b \leftarrow \texttt{not } a$$
$$r_4 : c \leftarrow \texttt{not } c, \texttt{not } a \quad r_5 : e \leftarrow \texttt{not } e, \texttt{not } b$$

This program has partial stable models $\mathcal{M}_1 = (\emptyset, \emptyset, HB_P)$, $\mathcal{M}_2 = (\{a\}, \{b\}, \{c, d, e\})$, $\mathcal{M}_3 = (\{b\}, \{a, e\}, \{c, d\})$; well-founded model is $\mathcal{M}_1$; regular models are $\mathcal{M}_2$ and $\mathcal{M}_3$; there are no stable models; and L-stable model is $\mathcal{M}_3$.

3 From *NLP*s to *BAF*s

In this section, we devise a translation from *NLP*s to *BAF*s. We wish to show that it preserves the main semantics considered in this paper, including the L-stable semantics. We start by creating arguments for derivations of rules of P, exactly as done in [7].

Definition 8. *[7] Let P be an NLP and $\mathcal{A}_P$ be the set of arguments recursively defined below.*

- *If $c \leftarrow \texttt{not } b_1, \ldots, \texttt{not } b_m$ is a rule in P, it is also an argument (say A) with $\texttt{Conc}(A) = c$, $\texttt{Rules}(A) = \{c \leftarrow \texttt{not } b_1, \ldots, \texttt{not } b_m\}$, $\texttt{Vul}(A) = \{b_1, \ldots, b_m\}$, and $\texttt{Sub}(A) = \{A\}$.*
- *If $c \leftarrow a_1, \ldots, a_n, \texttt{not } b_1, \ldots, \texttt{not } b_m$ is a rule (say r) in P and for each a_i $(1 \leq i \leq n)$ there exists an argument A_i with $\texttt{Conc}(A_i) = a_i$ and $r \notin \texttt{Rules}(A_i)$, then r derives the argument $c \leftarrow (A_1), \ldots, (A_n), \texttt{not } b_1, \ldots, \texttt{not } b_m$ (say A) with $\texttt{Conc}(A) = c$, $\texttt{Rules}(A) = \bigcup_{i=1}^{n} \texttt{Rules}(A_i) \cup \{r\}$, $\texttt{Vul}(A) = \bigcup_{i=1}^{n} \texttt{Vul}(A_i) \cup \{b_1, \ldots, b_m\}$, and $\texttt{Sub}(A) = \{A\} \cup \bigcup_{i=1}^{n} \texttt{Sub}(A_i)$.*

For any argument A, we say $\texttt{Conc}(A)$ is its conclusion; $\texttt{Rules}(A)$, its set of rules; $\texttt{Vul}(A)$, its set of vulnerabilities; and $\texttt{Sub}(A)$, its set of subarguments. We can intuitively see an argument as a tree-like structure representing a possible derivation of an atom from the rules of a program.

Example 3. Consider the *NLP* P below with rules $\{r_1, \ldots, r_8\}$:

$$r_1 : a \qquad r_2 : b \leftarrow a \qquad r_3 : c \leftarrow \texttt{not } c \qquad r_7 : c \leftarrow f, \texttt{not } g$$
$$r_4 : d \leftarrow b, \texttt{not } a \quad r_5 : d \leftarrow \texttt{not } c \quad r_6 : e \leftarrow b, c, \texttt{not } e \quad r_8 : f \leftarrow c, g$$

According to Definition 8, we can construct the following statements from P:

$$A_1 : a \qquad A_2 : b \leftarrow (A_1) \qquad A_3 : c \leftarrow \texttt{not } c$$
$$A_4 : d \leftarrow (A_2), \texttt{not } a \quad A_5 : d \leftarrow \texttt{not } c \quad A_6 : e \leftarrow (A_2), (A_3), \texttt{not } e$$

Next, we give the conclusions and vulnerabilities of each argument:

	A_1	A_2	A_3	A_4	A_5	A_6
$\texttt{Conc}(.)$	a	b	c	d	d	e
$\texttt{Vul}(.)$	$\emptyset$	$\emptyset$	$\{c\}$	$\{a\}$	$\{c\}$	$\{c, e\}$

The vulnerabilities of an argument A are associated with the negative literals found in the derivation of A. If not b is one of them, then b is a vulnerability of A. This means that if b is derived, then $\texttt{Conc}(A)$ cannot be obtained via the derivation represented by A. However, it can still be obtained via other derivations/arguments. For instance, in the program P of Example 3, the derivation of a suffices to prevent the derivation of d via argument A_4 (for that reason, $a \in \texttt{Vul}(A_4)$), but we still can derive d via A_5. Notice also that there are no arguments with conclusions f and g.

Now we are entitled to determine both the attack and support relations:

Definition 9. *Let P be an NLP and let A and B be arguments in the sense of Definition 8. We define the attack relation Att_P and the support relation Sup_P as follows: $(A, B) \in Att_P$ iff $\texttt{Conc}(A) \in \texttt{Vul}(B)$; $(A, B) \in Sup_P$ iff $A \neq B$ and $\texttt{Conc}(A) = \texttt{Conc}(B)$.*

Our definition of the set Att_P of attacks obtained from an *NLP* P is exactly as it has been characterised in [7]. The novelty here is the definition of Sup_P, which in our proposal is determined by arguments with the same conclusion (the condition $A \neq B$ is just to avoid incorporating the redundant case of an argument supporting itself). The failure in guaranteeing the correspondence between β-semi-stable and L-stable semantics, as discussed in [7], lies in the presence of arguments with distinct acceptabilities representing the derivations of the same atom. As Definition 9 ensures that such arguments support each other, they will have the same acceptability according to the β-semantics [1]. For the arguments of Example 3, A_1 attacks A_4; A_3 attacks itself, A_5 and A_6; A_6 attacks itself. We also have A_4 and A_5 supporting each other. Using the thus-defined concepts of arguments, attacks and supports, one can define the Bipolar Argumentation Framework corresponding to a particular *NLP*:

Definition 10. *Let P be an NLP. We define its corresponding BAF as $\mathcal{B}_P = (\mathcal{A}_P, Att_P, Sup_P)$, where $\mathcal{A}_P$ is the set of arguments in the sense of Definition 8 and Att_P is the attack relation in the sense of Definition 9.*

Although the internal structure of arguments constructed from Definition 8 is used to construct Att_P and Sup_P, any argument of $\mathcal{B}_P$ is abstract regarding *BAF* semantics, which depend exclusively on Att_P and Sup_P. As an example, the corresponding *BAF* of the *NLP* of Example 3 is depicted in Fig. 2.

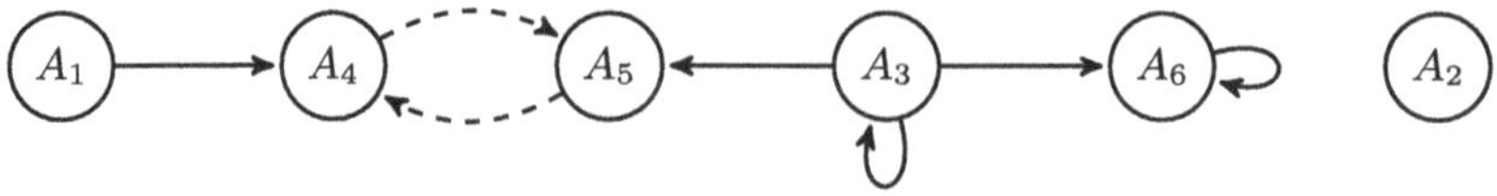

Fig. 2. A *BAF* $\mathcal{B}_P = (\mathcal{A}_P, Att_P, Sup_P)$

Once the *BAF* has been constructed, we show the equivalence between the semantics for P and their counterpart for the corresponding *BAF* $\mathcal{B}_P$. We prove the equivalence results by identifying connections between fundamental notions used in the definition of the semantics for *NLP*s and *BAF*s. For this purpose, we introduce two functions: $\mathcal{L}2\mathcal{I}_P$ associates an interpretation to each labelling, while $\mathcal{I}2\mathcal{L}_P$ associates a labelling to each interpretation. We then investigate the conditions under which they are each other's inverses and employ these results to prove the equivalence between the semantics. These functions essentially permit us to treat interpretations and labellings interchangeably.

Definition 11 (**$\mathcal{L}2\mathcal{I}_P$ and $\mathcal{I}2\mathcal{L}_P$ Functions**). *Let P be an NLP, $\mathcal{B}_P = (\mathcal{A}_P, Att_P, Sup_P)$ be its corresponding BAF, $\mathcal{I}nt$ be the set of all the 3-valued interpretations of P and $\mathcal{L}ab$ be the set of all labellings of $\mathcal{B}_p$.*

We introduce a function $\mathcal{L}2\mathcal{I}_P : \mathcal{L}ab \rightarrow \mathcal{I}nt$ where $\mathcal{L}2\mathcal{I}_P(\mathcal{L}) = (T, F, HB_P - (T \cup F))$, in which $T = \{c \in HB_P \mid \exists A \in \mathtt{in}(\mathcal{L})$ with $\mathtt{Conc}(A) = c\}$; and $F = \{c \in HB_P \mid \mathcal{L}(A) = \mathtt{out}$ for every $A \in \mathcal{A}_P$ such that $\mathtt{Conc}(A) = c\}$.

We introduce a function $\mathcal{I}2\mathcal{L}_P : \mathcal{I}nt \rightarrow \mathcal{L}ab$ such that for any $\mathcal{I} \in \mathcal{I}nt$ and any $A \in \mathcal{A}_P$, $\mathcal{I}2\mathcal{L}_P(\mathcal{I})(A) = \mathtt{in}$ if $\mathcal{I}(\mathtt{Conc}(A)) = \mathbf{t}$; $\mathcal{I}2\mathcal{L}_P(\mathcal{I})(A) = \mathtt{out}$ if $\mathcal{I}(\mathtt{Conc}(A)) = \mathbf{f}$; $\mathcal{I}2\mathcal{L}_P(\mathcal{I})(A) = \mathtt{undec}$ if $\mathcal{I}(\mathtt{Conc}(A)) = \mathbf{u}$.

In general, $\mathcal{I}2\mathcal{L}_P(\mathcal{L}2\mathcal{I}_P(\mathcal{L}))$ is not equal to $\mathcal{L}$. For instance, considering the labelling $\mathcal{L} = (\emptyset, \{A_4\}, \{A_1, A_2, A_3, A_5, A_6\})$ and the *BAF* $\mathcal{B}_P$ of Fig. 2, we have that $\mathcal{I}2\mathcal{L}_P(\mathcal{L}2\mathcal{I}_P(\mathcal{L})) = (\emptyset, \emptyset, \{A_1, A_2, A_3, A_4, A_5, A_6\})$. Such an inequality will always occur when two arguments with the same conclusion receive distinct labels. In our example, A_4 and A_5 have the same conclusion d; $\mathcal{L}(A_4) = \mathtt{out}$, but $\mathcal{L}(A_5) = \mathtt{undec}$. It holds $\mathcal{L}2\mathcal{I}_P(\mathcal{L})(d) = \mathbf{u}$, and consequently, $\mathcal{I}2\mathcal{L}_P(\mathcal{L}2\mathcal{I}_P(\mathcal{L}))(A_4) = \mathcal{I}2\mathcal{L}_P(\mathcal{L}2\mathcal{I}_P(\mathcal{L}))(A_5) = \mathtt{undec}$, i.e., $\mathcal{I}2\mathcal{L}_P(\mathcal{L}2\mathcal{I}_P(\mathcal{L}))$ is not equal to $\mathcal{L}$. However, if $\mathcal{L}$ is a β-complete labelling of $\mathcal{B}_P$, arguments with the same conclusion will always have the same label:

Theorem 1. *Let $\mathcal{B}_P = (\mathcal{A}_P, Att_P, Sup_P)$ be the corresponding BAF of P. If $\mathcal{L}$ is a β-complete labelling of $\mathcal{B}_P$ and $\mathtt{Conc}(A) = \mathtt{Conc}(B)$, then $\mathcal{L}(A) = \mathcal{L}(B)$.*

Proof. Let $A, B \in \mathcal{A}_P$ with $\mathtt{Conc}(A) = \mathtt{Conc}(B)$ and $\mathcal{L}$ be a β-complete labelling of $\mathcal{B}_P$. The result holds trivially for $A = B$. Assume $A \neq B$. By Definition 9, $(A, B) \in Sup_P$ and $(B, A) \in Sup_P$. As $\mathcal{L}$ is β-complete, $\mathcal{L}(A) = \mathcal{L}(B)$.

Thus, if $\mathcal{L}$ is a β-complete labelling, we have that $\mathcal{I}2\mathcal{L}_P(\mathcal{L}2\mathcal{I}_P(\mathcal{L})) = \mathcal{L}$:

Theorem 2. *Let $\mathcal{B}_P = (\mathcal{A}_P, Att_P, Sup_P)$ be the corresponding BAF of P. For any β-complete labelling $\mathcal{L}$ of $\mathcal{B}_P$, it holds $\mathcal{I}2\mathcal{L}_P(\mathcal{L}2\mathcal{I}_P(\mathcal{L})) = \mathcal{L}$.*

Proof. Let $A \in \mathcal{A}_P$. There are 3 cases: $\mathcal{L}(A) = \mathtt{in} \Rightarrow \mathcal{L}2\mathcal{I}_P(\mathcal{L})(\mathtt{Conc}(A)) = \mathbf{t} \Rightarrow \mathcal{I}2\mathcal{L}_P(\mathcal{L}2\mathcal{I}_P(\mathcal{L}))(A) = \mathtt{in}$; $\mathcal{L}(A) = \mathtt{out} \Rightarrow$ (Theorem 1) $\mathcal{L}(A') = \mathtt{out}$ for every $A' \in \mathcal{A}_P$ such that $\mathtt{Conc}(A') = \mathtt{Conc}(A) \Rightarrow \mathcal{L}2\mathcal{I}_P(\mathcal{L})(\mathtt{Conc}(A)) = \mathbf{f} \Rightarrow \mathcal{I}2\mathcal{L}_P(\mathcal{L}2\mathcal{I}_P(\mathcal{L}))(A) = \mathtt{out}$; $\mathcal{L}(A) = \mathtt{undec} \Rightarrow$ (Theorem 1) $\mathcal{L}(A') = \mathtt{undec}$ for every $A' \in \mathcal{A}_P$ such that $\mathtt{Conc}(A') = \mathtt{Conc}(A) \Rightarrow \mathcal{L}2\mathcal{I}_P(\mathcal{L})(\mathtt{Conc}(A)) = \mathbf{u} \Rightarrow \mathcal{I}2\mathcal{L}_P(\mathcal{L}2\mathcal{I}_P(\mathcal{L}))(A) = \mathtt{undec}$.

In general, $\mathcal{L}2\mathcal{I}_P(\mathcal{I}2\mathcal{L}_P(\mathcal{I}))$ is not equal to $\mathcal{I}$. Consider an atom $c \in HB_P$ that is not the conclusion of any argument in $\mathcal{A}_P$. Then, $\mathcal{L}2\mathcal{I}_P(\mathcal{I}2\mathcal{L}_P(\mathcal{I}))(c) = \mathbf{f}$, but $\mathcal{I}$ may assign any truth value to c. However, atom c is evaluated as false in any partial stable model. We obtain the following result:

Theorem 3. *Let $\mathcal{B}_P = (\mathcal{A}_P, Att_P)$ be the corresponding BAF of P and $\mathcal{M}$ be a partial stable model of P. It holds that $\mathcal{L}2\mathcal{I}_P(\mathcal{I}2\mathcal{L}_P(\mathcal{M})) = \mathcal{M}$.*

Proof. Let $\mathcal{M}$ be a partial stable model of P and $c \in HB_P$. It suffices to prove the following results:

- $\mathcal{M}(c) = \mathbf{t}$ iff $\mathcal{L}2\mathcal{I}_P(\mathcal{I}2\mathcal{L}_P(\mathcal{M}))(c) = \mathbf{t}$. ($\Rightarrow$) Assume $\mathcal{M}(c) = \mathbf{t}$. As $\Psi_P(\mathcal{M}) = \mathcal{M}$, there exists an argument A with $\texttt{Conc}(A) = c$ such that $\texttt{Vul}(A) \subseteq \mathbf{f}(\mathcal{M})$. This implies $\mathcal{I}2\mathcal{L}_P(\mathcal{M})(A) = \texttt{in}$ and $\mathcal{L}2\mathcal{I}_P(\mathcal{I}2\mathcal{L}_P(\mathcal{M}))(c) = \mathbf{t}$. ($\Leftarrow$) Assume $\mathcal{L}2\mathcal{I}_P(\mathcal{I}2\mathcal{L}_P(\mathcal{M}))(c) = \mathbf{t}$. Then there exists $A \in \mathcal{A}_P$ such that $\texttt{Conc}(A) = c$ and $\mathcal{I}2\mathcal{L}_P(\mathcal{M})(A) = \texttt{in}$. From Definition 11, $\mathcal{M}(c) = \mathbf{t}$.
- $\mathcal{M}(c) = \mathbf{f}$ iff $\mathcal{L}2\mathcal{I}_P(\mathcal{I}2\mathcal{L}_P(\mathcal{M}))(c) = \mathbf{f}$, which can be similarly proved.

This means that when restricted to β-complete labellings and partial stable models, $\mathcal{L}2\mathcal{I}_P$ and $\mathcal{I}2\mathcal{L}_P$ are each other's inverse. The function $\mathcal{L}2\mathcal{I}_P$ when applied to β-complete labellings can be characterised as follows:

$$\mathcal{L}2\mathcal{I}_P(\mathcal{L})(c) = \begin{cases} \mathbf{t} & \textit{if } \exists A \in \mathcal{A}_P \textit{ such that } \texttt{Conc}(A) = c \textit{ and } \mathcal{L}(A) = \texttt{in} \\ \mathbf{u} & \textit{if } \exists A \in \mathcal{A}_P \textit{ such that } \texttt{Conc}(A) = c \textit{ and } \mathcal{L}(A) = \texttt{undec} \\ \mathbf{f} & \textit{otherwise} \end{cases}$$

It is clear the relation between a β-complete labelling $\mathcal{L}$ and $\mathcal{L}2\mathcal{I}_P(\mathcal{L})$ for those atoms $c \in HB_P$ in which there is an argument A with $\texttt{Conc}(A) = c$. The distinction between them is for those atoms $c' \in HB_P$ without a corresponding argument A' with $\texttt{Conc}(A') = c'$. For this case, $\mathcal{L}2\mathcal{I}_P(\mathcal{L})(c') = \mathbf{f}$ and $\mathcal{L}$ is not defined for any argument A' with $\texttt{Conc}(A') = c'$.

From Theorems 2 and 3, we can obtain the following result:

Theorem 4. *Let $\mathcal{B}_P$ be the corresponding BAF of P. It holds $\mathcal{L}$ is a β-complete labelling of $\mathcal{B}_P$ iff $\mathcal{L}2\mathcal{I}_P(\mathcal{L})$ is a partial stable model of P; and $\mathcal{M}$ is a partial stable model of P iff $\mathcal{I}2\mathcal{L}_P(\mathcal{M})$ is a β-complete labelling of $\mathcal{B}_P$.*

Proof. 1. If $\mathcal{L}$ is a β-complete labelling, then $\mathcal{L}2\mathcal{I}_P(\mathcal{L})$ is a partial stable model:

Let $\mathcal{M} = \mathcal{L}2\mathcal{I}_P(\mathcal{L}) = (T, F, U)$. We show $\Psi_P(\mathcal{M}) = (T', F', U') = (T, F, U)$:

- $c \in T$ iff there exists $A \in \mathcal{A}_P$ such that $\texttt{Conc}(A) = c$ and $\mathcal{L}(A) = \texttt{in}$ iff there exists $A' \in \mathfrak{Sup}(A)$ such that $\texttt{Conc}(A') = c$ and for every $B \in Att(A')$, it holds $\mathcal{L}(B) = \texttt{out}$ iff there exists $A' \in \mathcal{A}_P$ such that $\texttt{Conc}(A') = c$ and for every $B \in Att(A')$, it holds $\mathcal{L}(B) = \texttt{out}$ iff there exists $A' \in \mathcal{A}_P$ such that $\texttt{Conc}(A') = c$ and $\texttt{Vul}(A') \subseteq F$ iff $c \in T'$.
- $c \notin F$ iff $c \notin F'$ can be similarly proved.

2. If $\mathcal{M}$ is a partial stable model, then $\mathcal{I}2\mathcal{L}_P(\mathcal{M})$ is a complete labelling:

   Let $\mathcal{M} = (T, F, U)$ be a partial stable model of P. Then $\Psi_P(\mathcal{M}) = (T, F, U)$. Let $A \in \mathcal{A}_P$. We will prove $\mathcal{L} = \mathcal{I}2\mathcal{L}_P(\mathcal{M})$ is a β-complete labelling of $\mathcal{B}_P$:
   - $\mathcal{L}(A) = \mathtt{in}$ iff $\mathtt{Conc}(A) \in T$ iff there exists $A' \in \mathcal{A}_P$ with $\mathtt{Conc}(A') = \mathtt{Conc}(A)$ and $\mathtt{Vul}(A') \subseteq F$ iff there exists $A' \in \mathfrak{Sup}(A)$ such that for every $B \in Att(A')$, it holds $\mathcal{L}(B) = \mathtt{out}$.
   - $\mathcal{L}(A) \neq \mathtt{out}$ iff $\mathtt{Conc}(A) \neq F$ iff there exists $A' \in \mathcal{A}_P$ with $\mathtt{Conc}(A') = \mathtt{Conc}(A)$ and $\mathtt{Vul}(A') \cap T = \emptyset$ iff there exists $A' \in \mathfrak{Sup}(A)$ such that for every $B \in Att(A')$, it holds $\mathcal{L}(B) \neq \mathtt{in}$.
3. $\mathcal{L}2\mathcal{I}_P(\mathcal{L})$ is a partial stable model of $P \Rightarrow$ according to item 2. above, $\mathcal{I}2\mathcal{L}_P(\mathcal{L}2\mathcal{I}_P(\mathcal{L}))$ is a β-complete labelling of $\mathcal{B}_P \Rightarrow$ (via Theorem 2) $\mathcal{L}$ is a β-complete labelling of $\mathcal{B}_P$.
4. $\mathcal{I}2\mathcal{L}_P(\mathcal{M})$ is a β-complete labelling of $\mathcal{B}_P \Rightarrow$ according to item 1. above, $\mathcal{L}2\mathcal{I}_P(\mathcal{I}2\mathcal{L}_P(\mathcal{M}))$ is a partial stable model of $P \Rightarrow$ (via Theorem 3) $\mathcal{M}$ is a partial stable model of P.

Theorem 4 is one of the main results of this paper. It plays a central role in ensuring the equivalence between the semantics for *NLP*s and their counterpart for *BAF*s:

Theorem 5. *Let P be an NLP and $\mathcal{B}_P = (\mathcal{A}_P, Att_P)$ be the corresponding BAF. It holds $\mathcal{L}$ is a β-grounded (resp. β-preferred, β-stable, β-semi-stable) labelling of $\mathcal{B}_P$ iff $\mathcal{L}2\mathcal{I}_P(\mathcal{L})$ is a well-founded (resp. regular, stable, L-stable) model of P.*

The following result is a direct consequence of Theorems 3 and 5:

Corollary 1. *Let P be an NLP and $\mathcal{B}_P = (\mathcal{A}_P, Att_P)$ its corresponding BAF. $\mathcal{M}$ is a well-founded (resp. regular, stable, L-stable) model of P iff $\mathcal{I}2\mathcal{L}_P(\mathcal{M})$ is a β-grounded (resp. β-preferred, β-stable, β-semi-stable) labelling of $\mathcal{B}_P$.*

Next, we consider the *NLP* exploited by Caminada et al. [7] as a counterexample to show that in general, *L*-stable models and semi-stable labellings do not coincide with each other in their translation from *NLP*s to *AAF*s:

Example 4. Let P and $\mathcal{B}_P$ be the *NLP* and corresponding *BAF* in Fig. 3, where the arguments of $\mathcal{B}_P$ are

	A	B	C_1	C_2	D	E
$\mathtt{Conc}(.)$	a	b	c	c	d	e
$\mathtt{Vul}(.)$	$\{b\}$	$\{a\}$	$\{a, c\}$	$\{c, d\}$	$\{d\}$	$\{b, e\}$

Note in $\mathcal{B}_P$ that there is both a mutual attack and a mutual support between arguments C_1 and C_2. As expected from Theorems 4 and 5, we obtain in Table 1 the equivalence between β-semantics of *BAF*s and 3-valued semantics of *NLP*s. We emphasise the coincidence between *L*-stable models and β-semi-stable labellings in Table 1 as it does not occur in [7]. In that reference, the

$a \leftarrow \texttt{not}\ b$
$b \leftarrow \texttt{not}\ a$
$c \leftarrow \texttt{not}\ c, \texttt{not}\ a$
$c \leftarrow \texttt{not}\ c, \texttt{not}\ d$
$d \leftarrow \texttt{not}\ d$
$e \leftarrow \texttt{not}\ e, \texttt{not}\ b$

(a) P

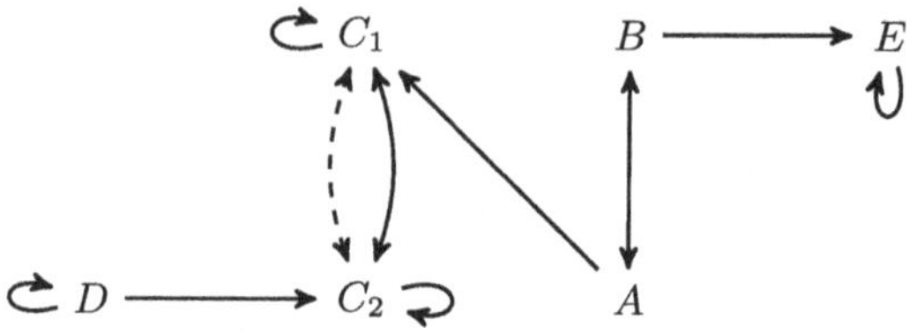

(b) $\mathcal{B}_P$

Fig. 3. *NLP* P and its corresponding *BAF* $\mathcal{B}_P$

Table 1. Semantics for P and $\mathcal{B}_P$

Partial Stable Models	β-Complete Labellings
$\mathcal{M}_1 = (\emptyset, \emptyset, \{a, b, c, d, e\})$	$\mathcal{L}_1 = (\emptyset, \emptyset, \{A, B, C_1, C_2, D, E\})$
$\mathcal{M}_2 = (\{a\}, \{b\}, \{c, d, e\})$	$\mathcal{L}_2 = (\{A\}, \{B\}, \{C_1, C_2, D, E\})$
$\mathcal{M}_3 = (\{b\}, \{a, e\}, \{c, d\})$	$\mathcal{L}_3 = (\{B\}, \{A, E\}, \{C_1, C_2, D\})$
Well-Founded Model	β-Grounded Labelling
$\mathcal{M}_1 = (\emptyset, \emptyset, \{a, b, c, d, e\})$	$\mathcal{L}_1 = (\emptyset, \emptyset, \{A, B, C_1, C_2, D, E\})$
Regular Models	β-Preferred Labellings
$\mathcal{M}_2 = (\{a\}, \{b\}, \{c, d, e\})$	$\mathcal{L}_2 = (\{A\}, \{B\}, \{C_1, C_2, D, E\})$
$\mathcal{M}_3 = (\{b\}, \{a, e\}, \{c, d\})$	$\mathcal{L}_3 = (\{B\}, \{A, E\}, \{C_1, C_2, D\})$
Stable Models	β-Stable Labellings
None	None
L-stable Models	β-Semi-stable Labellings
$\mathcal{M}_3 = (\{b\}, \{a, e\}, \{c, d\})$	$\mathcal{L}_3 = (\{B\}, \{A, E\}, \{C_1, C_2, D\})$

corresponding *AAF* possesses two semi-stable labellings in contrast with the unique L-stable model $\mathcal{M}_3$ of P.

Note that if the support relation was ignored in the framework of Fig. 2 as was done in [7], the β-complete labellings of $\mathcal{B}_P$ would be $\mathcal{L}_1, \mathcal{L}_2', \mathcal{L}_3$, where $\mathcal{L}_2' = (\{A\}, \{B, C_1\}, \{C_2, D, E\})$, and $\mathcal{L}_2'$ and $\mathcal{L}_3$ would be the semi-stable β-complete labellings and the correspondence with L-stable models would not sustain anymore.

4 From *BAF*s to *NLP*s

Now we will provide a translation for the opposite direction, i.e., from *BAF*s to *NLP*s. As in the previous section, this translation guarantees the equivalence between the semantics for *NLP*s and their counterpart for *BAF*s.

Definition 12 (Corresponding *NLP***).** *Let* $\mathcal{B} = (\mathcal{A}, Att, Sup)$ *be a BAF. For any* $A, B \in \mathcal{A}$, *define the rule* $r_{A,B}$ *such that* $head(r_{A,B}) = \mathfrak{Sup}(A)$, $body^+(r_{A,B}) = \emptyset$ *and* $body^-(r_{A,B}) = \{\texttt{not}\ \mathfrak{Sup}(X) \mid X \in Att(B)\}$. *The corresponding NLP of* $\mathcal{B}$ *is* $P_\mathcal{B} = \{r_{A,B} \mid A \in \mathcal{A}, B \in \mathfrak{Sup}(A)\}$.

Each argument is encoded in the corresponding *NLP* by its set of supporters, and the Herbrand Base of $P_{\mathcal{B}}$ is $HB_{P_{\mathcal{B}}} = \{\mathfrak{Sup}(A) \mid A \in \mathcal{A}\}$. Intuitively, with respect to any partial stable model of $P_{\mathcal{B}}$, we know $\mathfrak{Sup}(A)$ is true iff there exists $B \in \mathfrak{Sup}(A)$ such that $\mathfrak{Sup}(X)$ is false for every $X \in Att(B)$. This property follows directly from Definition 12, that was conceived to reflect the concept of acceptability in the β-complete semantics, where an argument $A \in \mathcal{A}$ is accepted iff there exists $B \in \mathfrak{Sup}(A)$ such that X is rejected for every $X \in Att(B)$.

Example 5. Let $\mathcal{B} = (\mathcal{A}, Att, Sup)$ be the *BAF* of Fig. 4a; its corresponding *NLP* $P_{\mathcal{B}}$ is shown in Fig. 4b.

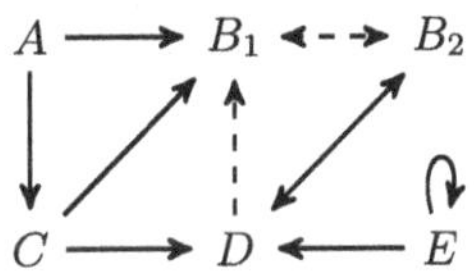

(a) $\mathcal{B}$

$$\{B_1, B_2, D\} \leftarrow \texttt{not}\ \{A\}, \texttt{not}\ \{C\}$$
$$\{B_1, B_2, D\} \leftarrow \texttt{not}\ \{D\}$$
$$\{B_1, B_2, D\} \leftarrow \texttt{not}\ \{B_1, B_2, D\}, \texttt{not}\ \{C\}, \texttt{not}\ \{E\}$$
$$\{C\} \leftarrow \texttt{not}\ \{A\}$$
$$\{D\} \leftarrow \texttt{not}\ \{B_1, B_2, D\}, \texttt{not}\ \{C\}, \texttt{not}\ \{E\}$$
$$\{E\} \leftarrow \texttt{not}\ \{E\}$$
$$\{A\}$$

(b) $P_{\mathcal{B}}$

Fig. 4. (a) *BAF* $\mathcal{B}$ and (b) its corresponding *NLP* $P_{\mathcal{B}}$

Observe that distinct arguments A, A' are associated to the same atom if $\mathfrak{Sup}(A) = \mathfrak{Sup}(A')$. For instance, the corresponding *NLP* of Example 4b encodes both arguments B_1 and B_2 as the atom $\mathfrak{Sup}(B_1) = \mathfrak{Sup}(B_2) = \{B_1, B_2, D\}$. From the supporters of B_1, we obtain the rules $r_{B_1,B_1} = \{B_1, B_2, D\} \leftarrow \texttt{not}\ \{A\}, \texttt{not}\ \{C\}$; $r_{B_1,B_2} = \{B_1, B_2, D\} \leftarrow \texttt{not}\ \{D\}$; and $r_{B_1,D} = \{B_1, B_2, D\} \leftarrow \texttt{not}\ \{B_1, B_2, D\}, \texttt{not}\ \{C\}, \texttt{not}\ \{E\}$. The intuition is that B_1 is accepted iff either (i) A and C are rejected or (ii) D is rejected or (iii) B_1, B_2, C, D, E are rejected. Condition (iii) seems contradictory, but it is similar to what occurs when an argument attacks itself (e.g., argument E is accepted iff E is rejected).

As done in Sect. 3, we want to establish a bijection between (some class of) labellings and interpretations, such that β-complete labellings correspond to partial stable models. However, we can have more labellings than interpretations.

Example 6. Let $\mathcal{B}$ be the *BAF* $\mathcal{B} = (\{A, B, C\}, \{(C, B)\}, \{(A, B), (B, A)\})$. Its corresponding *NLP* $P_{\mathcal{B}}$ has 3 rules: $\{A, B\} \leftarrow \{C\}$ and $\{A, B\}$ and $\{C\}$. There are $3^{|\mathcal{A}|} = 27$ labellings of $\mathcal{B}$ and $3^{|HB_{P_{\mathcal{B}}}|} = 9$ interpretations of $P_{\mathcal{B}}$. Hence, there is no bijection between labellings of $\mathcal{B}$ and interpretations of $P_{\mathcal{B}}$. Notice that a labelling may assign different labels to A and B, whereas an interpretation assigns only one truth value to the atom $\{A, B\}$.

Although there is no bijection between labellings and interpretations for the example above, there is always a bijection if we restrict our translations to labellings that assign the same label to arguments with identical sets of supporters.

Definition 13. *Let $\mathcal{B} = (\mathcal{A}, Att, Sup)$ be a BAF. A labelling $\mathcal{L}$ of $\mathcal{B}$ respects $\mathfrak{Sup}$ iff for every $A, B \in \mathcal{A}$ with $\mathfrak{Sup}(A) = \mathfrak{Sup}(B)$ it holds $\mathcal{L}(A) = \mathcal{L}(B)$.*

For the *BAF* $\mathcal{B}$ of Example 6, there are 9 labellings respecting $\mathfrak{Sup}$, since a labelling that respects $\mathfrak{Sup}$ assigns the same label to A and B. The representation of possibly many arguments by the same atom is motivated by the fact that every β-complete labelling satisfies the following property:

Proposition 1. *If $\mathcal{L}$ is a β-complete labelling of $\mathcal{B}$, then $\mathcal{L}$ respects $\mathfrak{Sup}$.*

By restricting labellings to those that respect $\mathfrak{Sup}$, the relationship between labellings and interpretations is straightforward:

Definition 14 ($\mathcal{L}2\mathcal{I}_{\mathcal{B}}$ and $\mathcal{I}2\mathcal{L}_{\mathcal{B}}$ Functions). *Let $\mathcal{B} = (\mathcal{A}, Att, Sup)$ be a BAF and $P_{\mathcal{B}}$ be its corresponding NLP. Denote the set of all labellings of $\mathcal{B}$ respecting $\mathfrak{Sup}$ as $\mathcal{L}ab^*$ and the set of all interpretations of $P_{\mathcal{B}}$ as $\mathcal{I}nt$. We introduce the functions (i) $\mathcal{L}2\mathcal{I}_{\mathcal{B}} : \mathcal{L}ab^* \rightarrow \mathcal{I}nt$, in which for every $\mathcal{L} \in \mathcal{L}ab^*$, $\mathcal{L}2\mathcal{I}_{\mathcal{B}}(\mathcal{L})(\mathfrak{Sup}(A)) = \mathbf{t}$ (resp. $\mathbf{f}, \mathbf{u}$) if $\mathcal{L}(A) = \mathtt{in}$ (resp. $\mathtt{out}, \mathtt{undec}$); and (ii) $\mathcal{I}2\mathcal{L}_{\mathcal{B}} : \mathcal{I}nt \rightarrow \mathcal{L}ab^*$, in which for every $\mathcal{M} \in \mathcal{I}nt$, $\mathcal{I}2\mathcal{L}_{\mathcal{B}}(\mathcal{M})(A) = \mathtt{in}$ (resp. $\mathtt{out}, \mathtt{undec}$) if $\mathcal{M}(\mathfrak{Sup}(A)) = \mathbf{t}$ (resp. $\mathbf{f}, \mathbf{u}$).*

In contrast with $\mathcal{L}2\mathcal{I}_P$ and $\mathcal{I}2\mathcal{L}_P$, the functions $\mathcal{L}2\mathcal{I}_{\mathcal{B}}$ and $\mathcal{I}2\mathcal{L}_{\mathcal{B}}$ are each other's inverse in the general case:

Theorem 6. *Let $\mathcal{B} = (\mathcal{A}, Att, Sup)$ be a BAF and $P_{\mathcal{B}}$ be its corresponding NLP. For any labelling $\mathcal{L}$ of $\mathcal{B}$ respecting $\mathfrak{Sup}$, it holds $\mathcal{I}2\mathcal{L}_{\mathcal{B}}(\mathcal{L}2\mathcal{I}_{\mathcal{B}}(\mathcal{L})) = \mathcal{L}$; and for any interpretation $\mathcal{I}$ of $P_{\mathcal{B}}$, it holds $\mathcal{L}2\mathcal{I}_{\mathcal{B}}(\mathcal{I}2\mathcal{L}_{\mathcal{B}}(\mathcal{I})) = \mathcal{I}$.*

The next lemma is essential for the correspondence between β-complete labellings of $\mathcal{B}$ and partial stable models of $P_{\mathcal{B}}$. Its proof follows straightforwardly from Definitions 12 and 14.

Lemma 1. *Let $\mathcal{B} = (\mathcal{A}, Att, Sup)$ be a BAF and $P_{\mathcal{B}}$ be its corresponding NLP. Let $\mathcal{L}$ be a labelling of $\mathcal{B}$ respecting $\mathfrak{Sup}$ and $\mathcal{M}$ be an interpretation of $P_{\mathcal{B}}$. If $\mathcal{L} = \mathcal{I}2\mathcal{L}_{\mathcal{B}}(\mathcal{M})$ or $\mathcal{M} = \mathcal{L}2\mathcal{I}_{\mathcal{B}}(\mathcal{L})$, then for any $A \in \mathcal{A}$, (i) $Att(B) \subseteq \mathtt{out}(\mathcal{L})$ for some $B \in \mathfrak{Sup}(A)$ iff $\{\mathfrak{Sup}(X) \mid \mathtt{not}\ \mathfrak{Sup}(X) \in body^-(r)\} \subseteq \mathbf{f}(\mathcal{M})$ for some $r \in P_{\mathcal{B}}$ with $head(r) = \mathfrak{Sup}(A)$; and (ii) $Att(B) \cap \mathtt{in}(\mathcal{L}) \neq \emptyset$ for every $B \in \mathfrak{Sup}(A)$ iff $\{\mathfrak{Sup}(X) \mid \mathtt{not}\ \mathfrak{Sup}(X) \in body^-(r)\} \cap \mathbf{t}(\mathcal{M}) \neq \emptyset$ for every rule $r \in P_{\mathcal{B}}$ with $head(r) = \mathfrak{Sup}(A)$.*

From Lemma 1, we obtain a similar result to Theorem 4:

Theorem 7. *Let $\mathcal{B} = (\mathcal{A}, Att, Sup)$ be a BAF with corresponding NLP $P_{\mathcal{B}}$, $\mathcal{L}$ be a labelling of $\mathcal{B}$ respecting $\mathfrak{Sup}$ and $\mathcal{M}$ be an interpretation of $P_{\mathcal{B}}$. $\mathcal{L}$ is a β-complete labelling of $\mathcal{B}$ iff $\mathcal{L}2\mathcal{I}_{\mathcal{B}}(\mathcal{L})$ is a partial stable model of $P_{\mathcal{B}}$; and $\mathcal{M}$ is a partial stable model of $P_{\mathcal{B}}$ iff $\mathcal{I}2\mathcal{L}_{\mathcal{B}}(\mathcal{M})$ is a β-complete labelling of $\mathcal{B}$.*

Proof. Owing to space restrictions, only part of the proof is shown. Assume $\mathcal{L}$ is a β-complete labelling of $\mathcal{B}$. It holds $\mathfrak{Sup}(A) \in \mathtt{t}(\mathcal{I}) \iff \mathcal{L}(A) = \mathtt{in} \iff \exists B \in \mathfrak{Sup}(A) : Att(B) \subseteq \mathtt{out}(\mathcal{L}) \overset{\text{Lemma 1}}{\iff} \exists r \in P_{\mathcal{B}} : head(r) = \mathfrak{Sup}(A) \wedge \{\mathfrak{Sup}(X) \mid \mathtt{not}\ \mathfrak{Sup}(X) \in body^{-}(r)\} \subseteq \mathtt{f}(\mathcal{I}) \iff \exists r \in P_{\mathcal{B}}/\mathcal{I} : head(r) = \mathfrak{Sup}(A) \wedge body^{+}(r) = \emptyset \iff \exists r \in P_{\mathcal{B}}/\mathcal{I} : head(r) = \mathfrak{Sup}(A) \wedge body^{+}(r) \subseteq \mathtt{t}(\psi_{P_{\mathcal{B}}}(\mathcal{I})) \iff \mathfrak{Sup}(A) \in \mathtt{t}(\psi_{P_{\mathcal{B}}}(\mathcal{I}))$. Similarly, we can prove $\mathfrak{Sup}(A) \in \mathtt{f}(\mathcal{I})$ iff $\mathfrak{Sup}(A) \in \mathtt{f}(\Psi_{P\mathcal{B}}(\mathcal{I}))$.

From Theorem 7, we can ensure the equivalence between the semantics for *BAF*s and their counterpart for *NLP*s:

Theorem 8. *Let $\mathcal{B}$ be a BAF and $P_{\mathcal{B}}$ be its corresponding NLP. For any labelling $\mathcal{L}$ of $\mathcal{B}$ respecting $\mathfrak{Sup}$, it holds $\mathcal{L}$ is a β-grounded (resp. β-preferred, β-stable, β-semi-stable) labelling of $\mathcal{B}$ iff $\mathcal{L}2\mathcal{I}_{\mathcal{B}}(\mathcal{L})$ is a well-founded (resp. regular, stable, L-stable) model of $P_{\mathcal{B}}$.*

The following result is a direct consequence of Theorems 6 and 8:

Corollary 2. *Let $\mathcal{B}$ be a BAF and $P_{\mathcal{B}}$ be its corresponding NLP. $\mathcal{M}$ is a well-founded (resp. regular, stable, L-stable) model of $P_{\mathcal{B}}$ iff $\mathcal{I}2\mathcal{L}_{\mathcal{B}}(\mathcal{M})$ is a β-grounded (resp. β-preferred, β-stable, β-semi-stable) labelling of $\mathcal{B}$.*

Recalling the *BAF* $\mathcal{B}_P$ of Example 4, its corresponding *NLP* $P_{\mathcal{B}_P}$ is structurally identical (i.e., equal under renaming) to the *NLP* P of Example 4. We obtain the expected equivalence results related to their semantics (see Table 2).

Table 2. Semantics for $\mathcal{B}_P$ and $P_{\mathcal{B}_P}$

β-Complete Labellings	Partial Stable Models
$\mathcal{L}_1 = (\emptyset, \emptyset, \{A, B, C_1, C_2, D, E\})$	$\mathcal{M}_1 = (\emptyset, \emptyset, \{\{A\}, \{B\}, \{C_1, C_2\}, \{D\}, \{E\}\})$
$\mathcal{L}_2 = (\{A\}, \{B\}, \{C_1, C_2, D, E\})$	$\mathcal{M}_2 = (\{\{A\}\}, \{\{B\}\}, \{\{C_1, C_2\}, \{D\}, \{E\}\})$
$\mathcal{L}_3 = (\{B\}, \{A, E\}, \{C_1, C_2, D\})$	$\mathcal{M}_3 = (\{\{B\}\}, \{\{A\}, \{E\}\}, \{\{C_1, C_2\}, \{D\}\})$
β-Grounded Labelling	Well-Founded Model
$\mathcal{L}_1 = (\emptyset, \emptyset, \{A, B, C_1, C_2, D, E\})$	$\mathcal{M}_1 = (\emptyset, \emptyset, \{\{A\}, \{B\}, \{C_1, C_2\}, \{D\}, \{E\}\})$
β-Preferred Labellings	Regular Models
$\mathcal{L}_2 = (\{A\}, \{B\}, \{C_1, C_2, D, E\})$	$\mathcal{M}_2 = (\{\{A\}\}, \{\{B\}\}, \{\{C_1, C_2\}, \{D\}, \{E\}\})$
$\mathcal{L}_3 = (\{B\}, \{A, E\}, \{C_1, C_2, D\})$	$\mathcal{M}_3 = (\{\{B\}\}, \{\{A\}, \{E\}\}, \{\{C_1, C_2\}, \{D\}\})$
β-Stable Labellings	Stable Models
None	None
β-Semi-stable Labellings	L-Stable Models
$\mathcal{L}_3 = (\{B\}, \{A, E\}, \{C_1, C_2, D\})$	$\mathcal{M}_3 = (\{\{B\}\}, \{\{A\}, \{E\}\}, \{\{C_1, C_2\}, \{D\}\})$

5 Conclusions

In this paper, we extend the work of Caminada et al. [7] by considering argumentation frameworks with support under the β-semantics [1]. We observe that β-complete, β-grounded, β-preferred, β-stable and β-semi-stable semantics of Bipolar Argumentation Frameworks (BAFs) respectively coincide with the partial stable, well-founded, regular, stable and L-stable semantics of Normal Logic Programs (*NLP*s). In particular, we note that the inclusion of support in argumentation allowed the connection between β-semi-stable and L-stable semantics, whereas semi-stable and L-stable semantics did not coincide for attack-only argumentation frameworks [7].

Each formalism encodes knowledge according to its own perspective: argumentation is concerned about arguments and their interaction, whereas *NLP*s are interested in deriving atoms from rules. We provide explicit translations between these formalisms and their semantics, allowing them to be used interchangeably and enhancing our understanding about how they relate to each other. Future works include exploring the connection of *NLP*s with other extensions of AAFs.

Acknowledgments. This work was supported by the Brazilian research agencies FUNCAP and CNPq.

Disclosure of Interests. The authors have no competing interests to declare that are relevant to the content of this article.

References

1. Alcântara, J., Cordeiro, R.: Bipolar argumentation frameworks with a dual relation between defeat and defence. J. Logic Comput., exae006 (2024)
2. Alcântara, J., Sá, S., Acosta-Guadarrama, J.: On the equivalence between abstract dialectical frameworks and logic programs. Theory Pract. Logic Program. **19**(5–6), 941–956 (2019)
3. Alcântara, J., Cordeiro, R., Sá, S.: On the equivalence between logic programming and SETAF (2024), (submitted)
4. Alfano, G., Greco, S., Parisi, F., Trubitsyna, I.: On the semantics of abstract argumentation frameworks: a logic programming approach. Theory Pract. Logic Program. **20**(5), 703–718 (2020)
5. Amgoud, L., Cayrol, C., Lagasquie-Schiex, M.C., Livet, P.: On bipolarity in argumentation frameworks. Int. J. Intell. Syst. **23**(10), 1062–1093 (2008)
6. Brewka, G., Woltran, S.: Abstract dialectical frameworks. In: Twelfth International Conf. on the Principles of Knowledge Representation and Reasoning, pp. 102–111. AAAI Press (2010)
7. Caminada, M., Sá, S., Alcântara, J., Dvořák, W.: On the equivalence between logic programming semantics and argumentation semantics. Int. J. Approximate Reasoning **58**, 87–111 (2015)
8. Caminada, M., Schulz, C.: On the equivalence between assumption-based argumentation and logic programming. J. Artifi. Intell. Res. **60**, 779–825 (2017)

9. Čyras, K., Schulz, C., Toni, F.: Capturing bipolar argumentation in non-flat assumption-based argumentation. In: An, B., Bazzan, A., Leite, J., Villata, S., van der Torre, L. (eds.) PRIMA 2017. LNCS (LNAI), vol. 10621, pp. 386–402. Springer, Cham (2017). https://doi.org/10.1007/978-3-319-69131-2_23
10. Dung, P.: On the acceptability of arguments and its fundamental role in nonmonotonic reasoning, logic programming and n-person games. Artif. Intell. **77**(2), 321–357 (1995)
11. Eiter, T., Leone, N., Saccá, D.: On the partial semantics for disjunctive deductive databases. Ann. Math. Artif. Intell. **19**(1–2), 59–96 (1997)
12. Przymusinski, T.: The well-founded semantics coincides with the three-valued stable semantics. Fund. Inform. **13**(4), 445–463 (1990)
13. Rocha, V.H.N., Cozman, F.G.: Bipolar argumentation frameworks with explicit conclusions: Connecting argumentation and logic programming. In: NMR, pp. 49–60 (2022)
14. Strass, H.: Approximating operators and semantics for abstract dialectical frameworks. Artif. Intell. **205**, 39–70 (2013)
15. Toni, F., et al.: Just a matter of perspective. Comput. Models Argument: Proc. COMMA 2022 **353**, 212 (2022)
16. Wu, Y., Caminada, M., Gabbay, D.M.: Complete extensions in argumentation coincide with 3-valued stable models in logic programming. Stud. Logica. **93**(2–3), 383 (2009)

Optimizing CleanUNet Architecture Parameters for Enhancing Speech Denoising

Matheus Vieira da Silva[1], João Fernando Mari[2], and André Ricardo Backes[3](✉)

[1] School of Computer Science, Federal University of Uberlandia, Uberlandia, Brazil
matheus.silva007@ufu.br

[2] Institute of Exacts and Technological Sciences, Federal University of Viçosa, Rio Paranaíba, Brazil
joaof.mari@ufv.br

[3] Departament of Computing, Federal University of São Carlos, São Carlos, Brazil
arbackes@yahoo.com.br

Abstract. Speech enhancement refers to a set of techniques aiming to recover clean speech from a corrupted signal. One way to corrupt a signal is through noise addition. Noise comes in a variety of ways. Suboptimal acoustic conditions can cause background noise and echo, hampering speech clarity and making denoising techniques necessary to enhance the speech. In this work, we optimized CleanUNet, a convolutional neural network (CNN) architecture proposed specifically for causal speech denoising tasks. We explored alternatives for the transformer bottleneck, such as Mamba architecture, capable of handling encoder outputs more efficiently with linear complexity, we also reduced the number of hidden channels in the convolutional layers. This decreases the model's parameter count and improves training and inference speed on a single GPU, offering a streamlined approach for enhanced performance. To our understanding, this is the first attempt to incorporate Mamba as a replacement for the vanilla transformer in the CleanUnet architecture.

Keywords: CleanUNet · Audio processing · Mamba architecture · Convolutional neural network

1 Introduction

The COVID-19 pandemic and lockdowns have led to significant changes in how people work and interact with technology, exemplified by Zoom's exponential growth from 10 million daily participants in December 2019 to over 300 million by April 2020 [8]. The pandemic led to a sudden shift to remote work and virtual meetings, revealing a significant issue: poor home office acoustics negatively impact videoconferencing. Suboptimal acoustic conditions caused more background noise and echo, hampering speech clarity and increasing the effort

A. Paes and F. A. N. Verri (Eds.): BRACIS 2024, LNAI 15413, pp. 254–264, 2025.
https://doi.org/10.1007/978-3-031-79032-4_18

required for video calls. Inadequate acoustic environments at home also hurt audio quality and workforce productivity. It's worth noting that the problem of insufficient acoustics isn't limited to home offices during videoconferencing. Challenges of background noise disrupting communication have been a concern from the beginning of the digital revolution [20].

Advanced techniques like Spectral Subtraction [2] and Wiener filtering [15] have been developed to address the challenges in speech enhancement. These methods use advanced algorithms and filters to reduce background noise, enhance speaker voices, and improve audio quality. The main drawback of these techniques is that they assume stationary noise, meaning that the background noise features are expected to remain relatively constant during the entire audio recording. In real-world scenarios, like Zoom meetings, noise can often be non-stationary, fluctuating in intensity and spectral content over time. In recent years, thanks to advancements in deep learning, the efficient removal of non-stationary noise from speech has been significantly enhanced [26]. There are two primary methods for speech enhancement: the time-frequency domain and the waveform domain. Some cross-domain studies explore hybrid approaches that combine both methods [21,25].

Recent advancements in deep learning, such as convolutional networks [10], generative adversarial networks [5], U-Net for segmentation [19], and transformer architecture [24] have facilitated the attainment of state-of-the-art performance by strategically harnessing the merits and limitations inherent to each architectural paradigm.

An architecture that attains state-of-the-art results is known as CleanUNet [9]. This model draws inspiration from [17] and combines transformers and U-Net to enhance noisy speech within the waveform domain. However, the CleanUNet architecture presents a formidable challenge: the size of the model and its vulnerability to quadratic complexity within the transformer bottleneck. These challenges lead to three main implications: 1) Constrained Computational Resources: the training and inference processes necessitate substantial computational resources due to the architecture's computational intensity and its dependency on audio samples that vary with the sample rate; 2) Environmental Concerns: the first implication raises environmental concerns, including issues such as water and carbon footprint, which are exacerbated by the extensive computational demands when operating at scale. 3) Real-Time Performance: the model's adaptability to resource-limited hardware is inadequate for real-time denoising applications.

In this work, we address the three primary challenges of CleanUNet by optimizing its architecture. Our contributions include introducing alternatives to the transformer bottleneck to reduce the model's parameter count for efficiency and enhancing inference speed to facilitate real-time applications. Our revised model enhances efficiency by replacing the transformer bottleneck with alternatives that handle encoder outputs with linear complexity O(n). Additionally, we empirically optimized the number of hidden channels in the convolutional layers.

These modifications reduce the model's parameter count and significantly improve both training and inference speeds on a single GPU.

The remainder of this paper is organized as follows: Sect. 2 presents the related work. Section 3 describes the material and methods used in our work. Section 4 presents the experiments and results, while Sect. 5 concludes the paper.

2 Related Works

Spectral subtraction is a simple yet effective method for reducing background noise [23]. This method immediately reduces the amplitude of noise components while preserving speech signals by estimating the average noise spectrum during silent periods and subtracting it from the input signal spectrum. However, the randomness of noise may result in negative values and distortions.

Based on statistical estimates of the underlying signal and noise, Wiener filtering dynamically creates a linear filter capable of suppressing existing noise [18]. The signal and noise are modeled as stochastic processes with well-known spectral properties. Its self-adjusting and statistically-based design make the Wiener filtering very effective at removing real-world speech noise, especially in home office environments.

The statistical characteristics of stationary noises are stable and do not change significantly over time. Examples include constant motor hum, continuous background conversation, and consistent background noise. Constant background noise can be approximated by updating silent periods, techniques like spectral subtraction work best. Non-stationary noises that change rapidly are hard to estimate and suppress. Non-stationary sounds have time-varying statistics, meaning their characteristics change rapidly. Examples include abrupt sounds like breaking glass, sporadic construction noises, and sudden horns.

One of the early deep learning methods used for speech denoising was Recurrent Neural Networks (RNNs) [14]. RNNs process input sequences incrementally while preserving an internal state to capture temporal context, making them ideal for speech processing. RNN training gradually maps noisy speech inputs to clean target outputs and aims to reduce audio noise. However, RNNs have limitations, such as challenging inference and high computational cost.

As an alternative, Convolutional Neural Networks (CNNs) [12] have become popular. 1D CNNs use local patterns and work directly on raw audio for noise reduction. 2D CNNs operate using spectrogram representations. In both cases, the network captures local speech features using small convolutional filters to separate them from background noise. Pooling and downsampling aggregate information over time. Multiple convolutional layers learn higher-level features, recognizing speech components even in noisy conditions. One of the main advantages of CNNs is weight sharing, significantly reducing parameters compared to RNNs, thus resulting in more efficient training and inference. While RNNs leverage temporal context, CNNs employ localized filters to effectively isolate speech from noise in the time-frequency domain for effective noise reduction. Ongoing research is improving the performance of CNN-based speech denoising [26].

Generative Adversarial Networks (GANs) [5] employ a generator to enhance noisy speech and a discriminator that distinguishes between enhanced and clean speech. Adversarial training improves the quality of the output. CNNs are commonly used in the generator and discriminator models in GAN architectures.

Recent advancements in deep learning have incorporated hybrid solutions involving U-Net [19] and Transformer [24] architectures. Transformers can capture long-term speech contexts and precise interactions between noisy inputs for denoising. Self-attention layers prioritize important speech components over background noise. Although the use of Transformers for speech processing is still being investigated, early results are promising [11].

3 Materials and Methods

3.1 Dataset

The database employed in this work was obtained from [22] and is composed of two separate datasets from 84 speakers with 28.4 h of clean and noisy speech pairs, designed for training and testing speech enhancement models that operate at 48 kHz. The training set consists of 34,647 pairs of audio samples, while the test set comprises 824 pairs. Due to variations in the lengths of the audio samples within each pair, a preprocessing step was undertaken to normalize their sizes. This involved concatenating all the files and subsequently cropping a new file to a specified length, denoted as L. If the last file is smaller than L, it will be padded with the last valid value.

3.2 CleanUNet

CleanUNet [9] is a CNN architecture proposed specifically for causal speech denoising tasks. It follows the principles of the U-Net architecture, a popular model in image segmentation tasks, and it is similar to the model presented in [3]. As in U-Net, CleanUNet consists of an encoder-decoder structure with skip connections, enabling the network to preserve fine-grained details during upsampling. Each convolutional block usually contains multiple convolutional layers followed by max-pooling layers and ReLU activation. CleanUNet adapts this architecture to the domain of audio processing on the raw waveform. It also uses self-attention blocks in the bottleneck to improve its ability to separate noise from the clean signal. Each self-attention block is composed of a multi-head self-attention layer and a position-wise fully connected layer [24].

One notable aspect of CleanUNet is its ability to handle varying levels and types of noise present in audio signals. It outperforms state-of-the-art models in terms of speech quality, according to several objective and subjective evaluation metrics.

3.3 Evaluation Metrics

We employed two objective evaluation metrics to assess the performance of our trained model: the Perceptual Evaluation of Speech Quality (PESQ) and the Short-Time Objective Intelligibility (STOI) [13].

PESQ. This metric focuses on the perceived quality of the speech, and it involves a psychoacoustic algorithm to simulate human hearing and compare the clean speech (original) and the enhanced speech (model output) to predict perceived quality. The calculation includes steps such as level alignment, filtering, and a detailed comparison process that considers various distortions and artifacts introduced by processing. The output score ranges from -0.5 to 4.5, where higher scores indicate better quality.

STOI. Designed to predict the intelligibility of speech signals, the method involves calculating the short-time temporal correlation between the clean (original) and enhanced speech (model output) signals across various frequency bands. The process includes segmenting the signals into short frames, applying a Fourier transform to analyze the frequency content, and then computing the correlation between the corresponding frames of the clean and enhanced signals. The STOI score ranges from 0 to 1, where 0 indicates poor intelligibility and 1 indicates excellent intelligibility.

3.4 Transformer Architecture

First introduced by [24], the transformer architecture uses self-attention mechanisms [1] to capture the dependencies between different words or tokens in a sequence. It allows the model to weigh the importance of each token in the sequence concerning every other token. The key component of self-attention is the calculation of attention scores based on queries (Q), keys (K), and values (V). The mathematical representation of self-attention is described below:

$$\text{Attention}(Q, K, V) = \text{softmax}\left(\frac{QK^T}{\sqrt{d_k}}\right)V \tag{1}$$

where Q, K, V represent the queries, keys, and values matrices derived from the input matrix. $\sqrt{d_k}$ is a scaling factor, where d_k has the dimensionality of K and V. The softmax function normalizes the attention scores, and the result is used to compute a weighted sum of the V vectors.

Inspired by CNNs, the architecture employs a self-attention technique variant called multi-headed attention to process input sequences simultaneously. The following expression illustrates the process:

$$\text{MultiHead}(Q, K, V) = \text{Concat}(\text{head}_1, \text{head}_2, \ldots, \text{head}_h)W^O \tag{2}$$

$$\text{head}_i = \text{Attention}(QW_i^Q, KW_i^K, VW_i^V) \tag{3}$$

where $head_i$ is the attention score, and W_i^Q, W_i^K, W_i^V and W_i^O are parameter matrices. During training, the computational complexity of computing attention scores for each token pair of results has a quadratic growth of $O(n^2)$ with an input sequence of length n. Researchers have proposed optimizations and approximations to address this challenge to reduce computational overhead [7,16].

3.5 Longnet

The Longnet architecture was introduced by [4]. This new Transformer variant achieves linear computational complexity and logarithmic token dependency, optimizing efficiency for extensive sequence processing. The core innovation of the dilated attention mechanism lies in its ability to expand the attentive field exponentially with increasing distance. This mechanism achieves this by dividing the input (Q, K, V) into equally sized segments with a predetermined segment length W. Each segment is then made sparse along the sequence dimension by selecting rows at specified intervals r. The segments are processed in parallel, after which they are scattered and concatenated to produce the output. Incorporating the concept of multi-head attention, the innovative method further enhances the dilated attention mechanism by diversifying the computation across different heads. Specifically, the computation varies among heads by sparsifying distinct segments of the (Q, K, V) pairs. This architecture is fully compatible with Transformers with an introduction of segment sizes r and dilation rates d hyperparameters.

3.6 Mamba Architecture

Mamba [6] is an innovative architecture designed to mitigate the quadratic complexity challenges faced by conventional attention models. This model addresses the scalability constraints of traditional Transformer models in processing long sequences and significantly enhances efficiency and flexibility with linear-time complexity for sequence modeling. It introduces two significant State Space Models (SSM) innovations: a selection mechanism and a hardware-aware algorithm.

A SSM can be simplified by the parameters (Δ, A, B, C), where each parameter represents a diagonal matrix. The discretization of (Δ, A, B) yields $(\overline{A}, \overline{B})$. This procedure translates the model from a continuous domain to a discrete one, enabling computation within a digital framework. The discretization process can be done through fixed formulas $\overline{A} = F_a(\Delta, A)$ and $\overline{B} = F_b(\Delta, A, B)$, where (F_a, F_b) pair is a discretization rule such as zero-order hold (ZOH).

One essential property of SSMs is Linear Time Invariance (LTI). In this case, all time steps in the dynamics represented by the matrices (Δ, A, B, C) are the same. This characteristic enables parallelizing training methods, using convolutional or recursive algorithms to efficiently compute the model's output.

The Mamba architecture introduces a selection mechanism similar to a gating mechanism of RNNs that can focus or filter out different parts of a context allowing SSMs to adapt over time and being input-dependent. However, this modification compromises the LTI property, resulting in the inability to apply convolution methods as previously done. To overcome this, the authors implement a hardware-aware method to take advantage of the memory hierarchy of modern GPUs.

The hardware-aware algorithm incorporates three key techniques: kernel fusion, parallel scan, and recomputation. Kernel fusion reduces memory I/O by combining multiple operations into a single GPU operation. Parallel scan algorithms are applied to enable parallel computation despite the non-linear nature of the SSMs. Recomputation is used during backpropagation to minimize memory usage by recalculating intermediate states instead of storing them.

4 Experiments and Results

Experiments were performed on a PC with an AMD Ryzen 7 7800X3D processor, 32GB of RAM, and an RTX 3090 24GB GPU running Linux Ubuntu 20.04 OS. We modified the original CleanUNet[1] code to run on Pytorch 2.1.1[2], CUDA 12.3, and Python 3.10[3], along with the modifications needed to implement Mamba architecture using mamba-ssm[4].

We proposed and evaluated two variations of CleanUNet, both designed to reduce the number of learnable parameters, reducing complexing, training, and prediction time, and looking to maintain the quality. The first variation consists of applying dilated attention mechanisms, while the second involves replacing the transformer architecture with the Mamba architecture. Both models were compared with CleanUnet described in [9], which we considered the Baseline.

The baseline CleanUnet [9] consists of an encoder and a decoder, both with D layers and a bottleneck composed of N self-attention blocks (bottleneck layers). Each encoder layer is composed of a 1D convolution layer (Conv1D) followed by a ReLu layer and a 1×1 convolution (Conv1×1) followed by a gated linear unit (GLU). Each Conv1D has kernel size K and stride $S = K/2$. The first Conv1D has H kernels, and the other layers double the number of channels. The Conv1x1 doubles the number of channels while the GLU reduces it by half. Each decoder is composed of a 1×1 convolution followed by a glu and a transposed 1D convolution layer (ConvTranspose1D). Following the U-Net architecture, each encoder layer is connected to one encoder layer by a skip connection in the reverse order. The bottleneck is composed of K self-attention blocks. Each self-attention block comprises a multi-head self-attention layer (8 heads) and a fully connected layer (input/output size of 512 and inner-layer size of 2048) as described in [9]. For this study we considered a CleanUNet model with $D = 8$, $K = 4$, and $N = 5$ [9]

The proposed dilated attention base transformers follow the UNet structure from CleanUNet, which is considered a baseline with a bottleneck composed of

[1] https://github.com/NVIDIA/CleanUNet.
[2] https://pytorch.org/.
[3] https://www.python.org/.
[4] https://pypi.org/project/mamba-ssm/1.1.0/.

dilated attention layers. For this configuration, we considered a dilated attention-based CleanUNet model with $D = 4$, $K = 3$, and $N = 1$ for training and prediction in GPU. The self-attention block comprises a dilated multi-head self-attention layer (32 heads) and a fully connected layer (input/output size of 1024 and inner-layer size of 2048). The list of segment sizes {10, 20, 30, 60} was provided with its respective dilation rates {2, 4, 8, 16}.

The proposed CleanUNet with a bottleneck composed of Mamba architecture also follows the configuration of the baseline CleanUNet but with changes in the number of encoder/decoder layers $D = 10$, and hidden channels starting from 32 from the first layers to max hidden channels 256. A Mamba architecture replaced the transformer bottleneck with $N = 1$. The dimensionality of the input vector in this bottleneck was set to 512 (input/output) with 16 layers.

We used a training configuration similar to [9]. The models were trained using Adam optimizer with $\beta_1 = 0.9$ and $\beta_2 = 0.999$. The learning rate decreases along the training using a cosine annealing rate scheduler with a maximum learning rate of 0.00025 and a warmup ratio of 5%. In this work, we employ the random shift data augmentation technique within a range of 0 to S seconds and BandMask augmentation [3].

The CleanUNet loss function described in [9] combines a multi-resolution STFT (Short-Time Fourier Transform) loss with a waveform-based loss to maximize speech-denoising performance. It is composed of two parts: the first is the $l1$ loss applied directly to the waveform, which encourages the denoised output to resemble the time domain clean speech waveform closely, and the second part is the multi-resolution STFT loss, which modifies the spectrogram's magnitude at various resolutions.

Table 1 presents the optimal outcomes and configurations for the speech enhancement model, highlighting the efficiency of implementing the Mamba architecture as a bottleneck. This approach significantly reduces the model's parameters, enabling increased data throughput and larger batch sizes. Consequently, it facilitates training the model on a single GPU. Given the linear complexity of Mamba's input handling, the model performs computations more quickly in the bottleneck. This design also supports larger vector sequences. Similarly, LongNet claims linear complexity, theoretically supporting longer sequences as well. According to the table results, LongNet uses fewer convolutional encoders to fit within memory constraints. This efficiency allows the preservation and utilization of features that might otherwise be discarded during the convolution process, thereby enhancing the model's performance.

The results presented in Table 1 are visually detailed by the scatter graphs in Fig. 1, in which we show the number of parameters on the x-axis versus the audio quality metrics on the y-axis. In the first row, we consider the total number of model parameters, including encoder, decoder, and bottleneck, and in the last row, we consider only the number of parameters in the bottleneck.

The proposed CleanUNet with a Mamba-based bottleneck significantly reduces the number of trainable parameters with a slight reduction in the audio quality indexes. While the total number of parameters is reduced by 69.59%,

Table 1. CleanUNet model results and settings

Audio Quality	Baseline	D. Attention	Mamba
PESQ (Wideband)	2.905	1.537	2.730
PESQ (Narrowband)	3.408	2.156	3.284
STOI	0.956	0.822	0.912
Training Performance			
Total Params.	46.082177M	11.882497M	14.012545M
Bottleneck Params.	16.550656M	8.401920M	6.666240M
Network Configurations			
Conv. Encoders (D)	8	4	10
Kernel Size (K)	(4×4)	(3×3)	(5×5)
Stride (S)	2	2	2
Bottleneck Layers (N)	5	1	1
Training Configurations			
Learning Rate	0.0002	0.00025	0.00025
Batch Size	64	24	32
Crop Length	1.33 s and 1.5 s	3 s	3 s
Split	-	Train/Test	Train/Test

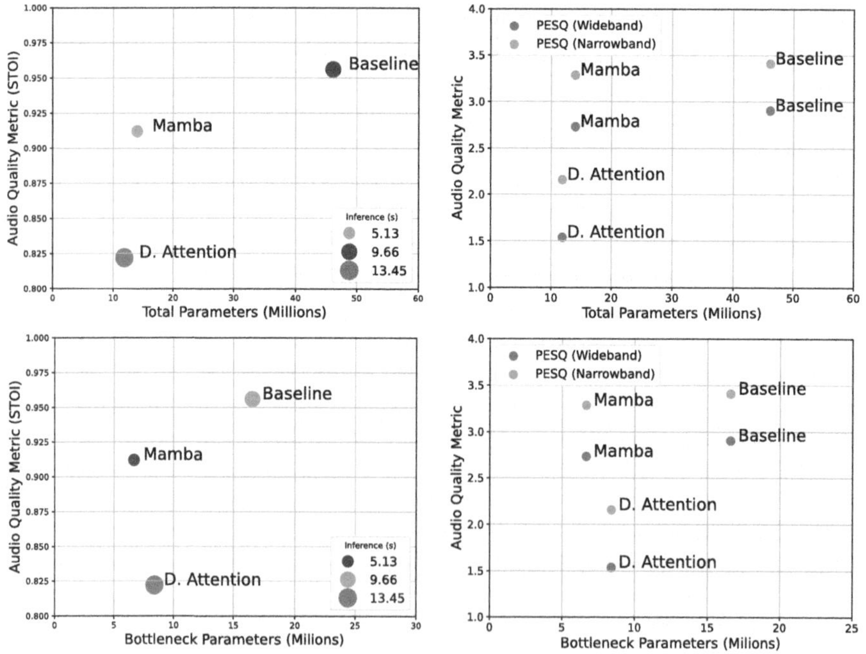

Fig. 1. Comparing the models considering the number of parameters Vs audio quality metrics. In the first row, we show the total number of parameters. In the last row, we show the number of parameters in the bottleneck. The first column also includes the inference time in seconds (s).

the reduction in the STOI, PESQ-wideband, and PESQ-narrowband is only 4.60%, 3.63%, and 6.02%, respectively. It is important to note that, besides replacing the bottleneck with a Mamba architecture, the number of layers in the encoders/decoders was increased, however, the number of channels was reduced to enable the model to perform faster operations while saving memory. During inference using the entire test set, the proposed model performed 46.9% faster than the baseline model.

5 Conclusion

This paper addressed some challenges related to using CleanUNet for speech denoising. Its good performance comes from a deep convolutional network that combines transformers and U-Net to enhance noisy speech within the waveform domain. Training requires a huge amount of resources, and its architecture limits the model's adaptability to resource-limited hardware, making it inadequate for real-time denoising applications. We explored alternatives for the transformer bottleneck, such as Mamba architecture, capable of handling encoder outputs more efficiently with linear complexity, reducing the need for multiple convolutional layers. As a result, we drastically reduced the number of trainable parameters while preserving its performance, thus improving training and inference speed on a single GPU.

Acknowledgments. André R. Backes gratefully acknowledges the financial support of CNPq (National Council for Scientific and Technological Development, Brazil) (Grant #307100/2021-9). This study was financed in part by the Coordenação de Aperfeiçoamento de Pessoal de Nível Superior - Brazil (CAPES) - Finance Code 001.

References

1. Bahdanau, D., Cho, K., Bengio, Y.: Neural machine translation by jointly learning to align and translate. arXiv preprint arXiv:1409.0473 (2014)
2. Boll, S.: Suppression of acoustic noise in speech using spectral subtraction. IEEE Trans. Acoust. Speech Signal Process. **27**(2), 113–120 (1979)
3. Defossez, A., Synnaeve, G., Adi, Y.: Real time speech enhancement in the waveform domain (2020)
4. Ding, J., et al.: Longnet: Scaling transformers to 1,000,000,000 tokens. arXiv preprint arXiv:2307.02486 (2023)
5. Goodfellow, I., et al.: Generative adversarial nets. Adv. Neural Inf. Process. Syst. **27** (2014)
6. Gu, A., Dao, T.: Mamba: Linear-time sequence modeling with selective state spaces. arXiv preprint arXiv:2312.00752 (2023)
7. Hassani, A., Shi, H.: Dilated neighborhood attention transformer. arXiv preprint arXiv:2209.15001 (2022)
8. Karl, K.A., Peluchette, J.V., Aghakhani, N.: Virtual work meetings during the covid-19 pandemic: the good, bad, and ugly. Small Group Res. **53**(3), 343–365 (2022)

9. Kong, Z., Ping, W., Dantrey, A., Catanzaro, B.: Speech denoising in the waveform domain with self-attention. In: ICASSP – IEEE International Conference on Acoustics, Speech and Signal Processing, pp. 7867–7871. IEEE (2022)
10. Krizhevsky, A., Sutskever, I., Hinton, G.E.: Imagenet classification with deep convolutional neural networks. Adv. Neural Inform. Process. Syst. **25** (2012)
11. Latif, S., Zaidi, A., Cuayahuitl, H., Shamshad, F., Shoukat, M., Qadir, J.: Transformers in speech processing: a survey. arXiv preprint arXiv:2303.11607 (2023)
12. LeCun, Y., Bottou, L., Bengio, Y., Haffner, P.: Gradient-based learning applied to document recognition. Proc. IEEE **86**(11), 2278–2324 (1998)
13. Loizou, P.C.: Speech enhancement: theory and practice. CRC press (2013)
14. Maas, A., Le, Q.V., O'neil, T.M., Vinyals, O., Nguyen, P., Ng, A.Y.: Recurrent neural networks for noise reduction in robust ASR (2012)
15. McAulay, R., Malpass, M.: Speech enhancement using a soft-decision noise suppression filter. IEEE Trans. Acoust. Speech Signal Process. **28**(2), 137–145 (1980)
16. Pagliardini, M., Paliotta, D., Jaggi, M., Fleuret, F.: Faster causal attention over large sequences through sparse flash attention. arXiv preprint arXiv:2306.01160 (2023)
17. Petit, O., Thome, N., Rambour, C., Themyr, L., Collins, T., Soler, L.: U-Net transformer: self and cross attention for medical image segmentation. In: Lian, C., Cao, X., Rekik, I., Xu, X., Yan, P. (eds.) MLMI 2021. LNCS, vol. 12966, pp. 267–276. Springer, Cham (2021). https://doi.org/10.1007/978-3-030-87589-3_28
18. Robinson, E.A., Treitel, S.: Principles of digital wiener filtering. Geophys. Prospect. **15**(3), 311–332 (1967)
19. Ronneberger, O., Fischer, P., Brox, T.: U-Net: convolutional networks for biomedical image segmentation. In: Navab, N., Hornegger, J., Wells, W.M., Frangi, A.F. (eds.) MICCAI 2015. LNCS, vol. 9351, pp. 234–241. Springer, Cham (2015). https://doi.org/10.1007/978-3-319-24574-4_28
20. Shannon, C.E.: A mathematical theory of communication. Bell Syst. Technical J. **27**(3), 379–423 (1948)
21. Tang, C., Luo, C., Zhao, Z., Xie, W., Zeng, W.: Joint time-frequency and time domain learning for speech enhancement. In: 29th International Conference on International Joint Conferences on Artificial Intelligence, pp. 3816–3822 (2021)
22. Valentini-Botinhao, C., et al.: Noisy speech database for training speech enhancement algorithms and tts models. University of Edinburgh. School of Informatics, Centre for Speech Technology Research (2017)
23. Vaseghi, S.V., Vaseghi, S.V.: Spectral subtraction. Adv. Signal Process. Digital Noise Reduct., 242–260 (1996)
24. Vaswani, A., et al.: Attention is all you need. Adv. Neural Inform. Process. Syst. **30** (2017)
25. Wang, H., Wang, D.: Cross-domain speech enhancement with a neural cascade architecture. In: ICASSP – IEEE International Conference on Acoustics, Speech and Signal Processing, pp. 7862–7866. IEEE (2022)
26. Yuliani, A.R., Amri, M.F., Suryawati, E., Ramdan, A., Pardede, H.F.: Speech enhancement using deep learning methods: a review. Jurnal Elektronika dan Telekomunikasi **21**(1), 19–26 (2021)

Portuguese Emotion Detection Model Using BERTimbau Applied to COVID-19 News and Replies

Francisco Bráulio Oliveira(✉) and Jaime Simão Sichman

Laboratório de Técnicas Inteligentes (LTI), Escola Politécnica (EP), Universidade de São Paulo (USP), São Paulo, Brazil
{fcobraulio,jaime.sichman}@usp.br

Abstract. In this study, we developed an Emotion Detection (ED) model for Brazilian Portuguese social media comments and used it to analyze emotion distribution in replies to COVID-19-related news on Twitter. We translated the GoEmotions dataset into Portuguese using a Large Language Model, preserving the original social media writing style, and fine-tuned the BERTimbau model for detecting Ekman's six basic emotions plus neutral emotion. We collected tweets related to the COVID-19 pandemic from major Brazilian news portals and analyzed user replies for expressed emotions. Chi-square tests indicated that emotion distribution depended on both the news topic and the media outlet, while the Augmented Dickey-Fuller (ADF) test showed that some emotions' distributions were influenced by the publication date. Among replies with identifiable emotions, anger (21.1%), joy (17.2%), and surprise (11.6%) were most prevalent. The Brazilian pandemic inquiry (CPI) prompted the highest anger prevalence (41.9%), while news from UOL Notícias had the highest anger (40.0%) and g1 led in sadness (13.6%).

Keywords: NLP · Emotion Detection · Deep Learning · Topic Modeling · Covid-19 · News

1 Introduction

Emotions constitute a fundamental aspect of human cognition and behavior. They play a critical indicator of psychological state and social interaction. Understanding the emotional expressions within text not only provides insights into individual sentiment, but also unveils collective trends and behavioral patterns within communities. Therefore, analysis of textual data using Artificial Intelligence (AI) techniques has became an important activity across multiple domains. In recent years, the intersection of natural language processing (NLP) and machine learning has facilitated significant advancements in emotion analysis. Notably, the utilization of deep learning architectures, such as BERT (Bidirectional Encoder Representations from Transformers), achieve significant results.

A. Paes and F. A. N. Verri (Eds.): BRACIS 2024, LNAI 15413, pp. 265–280, 2025.
https://doi.org/10.1007/978-3-031-79032-4_19

Social media platforms provide a significant amount of data to be analyzed by text classification models. Also, sentiments and topics on social media can be a good proxy for public opinion. For instance, data from social media, such as Twitter, replicate consumer confidence and Presidential job approval polls [28] and contribute to understand the response to terrorist attacks [6]. Recently, it helps elucidate COVID-19-related discourse, providing valuable insights into public sentiment, reactions to health policies, and the dissemination of information during crises.

However, while numerous studies have explored emotion analysis in English-language contexts, research in Brazilian Portuguese remains limited. This gap in the literature restricts comprehensive analyses of emotions expressed by Portuguese-speaking communities, particularly in response to significant events like the COVID-19 pandemic. Addressing this gap is crucial for promoting inclusive research and understanding the unique cultural nuances present in Brazilian Portuguese discourse.

In this study, we aim to contribute to the advancement of Natural Language Processing (NLP) research in Brazilian Portuguese by developing and applying a BERT-based emotion classification model, using the BERTimbau language model. Specifically, we seek to answer the following research questions:

RQ1: Given an Ekman-based labeled emotion corpus in English, is it feasible to use LLMs to generate a similarly labeled corpus in Brazilian Portuguese while maintaining the language style? Moreover, can we build an efficient emotion transformer-based classifier in Brazilian Portuguese using this new corpus?
RQ2: What are the predominant emotions expressed in comments on Brazilian media news about the COVID-19 pandemic on Twitter?
RQ3: How do these emotions distribute across different press outlets, news themes, and over time?

To address these questions, we employ a multi-step methodology involving corpus translation, model fine-tuning, dataset collection, and emotion detection. This study also aims to provide insights for policymakers, healthcare experts, and marketers by explaining the emotional landscape of COVID-19 discourse in Brazilian Portuguese. It hopes to enable more informed decision-making and communication strategies in the context of a major event. The remainder of the paper is organized as follows. Section 2 gives an overview of the related work. Section 3 describes our methodology, and Sect. 4 presents our findings. We conclude the paper in Sect. 5 with some general remarks.

2 Background and Related Work

Human emotions are expressed in a multi-modal fashion, encompassing facial expressions, voice, and language [23]. However, classification strategies focusing only on textual data exhibit performance comparable to multimodal approaches [1]. A crucial aspect in developing emotion classifiers is the establishment of an emotion model, defining how emotions are represented. The main kinds of emotion models include discrete and dimensional models [46]. Discrete models categorize emotions into a limited set of categories. Paul Ekman's study suggests six fundamental emotions—anger, sadness, disgust, joy, surprise, and fear—originating from distinct neural systems [11]. In contrast, Robert Plutchik's model suggests that emotions occur in opposing pairs at varying intensities, augmenting Ekman's set with acceptance/trust and anticipation [32]. Meanwhile, dimensional models propose emotions as non-independent entities, represented within a continuous multi-dimensional space. Plutchik's two-dimensional wheel [32] and Russell's circumplex model [34] differentiate emotions based on arousal (activation vs. deactivation) and valence (pleasantness vs. unpleasantness). Furthermore, the Pleasure-Arousal-Dominance (PAD) model delineates emotions across three axes [22].

The field of research known as Affective Computing emerged with the recognition that computational systems could potentially classify a user's emotional state [30], from where emerged the Text-Based Emotional Detection (TBED). Annotated collections of documents, or texts, with distinct emotions facilitated the development of artificial intelligence models capable of supervised emotion detection in new documents, as well as performance evaluation and comparison with other models. The first annotated corpus for emotions was the ISEAR dataset, encompassing 7 distinct emotions [35]. Subsequently, Wordnet was developed as one of the primary lexical databases specifically constructed for Natural Language Processing (NLP) in English [12].

Lexical databases consist of structured sets of terms aiming to approximate the terms native speakers associate with a given concept. Early NLP applications focused on emotion detection used lexical databases to identify the predominant emotion in a document [13,38,40,45]. Following this, word2vec [25] enabled the usage of word embeddings, opening the way for neural networks and deep learning in a variety of NLP tasks, including emotion detection. In 2017, the introduction of Transformers [42] enabled the development of language models, including BERT [10], representing a qualitative leap in emotion analysis tasks. Presently, TBED research indicates that lexicon-based applications are still in use; however, classifiers based on machine learning and deep learning are becoming more prevalent [2,15,16]. Recent works have employed Large Language Models (LLMs) for TBED activities [3,44].

Among works focused on the Portuguese language, notable efforts include the development of WordnetAffectBR, one of the earliest and most significant lexical databases in Brazilian Portuguese [29]. Another contribution constructed a corpus of short news items (headlines and brief descriptions) and utilized Latent Semantic Analysis to classify documents among Ekman's six emotions, along with neutral emotion. More recently, a BERT-based language model named BERTimbau was developed [37]. BERTimbau underwent fine-tuning using a version of the GoEmotions corpus [9] translated by Google Translate to generate a model specialized in emotion classification in Portuguese [14]. This model was utilized to evaluate the polarity of comments related to political discourse, and it was compared to ChatGPT, demonstrating superior performance [36]. In contrast, other techniques, such as classification models based on lexical databases, were employed to identify emotions in tweets related to the COVID-19 pandemic in Portuguese [4]. Regarding tweets related to the COVID-19 pandemic written in English, multiple studies used machine learning and deep learning techniques to investigate the distribution of emotions among social media users, finding a prevalence of negative emotions such as anger [18,27,47].

3 Methodology

In this study, we construct a model for classifying emotions in social media comments, applied to a dataset of COVID-19-related tweets collected from Twitter. We adapt the GoEmotions dataset [9] to Portuguese using ChatGPT to ensure that we keep the writing style of social media comments. Then, we fine-tune a pre-trained Portuguese language model, BERTimbau, for Emotion Detection (ED), employing Ekman's model of six basic emotions plus neutral emotion. Following this, we gather tweets related to the COVID-19 pandemic from major Brazilian news portals on Twitter, and their user replies. The fine-tuned model is used to detect the emotions expressed in these replies. Finally, we conduct statistical analysis on the distribution of these emotions relating them to the news topic, the news media, and the date when the news was published.

3.1 Model Training

Table 1. Sample comments from GoEmotions, their translation to Portuguese using Google Translate, and chatGPT (GoEmotions-BR).

GoEmotions [9]	Google Translate	GoEmotions-BR
I brought my love a cherrrryyyyy....	Trouxe uma cereja para meu amor....	Eu trouxe umaaaa cerreeeejj-jaaaaa pro meu amoooorrrrrr....
*tomorrow Thank you for this.	*amanhã Obrigado por isso.	amanhã vlw por isso.
Thats super fricking cool I love it	Isso é super legal, eu adorei	Isso é muito legal pra caramba, adoro!
Lucky, when I did it my nickname ended up being [NAME].	Sorte, quando fiz isso meu apelido acabou sendo [NOME].	Sortudo, quando fiz isso, meu apelido acabou sendo [NOME].
Yeah until the last one has creepily wed and reproduced.	Sim, até que o último se case e se reproduza assustadoramente.	Sim, até que o último tenha se casado de forma assustadora e reproduzido.
Looks better imo	Parece melhor, imo	Parece melhor na minha opinião.

Language Model (LM). Language Models are artifacts capable of performing language-related tasks automatically. Typically, they are generated through machine learning or statistical modeling. Recently, LMs have been build using Deep Neural Network (DNN) models trained on large volumes of data, utilizing a transformer architecture [42]. One of the most commonly used LM architectures is the Bidirectional Encoder Representations from Transformers (BERT) [10]. This modeling approach provides contextual language features, enabling LMs to have generalist capabilities. That is, they can undergo fine-tuning processes, generating new models specialized in specific tasks, including sentiment analysis, machine translation, question-answering systems, and more. In this study, we utilize a LM called BERTimbau [37], trained using the BERT architecture on a dataset collected in Brazilian Portuguese. BERTimbau was trained on the brWaC corpus [43], compiled from web pages, consisting of 3.53 million documents and a total of 2.68 billion tokens.

Fine-Tuning. Fine-tuning a language model (LM) typically involves retraining one of its layers using an annotated dataset tailored to a specific task. This study aimed to develop an LM specialized in Emotion Detection (ED) using the GoEmotions dataset, which includes about 58,000 Reddit comments manually annotated into 27 emotions, later aggregated into Ekman's six basic emotions plus neutral. Hammes and Freitas [14] previously translated this dataset using Google Translate but noted its inability to preserve the social media writing style.

Recent studies [8,17] have used ChatGPT for translating corpora to fine-tune LMs for underrepresented languages. Following this approach, we created GoEmotions-BR by translating the GoEmotions corpus with chatGPT-3.5-turbo, maintaining the original writing style and localizing slang (see Table 1).

The translation prompt used was: "Translate the following message to Portuguese. Preserve the writing style, using equivalent slang and abbreviations, if needed." For example, "Looks better imo"' was accurately translated to "Parece melhor na minha opinião," with chatGPT recognizing "imo" as "na minha opinião". Similarly, "I brought my love a cherrrryyyyy" was translated as "Eu trouxe umaaaa cerreeeejjjaaaaa pro meu amoooorrrrrr", preserving the original's expressive style, unlike Google Translate.

The GoEmotions-BR corpus was then used to fine-tune the BERTimbau-base model, creating BERTimbau-Ekman-BR, specialized in ED. We followed the code and configurations provided by Hammes and Freitas [14].

3.2 Emotion Distribution

The Corpus. We utilize the Reuters Institute's Digital News Report 2023 [26] to identify the most prominent digital media outlets in Brazil. Then, we search for the highly active Twitter accounts that best represent each media outlet, as presented in Table 2. We utilize the Twitter API to collect tweets related to

Table 2. Statistics of the twitter accounts of the selected media outlets.

Media	Account	Created at	Followers Count	Tweets Count	COVID Tweets Count	Replies Count
Globo News	g1	2007-09-11	13,503,670	728,050	13,636	1,139,807
Folha de S. Paulo	folha	2008-04-30	8,138,207	554,220	20,558	872,361
UOL	UOLNoticias	2008-04-30	4,208,743	526,127	25,628	681,007
Estadão	Estadao	2007-10-08	7,201,376	453,548	16,890	585,150
CNN Brasil	CNNBrasil	2016-12-14	1,551,680	89,129	24,262	575,504
O Globo	JornalOGlobo	2009-07-06	6,911,678	442,766	17,682	454,334
Veja	VEJA	2008-11-28	8,768,850	339,102	8,042	202,266
BBC News	bbcbrasil	2007-02-23	3,092,022	254,376	8,504	108,766
Record News	portalR7	2009-08-13	4,950,294	500,451	12,035	89,059
Jornal Extra	jornalextra	2009-03-17	1,094,180	211,136	8,088	74,143
BandNews	radiobandnewsfm	2009-03-25	1,511,125	200,237	7,229	46,498
Terra.com.br	Terra	2009-10-08	3,011,485	199,083	4,620	20,414
TV SBT	SBTonline	2009-02-11	5,018,484	109,309	2,107	12,395

the COVID-19 pandemic from these accounts. The tweets are identified through contextual annotations provided by Twitter itself [41], a tool that employs artificial intelligence and manual annotations to determine the topic of posts on the social network. In total, 169,281 news-related tweets and 4,861,704 replies on these news items were collected, originated from 971,176 distinct users.

Topic Modeling. We employ a technique called Latent Dirichlet Allocation (LDA) [5] to identify the topics present in the news tweets. LDA is a statistical method that discovers latent topics from a corpus of documents. Before training, we exclude stopwords (i.e., very common words such as articles, prepositions, and others), web links, punctuation marks, numbers, and named entities of geographical type (e.g., city names, countries). Additionally, other terms that are either too common (i.e., appearing in more than one-third of the documents) or too rare (i.e., appearing in less than 0.05% of the documents) are removed. We also create tokens from bigrams—pairs of words that occur frequently together. At the end, the dictionary have 56,444 unique tokens, used in 169,281 documents, with an average of 7.56 tokens per document. We use the implementation of the MAchine Learning for LanguagE Toolkit (MALLET) library [20] for the LDA algorithm to topic modeling.

We identify the appropriate number of topics (as the LDA input) by analyzing the coherence scores ranging from 3 to 39, with a step of 3. The coherence of a topic, used as an indicator for topic quality, is based on the distributional hypothesis that words with similar meaning tend to co-occur within a similar context [39]. The coherence score of a model is the average of the coherence scores of all topics. A coherence score of 0.38 was achieved for 12 topics, in our dataset.

Statistical Analysis. We use the chi-square test of independence [21] to assess whether a pair of categorical variables (e.g., emotion and media outlet) are related (alternative hypothesis) or independent (null hypothesis). This is a non-parametric test, appropriate for our dataset because some of the variables (i.e., emotion, topic, and media outlet) are nominal in nature. By analyzing the test statistic and the degrees of freedom, we can determine the p-value. When the p-value is greater than the significance level ($p > 0.05$), we do not reject the null hypothesis, indicating insufficient evidence to conclude that the categorical variables are related. Additionally, the effect size measures the strength of the relationship between the variables; larger effect sizes suggest a stronger association.

We also use the Augmented Dickey-Fuller (ADF) Test [19] to assess whether a time series (e.g., the monthly prevalence of emotion in replies) is stationary (alternative hypothesis) or non-stationary (null hypothesis). A non-stationary time series exhibits some time-dependent structure (e.g., variance that changes over time). When the test statistic exceeds the critical value at a 5% significance level, this is reflected in the p-value ($p > 0.05$), leading us not to reject the null hypothesis, indicating insufficient evidence to conclude that the time series is stationary.

4 Results

In this work, our aim is to train a model for Emotion Detection (ED) in Brazilian Portuguese. We useCchatGPT for dataset translation, enabling the model to classify social media comments within the scope of Ekman's basic emotions model. We then utilize this model to analyze the distribution of emotions in replies to Brazilian news tweets related to the COVID-19 pandemic, considering variations related to the news outlet, the topic of the news, and the time of publication.

4.1 Emotion Detection Model

Table 3 presents the F1-score results for each emotion detected by the BERT model used in the GoEmotions study [9], as well as the performance achieved with our model for the GoEmotions-BR corpus. Although the results are not directly comparable, since they were calculated on different corpora, they provide an indicative of expected outcome. Our results demonstrate a similar order of magnitude and consistency across emotion classes, indicating that the developed model satisfactorily meets the conditions for analyzing emotion distribution in responses to COVID-19-related news tweets in Brazil.

4.2 Emotion Distribution

We conducted a multi-stage statistical analysis to investigate how the distribution of emotions varies according to different factors. Firstly, we examine the

Table 3. F1-score comparison of Ekman Emotion for GoEmotions and BERTimbau-Ekman-BR models.

Emotion	GoEmotions	BERTimbau-Ekman-BR
anger	0.57	0.50
disgust	0.53	0.49
fear	0.68	0.60
joy	0.82	0.80
sadness	0.59	0.54
surprise	0.61	0.55
neutral	0.66	0.64
macro-average	0.64	0.59
std	0.10	0.11

Table 4. Summary of chi-square test of independence results for multiple pairs of categorical variables. The test statistics is presented in thousands. A p-value lower than 5% rejects the null hypothesis that the two variables are independent. The table presents the rule of thumb over Cohen's ω measure [7] as the Effect Size result.

Variables	Test Statistic	P-Value	Effect Size	Test Result
Emotion x Topic	35	0.0%	Small	Reject the null hypothesis
Emotion x Media	26	0.0%	Small	Reject the null hypothesis

global distribution of emotions in replies to news tweets, how reply volume varies over time, and how the distribution of emotions also changes over time. Subsequently, we used Latent Dirichlet Allocation (LDA) to identify latent topics within the corpus of COVID-19-related news tweets. Then, we assess how emotions are distributed among these topics. Finally, we evaluate how emotions are distributed among different news outlets. With that, we are able to map different scenarios impacting the distribution of emotions in replies, and statistically assess their impact. The results for multiple pairs of variables are summarized in Table 4, along with the effect size estimated using Cohen's ω [7]. Both the topic and the news outlet are hypothesized to influence the distribution of emotions, as the statistical test fails to reject the null hypothesis that they are independent in both cases. However, the effect in both instances is small.

Distribution over Time. In total, 4,861,704 replies to news tweets were collected over a three-year period, from January 2020 to December 2022. The volume of replies varied considerably throughout the observed period, peaking at over 300,000 posts per month in the first quarters of 2020 and 2021, but becoming virtually irrelevant from February 2022 onward. Conversely, Fig. 1 shows the distribution of emotions across the entire corpus of news tweet replies. Just under half (40.9%) of the replies exhibit no detectable emotion, categorized as

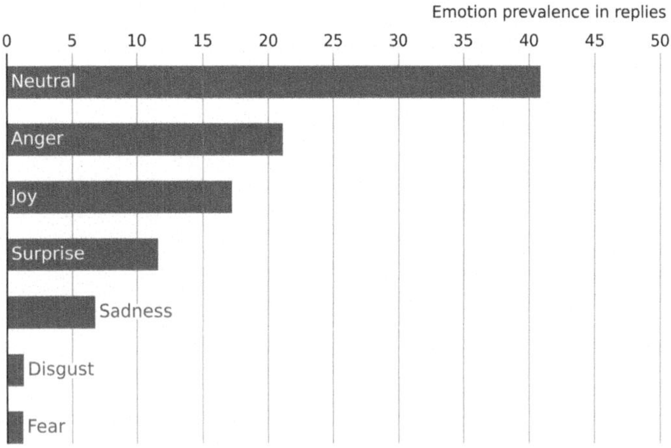

Fig. 1. The prevalence of emotions in user replies.

neutral. Among the replies with detectable emotions, anger (21.1%) emerges as the most prevalent, followed by joy (17.2%), surprise (11.6%), sadness (6.7%), disgust (1.3%), and fear (1.2%), which represents the least prevalent emotion.

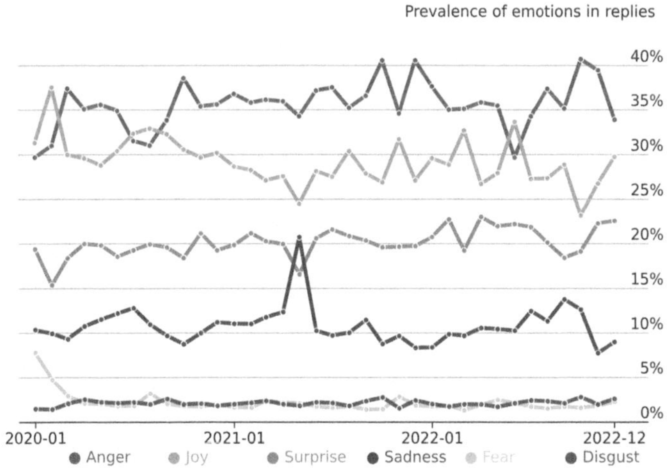

Fig. 2. Monthly distribution of emotions in the corpus.

Furthermore, Fig. 2 presents the distribution of emotions over time in replies. This figure considers only the replies with detectable emotions, thus excluding those categorized as neutral. Over the course of time, slightly more than a third of the replies with detectable emotions were associated with anger, while just under a third were linked to joy. The remainder was distributed among the other emotions, with sadness experiencing a significant peak in May 2021, surpassing 20% of the replies with detectable emotions.

Table 5. Augmented Dickey-Fuller test for monthly prevalence of emotions in user comments. A p-value higher than 5% fails to reject the null hypothesis that the time series of the daily amount of user comments with each emotion is non-stationary, thus indicating that the time series probably have time-dependent structures.

Emotion	Test Statistics	P-Value	Test Result
Anger	-4.7	0.0%	Reject the null hypothesis
Disgust	-5.8	0.0%	Reject the null hypothesis
Fear	-4.0	0.1%	Reject the null hypothesis
Joy	-1.7	41.1%	Fails to reject the null hypothesis
Sadness	-4.5	0.0%	Reject the null hypothesis
Surprise	-3.1	2.6%	Fails to reject the null hypothesis
Neutral	-3.6	8.9%	Reject the null hypothesis

Table 5 indicates that joy and neutral are likely the only emotions that are non-stationary, whereas anger, disgust, fear, surprise, and sadness are likely stationary. This suggests that the monthly prevalence of these emotions in user replies is not dependent on time. A more negative test statistic indicates a stronger rejection of the null hypothesis. Consequently, we can infer with greater confidence that joy is non-stationary, while disgust is stationary.

Distribution over Topics. Using LDA, we identify 12 topics in the news tweets, shown in Table 6. Each topic is represented by the ten most related words, a sample tweet, its prevalence among the news tweets, coherence score, and label (e.g., Cases & Deaths"). For instance, the first topic is more likely to be chosen by the LDA model if a news tweet contains keywords like coronavírus," saúde," or pacientes." This topic has a coherence score of 38.94%, appears in 15% of news tweets, and is labeled as "Impact on health system."

The chi-square test (χ^2) shows a significant relationship between topics and the prevalence of emotions in user comments (Table 4), although with a small effect size. Table 7 provides χ^2 results and the prevalence of emotions in comments according to the topic. A higher χ^2 value indicates a greater deviation from the mean (e.g., anger in responses to vaccination news).

The highest χ^2 is observed for anger in replies to Pandemic Investigation news (41.9%), higher than the mean anger prevalence (35.7%). This topic also has the lowest joy (25.4%) and sadness (9.2%). Replies to cases & deaths news show the lowest anger (29.4%) but the highest sadness (16.3%), surprise (21.5%), and fear (3.5%). News on preventive measures have the second-highest sadness (14.3%) and lowest surprise (17.6%). Joy is prevalent in replies to almost any topic, with the highest rates in vaccine distribution (34.6%) and international impact news (32.3%).

Table 6. The topics identified by LDA.

Keywords	Sample Tweets	Prevalence	Coherence	Label
coronavírus, saúde, pacientes, dias, pessoas, doença, hospital, teste, médico, via_user	O enfermeiro Marcos (...) contraiu doença e acabou internado na mesma UTI onde trabalha.	0.15	38.94	Impact on health system
bolsonaro, cpi, presidente, disse, pazuello, afirmou, user, stf, queiroga, governo_federal	Pazuello deve afirmar à PF em depoimento que recebeu pedido do presidente Bolsonaro de investigar supostas irregularidades (...)	0.13	53.04	Pandemic investigation
vacinas, brasil, oms, mundo, afirma, países, testes, índia, estados_unidos, ministério	Covid-19: 172 países já aderiram a acordo mundial de vacinas; OMS diz ter iniciado diálogo com autoridades russas	0.10	34.77	Vaccination debate
pandemia, brasil, ano, semana, será, alta, brasileiros, apesar, abril	Desigualdade bate recorde na pandemia e renda média cai a pior nível desde 2012	0.09	25.93	Economic impact
quarentena, bolsonaro, anos, máscara, fala, morre, gt, isolamento, casa, vídeo	Alexandre Garcia se descuida com máscara em evento com Bolsonaro.	0.09	31.45	Preventive measures
sp, governo, rio, lockdown, estados, paulo, doria, volta, rj, anuncia	Confira as novas regras da prefeitura para reabertura do Rio: do comércio e salões de rua a escolas e bares	0.08	27.37	Governmental actions
vacina, ministério_saúde, anvisa, coronavac, pfizer, crianças, fiocruz, doses, dose, butantan	O Instituto Butantan enviou (...) uma solicitação à Anvisa de início dos testes clínicos da vacina contra a Covid-19 Butanvac	0.08	50.48	Vaccine production
coronavírus, eua, país, china, causa, trump, onda, itclia, surto, http_via	Coronavírus na Itália: como país conseguiu conter segunda onda de contágios e se tornou exceção na Europa	0.06	34.71	International impact
casos, mortes, brasil, dia, brasil_registra_mortes, país, horas, quase, mortos, doença	Brasil ultrapassa mil mortes e chega a quase 300 mil casos de Covid em 24 h. (...)	0.08	64.20	Cases & Deaths
pandemia, combate, durante_pandemia, crise, devido, auxílio_emergencial, medidas, empresas	Projeto tem como objetivo criar um fundo emergencial para técnicos e artistas do segmento teatral afetados pela pandemia	0.05	23.14	Political response
vacinação, paulo, rio, governo, hoje, população, cidade, imunização, segundafeira	São Paulo inicia vacinação de pessoas de 72 a 74 anos na segunda-feira (22)	0.04	36.20	Vaccine distribution
coronavírus, estudo, pessoas, reino_unido, ômicron, caso, vírus, variante, risco, manaus	Pesquisadores da UFMG divulgaram descoberta de 18 mutações no Sars-Cov-2 (...)	0.04	35.69	Research & variants

Distribution over Media Outlet. Media outlet and emotion are also two categorical variables that hold a significant relationship, according to the χ^2 test result (Table 4), although with a small effect size. Table 8 displays the prevalence of each emotion in user comments as well as the $\chi 2$ value for a given media outlet. The coverage of UOL has the highest prevalence of anger (40.0%) and the lowest of joy (26.1%) among user replies. Meanwhile, the highest prevalence of sadness (13.6%) and the lowest of surprise (18.0%) are observed in user replies to the news from g1, with a significant deviation from the global mean. The highest

Table 7. Prevalence and chi-square values (in thousands) of emotions in user comments per topic. A higher chi-square value indicates a more significant deviation from the expected (random) prevalence. For example, anger appears in 20.5% of comments on cases & deaths news, with a chi-square value of $5.9x10^3$.

Topic	Anger	Disgust	Fear	Joy	Sadness	Surprise
Impact on health system	33.5% (0.6)	2.0% (0.0)	2.2% (0.1)	30.2% (0.2)	11.5% (0.0)	20.5% (0.2)
Pandemic investigation	**41.9% (6.2)**	**2.5% (0.5)**	1.9% (0.1)	25.4% (3.1)	9.2% (2.6)	19.2% (0.1)
Vaccination debate	35.4% (0.0)	1.8% (0.1)	2.0% (0.0)	29.7% (0.0)	10.3% (0.3)	20.8% (0.1)
Economic impact	34.5% (0.2)	2.1% (0.0)	2.3% (0.1)	29.9% (0.0)	11.4% (0.0)	19.8% (0.0)
Preventive measures	34.9% (0.0)	**2.5% (0.5)**	1.8% (0.1)	28.9% (0.1)	**14.3% (4.5)**	**17.6% (0.4)**
Governmental actions	35.5% (0.0)	2.0% (0.0)	1.6% (0.2)	30.4% (0.1)	9.4% (0.7)	21.2% (0.2)
Vaccine production	32.4% (0.7)	1.6% (0.3)	2.0% (0.0)	**32.3% (0.8)**	13.6% (1.0)	18.2% (0.2)
International impact	32.0% (0.9)	1.8% (0.1)	2.3% (0.0)	32.3% (0.2)	10.1% (0.3)	21.5% (0.1)
Cases & Deaths	29.4% (1.4)	2.3% (0.0)	**3.5% (1.4)**	27.0% (0.2)	16.3% (2.9)	21.5% (0.3)
Political response	37.7% (0.1)	2.2% (0.0)	1.6% (0.2)	29.5% (0.0)	9.8% (0.3)	19.2% (0.0)
Vaccine distribution	32.6% (0.3)	1.4% (0.2)	1.7% (0.1)	34.6% (0.7)	9.3% (0.3)	20.5% (0.0)
Research & variants	36.3% (0.0)	1.8% (0.1)	2.6% (0.1)	29.3% (0.0)	11.1% (0.0)	18.8% (0.0)
Mean	**35.7%**	**2.1%**	**2.0%**	**29.1%**	**11.4%**	**19.6%**

Table 8. Prevalence and chi-square values (in thousands) of emotions in user comments per media outlet. A higher chi-square value indicates a more significant deviation from the expected prevalence. For example, anger appears in 29.7% of comments on BBC News Brasil, with a chi-square value of $1.1x10^3$.

Media outlet	Anger	Disgust	Fear	Joy	Sadness	Surprise
BBC News Brasil	29.7% (1.1)	2.0% (0.0)	3.0% (0.2)	32.9% (0.1)	12.7% (0.0)	19.7% (0.0)
CNN Brasil	33.4% (0.2)	1.8% (0.1)	1.8% (0.1)	28.0% (0.0)	13.1% (1.4)	**21.9% (1.5)**
Estadão	37.0% (0.1)	2.1% (0.0)	1.8% (0.1)	28.5% (0.1)	10.1% (0.5)	20.4% (0.1)
Folha de S.Paulo	39.0% (1.5)	**2.5% (0.3)**	1.8% (0.1)	27.7% (0.4)	10.0% (0.9)	19.0% (0.1)
Jornal Extra	30.0% (0.5)	1.6% (0.1)	1.9% (0.0)	36.9% (0.7)	11.2% (0.0)	18.4% (0.1)
Jornal O Globo	35.8% (0.0)	1.9% (0.1)	2.2% (0.0)	29.8% (0.0)	9.8% (0.7)	20.5% (0.0)
Portal R7.com	33.6% (0.1)	1.8% (0.0)	2.0% (0.0)	30.9% (0.0)	9.3% (0.2)	22.3% (0.1)
Rádio BandNews FM	31.7% (0.3)	1.8% (0.0)	2.5% (0.0)	30.2% (0.0)	12.3% (0.0)	21.6% (0.0)
SBT	36.0% (0.0)	3.0% (0.0)	1.6% (0.0)	30.6% (0.0)	12.0% (0.0)	16.8% (0.0)
Terra	36.4% (0.0)	2.1% (0.0)	1.9% (0.0)	27.7% (0.0)	11.0% (0.0)	20.9% (0.0)
UOL Notícias	**40.0% (1.7)**	2.3% (0.1)	1.9% (0.1)	26.1% (1.5)	10.2% (0.5)	19.5% (0.0)
VEJA	36.2% (0.0)	2.1% (0.0)	1.9% (0.0)	29.7% (0.0)	10.2% (0.2)	19.9% (0.0)
g1	32.4% (1.4)	2.1% (0.0)	**2.4% (0.6)**	**31.5% (1.9)**	**13.6% (3.6)**	18.0% (0.6)
Mean	**35.7%**	**2.1%**	**2.0%**	**29.1%**	**11.4%**	**19.6%**

prevalence of joy is observed in user replies to BBC News Brasil (32.9%), and surprise reach the highest prevalence in user replies to the news published by SBT (22.3%).

5 Conclusions and Future Work

In this study, we developed a specialized Emotion Detection (ED) model for Brazilian Portuguese social media comments. While GoEmotions [9] is widely used, there have been no prior efforts to create emotion classifiers in Brazilian Portuguese that preserve the original social media writing style. To address this, we translated GoEmotions into GoEmotions-BR, with chatGPT-3.5-turbo effectively capturing nuanced expressions. The BERTimbau-base model, fine-tuned on GoEmotions-BR, enabled the detection of Ekman's six basic emotions and neutral emotion in Portuguese.

Our model achieved a F1-score in the 0.5–0.6 range, close to state-of-the-art results, with consistent performance across emotions, indicating evenly distributed errors and minimal bias. Although there is room for improvement, the model's performance is significantly better than chance, given the number of emotion classes.

We analyzed factors influencing emotion distribution in replies to COVID-19-related news, using 169,281 tweets and 4,861,704 replies. LDA identified the top 12 topics, and chi-squared tests confirmed that emotion distribution depended on both the news topic and the publishing outlet. The ADF test revealed non-stationarity in joy prevalence, while other emotions were stationary.

Most replies were neutral, with anger (21.1%) being the most prevalent identifiable emotion, followed by joy (17.2%) and surprise (11.6%). The 2021 Brazilian pandemic inquiry had the highest prevalence of anger (41.9%), preventive measures news led to the most sadness (14.3%), and vaccine production news elicited the most joy (32.3%). UOL Notícias had the highest anger (40.0%) and lowest joy (26.1%), while g1 had the highest sadness (13.6%).

Future research should focus on creating ED datasets that reflect Brazilian cultural nuances and evaluating emotion classifiers with other language models, such as Albertina [33], peLLE [24], and Sabiá [31]. Additionally, comparing the results on GoEmotions-BR with other strategies, like bag-of-words, could help assess the proposed model's quality. Finally, the methodology could be extended to other tasks, such as hate speech detection.

Acknowledgments. This study was financed in part by the Coordenação de Aperfeiçoamento de Pessoal de Nível Superior - Brasil (CAPES) - Finance Code 001.

Data Availability Statement. The data and models developed in this study will be made available upon request.

Disclosure of Interests. The authors have no competing interests to declare that are relevant to the content of this article.

References

1. Akhtar, M.S., Chauhan, D.S., Ghosal, D., Poria, S., Ekbal, A., Bhattacharyya, P.: Multi-task learning for multi-modal emotion recognition and sentiment analysis. arXiv preprint arXiv:1905.05812 (2019)

2. Al Maruf, A., Khanam, F., Haque, M.M., Jiyad, Z.M., Mridha, F., Aung, Z.: Challenges and opportunities of text-based emotion detection: a survey. IEEE Access (2024)
3. Amin, M.M., Mao, R., Cambria, E., Schuller, B.W.: A wide evaluation of chatgpt on affective computing tasks. arXiv preprint arXiv:2308.13911 (2023)
4. Bernardi, F.A., Lima, V.C., Rijo, R.P.C.L., Alves, D.: Mais do que palavras: uma análise das emoções brasileiras durante a covid-19. Revista Ibérica de Sistemas e Tecnologias de Informação **E40**, 526–541 (2021)
5. Blei, D.M., Ng, A.Y., Jordan, M.I.: Latent Dirichlet allocation. J. Mach. Learn. Res. **3**, 993–1022 (2003)
6. Cheong, M., Lee, V.C.: A microblogging-based approach to terrorism informatics: exploration and chronicling civilian sentiment and response to terrorism events via twitter. Inf. Syst. Front. **13**, 45–59 (2011). https://doi.org/10.1007/s10796-010-9273-x
7. Cohen, J.: Statistical power analysis for the behavioral sciences. Routledge (2013)
8. Cui, Y., Yang, Z., Yao, X.: Efficient and effective text encoding for chinese llama and alpaca. arXiv preprint arXiv:2304.08177 (2023)
9. Demszky, D., Movshovitz-Attias, D., Ko, J., Cowen, A., Nemade, G., Ravi, S.: Goemotions: a dataset of fine-grained emotions. arXiv preprint arXiv:2005.00547 (2020)
10. Devlin, J., Chang, M.W., Lee, K., Toutanova, K.: Bert: Pre-training of deep bidirectional transformers for language understanding. arXiv preprint arXiv:1810.04805 (2018)
11. Ekman, P.: An argument for basic emotions. Cognit. Emotion **6**(3–4), 169–200 (1992)
12. Fellbaum, C.: Wordnet. In: Theory and Applications of Ontology: Computer Applications, pp. 231–243. Springer (2010)
13. Gill, A.J., French, R.M., Gergle, D., Oberlander, J.: Identifying emotional characteristics from short blog texts. In: Proceedings for the 30th Annual Meeting of the Cognitive Science Society, pp. 2237–2242. Cognitive Science Society (2008)
14. Hammes, L.O.A., de Freitas, L.A.: Utilizando bertimbau para a classificação de emoções em português. In: Anais do XIII Simpósio Brasileiro de Tecnologia da Informação e da Linguagem Humana, pp. 56–63. SBC (2021)
15. Hung, L.P., Alias, S.: Beyond sentiment analysis: a review of recent trends in text based sentiment analysis and emotion detection. J. Adv. Comput. Intell. Intell. Inform. **27**(1), 84–95 (2023)
16. Kusal, S., Patil, S., Choudrie, J., Kotecha, K., Vora, D., Pappas, I.: A systematic review of applications of natural language processing and future challenges with special emphasis in text-based emotion detection. Artif. Intell. Rev. **56**(12), 15129–15215 (2023)
17. Larcher, C., Piau, M., Finardi, P., Gengo, P., Esposito, P., Caridá, V.: Cabrita: closing the gap for foreign languages. arXiv preprint arXiv:2308.11878 (2023)
18. Lwin, M.O., et al.: Global sentiments surrounding the covid-19 pandemic on twitter: analysis of twitter trends. JMIR Public Health Surveill. **6**(2), e19447 (2020)
19. MacKinnon, J.G.: Approximate asymptotic distribution functions for unit-root and cointegration tests. J. Bus. Econ. Statist. **12**(2), 167–176 (1994)
20. McCallum, A.K.: Mallet: A machine learning for language toolkit (2002). http://mallet.cs.umass.edu
21. McHugh, M.L.: The chi-square test of independence. Biochemia Medica **23**(2), 143–149 (2013)

22. Mehrabian, A.: Basic dimensions for a general psychological theory: implications for personality, social, environmental, and developmental studies (1980)
23. Mehrabian, A.: Communication without words. In: Communication Theory, pp. 193–200. Routledge (2017)
24. de Mello, G.L., Finger, M., Carpi, M.d.M., Jose, M.M., Domingues, P.H., Cavalim, P., et al.: Pelle: encoder-based language models for brazilian portuguese based on open data. arXiv preprint arXiv:2402.19204 (2024)
25. Mikolov, T., Chen, K., Corrado, G., Dean, J.: Efficient estimation of word representations in vector space. arXiv preprint arXiv:1301.3781 (2013)
26. Newman, N., Fletcher, R., Eddy, K., Robertson, C.T., Nielsen, R.K.: Digital news report 2023 (2023)
27. Oliveira, F.B., Haque, A., Mougouei, D., Evans, S., Sichman, J.S., Singh, M.P.: Investigating the emotional response to covid-19 news on twitter: a topic modeling and emotion classification approach. IEEE Access **10**, 16883–16897 (2022)
28. O'Connor, B., Balasubramanyan, R., Routledge, B., Smith, N.: From tweets to polls: linking text sentiment to public opinion time series. In: Proceedings of the International AAAI Conference on Web and Social Media (ICWSM), vol. 4(1), pp. 122–129 (2010). https://doi.org/10.1609/icwsm.v4i1.14031
29. Pasqualotti, P.R., Vieira, R.: Wordnetaffectbr: uma base lexical de palavras de emoções para a língua portuguesa. Revista Novas Tecnologias na Educação **6**(1) (2008)
30. Picard, R.W.: Affective computing. MIT press (2000)
31. Pires, R., Abonizio, H., Almeida, T.S., Nogueira, R.: Sabiá: Portuguese large language models. In: Brazilian Conference on Intelligent Systems, pp. 226–240. Springer (2023)
32. Plutchik, R.: Emotions: a general psychoevolutionary theory. Approach. Emot. **1984**, 197–219 (1984)
33. Rodrigues, J., et al.: Advancing neural encoding of portuguese with transformer albertina pt. In: EPIA Conference on Artificial Intelligence, pp. 441–453. Springer (2023). https://doi.org/10.1007/978-3-031-49008-8_35
34. Russell, J.A.: A circumplex model of affect. J. Pers. Soc. Psychol. **39**(6), 1161 (1980)
35. Scherer, K.: Profiles of emotion-antecedent appraisal: testing theoretical predictions across cultures. Cogn. Emot. **11**(2), 113–150 (1997)
36. Silva, L.G., Caseli, H.M.: Aspect-based sentiment analysis in comments on political debates in Portuguese: evaluating the potential of chatgpt. PROPOR **2024**, 312 (2024)
37. Souza, F., Nogueira, R., Lotufo, R.: BERTimbau: pretrained BERT models for Brazilian Portuguese. In: Cerri, R., Prati, R.C. (eds.) BRACIS 2020. LNCS (LNAI), vol. 12319, pp. 403–417. Springer, Cham (2020). https://doi.org/10.1007/978-3-030-61377-8_28
38. Strapparava, C., Mihalcea, R.: Learning to identify emotions in text. In: Proceedings of the 2008 ACM symposium on Applied Computing, pp. 1556–1560 (2008)
39. Syed, S., Spruit, M.: Full-text or abstract? examining topic coherence scores using latent Dirichlet allocation. In: 2017 IEEE International Conference on Data Science and Advanced Analytics (DSAA), pp. 165–174 (2017). https://doi.org/10.1109/DSAA.2017.61
40. Sykora, M., Jackson, T., O'Brien, A., Elayan, S.: Emotive ontology: Extracting fine-grained emotions from terse, informal messages (2013)
41. Twitter, I.: Twitter context annotations: List of entities (2022). https://github.com/twitterdev/twitter-context-annotations

42. Vaswani, A., et al.: Attention is all you need. Adv. Neural Inform. Process. Syst. **30** (2017)
43. Wagner Filho, J.A., Wilkens, R., Idiart, M., Villavicencio, A.: The brwac corpus: a new open resource for brazilian portuguese. In: Proceedings of the Eleventh International Conference on Language Resources and Evaluation (LREC 2018) (2018)
44. Wake, N., Kanehira, A., Sasabuchi, K., Takamatsu, J., Ikeuchi, K.: Bias in emotion recognition with chatgpt. arXiv preprint arXiv:2310.11753 (2023)
45. Wang, X., Zheng, Q.: Text emotion classification research based on improved latent semantic analysis algorithm. In: Conference of the 2nd International Conference on Computer Science and Electronics Engineering (ICCSEE 2013), pp. 210–213. Atlantis Press (2013)
46. Wang, Y., et al.: A systematic review on affective computing: emotion models, databases, and recent advances. Inform. Fusion **83**, 19–52 (2022)
47. Xue, J., et al.: Twitter discussions and emotions about the covid-19 pandemic: machine learning approach. J. Med. Internet Res. **22**(11), e20550 (2020)

Predicting Bull and Bear Markets: A Deep Learning and Linear Regression Study in Cryptocurrencies

João Paulo Costa e Souza[1(✉)], Rodolfo I. Meneguette[2], Vinícius P. Gonçalves[1], Fábio L. L. de Mendonça[1], Francisco Airton Silva[3], and Geraldo P. Rocha Filho[1,4]

[1] University of Brasilia, Brasilia, Brazil
joaopaulo.csz@outlook.com, vpgvinicius@unb.br, fabio.mendonca@redes.unb.br

[2] University of São Paulo, São Carlos, Brazil
meneguette@icmc.usp.br

[3] Federal University of Piauí, Picos, Brazil
faps@ufpi.edu.br

[4] State University of Southwest Bahia, Vitória da Conquista, Brazil
geraldo.rocha@uesb.edu.br

Abstract. Despite the growing popularity and increasingly widespread use of cryptocurrencies in contemporary financial markets, understanding market trends and predicting their future movements is a formidable challenge in financial analysis. Regardless of the numerous tools and techniques available, accurately forecasting market behavior remains complex due to financial data's inherent volatility and noise. This work proposes a novel hybrid model for market trend classification and trend slope regression to address this gap. The proposed approach leverages technical indicators, autoencoders, a BiLSTM-CNN architecture, and the use of linear regression to label the trends, enhancing prediction accuracy. The core innovation lies in the integration of these advanced techniques to extract meaningful and useful patterns from the data. This methodology allows the model to better capture the complexities of financial time series, thus improving its predictive capabilities. As a proof of concept, the proposed model was evaluated using historical Bitcoin data spanning two years, with performance metrics such as accuracy, precision, recall for trend classification, and mean squared error (MSE), mean absolute error (MAE), and mean absolute percentage error (MAPE) for trend slope regression. The results are compelling, demonstrating that the combination of linear regression labeling and the architecture proposed achieves a remarkable level of accuracy both in-sample and out-of-sample. This accuracy significantly surpasses the 50% random benchmark in the predictions of bull and bear markets. Ultimately, this research presents a promising approach to enhancing the precision of trend predictions in the volatile and unpredictable cryptocurrency market.

A. Paes and F. A. N. Verri (Eds.): BRACIS 2024, LNAI 15413, pp. 281–295, 2025.
https://doi.org/10.1007/978-3-031-79032-4_20

Keywords: Artificial Intelligence · Cryptocurrency · Linear Regression · CNN · LSTM · BiLSTM · Machine Learning · Deep Learning · Time Series Forecasting

1 Introduction

Cryptocurrency refers to electronic or virtual currency used for online transactions and asset transfers, secured through cryptographic techniques that regulate the creation of new units and safeguard transactions, ensuring secure transfers [3,18]. Unlike conventional currencies, cryptocurrencies operate on a decentralized framework without central banking systems. Due to their uncontrolled and untraceable nature, the cryptocurrency market has experienced nearly exponential growth [4,12,14].

As cryptocurrencies gain recognition as a new electronic option for currency exchange, they have become increasingly important in emerging economies and the global financial landscape [11]. They are now integral to numerous financial transactions, making them a promising investment avenue. However, the market is characterized by significant volatility, price fluctuations, and significant randomness. Predicting cryptocurrency trends is a formidable challenge in time-series forecasting due to the multitude of unpredictable variables, pronounced volatility in pricing, the lack of stationarity, and the fact that they are close to a random process [6,10].

Researchers have proposed various market rules to guide investors in their decisions, using historical price movements and trading volume in the form of technical analysis. This analysis helps forecast the continuation or reversion of market trends by identifying patterns in past price data. Nevertheless, manual trading remains difficult due to market uncertainty, emotional influences on traders [16], and the infinitude of indicators and other financial data [1]. Algorithmic trading, incorporating machine learning techniques, offers a solution. Deep learning techniques, particularly Convolutional Neural Networks (CNN) and long short-term memory (LSTM) layers, have proven effective for financial forecasting in the cryptocurrency market. LSTM layers efficiently capture sequential patterns in both long- and short-term dependencies, whereas convolutional layers serve to eliminate noise from intricate time-series datasets and extract valuable patterns [9].

The literature presents a variety of models aimed at predicting the dynamics of financial markets, often achieving impressive results in price prediction. For instance, Selvin et al. compare classical econometric methods with cutting-edge deep learning techniques, concluding that the latter are superior [15]. Similarly, Zhanhong He et al. study different architectures for predicting gold prices, identifying a combination of CNN, LSTM, and attention mechanisms as the most effective [5]. Zhuorui Zhang et al. develop a robust price prediction regressor using memory and convolutional layers that consider the interrelations between cryptocurrencies [19]. In another study, Ioannis E. Livieris et al. create a CNN-LSTM based neural network to predict gold prices, which shows promising results in regression but falls short in trend classification, with accuracy close to randomness [9]. Livieris et al. also propose a non-sequential CNN-LSTM neural network for price regression and trend classification, again finding trend classification results near randomness [8]. Yanhui Liang et al. take a different approach by decomposing gold prices into various frequencies before feeding them into a CNN-LSTM structure, though they do not test the model in trend classification [7]. Sima Siami-Namini et al. compare Bidirectional Long Short Term Momory (BiLSTM) and LSTM for price prediction, showing that BiLSTM achieves lower errors, but they do not test the model in trend classification [17]. Iromie K. Samarasekara et al. focus on risk management rather than trading, developing a dynamic Stop-Loss tool that uses a CNN-LSTM trend classifier to feed a price regressor, the study showed good results compared to other tools of risk management [13]. Faraz et al. enhance an LSTM price regressor with an LSTM autoencoder, achieving good results in price prediction but not testing it in trend classification [2]. Despite these successes in price prediction, trend prediction often yields poor results. To address this gap, the current study proposes a novel trend classification method using a robust BiLSTM-CNN model and suggests linear regression for better sample labeling in trend slope regression and classification tasks.

To effectively manage the inherent noise in financial markets, reduce input dimensionality, and eliminate outliers, an encoding structure is applied to the features. This compression process, handled by BiLSTM layers, forces market noise filtering, which is essential when dealing with financial data. Once encoded, the features are fed into a BiLSTM-CNN neural network. The BiLSTM layers track short- and long-term dependencies in both directions of the time series, while the CNN layers capture significant patterns. The model is then trained using labels produced by linear regression over past and future price candles. Trends are identified by their slope, with an ascending slope indicating a bull market and a descending one signaling a bear market. This approach enhances accuracy in identifying and predicting market trends. The point-by-point contributions are listed below:

- A novel autoencoder architecture to filter market noise based on BiLSTM.
- Proposed a BiLSTM-CNN to handle trend prediction, both in market classification and slope regression tasks.
- A novel trend labeling system based on linear regression slope.

The rest of this paper is organized as: Sect. 2 describes the related work related to financial forecasting, both in price and trend prediction. Section 3 shows the proposed model. Section 4 discusses the results. Section 5 brings a conclusion, followed by references.

2 Related Work

This session aims to discuss several models and strategies proposed by various researchers. These proposals cover various architectures as well as methods for pre-processing the data and variables that feed the input layer of the various models.

Sreelekshmy Selvin et al. [15] conducted a comparative analysis of three machine learning approaches: Recurrent Neural Network (RNN), Long Short-Term Memory, and Convolutional Neural Network, focusing on their performance in predicting financial market price. Employing a sliding window methodology with a window size of 100 min, wherein 90 min of historical data is utilized to forecast the price of the subsequent 10 min. The method used to evaluate the models is MAPE. Their study revealed that CNN outperformed RNN and LSTM. The superior performance of CNN is attributed to its ability to adapt to the dynamic nature of the market by prioritizing current information over past patterns, which may not always be indicative of future trends. Unlike RNN and LSTM, which rely heavily on historical data, CNN's emphasis on current information enables it to identify sudden shifts in market trends more effectively. This study underscores the efficacy of CNN in financial forecasting, particularly in scenarios characterized by rapid market changes and volatility.

In the research conducted by Zhanhong He et al. [5], a novel deep learning model for predicting gold prices is introduced, integrating Long Short-Term Memory and Convolutional Neural Networks with an attention mechanism. Their study reveals that arranging the model with the LSTM layer preceding the CNN layer yields superior results compared to the reverse order. Additionally, incorporating an attention mechanism after the LSTM layer and before the CNN layer further enhances predictive accuracy. The author attributes this improvement to the prevention of information loss in the attributes, which would occur if CNN were placed before LSTM. Furthermore, the performance evaluation of the LSTM-CNN hybrid model against CNN and LSTM individually demonstrates a reduction in error, highlighting the efficacy of the combined approach. The author proposes that employing bidirectional LSTM (BiLSTM) could potentially yield even more favorable outcomes, indicating avenues for future research and refinement in forecasting gold prices.

Introducing a new approach to cryptocurrency price prediction, the work of Zhuorui Zhang et al. [19] emphasizes volatility and strong intercorrelations among different cryptocurrencies. Their predictive system comprises three key modules: the attentive memory module, leveraging a Gated Recurrent Unit (GRU) layer and self-attention mechanism to capture both short- and long-term dependencies; the channel weighting module, explicitly modeling interdependencies between cryptocurrency channels to account for their correlated nature; and the convolutional and pooling module, identifying attribute forms crucial for prediction. While the attentive memory module effectively captures dependencies, it struggles to assess the true importance of temporally delayed instances, addressed by the self-attention mechanism, which selects significant information across various time intervals. By explicitly modeling interdependencies between cryptocurrency channels, the channel weighting module proves essential, given the inherent correlations among cryptocurrency prices. The convolutional module further enhances predictive accuracy by identifying attribute forms. Evaluation metrics, including average absolute percentage error and accuracy, demonstrate superior performance of the proposed model over traditional methods like CNN, LSTM, or GRU, indicating its effectiveness in cryptocurrency price prediction.

In the research conducted by Ioannis E. Livieris et al. [9], the focus lies on leveraging convolutional neural networks and long short-term memory neural networks for gold price and movement prediction. The proposed model capitalizes on CNN's capability to extract crucial information and learn internal representations from time series data, complemented by LSTM's proficiency in capturing both long- and short-term dependencies. While the study yielded highly accurate results for price prediction with minimal errors, the performance for trend prediction was noted to be average. This observation suggests that while the model effectively captures and predicts the price movements of gold, it may encounter challenges in accurately forecasting the directional trends over time. Further refinements or alternative approaches may be necessary to enhance the model's efficacy in trend prediction, potentially through additional feature engineering, model adjustments, or the incorporation of complementary techniques.

In another study by Ioannis E. Livieris et al. [8], an advanced non-sequential neural network architecture is proposed to predict cryptocurrency prices. This model capitalizes on the intricate interrelations between different cryptocurrencies by using data from multiple cryptocurrencies as inputs. The research focuses on the three leading cryptocurrencies by market capitalization: Bitcoin, Ethereum, and Ripple. The model is applied to both price regression and trend classification tasks, with its performance evaluated using a range of metrics: Mean Absolute Error (MAE), Root Mean Squared Error (RMSE), and R squared (R^2) for regression tasks, as well as Accuracy, Geometric Mean, Sensitivity, and Specificity for classification tasks. While the model achieved relatively low errors in regression, its accuracy in trend classification was unfortunately close to randomness.

Yanhui Liang et al. [7] also concentrate on predicting gold prices through a multi-step approach. Initially, they decompose gold prices into distinct frequency components utilizing the Improved Complete Ensemble Empirical Mode Decomposition with Adaptive Noise (ICEEMDAN) algorithm. Subsequently, each of these decomposed components undergoes neural network layers such as CNN and LSTM networks to refine the prediction. The study reveals that decomposing prices into various frequencies significantly enhances prediction accuracy, outperforming alternative decomposition methods. This approach underscores the importance of considering diverse frequency components in price prediction models, highlighting its potential to improve forecasting accuracy in financial markets, particularly for assets like gold characterized by complex and multifaceted price dynamics.

Sima Siami-Namini et al. [17]conducted a comparative study on the performance of LSTM and BiLSTM in price prediction tasks. While BiLSTM is known to excel in tasks like predicting the next word in a sentence, its superiority in time series prediction was uncertain. The study revealed that BiLSTM outperformed LSTM by 37.78% in MSE for price prediction tasks. However, biLSTM requires more effort to train compared to LSTM. The key advantage of BiLSTM in financial data analysis is its bidirectional training, allowing the model to process information both forward and backward, enhancing its predictive capabilities.

Iromie K. Samarasekara et al. [13] deviate from conventional approaches in financial forecasting by focusing their research on stop-loss dynamics. While traditional studies center on developing price prediction systems to aid buying and selling decisions, this work directs attention toward creating a prediction system to optimize stop-loss placement. The authors employ convolutional neural networks and long short-term memory networks to forecast future trends, utilizing these predictions as inputs for a subsequent price prediction system. Subsequently, both the future trend predictor and price predictor are fed into a regression neural network, also leveraging CNN and LSTM architectures, tasked with determining optimal stop-loss positions. This neural network is trained according to a specific algorithm. Results demonstrate that losses incurred using the proposed strategy are substantially lower than losses experienced with traditional strategies employing fixed stop-loss and trailing stop mechanisms.

Faraz et al. [2] employed an autoencoder in their study to compress information extracted from technical indicators. Both the encoder and decoder components of the autoencoder utilized Long Short-Term Memory cells. Following the encoding process and reduction in input dimensionality, the data was then fed into a neural network consisting of LSTM cells for price prediction. Comparative analysis with the Generative Adversarial Networks (GAN) model demonstrated a reduction in error, indicating the superior performance of the LSTM-based autoencoder approach. This suggests the effectiveness of leveraging LSTM cells within the autoencoder framework for information compression and subsequent price prediction tasks, showcasing its potential for enhancing accuracy in financial forecasting models.

3 Proposed Method

In this section, we elucidate the proposed models and main concepts about trend classification and trend slope regression, which are the aims of this study.

Figure 1 illustrates the pipeline of the models of trend prediction. First, the technical indicators—which contain valuable trend information—go through the normalization process. Then, the normalized data passes through an autoencoder to reduce the dimension of features, filter out some inherent market noise, and extract outliers. Finally, the data is sent to a BiLSTM-CNN neural network to forecast the trends. In this step, linear regression is used to train the neural net both in the trend classifier and trend slope regressor.

A set of candles is used in the trend forecasters' training. Candles are a type of financial instrument that displays four different prices during a given period: the opening, closing, minimum, and maximum prices. Half of this candle set is composed of past candles, and the other half is composed of future candles. The closing prices of this set of candles are then used to perform a linear regression. In conclusion, regarding the trend classifier, a positive linear regression slope indicates a bull market, while a negative linear regression slope indicates a bear market. Regarding the trend slope regressor, the linear regression slope measure provides an approximation of the trend's strength.

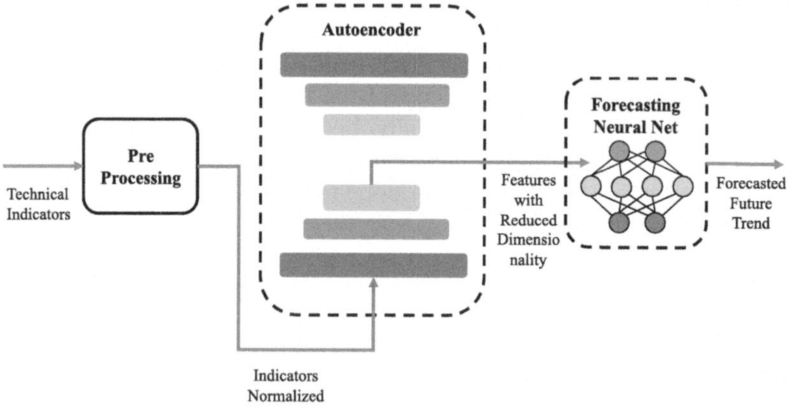

Fig. 1. Pipeline of the model, showing the flow of market information carried out by technical indicators from the preprocessing stage, passing through encoding, and reaching the forecasting neural network.

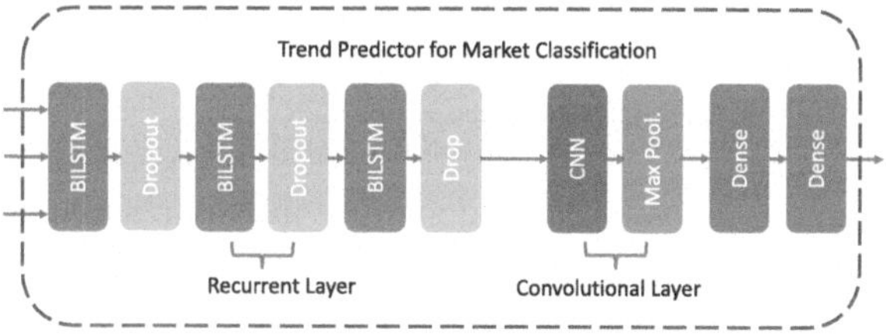

Fig. 2. Architectural scheme of the neural network used for trend market classification. The model uses three BiLSTM layers, one convolutional layer, and two dense layers.

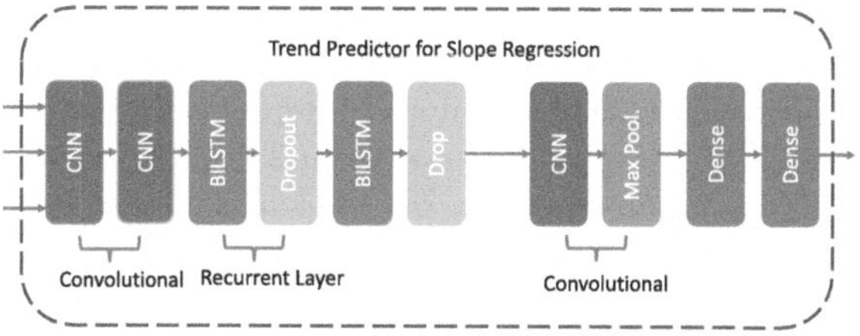

Fig. 3. Architectural scheme of the neural network used for trend slope regression. The model uses a prior convolutional layer, two BiLSTM layers, one convolutional layer, and two dense layers.

The trend classifier's architecture is depicted in Fig. 2. It is made up of three BiLSTM layers with a dropout layer to prevent overfitting, a convolutional layer, a Max Pooling layer, and two dense layers. BiLSTM layers have the advantage of being able to obtain time dependences from the past to the future as well as from the future to the past. Additionally, because they come before the CNN layer, they ensure that information crucial to classification is not improperly pruned. Furthermore, the design of the slope regressor is depicted in Fig. 3. Two CNN layers, two BiLSTM layers, a CNN layer, a Max Pooling layer, and two dense layers make up this structure.

3.1 Normalization

Neural networks are sensitive to the size of the input data, so learning may be dominated by features with wider numerical ranges if normalization isn't applied. This could lead to the model skewing in favor of some features and possibly ignoring others. In that regard, Normalization was used to stabilize the training process by scaling the data to a common range, $[-1, 1]$. This accelerates the convergence of the gradient descent technique and increases its efficiency. Additionally, the uniform scaling of normalization helps to achieve better and more balanced weight adjustments during training, which improves the model's performance and generalization on fresh data. It also prevents numerical instability.

$$Norm(x) = \frac{x - \min(X)}{\max(X) - \min(X)} \tag{1}$$

In Eq. 1, x represents the value of the sample in a particular feature, $\min(x)$ represents the minimum value the feature takes, whereas $\max(x)$ represents its maximum value.

3.2 Autoencoder

An autoencoder is a kind of artificial neural network that is designed to automatically learn effective coding from input data without supervision. An autoencoder is a structure that employs two structures to learn a compressed representation of data: an encoder that codifies the information using the function $h = f(x)$, and a decoder that tries to match the encoder's input with its output $x' = g(h)$.

BiLSTM layers were used to build the autoencoder, which was based on the premise of dimensionality reduction and filtering outliers and market noise. The number of BiLSTM layers in the encoder and decoder components is equal, and each layer has an innate dropout mechanism that keeps the model from overfitting.

3.3 Convolutional Neural Networks

Convolutional neural networks (CNNs) are engineered to autonomously discern and internalize intricate patterns from single- or multi-dimensional datasets. The distinctive feature of CNNs lies in their convolutional layer, where convolutional operations are applied, enabling the identification of patterns such as edges and shapes. The application of CNNs in financial time series analysis offers significant potential for generating valuable insights into market forecasting, identifying investment opportunities, and recognizing price patterns. These networks excel in handling complex sequential data and extracting pertinent information, rendering them indispensable tools in financial analytics. Unlike densely connected neural networks, CNNs leverage weight sharing in convolutional layers, enhancing scalability and enabling the network to learn specific types of patterns more efficiently. This characteristic contributes to the adaptability and effectiveness of CNNs in capturing and interpreting intricate patterns inherent in financial data, empowering traders and analysts with enhanced decision-making capabilities in dynamic market environments.

$$f * g(x) = \sum_{-\infty}^{\infty} f(k)g(x-k) \tag{2}$$

Equation 2 shows the convolution operation applied to one dimension, where f is the function to pass through the operation, g is the convolution filter and $(*)$ represents the convolution operation.

3.4 LSTM Neural Networks

Long Short-Term Memory networks (LSTMs) represent a specialized variant of recurrent neural networks (RNNs) designed specifically for processing sequential information. LSTMs incorporate three essential channels for managing and processing sequential data. The first channel, termed the "forgetting gate," regulates the extent to which information is discarded from the long memory, enabling the network to forget irrelevant or outdated information. The second channel, known as the "input gate," receives new data and assesses its significance in the context of previous information, determining what information to retain and what to discard. This gate plays a crucial role in updating the network's internal state based on incoming data. Finally, the third channel, the "output gate," determines which information stored in the long-term memory should be utilized to predict the current output, effectively filtering and selecting relevant memory components for the current context. These three channels interact dynamically, allowing LSTMs to capture and preserve long-term temporal dependencies within sequential data (which financial prices are an example) while mitigating issues such as gradient vanishing or exploding, which are common in traditional RNN architectures.

$$i_t = \sigma_g(W_i[x_t, h_{t-1}] + b_i) \tag{3}$$

$$f_t = \sigma_g(W_f[x_t, h_{t-1}] + b_f) \tag{4}$$

$$o_t = \sigma_g(W_o[x_t, h_{t-1}] + b_o) \tag{5}$$

$$c_t = f_t \odot c_{t-1} + i_t \odot \sigma_c(W_c[x_t, h_{t-1}] + b_c) \tag{6}$$

$$h_t = o_t \odot \sigma_c(c_t) \tag{7}$$

In the equations above, i_t, f_t and o_t represent respectively the input, forget and output gates at the time step t; W and b represent which one the weights and the bias of the LSTM cell, it's noted that each gate has their own weights and biases; x_t is the input of the cell, and h_{t-1} the hidden state at the previous time step. The operation $[y, z]$ represents concatenation. σ_g is the sigmoid function, whereas σ_c is the hyperbolic tangent function. Thus, memory c_t is formed taking the value of previous memory c_{t-1} weighted by the forget gate f_t added the new information given by $\sigma_c(W_c[x_t, h_{t-1}] + b_c)$ weighted by the input gate i_t. The symbol $\odot$ represents the Hadamard's multiplication. Finally, the actual hidden state is formed by the memory c_t transformed by the function σ_c.

4 Experiments and Results

This section assesses the effectiveness of our suggested market trend classifier and trend slope regressor models using various parameter values. While the accuracy, precision, and recall metrics are used to assess the market trend classifier's performance, the Mean Squared Error, Mean Absolute Error, and Mean Absolute Percentage Error are used to evaluate the trend slope regressor's performance.

In this study, Bitcoin data was used to perform the training and testing of the trend predictors. Bitcoin was chosen as it's the first and most widely recognized cryptocurrency, and holds significant representativity in the market, often serving as the benchmark for other digital assets. It commands the largest market capitalization, influencing market trends and investor sentiment. Bitcoin's dominance also reflects its role as a store of value and a leading indicator of the cryptocurrency sector's health and maturity. Said that, the dataset consists of the last 30'000 candles of the BTCUSDT (Bitcoin price given in dollars) in the period between January 01, 2022, and January 01, 2024. Opening, high, low, close, and volume of each candle were used to calculate technical indicators.

The technical indicators were: Exponential Moving Average(EMA), Weighted Moving Average(WMA), Moving Average Convergence Divergence(MACD), Momentum (MOM), Commodities Channel Index(CCI), Detrended Price Oscillator(DPO), Relative Strength Index(RSI), Average True Range(ATR), Stochastic K% and Stochastic D%.

The results for training and test of the market trend classifier are shown in Table 1. The number of candles taken for the linear regression calculation was varied between two to twelve, where half of the candles are future candles and half are past candles. The greater the number of candles, the more the trend is consolidated. It's clear from the data that market forecasting is possible, given that if the market was all random, the accuracy of all models would be near 50%. Furthermore, as shown in the table, the models trained with more candles show a better performance than the models trained with fewer candles, which suggests that market forecasting is more sympathetic to more consolidated trends.

Table 1. Metrics for the Market Trend Classifier

Trend Predictor for Bull and Bear Market Classifier						
Train						
Metrics	2C	4C	6C	8C	10C	12C
Accuracy	0.624	0.685	0.707	0.709	0.713	0.720
Bull precision	0.649	0.655	0.751	0.692	0.687	0.731
Bear precision	0.593	0.752	0.669	0.732	0.754	0.709
Precision mean	0.621	0.704	0.710	0.712	0.721	0.720
Bull Recall	0.659	0.858	0.663	0.789	0.810	0.722
Bear Recall	0.582	0.488	0.756	0.621	0.602	0.718
Recall Mean	0.621	0.673	0.710	0.705	0.709	0.720
Test						
Accuracy	0.590	0.695	0.683	0.731	0.705	0.725
Bull Precision	0.612	0.709	0.767	0.743	0.769	0.770
Bear Precision	0.550	0.675	0.617	0.715	0.645	0.678
Precision Mean	0.581	0.692	0.692	0.729	0.707	0.724
Bull Recall	0.707	0.751	0.619	0.793	0.673	0.712
Bear Recall	0.440	0.626	0.772	0.654	0.746	0.741
Recall Mean	0.576	0.689	0.691	0.723	0.709	0.726

Table 2. Metrics for the Trend Slope Regressor

Trend Slope Regressor						
Train						
Metrics	2C	4C	6C	8C	10C	12C
MSE	0.0006	0.0006	0.00007	0.0008	0.0008	0.0009
MAE	0.0133	0.0162	0.0174	0.0197	0.0196	0.0213
MAPE	301	985	479	578	358	536
Test						
MSE	0.0010	0.0021	0.0026	0.0033	0.0045	0.0080
MSE	0.0189	0.0311	0.0360	0.0406	0.0453	0.0562
MAPE	390	344	628	591	454	431

The results for training and testing of the trend slope regressor are shown in Table 2 and in Fig. 4. Visually, the graphs show, like in the case of the market trend classifier, that the greater the number of candles used in linear regression, the more likely the curves of true and predicted values will match. This phenomenon can be explained by the higher volatility of linear regression calculated with fewer candles. Nevertheless, Table 2 shows bigger values of errors as the number of candles increases. The smallest values of errors come from the

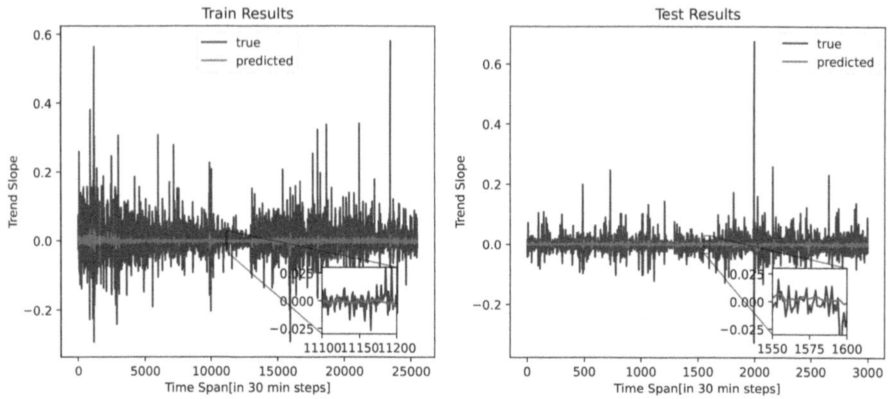

(a) 2 candles

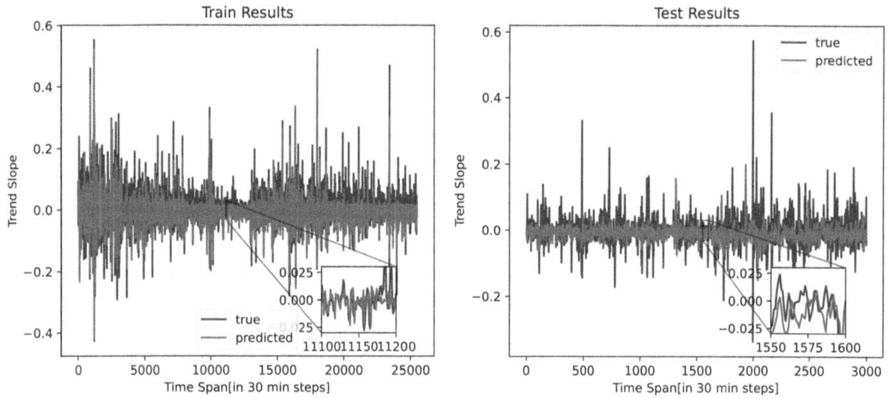

(b) 4 candles

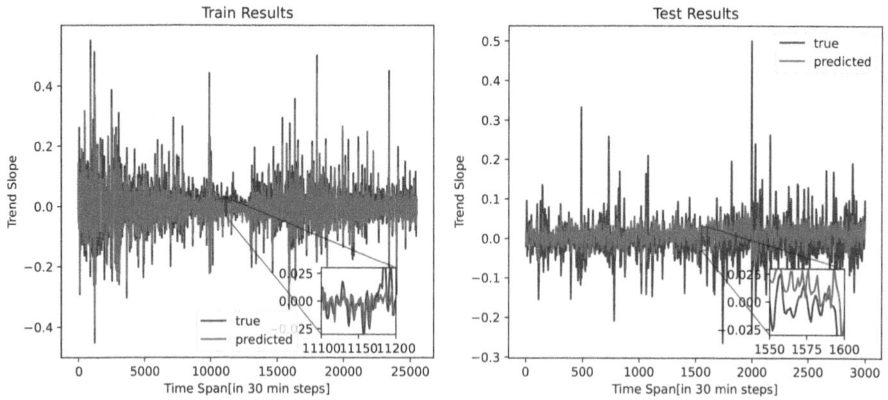

(c) 6 candles

Fig. 4. Trend slope over time, for the trend slope predictor built with (a) 2 candles, (b) 4 candles, (c) 6 candles, (d) 8 candles, (e) 10 candles, and (f) 12 candles. The curves in purple show the true values of the slope, whereas the curves in blue show the predicted values. The slopes are shown for train and test samples (Color figure online)

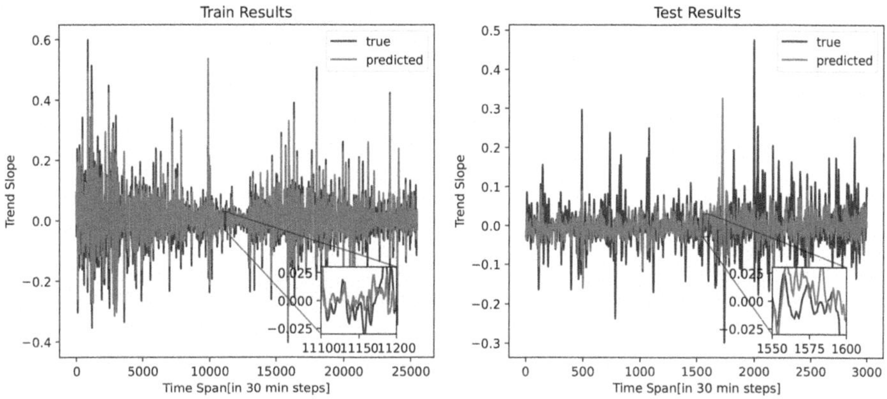

(d) 8 candles

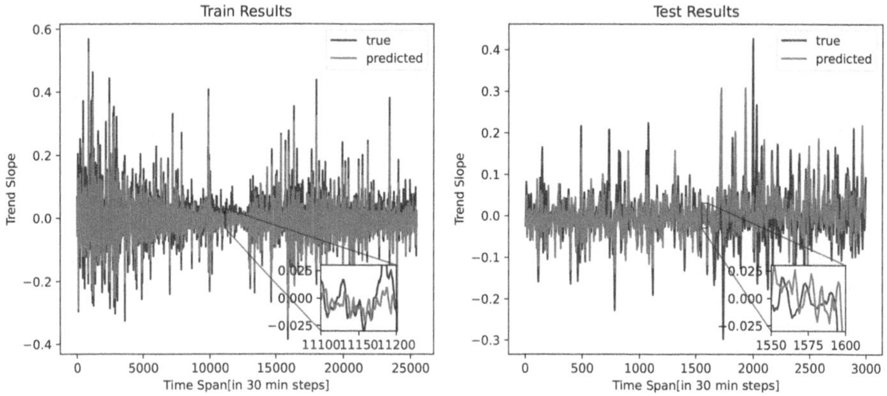

(e) 10 candles

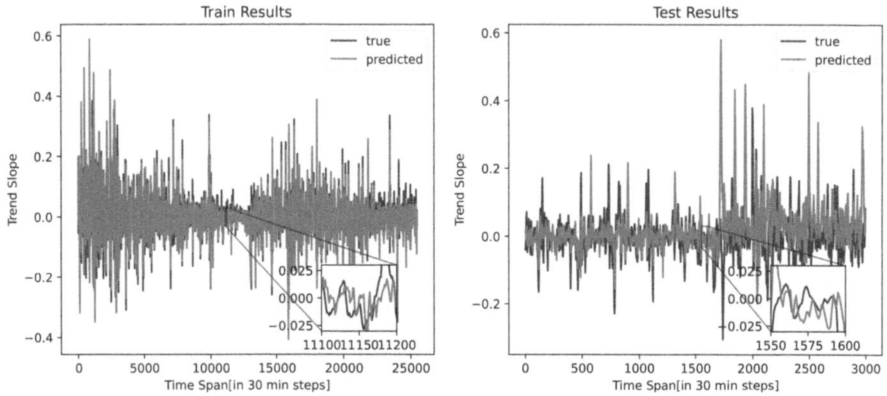

(f) 12 candles

Fig. 4. (*continued*)

case in which only 2 candles were used for linear regression calculation, and we see clearly from the pictures that was the worst scenario. This occurs because the model can be stuck in local minima in the gradient descent part of training, delivering small values of error, whereas the overall predicted curve does not match the real curve.

5 Conclusion

This paper presents a novel strategy for training financial predictors using linear regression, focusing on cryptocurrencies, especially Bitcoin, due to its market significance and trend forecasting complexities. The proposed architecture employs Bidirectional Long Short-Term Memory (BiLSTM) and Convolutional layers to capture spatial and temporal patterns in the data. Unlike traditional methods, this approach uses linear regression over a variable set of 2 to 12 candles -half future and half past candles- to better represent market trends.

The trend classifier shows promising potential, achieving significantly higher accuracy than the 50% random benchmark. Increasing the number of candles in the linear regression set enhances accuracy by providing more consolidated trends, which are easier to forecast. Consolidated trends exhibit stable patterns, while less consolidated ones are more volatile and random. Thus, using more candles in the linear regression set leads to more reliable predictions by effectively capturing market dynamics. The trend slope regressor shows discrepancies between predicted trend slopes and error metrics, with better matches for more consolidated trends.

Future research could explore a broader set of indicators and the maximum number of candles predictors can handle before accuracy degrades. Such advancements could enhance the model's predictive capabilities. The trend classification and slope regression techniques in this study offer valuable tools for investors, enabling precise identification of trend reversals and continuations, potentially leading to greater profits.

Disclosure of Interests. The authors have no competing interests.

References

1. Derbentsev, V., Matviychuk, A., Soloviev, V.N.: Forecasting of cryptocurrency prices using machine learning. Advanced studies of financial technologies and cryptocurrency markets, pp. 211–231 (2020)
2. Faraz, M., Khaloozadeh, H., Abbasi, M.: Stock market prediction-by-prediction based on autoencoder long short-term memory networks. In: 2020 28th Iranian Conference on Electrical Engineering (ICEE), pp. 1–5. IEEE (2020)
3. Garcia, D., Tessone, C.J., Mavrodiev, P., Perony, N.: The digital traces of bubbles: feedback cycles between socio-economic signals in the bitcoin economy. J. R. Soc. Interface **11**(99), 20140623 (2014)

4. Goundar, S.: Introduction to blockchains and cryptocurrencies. In: Blockchain Technologies, Applications and Cryptocurrencies: Current Practice and Future Trends, pp. ix–xix. World Scientific (2021)
5. He, Z., Zhou, J., Dai, H.N., Wang, H.: Gold price forecast based on lstm-cnn model. In: 2019 IEEE Intl Conf on Dependable, Autonomic and Secure Computing, Intl Conf on Pervasive Intelligence and Computing, Intl Conf on Cloud and Big Data Computing, Intl Conf on Cyber Science and Technology Congress (DASC/PiCom/CBDCom/CyberSciTech), pp. 1046–1053. IEEE (2019)
6. Lamberton, D., Lapeyre, B.: Introduction to stochastic calculus applied to finance. Chapman and Hall/CRC (2011)
7. Liang, Y., Lin, Y., Lu, Q.: Forecasting gold price using a novel hybrid model with iceemdan and lstm-cnn-cbam. Expert Syst. Appl. **206**, 117847 (2022)
8. Livieris, I.E., Kiriakidou, N., Stavroyiannis, S., Pintelas, P.: An advanced cnn-lstm model for cryptocurrency forecasting. Electronics **10**(3), 287 (2021)
9. Livieris, I.E., Pintelas, E., Pintelas, P.: A cnn-lstm model for gold price time-series forecasting. Neural Comput. Appl. **32**, 17351–17360 (2020)
10. Livieris, I.E., Stavroyiannis, S., Pintelas, E., Pintelas, P.: A novel validation framework to enhance deep learning models in time-series forecasting. Neural Comput. Appl. **32**(23), 17149–17167 (2020)
11. Nasir, M.A., Huynh, T.L.D., Nguyen, S.P., Duong, D.: Forecasting cryptocurrency returns and volume using search engines. Financial Innovation **5**(1), 1–13 (2019)
12. Nobre, R.A., et al.: Aurora: an autonomous agent-oriented hybrid trading service. Neural Comput. Appli., 1–16 (2022)
13. Samarasekara, I.K., Mendis, O.K., Ahangama, S., Atukorale, A.S.: Dynamic stop-loss approach for short term trades using deep learning. In: 2022 IEEE International Conference on Big Data (Big Data), pp. 3537–3546. IEEE (2022)
14. Schmitz, M., Immich, R., Pessin, G., Rocha Filho, G.P.: Towards the categorization of brazilian financial market headlines. IEEE Lat. Am. Trans. **20**(2), 344–351 (2021)
15. Selvin, S., Vinayakumar, R., Gopalakrishnan, E., Menon, V.K., Soman, K.: Stock price prediction using lstm, rnn and cnn-sliding window model. In: 2017 International Conference on Advances in Computing, Communications and Informatics (ICACCI), pp. 1643–1647. IEEE (2017)
16. Shefrin, H., Statman, M.: The disposition to sell winners too early and ride losers too long: theory and evidence. J. Financ. **40**(3), 777–790 (1985)
17. Siami-Namini, S., Tavakoli, N., Namin, A.S.: The performance of lstm and bilstm in forecasting time series. In: 2019 IEEE International Conference on Big Data (Big Data), pp. 3285–3292. IEEE (2019)
18. Souza, F.M.d.C., Filho, G.P.R., Guimarães, F.G., Meneguette, R.I., Pessin, G.: Navigating market sentiments: a novel approach to iron ore price forecasting with weighted fuzzy time series. Information **15**(5), 251 (2024)
19. Zhang, Z., Dai, H.N., Zhou, J., Mondal, S.K., García, M.M., Wang, H.: Forecasting cryptocurrency price using convolutional neural networks with weighted and attentive memory channels. Expert Syst. Appl. **183**, 115378 (2021)

Predicting Energy Consumption Data Using Deep Learning: An LSTM Approach

Julio Macedo Chaves[1,2](✉), Elene Firmeza Ohata[1,3], Matheus Araujo dos Santos[1,3], José Daniel de Alencar Santos[1,2], Matheus Jardim Bernardes[4], Daniel Seleme Dora[4], and Pedro Pedrosa Rebouças Filho[1,2,3]

[1] Laboratory for Processing Images, Signals and Computer Science (LAPISCO), Fortaleza, Ceará, Brazil
{elene.ohata,matheus.santos,julio.chaves}@lapisco.ifce.edu.br

[2] Federal Institute of Education, Science and Technology of Ceará (IFCE), Fortaleza, Ceará, Brazil
{jdaniel,pedrosarf}@ifce.edu.br

[3] Federal University of Ceará (UFC), Fortaleza, Ceará, Brazil

[4] Pontifical Catholic University of Rio Grande do Sul (PUCRS), Porto Alegre, Brazil
daniel.dora@ddora.com.br

https://lapisco.ifce.edu.br/

Abstract. Nowadays, most of the energy produced globally comes from fossil fuels. However, this type of energy generation harms the environment by emitting various toxic residues in natural bodies. In recent years, an area that has gained strength is producing and consuming clean energy. One that stands out is solar energy production because it is easy to implement and relatively cheap compared to other clean energy production. Therefore, several projects were created focusing on generating energy from sunlight, and using it. Due to this growing number of enterprises in the area, the creation of applications to help manage production and its use has increased. The use of Deep Learning techniques to help this industry has also gained strength; predictive models for energy consumption have been widely studied to help enterprises in future planning. In this work, we developed a deep learning model using Long Short-Term Memory (LSTM) capable of predicting energy consumption from solar power plant data, using the data to train the model and make inferences in the future. We explored configuration combinations such as data filtering with smoothing techniques, model hyperparameters, number of layers, number of neurons, and optimal prediction horizon. The achieved results demonstrate the validity and effectiveness of the implemented methodology.

Keywords: Solar energy · Deep learning · LSTM

1 Introduction

The growth of the global population has generated an increasing demand for energy [24]. However, most of the energy currently generated comes from fossil

A. Paes and F. A. N. Verri (Eds.): BRACIS 2024, LNAI 15413, pp. 296–308, 2025.
https://doi.org/10.1007/978-3-031-79032-4_21

fuels [14], which are known for their high emissions of toxic elements into the environment. This results in several problems, such as global warming, pollution of water bodies and the atmosphere, and the decrease of green areas [7,21]. In response to this situation, there has been a significant increase in the search for clean energy sources in recent years [24].

Data collected by a study by the International Renewable Energy Agency in 2020 point to a 7% growth in the use of clean energy sources worldwide. Of this increase, 90% is related to the increased use of solar energy [13]. Therefore, solar power plays a crucial role in the search for energy sources that do not harm the environment. Another point that draws attention to this form of energy generation is that it is relatively cheap when compared to other sources [31], which is why the industry field has turned its eyes to the horizon of investments in this area.

Another area highlighted in the industry field is Machine Learning [3]. Machine learning techniques have been used in several ways to solve various problems in the industry, such as fault diagnosis [12], quality detection [15,22], performance optimization [5,18], process control [20], signal processing [25], outcome prediction [16].

Among these applications, one that stands out in solar energy is the forecast of the amount of energy generated in a given period [17,33]. Predictions can be used for financial planning, low-generation detection, electrical crisis management, and energy distribution planning. This information can be used for the general planning of power generation plants, providing a service with excellent customer stability and higher quality [17].

Furthermore, another point investigated in the field of solar energy is the prediction of energy consumption [1]. This type of application, as well as generation, brings insights and better planning in the management of a power plant, mainly in managing and avoiding energy crises [26]. Therefore, consumption prediction applications are of paramount importance for the field of solar energy producers.

Based on that, we proposed a methodology using Deep Learning techniques to predict the amount of energy consumed by a solar plant based on time. We use data transformation techniques, search for optimal hyperparameters, and prediction horizons to create a robust tool with a clear purpose. With this tool, we aim to provide an artifact that brings valuable information to industries related to solar energy consumption. Based on the data obtained with a device, enterprises can get insights and manage energy crises in addition to other business planning.

2 Related Works

This section will examine previous studies and works related to this research topic, providing a broader and deeper view of the subject addressed.

Predicting future energy consumption is an issue frequently explored in the scientific community. This is evident in the study conducted by Amasyali *et al.* [2], providing a comprehensive analysis of the most commonly employed methods in temporal forecasting of energy consumption. Their conclusions highlight the strong results obtained by various forecasting approaches.

In the literature on the subject, Ekononomou *et al.* [9] stands out for using a Multi-Layer Perceptron (MLP) and an Artificial Neural Network (ANN) to predict energy consumption, incorporating factors such as temperature and Gross Domestic Product (GDP). In contrast, Hu *et al.* [11] presented the Adaboost-ESN model, developed to predict the monthly consumption of industrial electricity. In addition to considering previous consumption, this model included factors such as sales volume and import value, achieving promising results.

A notable approach was proposed by Du *et al.* [8]. They introduced a deep learning framework based on an encoder-decoder framework with BI-LSTM to predict multivariate energy consumption time series. The experiments, controlled on five different datasets, revealed that the model achieved robust performance in short- and long-term predictions.

Furthermore, it is essential to highlight studies such as Mahjoub *et al.* [19] and Olu *et al.* [23], which go beyond conventional approaches by exploring Deep Learning for time series forecasting. These studies stand out notably in the results achieved, showing the substantial potential of these techniques to address this problem.

Within this scenario, the long short-term memory (LSTM) architecture has gained prominence, as evidenced by the study by Wang *et al.* [29]. This work concludes that the LSTM model has a significant potential to predict future energy consumption. Such a model stands out for its ability to accurately capture the data's periodicity and exhibits a predictive performance superior to traditional models.

In the literature, smoothing techniques have been explored to improve the accuracy of time series predictions. A notable example is the study by Wibawa *et al.* [30], in which the application of smoothing in the data significantly enhanced the performance of his Convolutional Neural Network (CNN), surpassing several state-of-the-art models.

In a study conducted by Xiao *et al.* [32], the impact of data pre-processing and smoothing techniques on predicting time series regarding energy consumption in buildings was investigated. The results showed that the smoothing method resulted in a 0.91% reduction in the evaluated metric. Furthermore, a gradual decline in model performance over time was observed, suggesting the need for updating every seven weeks.

3 Methodology

In this section, we detail the methodology that supports our study. Our proposed approach is visually captured in Fig. 1, which encompasses three main blocks: input data, pipeline, and the model evaluation. The following subsections will describe each of the steps in this process.

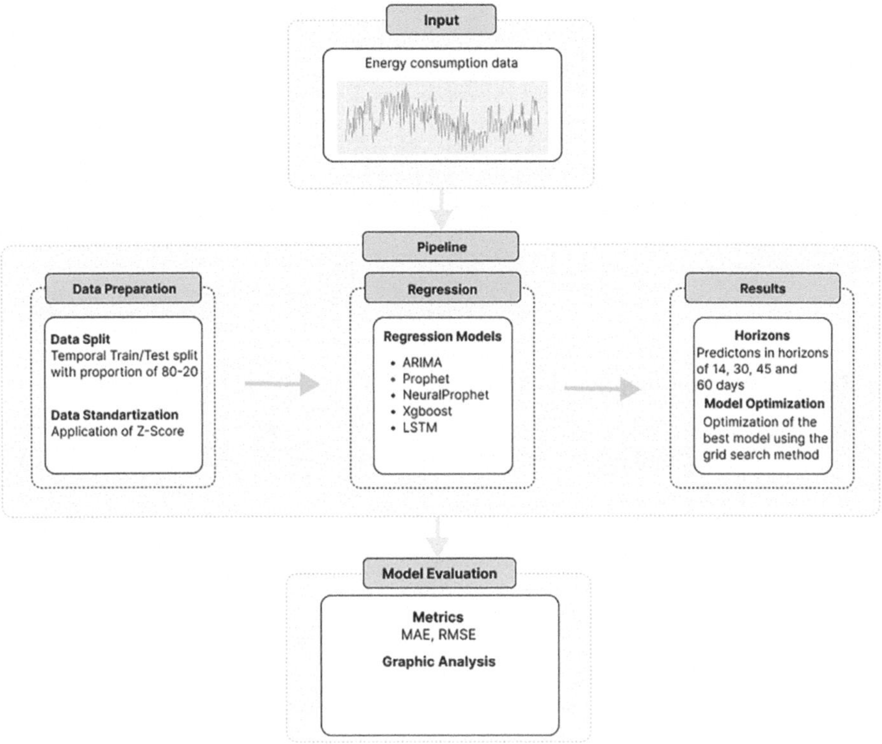

Fig. 1. Methodology applied in this work.

3.1 Dataset

The data used in this study was collected from a power plant, which we refer to by its pseudonym: SRI. The information was collected for approximately 390 days. These data were registered daily by specialized equipment, reflecting the energy consumption of each station during this period. The data used is in Fig. 2.

Figure 3 illustrates the decomposition of consumption data into its seasonality, trend, and residual components. We observed that the data trend presents variations, not following a continuous linear trajectory. These oscillations suggest the existence of patterns of growth followed by periods of decline in the data, possibly influenced by external factors such as climate change or economic events.

Furthermore, in Fig. 3, it is possible to notice a marked seasonality repeated approximately every 30 days, indicating a monthly seasonality. It is also possible to observe many residuals, showing that we could not solve the problem with linear models.

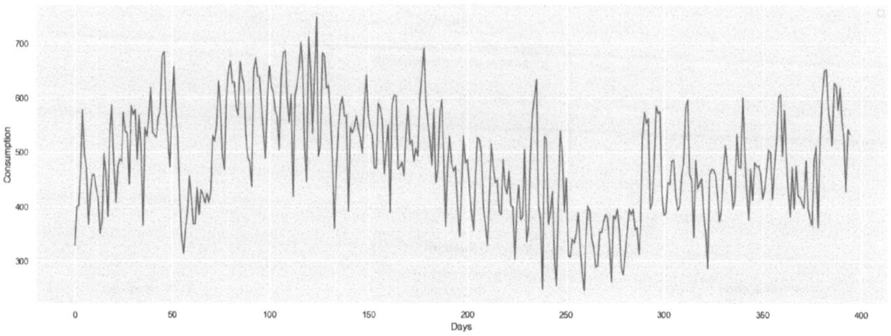

Fig. 2. Power consumption data collected from the station, covering 390 days.

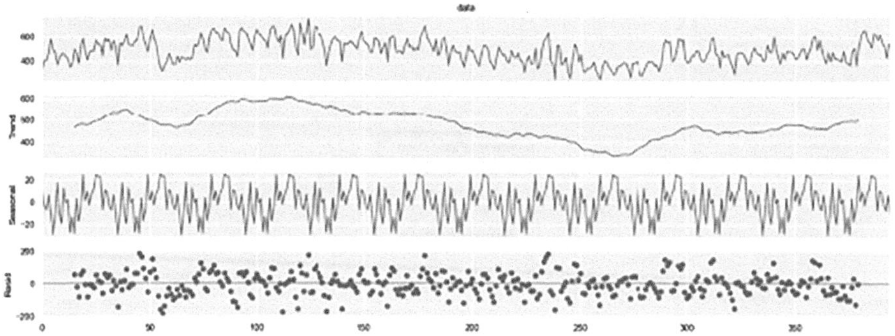

Fig. 3. Seasonal, trend, and residuals for the studied dataset.

3.2 Data Preparation

We split the dataset into two subsets, one for training and the other for testing. We performed this split temporally; we separated 80% of the dataset for training, equivalent to approximately 330 days, and we separated the remaining 20% for model validation, about 60 days.

For prediction purposes, we divided the test set into four groups, which covered the horizons of 14, 30, 45 and 60 days. We performed this division of the test set to evaluate the model's efficiency in different prediction horizons. The first 14-day horizon is chosen because it encompasses two complete weeks, which helps the model capture the typical patterns of weekdays and weekends. The second horizon, representing a whole month, allows the model to account for monthly variation, such as regular events that occur once per month. The third horizon represents a month and a half, making it possible to observe trends that may not be visible at shorter intervals. Finally, the fourth horizon represents two whole months, providing an even broader view and enabling the capture of more seasonal effects or periodic changes.

We used the Z-score method to standardize the data. This procedure recalibrates an attribute to align with 0, and its values are readjusted by dividing them by their standard deviation. As a result, the mean of the attribute is normalized to 0 and its standard deviation to 1. We apply the Eq. 1 to perform this transformation on the data.

$$x_{i_{std}} = \frac{x_i - mean(x)}{std(x)} \tag{1}$$

In Eq. 1, x denotes a value of a specific attribute taken from the training set. Here, mean(x) represents the mean of these attribute values, and std(x) indicates their standard deviation. Before the preprocessing step, this attribute's value is expressed as x_i. After undergoing standardization, it transitions to its standardized form, denoted by $x_{i_{std}}$.

It is important to emphasize that, while both training and test data undergo standardization, the parameters—specifically mean(x) and std(x)—used in Eq. 1 are derived solely from the training set.

3.3 Regression

Within the scope of this study, we evaluated different state-of-the-art models to find the architecture that best fits the problem. The models were AutoRegressive Integrated Moving Average (ARIMA) [4], Prophet [27], NeuralProphet [28], eXtreme Gradient Boosting (Xgboost) [6], and LSTM [10].

This model selection represents a comprehensive approach encompassing traditional methods and recent innovations. It aims to achieve a holistic understanding of these models' performance regarding the complexities of the analyzed time series.

We trained all models initially using default hyperparameters and tested their performance over a thirty-day prediction horizon. Then, we submit the model that obtained the best results for optimization by the grid search method.

3.4 Metrics

This subsection will present the metrics used to evaluate the trained models.

Mean Absolute Error (MAE). The MAE quantifies the mean error in terms of the magnitude of discrepancies. The corresponding mathematical formulation is given by the Eq. 2.

$$MAE = \frac{1}{n}\sum_{i=1}^{n} |y_i - \hat{y}_i| \tag{2}$$

In this context, n denotes the total number of observations, y_i refers to the observed value, and $\hat{y}_i$ refers to the value predicted by the model. The MAE, often used in regression problems, stands out for its direct interpretation: high values indicate a higher average of absolute errors, indicating a lower model performance. Furthermore, the MAE uses the same unit as the analyzed variable. In this study, for example, an MAE of 0.1 implies that, on average, the model estimates differ by 0.1 kWh from the actual values.

Root Mean Square Error (RMSE). The RMSE is calculated from the square root of the mean squared error, as defined by the Eq. 3.

$$RMSE = \sqrt{\frac{1}{n}\sum_{i=1}^{n}(y_i - \hat{y}_i)^2} \tag{3}$$

Unlike the MAE, the RMSE incorporates the squared error in its formula, amplifying the influence of more significant errors. This characteristic makes the RMSE especially valuable in situations where it is crucial to penalize more expressive deviations. Thanks to the inclusion of the square root, RMSE maintains the same unit as the original data. Like MAE, an RMSE of 0 represents a perfect fit where all model predictions are exact. Higher RMSE values indicate a declining performance of the model.

3.5 System Configuration

The device used for the tests has the following hardware configurations: Ryzen 5 5600x processor, 12 GB RTX 3060 graphics card, and 16 GB of RAM. Table 1 details the versions of libraries and software adopted in this study.

Table 1. Specifications of the softwares and libraries used in this study.

Software version
Ubuntu 22.04
Python 3.11.4
Torch 3.0.1
Tsmoothie 1.0.4
Pandas 2.0.2
Plotly 5.15.0
Scikit-learn 1.2.2
Pmdarima 2.0.3
Prophet 1.1.4
Statsmodels 0.14.0
Xgboost 1.7.6

4 Results and Discussion

In this section, we address the results achieved by the proposed methodology to forecast energy consumption; our proposal seeks to find the combination between the best method, use the appropriate pre-processing technique on the data, and adjust the hyperparameters.

4.1 Modeling of the Regression Method

In Table 2, we present the results of energy consumption prediction using different models; we used the default settings for all models. The model that demonstrated the best performance was the LTSM due to its lower values of MAE and RMSE, indicating its effectiveness in the prediction task. The ARIMA model showed the worst results, with an MAE of 84.40 kWh and RMSE of 96.67 kWh. The other models showed similar metrics but were significantly lower than the LSTM.

Table 2. Results of energy consumption prediction using different models.

	MAE (kWh)	RMSE (kWh)
ARIMA	84.40	96.67
Prophet	67.70	81.12
NeuralProphet	68.44	84.88
Xgboost	64.08	75.19
LSTM	**53.07**	**62.73**

4.2 Smoothing and Horizon

We set out for specific tests and adjustments for the LTSM model. We analysed two hypotheses: applying the smoothing technique to remove noise, and searching for an optimal prediction horizon threshold. For the smoothing step, we used a convolution-based smoothing, using a window of 30 days. The forecast horizon is analyzed to observe the limitation of how many steps (days) the model can predict without knowing more data, thus simulating a real environment. The results of applying the method in combination with different prediction horizons are shown in Table 3.

Table 3 shows that the data without smoothing showed a significant increase in the MAE in the 30-day and 45-day horizons. However, the MAE remained relatively stable when the prediction horizon was extended to 45 and 60 days. Based on these results, we used the 30-day threshold as the optimal prediction horizon. Beyond this point, accumulated noise and errors have a more significant impact on the non-smoothing data.

Table 3. Comparison of MAE and RMSE metrics for different prediction horizons, with and without data smoothing.

Days	Original		Smoothed	
	MAE (kWh)	RMSE (kWh)	MAE (kWh)	RMSE (kWh)
14	**52.84**	**60.55**	52.45	72.61
30	53.07	62.73	47.18	63.67
45	62.75	75.43	**45.59**	**60.91**
60	61.48	74.89	49.01	64.33

Furthermore, in Table 3, it is noted that data with smoothing were less affected by the increase in the prediction horizon, allowing the choice of longer prediction horizons. We, therefore, chose to use a 45-day forecast horizon with smoothed data, as they demonstrate the lowest MAE among all configurations.

4.3 Hyperparameters Selection

Later, as the last step, we searched for the best hyperparameters in the network; we used grid search techniques and calculated the model metrics for the 45-day horizon with smoothed data. The hyperparameters used were the number of layers and the number of neurons per hidden layer. The results are displayed in the Table 4. The configuration that obtained the best result was using five hidden layers with 100 neurons in each layer. This configuration achieved an MAE of 45.52 kWh and RMSE of 54.57 kWh, demonstrating greater accuracy than the other tested configurations.

Table 4. Model performance with hyperparameter variations.

No. of layers	No. of neurons	MAE (kWh)	RMSE (kWh)
5	100	**45.52**	**54.57**
5	50	45.58	59.5
5	25	47.69	62.29
20	50	48.29	60.5
10	50	50.37	63.37

Analyzing the other results, we found that the variation in the number of hidden layers and neurons per layer had different impacts on the model's performance. Configurations with five hidden layers and 25 neurons obtained results close to those of the highlighted layer. Meanwhile, the increase to 20 hidden layers resulted in lower performance, likely due to overfitting during the training step. Adding more layers to the model increases its learnability, which can lead to better adaptation to the training data. However, in some cases, this greater flexibility can cause the model to memorize the training data instead of learning more general patterns, impairing generalization to new data.

Based on these results, the configuration with five hidden layers and 100 neurons in each layer stood out as the most suitable for use in a 45-day horizon and smoothed data. This configuration has struck a balance between the network's learnability and generalization ability to new data, resulting in good forecasting performance for the specific scenario. Figure 4 shows the results obtained for the plant.

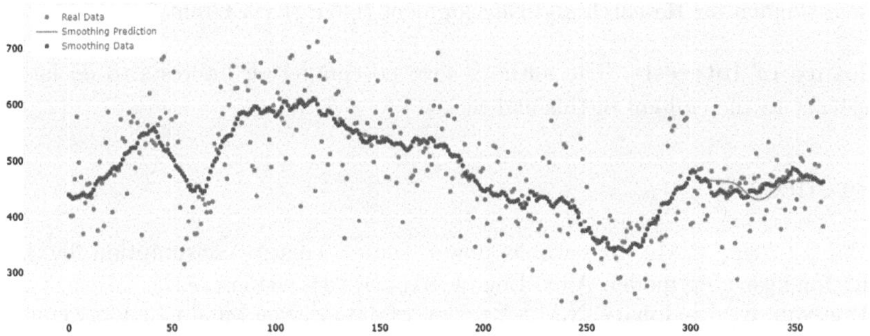

Fig. 4. Prediction results using the model with the best settings found in development. (Color figure online)

5 Conclusion

During the modeling development, we can explore different combinations of configurations such as Data transformation techniques, hyperparameter search, and search for an optimal prediction horizon. We built an artificial intelligence model from this exploration using Deep Learning techniques that can predict solar plants' energy consumption. The results found with the training from the database were satisfactory, with the model being able to generalize and make predictions, achieving good metrics.

The best model found has achieved an MAE of 45.52 kWh and an RMSE of 54.57 kWh and uses five hidden layers with 100 neurons. The optimal prediction horizon seen was 45 days, which provides a good generalization for a satisfactory time window. From this, a user can get good results for forecasting energy consumption up to 45 days into the future of the current day. Such a window provides a good window for insights and planning decisions for companies and industries.

Therefore, industries in the field of solar energy can take advantage of this artifact for future planning based on forecasts created by the model. Therefore, we hope that the modeling developed in this study can help the field of solar energy with helpful information.

In addition, studies with different databases and exploration of other solutions to the problem of predicting solar energy consumption are needed. For future work, we plan to use other databases to test the model, explore other modeling techniques, such as classical regression, and seek to use other Deep Learning models for better results.

Acknowledgments. This study was financed in part by the Coordenação de Aperfeiçoamento de Pessoal de Nível Superior - Brasil (CAPES) - Finance Code 001. Also Pedro Pedrosa Rebouças Filho acknowledges the sponsorship from the Brazilian National Council for Research and Development (CNPq) via Grant 301455/2022-8.

Disclosure of Interests. The authors have no competing interests to declare that are relevant to the content of this article.

References

1. Abd El-Aziz, R.M.: Renewable power source energy consumption by hybrid machine learning model. Alex. Eng. J. **61**(12), 9447–9455 (2022)
2. Amasyali, K., El-Gohary, N.M.: A review of data-driven building energy consumption prediction studies. Renew. Sustain. Energy Rev. **81**, 1192–1205 (2018)
3. Bertolini, M., Mezzogori, D., Neroni, M., Zammori, F.: Machine learning for industrial applications: a comprehensive literature review. Expert Syst. Appl. **175**, 114820 (2021)
4. Box, G.E., Pierce, D.A.: Distribution of residual autocorrelations in autoregressive-integrated moving average time series models. J. Am. Stat. Assoc. **65**(332), 1509–1526 (1970)
5. Chan, S.L., Lu, Y., Wang, Y.: Data-driven cost estimation for additive manufacturing in cybermanufacturing. J. Manuf. Syst. **46**, 115–126 (2018). https://doi.org/10.1016/j.jmsy.2017.12.001, https://www.sciencedirect.com/science/article/pii/S0278612517301577
6. Chen, T., Guestrin, C.: XGBoost: a scalable tree boosting system. In: Proceedings of the 22nd ACM SIGKDD International Conference on Knowledge Discovery and Data Mining, pp. 785–794 (2016)
7. Dincer, I.: Environmental impacts of energy. Energy Policy **27**(14), 845–854 (1999)
8. Du, S., Li, T., Yang, Y., Horng, S.J.: Multivariate time series forecasting via attention-based encoder-decoder framework. Neurocomputing **388**, 269–279 (2020)
9. Ekonomou, L.: Greek long-term energy consumption prediction using artificial neural networks. Energy **35**(2), 512–517 (2010)
10. Hochreiter, S., Schmidhuber, J.: Long short-term memory. Neural Comput. **9**(8), 1735–1780 (1997)
11. Hu, H., Wang, L., Peng, L., Zeng, Y.R.: Effective energy consumption forecasting using enhanced bagged echo state network. Energy **193**, 116778 (2020)
12. Iqbal, R., Maniak, T., Doctor, F., Karyotis, C.: Fault detection and isolation in industrial processes using deep learning approaches. IEEE Trans. Industr. Inf. **15**(5), 3077–3084 (2019)
13. Irena, A.: Renewable capacity highlights. Proc. Int. Renew. Energy Agency (IRENA), 1–8 (2020)

14. Jefferson, M.: World energy outlook to 2100. In: World Petroleum Congress, pp. WPC–26015. WPC (1994)
15. Jennings, C., Wu, D., Terpenny, J.: Forecasting obsolescence risk and product life cycle with machine learning. IEEE Trans. Compon. Packag. Manuf. Technol. **6**(9), 1428–1439 (2016). https://doi.org/10.1109/TCPMT.2016.2589206
16. Ji, S., Wang, X., Zhao, W., Guo, D.: An application of a three-stage XGBoost-based model to sales forecasting of a cross-border e-commerce enterprise. Math. Probl. Eng. **2019** (2019)
17. Lima, M.A.F., Carvalho, P.C., Fernández-Ramírez, L.M., Braga, A.P.: Improving solar forecasting using deep learning and portfolio theory integration. Energy **195**, 117016 (2020)
18. Maggipinto, M., Terzi, M., Masiero, C., Beghi, A., Susto, G.A.: A computer vision-inspired deep learning architecture for virtual metrology modeling with 2-dimensional data. IEEE Trans. Semicond. Manuf. **31**(3), 376–384 (2018). https://doi.org/10.1109/TSM.2018.2849206
19. Mahjoub, S., Chrifi-Alaoui, L., Marhic, B., Delahoche, L.: Predicting energy consumption using LSTM, multi-layer GRU and drop-GRU neural networks. Sensors **22**(11), 4062 (2022)
20. Mezzogori, D., Zammori, F.: An entity embeddings deep learning approach for demand forecast of highly differentiated products. Procedia Manuf. **39**, 1793–1800 (2019). https://doi.org/10.1016/j.promfg.2020.01.260, https://www.sciencedirect.com/science/article/pii/S2351978920303243, 25th International Conference on Production Research Manufacturing Innovation: Cyber Physical Manufacturing, 9–14 August 2019|Chicago, Illinois (USA)
21. Mirzaliev, S., Sharipov, K.: A review of energy efficient fluid power systems: fluid power impact on energy, emissions and economics, vol. 30 (2020)
22. Oh, Y., Ransikarbum, K., Busogi, M., Kwon, D., Kim, N.: Adaptive SVM-based real-time quality assessment for primer-sealer dispensing process of sunroof assembly line. Reliab. Eng. Syst. Saf. **184**, 202–212 (2019). https://doi.org/10.1016/j.ress.2018.03.020, https://www.sciencedirect.com/science/article/pii/S0951832017303861, Impact of Prognostics and Health Management in Systems Reliability and Maintenance Planning
23. Olu-Ajayi, R., Alaka, H., Sulaimon, I., Sunmola, F., Ajayi, S.: Building energy consumption prediction for residential buildings using deep learning and other machine learning techniques. J. Build. Eng. **45**, 103406 (2022)
24. Rabaia, M.K.H., et al.: Environmental impacts of solar energy systems: a review. Sci. Total Environ. **754**, 141989 (2021)
25. Scime, L., Beuth, J.: Using machine learning to identify in-situ melt pool signatures indicative of flaw formation in a laser powder bed fusion additive manufacturing process. Addit. Manuf. **25**, 151–165 (2019). https://doi.org/10.1016/j.addma.2018.11.010, https://www.sciencedirect.com/science/article/pii/S2214860418306869
26. Shamshirband, S., Rabczuk, T., Chau, K.W.: A survey of deep learning techniques: application in wind and solar energy resources. IEEE Access **7**, 164650–164666 (2019)
27. Taylor, S.J., Letham, B.: Forecasting at scale. Am. Stat. **72**(1), 37–45 (2018)
28. Triebe, O., Hewamalage, H., Pilyugina, P., Laptev, N., Bergmeir, C., Rajagopal, R.: Neuralprophet: Explainable forecasting at scale. arxiv 2021. arXiv preprint arXiv:2111.15397
29. Wang, J.Q., Du, Y., Wang, J.: LSTM based long-term energy consumption prediction with periodicity. Energy **197**, 117197 (2020)

30. Wibawa, A.P., Utama, A.B.P., Elmunsyah, H., Pujianto, U., Dwiyanto, F.A., Hernandez, L.: Time-series analysis with smoothed convolutional neural network. J. Big Data **9**(1), 44 (2022)
31. Wilberforce, T., Baroutaji, A., El Hassan, Z., Thompson, J., Soudan, B., Olabi, A.G.: Prospects and challenges of concentrated solar photovoltaics and enhanced geothermal energy technologies. Sci. Total Environ. **659**, 851–861 (2019)
32. Xiao, Z.: Impacts of data preprocessing and selection on energy consumption prediction model of HVAC systems based on deep learning. Energy Build. **258**, 111832 (2022)
33. Yagli, G.M., Yang, D., Srinivasan, D.: Automatic hourly solar forecasting using machine learning models. Renew. Sustain. Energy Rev. **105**, 487–498 (2019)

Pseudonymization in Legal Texts According to the LGPD: A Named Entity Recognition Approach

Marcelo Anselmo(✉) and Bruno César Ribas

Department of Computer Science, University of Brasilia, Brasilia, Brazil
marcelo.anselmo@aluno.unb.br, bruno.ribas@unb.br

Abstract. This study explores the application of Named Entity Recognition (NER) for the pseudonymization of data in legal texts, aiming to protect Personally Identifiable Information (PII) in compliance with Brazil's General Data Protection Law (LGPD). The research highlights the challenge of balancing data privacy and utility, presenting a methodology that uses artificial intelligence technologies to effectively identify and obscure PII in legal documents. In this study, we propose a Transformer model along with Regex techniques to identify entities in a text. To test the model, we created a new dataset from the existing LenerBR. We also used a function and prompt engineering applied to the Llama 8B version 3 model to generate synthetic data. Tests showed the need for further adjustments in the proposed new model. Future work will focus on improving the model's accuracy and efficiency, as well as enhancing the identification of sensitive data and learning from user interactions.

Keywords: Data Pseudonymization · Named Entity Recognition · Information Privacy · Legal Texts · LGPD · Transformer

1 Introduction

In today's digital age, where technology and computing power have grown exponentially, personal data has become an extremely valuable asset. Companies capture this information from online interactions, such as websites and social networks, to create detailed user profiles. These profiles range from consumer habits to sensitive information like health conditions and personal preferences. This practice not only allows for the personalization of online experiences, but it also exposes users to privacy risks, including the illegal sale of such data in dark markets, potentially for use in criminal activities [3].

This scenario highlights the critical importance of protecting personal data, not only for individual privacy, but also for the overall security of society. The solution to these issues has manifested through regulations and laws dedicated to data protection, such as the General Data Protection Regulation (GDPR) in the European Union and the General Data Protection Law (LGPD) in Brazil [9].

A. Paes and F. A. N. Verri (Eds.): BRACIS 2024, LNAI 15413, pp. 309–323, 2025.
https://doi.org/10.1007/978-3-031-79032-4_22

The main problem addressed in this work is the difficulty of balancing individual privacy with the need to use their information in legal and administrative texts. This issue deepens when dealing with identifying and hiding PII without compromising data integrity and utility. Using resources, such as Transformer models and Regex techniques, an effective balance between privacy and utility is sought, as already accomplished in other studies, such as the work of G. M. GR, S. Abhi, and R. Agarwal [5].

The aim of this study is to investigate and demonstrate the effectiveness of Named Entity Recognition (NER) applied to PII pseudonymization in legal documents. It proposes to develop and test a model capable of efficiently identifying and hiding named entities that correspond to PII. The Fig. 1 illustrates the pipeline adopted in our methodology. Another important note is the CNMP Resolution No. 281, dated December 12, 2023[1], in its Article 79, presents pseudonymization as an alternative to protecting the personal data of natural persons in procedures or processes within the Brazilian Public Prosecutor's Office.

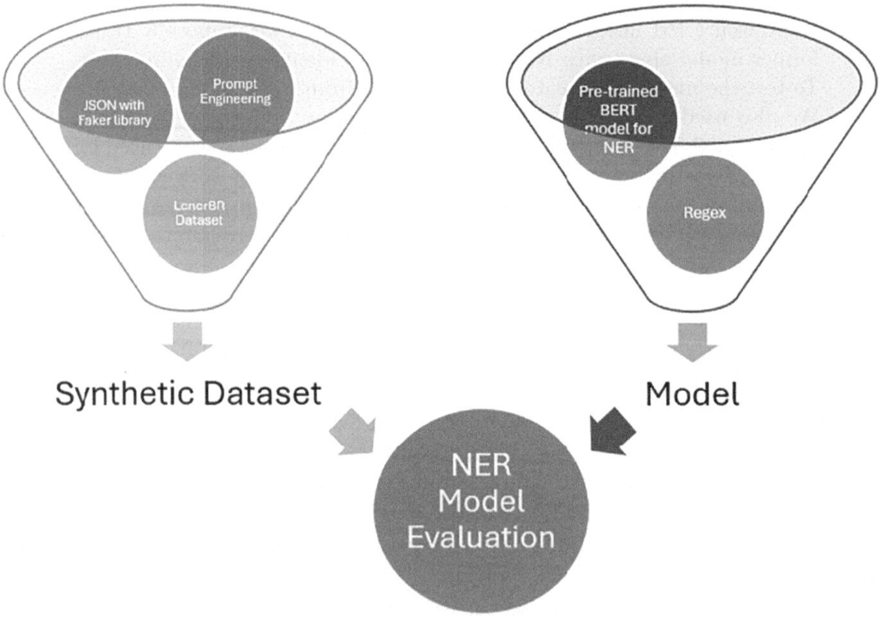

Fig. 1. Pipeline adopted in our methodology

Legal texts are often lengthy and complex, prioritizing formality over readability [12]. However, the implementation of advanced techniques, such as artificial intelligence models, often relies on large volumes of annotated data for

[1] https://www.cnmp.mp.br/portal/images/CALJ/resolucoes/Resoluo-281-de-2023.pdf.

training, which can be challenging in terms of cost and time [13]. With this in mind, we face a fundamental challenge: the need for data to be reliably and efficiently labeled. Traditionally, this labeling is done manually, a process that, despite improving the accuracy of artificial intelligence models, is notoriously time-consuming and expensive [4]. This scenario imposes a high cost in terms of time, money, and effort, limiting the scalability of solutions.

This paper is structured as follows: Sect. 2 presents the background concepts, covering fundamental concepts, like the related to personal data protection and pseudonymization. Section 3 describes related work, highlighting previous studies that addressed similar issues. Section 4 presents the proposed methodology, detailing the data used, tools, and terms to be found. Section 5 discusses the expected results for each test set. Finally, Sect. 6 presents the conclusions and suggestions for future work.

2 Background Concepts

2.1 General Data Protection Law (LGPD)

The General Data Protection Law (LGPD[2], Law No. 13,709, dated August 14, 2018) of Brazil is a regulation that establishes guidelines for the collection, use, processing, and storage of personal data. The law grants individuals more control over their personal information while balancing privacy rights with technological advancement. The LGPD requires that any operation involving personal data be based on clear legal justifications and respect strict principles, such as purpose, adequacy, and necessity [3].

2.2 Personal Data (PD)

Personal data (LGPD, Art. 5, I) refers to information that identifies or can directly or indirectly identify a natural person. This definition is broad and includes a variety of types of information, ranging from names and digital identities to physical and electronic characteristics, such as photos and location data. Such data is crucial in digital society, being used in various contexts, from identity verification to service personalization. However, handling it involves significant privacy risks, which requires robust protection measures to ensure data security and confidentiality[3].

2.3 Anonymization vs. Pseudonymization

Anonymization and pseudonymization are two fundamental concepts in data privacy management. Anonymization (LGPD, Art. 5, XI) refers to the process whereby personal data is processed to remove the ability to permanently associate this data with a specific individual, without using additional information.

[2] https://www.planalto.gov.br/ccivil_03/_ato2015-2018/2018/lei/l13709.htm.
[3] https://www.serpro.gov.br/lgpd/menu/protecao-de-dados/dados-pessoais-lgpd.

This means that once anonymized, the data cannot be linked back to the owner, even with the use of reasonable technical means. This process is irreversible, ensuring a high level of protection for the individual's identity.

On the other hand, pseudonymization (LGPD, Art. 13, ğ 4) is a technique that modifies personal data so that identification of the owner cannot be done without using additional information, which is maintained separately and under strict security measures by the data controller. Unlike anonymization, pseudonymization is reversible, provided that the additional information is made available[4]. This method will be adopted in this work as it allows the data to be protected without losing its utility for analysis and research.

2.4 Retrieval-Augmented Generation (RAG)

RAG is an advanced approach in natural language processing (NLP) models that integrates information retrieval with text generation [6]. This technique allows the model to retrieve relevant documents that are used to inform subsequent text generation. The unique ability of RAG to combine parametric and non-parametric memory enables it to adapt both retrieval and content generation for specific NLP tasks, improving the accuracy and relevance of the generated information[5]. By utilizing RAG, we can incorporate a vast amount of pre-existing legal information to enhance the quality of our generated dataset.

3 Related Work

In the article by Patsakis and Lykousas (2023) [11], the authors explore the challenge of effectively anonymizing text in the age of large language models. This study questions the effectiveness of these approaches and assesses their ability to prevent identification, especially with the use of AI on large datasets. An experiment is conducted using GPT on anonymized texts of well-known personalities to verify whether these language models can re-identify individuals. They argue that despite technological advances, there are still significant obstacles in protecting data privacy in texts processed by these tools.

In the work of de Andrade et al. (2023) [1], the authors present an approach called PromptNER, which focuses on NER in sensitive data using automatically labeled instances. The authors propose an approach that employs LLMs to identify entities in complaints and then trains simpler models like the SpERT method. The improved NER model shows substantial improvements in F-score, ranging from 41% to 129% compared to models using only manually labeled data. This study is crucial because it combines artificial intelligence methods to improve efficiency in identifying and handling personal data.

Mota et al. (2021) [7] investigate the use of neural networks for named entity recognition in legal documents in Portuguese. The authors used Spacy

[4] https://www.cnmp.mp.br/portal/images/CALJ/resolucoes/Resoluo-281-de-2023.pdf.

[5] https://huggingface.co/transformers/model_doc/rag.html.

and FLAIR libraries. Their results demonstrate the ability of these advanced technologies to handle the linguistic complexity of Portuguese, providing a technical foundation for the development of more effective solutions.

Nunes and dos Santos (2023) [9] discuss the application and impact of the LGPD in the Brazilian context, focusing on a technological and agnostic approach. They emphasize the need for organizations to adapt to legal requirements and the importance of implementing effective measures for protecting personal data. This study highlights that implementing the LGPD requires a comprehensive technical and organizational approach. Multiples tools, such as Data Management Systems (DMS), anonymization, pseudonymization, encryption, and auditing, are highlighted as essential for compliance. Techniques like Big Data and Machine Learning can improve compliance, while privacy by design and eDiscovery are emphasized as crucial for protecting data.

GR, Shinu, and Agarwal (2023) [5] presented a hybrid model that combines RegEx and NER for resume analysis and matching. This work illustrates the applicability of NER techniques along with regular expressions to efficiently extract and process information in structured documents like resumes. The proposed methodology is relevant to our study as it demonstrates the effectiveness of integrating NER and RegEx in data identification and pseudonymization tasks.

Luz de Araujo et al. (2018) [2] developed LeNER-BR, a dataset specifically for Named Entity Recognition in Brazilian legal texts. Their results provide an essential foundation for validating NER models in the Brazilian legal context. The dataset they proposed serves as a valuable resource for training and testing new methodologies, including the approach of our study.

Oliveira et al. (2024) [10] explored the combination of prompt-based language models and weak supervision to label the task of Named Entity Recognition (NER) in legal documents. They highlighted the effectiveness of their techniques in improving the accuracy and automation of the NER process, which is crucial for data pseudonymization in compliance with privacy regulations.

4 Methodology

4.1 Personal Data Used

We categorized terms according to some types of Personal Data (PD). It is important to note that the spectrum of terms is much broader than those discussed in this study, however, we highlight some popular terms. Note that the CPF is a Brazilian ID number. The terms used to generate synthetic data are presented in Table 1 below.

Table 1. Selected Terms for Identification of Personal Data (PD)

Term	Example
NAME	*João da Silva*
DATE	*12 de janeiro de 2013*
ADDRESS	*Rua do Alecrim, 0*
CPF	*123.456.789-00*
PHONE	*(11) 99999-9999*
EMAIL	*example@example.com*
MONEY	*R$ 1,000.00*
ZIP CODE	*12345-678*

4.2 Dataset

For this study, we used the LenerBR dataset derived from the study by Luz de Araujo and Pedro Henrique [2] and available at[6]. This dataset consists exclusively of legal documents. It includes labels for people, places, temporal entities, and organizations, as well as specific tags for legal entities and judicial processes.

We used the functions of the Faker library[7] to generate synthetic data. The *Faker* library is a Python tool that generates fake data as per the function used. It is useful for creating synthetic data for testing and prototyping. We created a Python function that generates random data for one to two items per term using the *Faker* library. Figure 2 below show examples of data generated by our function.


```
{
'NOME': ['Daniel Mendes'],
'DATA': ['dezembro de 1990'],
'CPF': ['490.183.567-10', '127.034.685-81'],
'TELEFONE': ['0800 170 6459', '61 6556 4995'],
'EMAIL': ['santosbarbara@example.net', 'frezende@example.net'],
'DINHEIRO': ['R$ 43,95', 'R$ 3,58'],
'CEP': ['28866-051', '29566719'],
'ENDERECO': ['Praia Antônio Caldeira, 4']
}
```

```
{
'NOME': ['Safe Moraes', 'Melissa Freitas'],
'DATA': ['26/05/2013', 'junho de 2021'],
'CPF': ['501.439.826-15'],
'TELEFONE': ['+55 (084) 5042-4202', '0900-372-6488'],
'EMAIL': ['ribeirobeatriz@example.net', 'tnovais@example.com'],
'DINHEIRO': ['R$ 8.691,11', 'R$ 48.931,48'],
'CEP': ['86471-581', '07788218'],
'ENDERECO': ['Rio Grande do Norte', 'Sítio Ana Júlia Campos, 5']
}
```


Fig. 2. Example of data generated by running our function twice.

The entities in portuguese and its respective translation to english, hidden the entities that has the same translation, are: "*NOME*" (NAME), "*DATA*" (DATE), "*ENDERECO*" (ADDRESS), "*TELEFONE*" (PHONE), "*DINHEIRO*" (MONEY), "*CEP*" (ZIP CODE).

As seen in the previous images, we generated synthetic data for the term CPF. For this term, being a personal identifier, we changed the values of the

[6] https://huggingface.co/datasets/lener_br.

[7] https://faker.readthedocs.io/en/master/.

check digit. According to the study by [8], the check digit is a number calculated from the other numbers of the CPF. It is used to verify whether the number is mathematically consistent with the standards defined by the competent Brazilian agencies. This action aims to ensure that invalid CPF is generated, thus preventing the exposure of sensitive data.

We applied prompt engineering techniques with RAG to create synthetic data. For this purpose, we used the Llama model version 3 with 8 billion parameters, a large-scale open-source language model from Meta. It is freely available for research purposes[8].

Some of the most relevant parameters for running the Llama model were: "*temperature = 0.2*" and "*max_new_tokens = 512*". The testing environment was a local machine with a GPU of 24 GB VRAM and an Intel I9 10th generation processor. For each generated text, the time was around 12 s. These parameters were used to generate the synthetic data. Our code is available at a github repository[9]. An example of a prompt to generate a synthetic text and the response to this prompt is shown in Fig. 3 below.

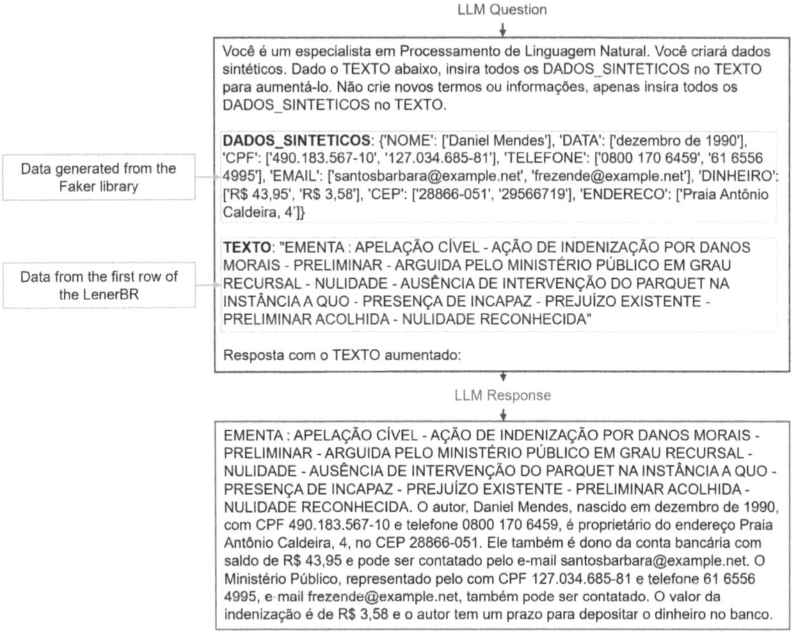

Fig. 3. Prompt Engineering to generate a synthetic example.

Using this prompt model from Fig. 3, we managed to generate 600 synthetic data of legal texts. These examples were used to evaluate the new proposed

[8] https://llama.meta.com/llama3.

[9] https://github.com/celiudos/paper_bracis_2024.

NER model for LGPD. We noticed that, even with the command passed to the prompt to use all synthetic data, the model did not use the value "*29566719*" for "*CEP*". Therefore, we created a function that removes these data from the generated JSON already displayed in Fig. 2.

4.3 NER Model for Evaluation

We used a pre-trained BERT model for named entity recognition in Portuguese, provided by Pierre[10]. This model leveraged the same dataset previously mentioned, the LenerBR. However, our study changed the names of some entities displayed in the original model, removed others, and also added new entities. The original model had 6 entities, while our model has 8 entities. The mentioned changes are shown in Table 2 below.

Table 2. Changes in entities for the new NER model.

Original Model	Current Model	Situation
PERSON	NAME	*Changed*
TIME	DATE	*Changed*
PLACE	ADDRESS	*Changed*
JURISPRUDENCE	–	*Removed*
LEGISLATION	–	*Removed*
ORGANIZATION	–	*Removed*
–	CPF	*Added*
–	PHONE	*Added*
–	EMAIL	*Added*
–	MONEY	*Added*
–	ZIP CODE	*Added*

For the new entities, we employed regular expressions (Regex) to enhance NER accuracy. Regex proved to be a valuable technique for identifying and replacing specific text patterns (highlighted in the study [5]), such as phone numbers, email addresses, and other personal identifiers, which are not always captured by standard NER methods. This complementary approach allowed for more granular filtering of added data, strengthening the effectiveness of identification when dealing with a wide range of formats.

5 Results

5.1 Evaluation of the New NER Model

Based on the generated dataset, we conducted evaluations using the F1-Score. The below Table 3 contains the results obtained.

Table 3. Evaluation Results

Entity	Precision	Recall	F1-Score	Support
NAME	0.73	0.83	0.777	948
DATE	0.913	0.994	0.952	856
ADDRESS	0.363	0.391	0.377	1135
CPF	0.988	1.0	0.994	887
PHONE	0.982	0.944	0.963	987
EMAIL	0.99	1.0	0.995	938
MONEY	0.971	1.0	0.985	965
ZIP CODE	1.0	0.526	0.69	851
Overall	0.837	0.826	0.832	–

We observed that the best-performing entity was "*EMAIL*" with an F1-Score of 0.995, while the worst was "*ADDRESS*" with an F1-Score of 0.377. It's important to note that the first is identified through Regex, while the latter by the NER model. With a simple analysis of the "*ADDRESS*", our assumption is that it contains text that could be confused with names of people, animals, among many others. There is also the fact that during the synthetic data generation process, the model may have generated addresses that are uncommon, which could have made correct identification more difficult.

The model achieved an overall F1-Score of 0.832, which is a satisfactory result for an NER model. The entity "*ZIP CODE*" had an F1-Score of 0.69, which is below expectations. This result can be attributed to the model's difficulty in correctly identifying ZIP codes, which consist of 8 digits and a hyphen. The hyphen was not captured by the initially created Regex pattern, as we had made the hyphen mandatory, hence a ZIP code like "*12345678*" was not recognized.

Table 4. Evaluation Results with Regex Corrections.

Entity	Precision	Recall	F1-Score	Support
NAME	0.73	0.83	0.777	948
DATE	0.913	0.994	0.952	856
ADDRESS	0.363	0.391	0.377	1135
CPF	0.988	1.0	0.994	887
PHONE	0.982	0.944	0.963	987
EMAIL	0.99	1.0	0.995	938
MONEY	0.971	1.0	0.985	965
ZIP CODE	1.0	0.999	0.999	851
Overall	0.845	0.879	0.862	–

[10] https://huggingface.co/pierreguillou/ner-bert-large-cased-pt-lenerbr.

To address this issue, we added a regular expression to also identify ZIP codes without a hyphen. The Table 4 contains the results obtained after this correction.

By comparing the Tables 3 and 4, we can see that the model correction with the addition of regular expressions for ZIP code identification significantly improved the F1-Score for this entity. The F1-Score for ZIP CODE increased from 0.69 to 0.99, which is an excellent result. The overall F1-Score of the model also improved, from 0.83 to 0.86. This demonstrates that adding regular expressions to correct entity identification errors can significantly enhance identification accuracy.

It is interesting to note that terms identified using Regex, such as "*CPF*", "*PHONE*", "*EMAIL*", "*MONEY*", and "*ZIP CODE*", achieved F1-Scores that were not always perfect. Upon further investigation, we observed some interesting behaviors. For this analysis, let's focus on the "*Entity*" column and the values identified with Regex.

Analyzing the "*Recall*" value for "*CPF*", we see a 100% score, meaning the model correctly identified all cases where "*CPF*" should not appear. However, for "*Precision*", the value was 98.2%, indicating the percentage correctly identified. Upon analyzing the cases where the model missed, we observed some instances like the one shown below in Fig. 4.

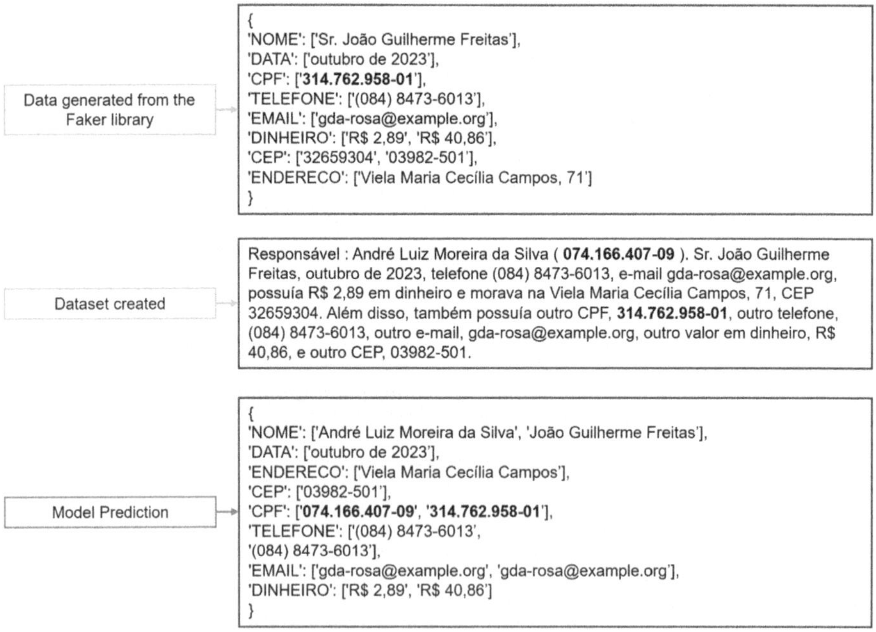

Fig. 4. Issues regarding the model's "*Precision*".

This indicates that, even though we had instructed the LLM to generate text without creating new terms, it did. This behavior repeats for the other Regex terms about "*Precision*". As for "*Recall*", the model correctly identified all cases, except for "*PHONE*", which scored 94.4%. In this case, our model made an error in just one instance, resulting in a false positive, as shown in the example in Fig. 5.

It is still valid to say that our proposed model also ended up correctly identifying an additional data entity. In this way, we see a kind of inversion of our proposal, where our model ends up correcting the new dataset, improving its quality. This is a positive point, as it demonstrates that the model can be used to enhance data quality, even if that is not its primary objective.

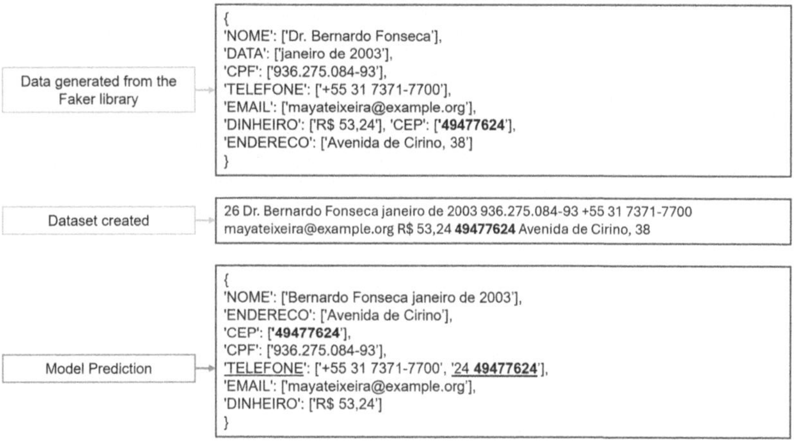

Fig. 5. Issues regarding the model's "*Recall*"

5.2 Usage Example of the New NER Model

For the following example, we used the synthetic text generated as shown in Fig. 3. We created an interface using Python to facilitate the pseudonymization of the text. Thus, we established one input field and two output fields. The input field is for the original text, and the output fields are for the identified terms and another for the pseudonymization secret. The Fig. 6 below contains the original text, while Figs. 7 and 8 contain data related to the generated outputs.

As shown in Figs. 7 and 8 below, here again is the translation of the entities from Portuguese to English: "*NOME*" (NAME), "*DATA*" (DATE), "*ENDERECO*" (ADDRESS), "*TELEFONE*" (PHONE), "*DINHEIRO*" (MONEY), "*CEP*" (ZIP CODE).

5.3 Limitations

As with any automated system, there are limitations that must be considered to ensure the effectiveness and accuracy of pseudonymization. One of these limitations is the need for human review. Our model may occasionally make errors

EMENTA : APELAÇÃO CÍVEL - AÇÃO DE INDENIZAÇÃO POR DANOS MORAIS - PRELIMINAR - ARGUIDA PELO MINISTÉRIO PÚBLICO EM GRAU RECURSAL - NULIDADE - AUSÊNCIA DE INTERVENÇÃO DO PARQUET NA INSTÂNCIA A QUO - PRESENÇA DE INCAPAZ - PREJUÍZO EXISTENTE - PRELIMINAR ACOLHIDA - NULIDADE RECONHECIDA. O autor, Daniel Mendes, nascido em dezembro de 1990, com CPF 490.183.567-10 e telefone 0800 170 6459, é proprietário do endereço Praia Antônio Caldeira, 4, no CEP 28866-051. Ele também é dono da conta bancária com saldo de R$ 43,95 e pode ser contatado pelo e-mail santosbarbara@example.net. O Ministério Público, representado pelo com CPF 127.034.685-81 e telefone 61 6556 4995, e-mail frezende@example.net, também pode ser contatado. O valor da indenização é de R$ 3,58 e o autor tem um prazo para depositar o dinheiro no banco.

Fig. 6. Input Data

NOME DATA CPF TELEFONE ENDERECO CEP DINHEIRO EMAIL

EMENTA : APELAÇÃO CÍVEL - AÇÃO DE INDENIZAÇÃO POR DANOS MORAIS - PRELIMINAR - ARG
UIDA PELO MINISTÉRIO PÚBLICO EM GRAU RECURSAL - NULIDADE - AUSÊNCIA DE INTERVENÇ
ÃO DO PARQUET NA INSTÂNCIA A QUO - PRESENÇA DE INCAPAZ - PREJUÍZO EXISTENTE - PRELI
MINAR ACOLHIDA - NULIDADE RECONHECIDA. O autor, <NOME> , nascido em <DATA> , co
m CPF <CPF> e telefone <TELEFONE> , é proprietário do endereço <ENDERECO> , 4, no
CEP <CEP> . Ele também é dono da conta bancária com saldo de <DINHEIRO> e pode ser
contatado pelo e-mail <NOME_2> @example.net. O Ministério Público, representado pelo com
CPF <CPF_2> e telefone <TELEFONE_2> , e-mail <EMAIL_2> , também pode ser contata
do. O valor da indenização é de <DINHEIRO_2> e o autor tem um prazo para depositar o dinhe
iro no banco.

Fig. 7. Output Data - Entity Recognition


```
{
  NOME: {
    <NOME>: "Daniel Mendes",
    <NOME_2>: "santosbarbara"
  },
  DATA: {
    <DATA>: "dezembro de 1990"
  },
  ENDERECO: {
    <ENDERECO>: "Praia Antônio Caldeira"
  },
  CEP: {
    <CEP>: "28866-051"
  },
  CPF: {
    <CPF>: "490.183.567-10",
    <CPF_2>: "127.034.685-81"
  },
  TELEFONE: {
    <TELEFONE>: "0800 170 6459",
    <TELEFONE_2>: "61 6556 4995"
  },
  EMAIL: {
    <EMAIL>: "santosbarbara@example.net",
    <EMAIL_2>: "frezende@example.net"
  },
  DINHEIRO: {
    <DINHEIRO>: "R$ 43,95",
    <DINHEIRO_2>: "R$ 3,58"
  }
}
```


Fig. 8. Output Data - Pseudonymization Secret

in the pseudonymization of named entities. This requires a subsequent human review process to correct possible errors and ensure that the data is properly hidden, according to legal requirements.

Another significant limitation is the lack of cross-referencing between terms previously identified in the document. For example, if the system identifies "*Maria Silva*" as a person in one part of the text, it may not recognize "*Maria*" in a subsequent reference as the same person, potentially labeling it differently, such as "*NAME_2*". This problem, which seems simple to solve, may involve complex issues, such as homonyms and the overall context of the text.

Additionally, the model can generate false positives, which are errors where non-personal data are erroneously categorized with a different label. For example, a street name may be incorrectly identified as a person's name.

Finally, the model struggles to identify and correct typographical errors in critical information, such as CPF or phone numbers. Such typographical errors, which may include spaces or incorrect characters inserted between numbers, are not recognized by the system, potentially leading to a failure in properly concealing these sensitive data. This issue is likely to be addressed in subsequent work through optimization of the transformer model.

6 Conclusion

This study addressed the challenge of pseudonymizing Personally Identifiable Information (PII) in legal texts to comply with Brazil's General Data Protection Law (LGPD). We employed artificial intelligence technologies, including a Transformer model and Regex techniques, to effectively identify and conceal PII. Tests revealed that although our model showed good initial results, adjustments are necessary to enhance its precision and effectiveness.

The conclusions highlight that the combined approach of NER and Regex is promising, allowing for more accurate entity identification. However, methodological limitations include the need for ongoing adjustments to the model to handle atypical cases and the reliance on high-quality training data to maintain accuracy. Moreover, the technique faces challenges in consistently identifying all PII categories, with some entities like addresses showing lower results compared to others, such as CPFs and emails.

In summary, while initial results on synthetic data were promising, tests on real data are necessary to ensure the model's effectiveness in practical environments. The main advantage of the proposed model is its lightweight nature, allowing it to operate on systems with less computational capacity. This feature makes the model particularly valuable for applications in resource-limited environments or where processing speed is crucial.

7 Future Work

For future work, one of the main intentions is to mitigate the limitations of the current model, especially in terms of consistency in entity identification and error reduction, such as false positives and the non-cross-referencing of named entities. The idea is to refine the existing model and incorporate more precise feedback in the model training phase to improve its accuracy and efficiency.

Plans also include developing an interactive model that learns from corrections made by users. This active learning approach will allow the system to continuously refine its identification and categorization of PII, adjusting to the peculiarities of legal text and the specific pseudonymization preferences of the user.

Lastly, one of the focuses will be on identifying new terms related to PII that are continually emerging with changes in data collection practices and legislation. Recognizing that new types of personal data may arise, the model needs to be prepared and ready to respond to these new demands, enabling all forms of PII to be adequately identified and protected in accordance with the latest legal and ethical standards.

Acknowledgments. The authors acknowledge the use of the Llama version 3 model for generating data to enhance the dataset utilized in this research. The synthetic data generation included sensitive information to ensure the robustness of the model. The authors take full responsibility for the content of the article, including ensuring the originality and correctness of all text, and have taken all necessary precautions to ensure that no real sensitive information is used or exposed.

References

1. de Andrade, C.M., et al.: Promptner: Uma abordagem para reconhecimento de entidades nomeadas em dados sensíveis a partir de instâncias rotuladas automaticamente. In: Anais do XXXVIII Simpósio Brasileiro de Bancos de Dados, pp. 269–281. SBC (2023)
2. Luz de Araujo, P.H., de Campos, T.E., de Oliveira, R.R.R., Stauffer, M., Couto, S., Bermejo, P.: LeNER-Br: a dataset for named entity recognition in Brazilian legal text. In: Villavicencio, A., et al. (eds.) PROPOR 2018. LNCS (LNAI), vol. 11122, pp. 313–323. Springer, Cham (2018). https://doi.org/10.1007/978-3-319-99722-3_32
3. de Dados Pessoais, P.: LGPD (2020)
4. Fredriksson, T., Mattos, D.I., Bosch, J., Olsson, H.H.: Data labeling: an empirical investigation into industrial challenges and mitigation strategies. In: Product-Focused Software Process Improvement: 21st International Conference, PROFES 2020, Turin, Italy, 25–27 November 2020, Proceedings, pp. 202–216. Springer, Heidelberg (2020). https://doi.org/10.1007/978-3-030-64148-1_13
5. Murthy, G., Abhi, S., Agarwal, R.: A hybrid resume parser and matcher using RegEx and NER. In: 2023 International Conference on Advances in Computation, Communication and Information Technology (ICAICCIT), pp. 24–29. IEEE (2023)
6. Lewis, P.S.H., et al.: Retrieval-augmented generation for knowledge-intensive NLP tasks. CoRR abs/2005.11401 (2020). 2005.11401
7. Mota, C.C., Nascimento, A.C., Miranda, P.B., Mello, R.F., Maldonado, I.W., Coelho Filho, J.L.: Reconhecimento de entidades nomeadas em documentos jurídicos em português utilizando redes neurais. In: Anais do XVIII Encontro Nacional de Inteligência Artificial e Computacional, pp. 130–140. SBC (2021)

8. Nascimento, B.A.R., Botto Filho, M., Gomes, V.G., Menezes, H.K.A., Silva, A.N.M., et al.: Breve histório do cadastro de pessoa física–cpf e sua relação com a teoria dos números. Caderno de Graduação-Ciências Exatas e Tecnológicas-UNIT-SERGIPE **2**(3), 125–135 (2015)
9. Nunes, L.F.P., dos Santos, J.C.F.: Lgpd-uma visão de tecnologia e agnóstica. Revista Direito & Paz **2**(49), 218–237 (2023)
10. Oliveira, V., Nogueira, G., Faleiros, T., Marcacini, R.: Combining prompt-based language models and weak supervision for labeling named entity recognition on legal documents. Artificial Intelligence and Law, pp. 1–21 (2024)
11. Patsakis, C., Lykousas, N.: Man vs the machine: the struggle for effective text anonymisation in the age of large language models. arXiv preprint arXiv:2303.12429 (2023)
12. Sakhaee, N., Wilson, M.C.: Information extraction framework to build legislation network. CoRR abs/1812.01567 (2018). 1812.01567
13. Zhang, S., He, L., Dragut, E., Vucetic, S.: How to invest my time: lessons from human-in-the-loop entity extraction. In: Proceedings of the 25th ACM SIGKDD International Conference on Knowledge Discovery & Data Mining, KDD 2019, pp. 2305–2313. Association for Computing Machinery, New York, NY, USA (2019). https://doi.org/10.1145/3292500.3330773

ptt5-v2: A Closer Look at Continued Pretraining of T5 Models for the Portuguese Language

Marcos Piau[1(✉)], Roberto Lotufo[1,2], and Rodrigo Nogueira[1,3]

[1] School of Electrical and Computing Engineering, State University of Campinas (UNICAMP), Campinas, Brazil
marcos.piau.vieira@gmail.com
[2] NeuralMind, Campinas, Brazil
[3] Maritaca AI, Campinas, Brazil

Abstract. Despite advancements in Natural Language Processing (NLP) and the growing availability of pretrained models, the English language remains the primary focus of model development. Continued pretraining on language-specific corpora provides a practical solution for adapting models to other languages. However, the impact of different pretraining settings on downstream tasks remains underexplored. This work introduces `ptt5-v2`, investigating the continued pretraining of T5 models for Portuguese. We first develop a baseline set of settings and pretrain models with sizes up to 3B parameters. Finetuning on three Portuguese downstream tasks (ASSIN2 STS, ASSIN2 RTE, and TweetSentBR) yields SOTA results on the latter two. We then explore the effects of different pretraining configurations, including quality filters, optimization strategies, and multi-epoch pretraining. Perhaps surprisingly, their impact remains subtle compared to our baseline. We release `ptt5-v2` pretrained checkpoints and finetuned MonoT5 rerankers on HuggingFace in their respective collections at https://huggingface.co/unicamp-dl.

Keywords: Portuguese T5 · Continued pretraining · Information Retrieval

1 Introduction

Transformer-based pretrained language models have established themselves as the core paradigm in the field of Natural Language Processing (NLP). Starting with the advent of BERT [12], which popularized the "pretrain, then fine-tune" approach and the use of the transformer architecture itself, these models acquire a general-purpose language representation by unsupervised pretraining on extensive corpora of unlabeled text. The dynamics of the pretraining process had been studied in depth by many works, like Raffel et al. [34], that introduced T5, and scaled up models to billions of parameters and set new SOTAs across many tasks. The trend towards increasing model sizes and datasets to improve performance

A. Paes and F. A. N. Verri (Eds.): BRACIS 2024, LNAI 15413, pp. 324–338, 2025.
https://doi.org/10.1007/978-3-031-79032-4_23

motivated studies like Kaplan et al. [23] on scaling laws and Hoffmann et al. [18], who demonstrated the importance of training data size relative to model size for compute-optimal training regimes; more recently work from Gadre et al. [14] specifically examined the influence of extended pretraining on downstream task performance.

Despite the extensive study of pretraining dynamics, the focus has predominantly been on English, leaving non-English languages less explored. Continued pretraining presents a strategic approach to adapting these models to additional languages and domains using significantly less data and computational resources than training from scratch. This method involves further pretraining on language-specific corpora, which has been shown to substantially enhance model performance on downstream tasks in the target language [5,6,11,25,32,42]. However, there is a lack of detailed investigations into how different settings during the continued pretraining phase influence on downstream tasks performance, with most studies merely aiming for benchmark-leading results without a thorough examination of the underlying factors.

In this work, we study the continued pretraining of T5 models for the Portuguese language, analyzing the impact of various settings on downstream task performance. Rather than solely focusing on achieving state-of-the-art results, our study also investigates how factors like model size, optimization schedules, and the application of quality filters in the pretraining dataset affect the performance. We continue the pretraning of Google's T5 with up to 3 billion parameters on Portuguese texts. By experimenting with different configurations in the pretraining stage, we observe nuanced effects on downstream tasks, with some settings only marginally outperforming the baselines. Our findings also suggest that while continued pretraining enhances model capabilities, the increments in performance diminish as model size increases.

T5 models [34] demonstrate adaptability across various natural language processing (NLP) tasks due to their encoder-decoder architecture. This structure enables them to process text for both understanding and generation, providing an advantage over encoder-only models like BERT. While not the focus of this work, T5's adaptability to instruction-based fine-tuning, as seen in FLAN-T5 [8], also enables effective zero-shot and few-shot applications. These factors, combined with the scarcity of Portuguese pretrained encoder-decoder models, motivate our choice to continue the investigation the T5 architecture in this study.

2 Related Work

The T5 model [34] is an encoder-decoder transformer, and one of its main innovations was to cast all tasks into a text-to-text format, allowing for a unified approach; scaling up models to 11B parameters, they consolidated the transfer learning approach, setting new SOTAs for GLUE [46], SuperGLUE [45], CNN/Daily Mail [17] benchmarks. It was pretrained using the "span corruption" objective over the C4 ("Colossal Clean Crawled Corpus") dataset, where random consecutive spans in the input are replaced by special mask tokens, and the model is trained to predict these corrupted tokens (models with up to 11B parameters). Building upon this foundation, mT5 [48] extended the T5 framework to multilingual settings, having been pretrained on mC4, a multilingual dataset covering 101 languages (models up to 13B parameters). PTT5 [6] further adapted T5 for Portuguese by continuing the pretraining of T5 models on the BrWac dataset [44]. This approach led to significant improvements in downstream Portuguese language tasks, which were further enhanced by a Portuguese tokenizer. For clarity, we'll refer to the work of Carmo et al. as `ptt5-v1`. Other notable international adaptations of T5/mT5 include it5 [40] (Italian), AfriTeVa [22] (low-resource African languages), AraT5 [29] (Arabic), and plT5 [7] (Polish).

Bertimbau [42], a popular adaptation of the BERT encoder model, remains influential within Portuguese language modeling. Others exploring encoder architectures include Albertina [37], DeBERTinha [5], the work of Gomes et al. [16] (which pretrains a Roberta model), and de Morais et al. [28]. Reflecting a broader trend, numerous recent Portuguese models prioritize decoder-only architectures, such as Sabiá [32], Glória [27], Bode [15], Cabrita [25], and Gervásio [39]. In the encoder-decoder space, the work of Carmo et al. (`ptt5-v1`) explored adapting T5 models for Portuguese. Beyond generic Portuguese models, several works specialize in custom domains: de Barros et al. [2] and BERTabaporu [10] were designed for Portuguese social media data, while Bertaú [13] focuses on financial language.

3 Methodology

This section describes the methodology for pretraining and evaluating our key experiments, covering the pretraining dataset, language-specific vocabulary, model architectures, optimization strategies, and finetuning and validation processes for downstream tasks.

3.1 Unsupervised Continued Pretraining

As the pretraining data, we utilized the Portuguese segment of the mC4 dataset (hereafter referred to as mC4-pt), comprising approximately 524 GB of uncompressed text across 169 million documents. This dataset is significantly larger than the one used for the pretraining of `ptt5-v1` models, which originated from the BrWac dataset [44] and consisted of around 15 GB of text from 7.4 million documents after preprocessing.

We adopted the Portuguese language vocabulary from ptt5-v1. This SentencePiece Unigram tokenizer [24], comprising 32,000 tokens, was trained over a corpus of 2 million documents from the Portuguese Wikipedia. This vocabulary shares the same number of tokens and control tokens as T5, facilitating the direct use of Google's model checkpoints.

As the pretraining objective, the span corruption task was employed, utilizing batches of 128 sequences of 512 tokens (65,536 tokens) - a methodology consistent with the baseline experiment by Raffel et al. [34]. Adafactor optimizer [41] with a constant learning rate of 0.001 and cross-entropy as loss was used during the entire pretraining process. Using these experimental settings, we started from Google's original checkpoints with sizes from `t5-small` (60M parameters) up to `t5-3B` (3B parameters), and performed a complete epoch of continued pretraining over the mC4-pt dataset. Considering these settings, a single epoch over the mC4-pt dataset comprises approximately 1,764,515 training steps and 116 billion training tokens. Additional pretraining experiments are detailed in Sect. 5.1.

Both pretraining and finetuning experiments utilized TPUv2-8 and TPUv3-8 devices, leveraging `t5` [34] and `seqio` [36] frameworks.

3.2 Supervised Finetuning on Downstream Tasks

We assess the impact of our pretraining on three Portuguese language downstream tasks: ASSIN2 RTE, ASSIN2 STS, and TweetSentBR. The ASSIN2 dataset [31] provides two tasks: RTE (Recognizing Textual Entailment), which involves determining whether one sentence entails another, and STS (Semantic Textual Similarity), which quantifies the semantic similarity between sentence pairs on a 1–5 scale. The TweetSentBR dataset [4] is a sentiment analysis task for Brazilian Portuguese tweets, classifying them as positive, negative, or neutral. Tables 1 and 2 show further details and examples for each task.

Table 1. Downstream tasks datasets.

Dataset	Task	Preferred metric	Train/validation/test	Random score	Possible outputs
ASSIN2 RTE	Binary classification	F1-macro	6,500/500/2,448	50	{Entailment, None}
ASSIN2 STS	Regression	Pearson	6,500/500/2,448	0	$[1, 5]$
TweetSentBR	Multiclass Classification (3)	F1-macro	11,525/1,281/1,982	32.4	{negativo, positivo, neutro}

Table 2. Downstream tasks inputs and targets

Dataset	Example inputs	Example targets
ASSIN2 RTE	`assin2_rte sentence1: Uma pessoa está escovando um gato sentence2: O pelo de um gato está sendo penteado por uma pessoa`	`Entailment`
ASSIN2 STS	`assin2_stsb sentence1: Uma mulher está cortando vegetais sentence2: Uma mulher está cortando brócolis`	`4.2`
TweetSentBR	`ttsbr_neg_pos_neu_sentiment_pt: adorando esse com dr dráuzio varela`	`positivo`

We finetuned the pretrained models over 100 epochs with batches of 128 sequences and a maximum length of 512 tokens, using Adafactor as the optimizer with a constant learning rate of 0.001. The model checkpoint yielding the best performance on the validation set was selected for testing, and greedy decoding was utilized as the decoding method. Because TweetSentBR lacks a validation set, we reserved 10% of the training data for validation and used the remaining 90% for training.

All tasks were approached using a text-to-text format. Specifically for the ASSIN2 STS task, which involves the prediction of continuous values in the range between 1 to 5, we adopted the strategy from Raffel et al. [34], by rounding the target scores to the nearest 0.2 increment and converting these to strings, thus framing it as a multiclass classification problem compatible with the text-to-text format.

To compare the quality of the new checkpoints against existing alternatives, we also use the same finetuning procedure on Google T5 and mT5 models.

3.3 MonoPTT5 Rerankers

To evaluate the adaptability of the ptt5-v2 models for information retrieval tasks, specifically passage reranking, we trained MonoT5 rerankers [30] using checkpoints generated as described in Sect. 3.1. We named these models MonoPTT5. MonoT5 rerankers are used for passage reranking, a two-stage process: first, a less computationally expensive method like BM25 generates an initial set of relevant documents for a given query; the reranker model then reranks a subset of these documents to improve relevance ordering. During training, the model learns in a supervised text-to-text manner to generate tokens corresponding to relevant and non-relevant labels. For inference, we greedy decode a single token and calculate the softmax over the logits of the two possible tokens, using the probability of the positive class as the relevance score.

We adapted the input and target format for the Portuguese language to the structure `"Pergunta: {query} Documento: {document} Relevante:"`, assigning the tokens "Sim" (relevant) and "Não" (non-relevant). This format is applied during both training and inference, regardless of the input language.

The training data originated from the mMARCO dataset [3], a translated version of the MS MARCO passage retrieval dataset [1], originally in English, to 13 languages, including Portuguese. The training subset consists of triples (query, relevant passage, non-relevant passage), which we split into two training example pairs, with each pair containing the query matched to one passage – either relevant or non-relevant – thus creating one example for each label. We created a bilingual Portuguese-English training dataset by randomly assigning one of the two languages to each training triplet. This "translate-train" approach [9,20, 48] leverages synthetic data augmentation by integrating machine translations with original text data to substantially expand the available training material in the target language. Prior research [3,9,38] has shown the effectiveness of this bilingual training strategy, motivating our adoption of this method.

The models were trained for 100k steps with batch sizes of 128 sequences and a maximum length of 512 tokens, utilizing Adafactor with a constant learning rate of 0.001 as the optimizer. Instances exceeding the maximum token length were excluded from the training dataset; these constituted approximately 0.01% of the training samples and were predominantly attributed to noisy translation data. Given the significant computational resources required for training these rerankers, we focused exclusively on models based on the main `ptt5-v2` checkpoints.

To evaluate the rerankers, we first used BM25 to generate an initial set of relevant documents[1] and then rerank the top-1000 documents. Retrieval metrics are calculated by comparing this ordered list with the relevance judgments from

[1] All applications of BM25 in this work use Pyserini's implementation [26] with default parameters $k_1 = 0.9$ and $b = 0.4$.

each dataset. We consider two retrieval scenarios: in-domain (using the "small dev" set of 6,980 queries from the mMARCO-pt dataset) and zero-shot (using the Portuguese subset of 249 annotated queries from mRobust [21]). Due to the longer document length in mRobust, we segmented documents into sliding sentence windows using a Spacy [19] sentencizer pipeline, with a maximum length of 8 and a stride of 4 sentences to mitigate truncation during reranking.

4 Main Results

Table 3 shows the results on the downstream tasks considered. In the ASSIN2 RTE task, our 3B sets a new SOTA, surpassing the current one by 0.61 F1-macro points. For the TweetSentBR dataset, we achieved better performance than current finetuned SOTAs with `ptt5-v2-large` and `ptt5-v2-3B`, by 0.52 and 1.54 F1-macro points, respectively, but our results are worse when comparing to GPT-4. We highlight that our `ptt5-v2` were trained exclusively on each task training data using the text-to-text framework without any data augmentation or adaptation to the model's architecture, unlike the works of [2] and [38], which held the SOTA for TweetSentBR and ASSIN2 RTE. In the ASSIN2 STS task, our models did not surpass the current SOTA; regardless, `ptt5-v2` still shows better performance than mT5 and T5 models with approximate sizes, and this is

Table 3. Main results after finetuning. For supervised SOTAs, model sizes are as follows: [38] (335M), [37] (900M), [2] (110M), [38], and [21] (580M). NPM results exclude retrieval tasks.

		Binary classification	Regression	Multiclass classification	←	Retrieval	
		ASSIN2 RTE	ASSIN2 STS	TweetSentBR	←	mMARCO-pt	mRobust-pt
Model	Parameters	F1-macro	Pearson	F1-macro	NPM	RR@10	nDCG@20
T5							
t5-small	60M	83.66	0.738	62.14	61.71	-	-
t5-base	220M	85.80	0.764	65.43	65.63	-	-
t5-large	770M	88.91	0.790	68.22	69.95	-	-
t5-3B	3B	90.78	0.827	72.58	74.56	-	-
mT5							
mt5-small	300M	75.36	0.688	61.81	54.35	-	-
mt5-base	580M	79.39	0.749	70.76	63.46	-	-
mt5-large	1.2B	88.25	0.753	61.11	64.76	-	-
mt5-xl	3.7B	91.81	0.827	77.05	77.45	-	-
ptt5-v2							
ptt5-v2-small	60M	87.14	0.782	70.99	69.86	0.273	0.344
ptt5-v2-base	220M	88.36	0.814	73.20	72.82	**0.311**	0.384
ptt5-v2-large	770M	91.73	0.839	**76.78**	77.68	**0.315**	**0.462**
ptt5-v2-3B	3B	**92.68**	0.829	**77.80**	78.48	**0.332**	**0.512**
Supervised SOTA	-	92.07 [38]	**0.868** [37]	76.26 [2]	-	0.306 [38]	0.391 [21]
Few shot baselines from [32]							
GPT-4	-	90.96	0.776	82.40	-	-	-
GPT-3.5-turbo	-	88.28	0.664	74.39	-	-	-

also the only task where a smaller `ptt5-v2` model (`ptt5-v2-large`) has better performance of a large one `ptt5-v2-3B`.

In addition to using individual metrics for each task, we also incorporate the Normalized Preferred Metric (NPM) [43] to facilitate the evaluation of the overall performance of a pretrained model across multiple tasks. The NPM normalizes a task's preferred metric (e.g., F1-macro for ASSIN2 RTE), assigning a value of 0 to represent random performance and 100 to denote maximum performance. Below is the equation used to calculating the NPM for a given model and set N of tasks:

$$\text{NPM} = \frac{100}{N} \sum_{i=1}^{N} \frac{[\text{raw preferred metric}]_i - [\text{random score}]_i}{[\text{maximum score}]_i - [\text{random score}]_i} \quad (1)$$

Given that MonoPTT5 rerankers were exclusively trained starting from `ptt5-v2` pretrained checkpoints, retrieval tasks were excluded from this evaluation. Therefore, we only considered ASSIN2 RTE, ASSIN2 STS, and TweetSentBR tasks. For each model, we calculate its aggregate performance by first determining the NPM for each task and then computing the average of these values.

Our `ptt5-v2` models have higher NPM values than T5 and mT5 models with considerably more parameters: for example, `ptt5-v2-base` is only surpassed by `t5-3B` ($\sim$13.6x larger) and `t5-xl` ($\sim$16.81x larger); this performance gap, however, is most pronounced in smaller models, narrowing as model size increases. A similar result was also observed by Xue et al. [48], which analyzed the performance of T5 and mT5 models performance on the SQuAD benchmark [35], observing a performance gap between `t5-small` and `t5-base` vs mT5 models of equivalent sizes, which is diminished starting from `t5-large`. This performance gap observed on smaller model sizes is advantageous when we consider environments constrained by computational resources, increasing the maximum attainable level of performance; additionally, a language-specific tokenizer reduces text splitting into fewer tokens, leading to lower latency and the potential to accommodate more text within the same maximum token context window. Interestingly, mT5 models tend to show lower NPM values, except in the 3 billion parameter range, where they slightly outperform T5 models.

For the retrieval tasks, our MonoPTT5 rerankers were able to set new SOTAs for both mMARCO-pt and mRobust-pt. For the mMARCO-pt dataset, models starting from `t5-base` size were able to surpass the current SOTA; the 3 billion parameters reranker obtained a gain of +0.026 points of MRR@10. In the mRobust-pt, our large and 3B rerankers surpassed the current SOTA by +0.071 and +0.121 in terms of nDGC@20, respectively. A more in-depth analysis of the retrieval tasks is conducted on Sect. 5.2.

5 Ablations

5.1 Additional Pretraining Experiments

This sections includes pretraining experiments additional to those described in Sect. 3.1.

Comparison with ptt5-v1: Given the title of our work, a pertinent question arises: *How do* `ptt5-v2` *models compare with the work in* `ptt5-v1`? A few key differences exist in the pretraining of `ptt5-v1`. Notably, it utilized BrWac, a considerably smaller dataset, and a slightly different pretraining objective (denoising, where some input tokens are masked and the model is trained to predict the original text, rather than span corruption). Additionally, `ptt5-v1` employed models ranging from `t5-small` to `t5-large` with an Adafactor optimizer using a learning rate of 0.003 (threefold larger than our setting). `ptt5-v1` also explored both T5's original English vocabulary and a Portuguese language-specific tokenizer. In contrast, ptt5-v2 exclusively uses the latter.

To ensure a fair comparison, we finetuned the `ptt5-v1` checkpoints following the methodology outlined in Sect. 3.2. Figure 1 presents the NPM values for both `ptt5-v1` and `ptt5-v2`, alongside comparisons to mT5 and T5 models. The data corroborates the enhancements achievable through monolingual pretraining on the target language, which is further augmented by employing a dedicated tokenizer. In comparison between the two `ptt5` iterations, a performance disparity favoring `ptt5-v2` is apparent for the `small` and `large` sizes, with the largest gap observed for `large`; however, models of the `base` size exhibit marginally superior performance in the `ptt5-v1` variant. Surprisingly, mT5's performance lags behind all other models, including the monolingual English T5, except in the 3B parameter range, where it closely matches `ptt5-v2` and surpasses T5.

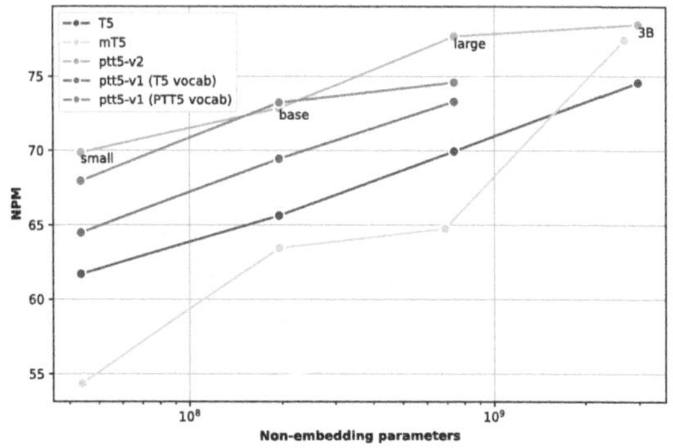

Fig. 1. NPM vs parameters for varying T5 configurations.

Quality Filters: In our primary experiments, we utilized the entire mC4-pt dataset, containing approximately 116 billion training tokens, in the pretraining phase. In this additional experiment, we consider the MassiveText quality filters [33] to investigate the impact of this filtering process on downstream tasks. Applying these filters to mC4-pt reduces the number of training tokens to approximately 82 billion, a reduction of about 30%. This experiment was restricted to models of `t5-base` size, keeping the same batch size and optimization strategy used in the main experiments. Figure 2 shows the effect of using MassiveText's quality filters on downstream task performance, measured in terms of NPM. Pretraining with the filtered dataset shows an upward trend in performance, which continues without saturation up to the duration of one mC4-pt epoch, which is the last point in the plot. It's important to note that one epoch in the filtered dataset has fewer steps than the full dataset, so at one full dataset epoch, data is repeated for the filtered dataset. Despite the upward trend favoring the pretraining with the filtered dataset, the difference in terms of NPM is small at the one mC4-pt epoch mark.

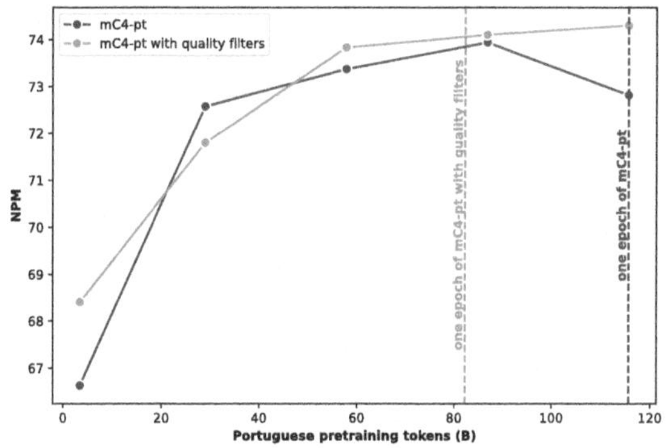

Fig. 2. Effect of using quality filters on mC4-pt. Vertical lines indicate the number of pretraining tokens for each dataset considered.

Pretraining Optimization Strategy: In our exploration of optimization strategies, we initially employed Adafactor with a constant learning rate. This ablation extends our investigation to the "inverse square root" learning rate schedule, as utilized by [34] in their final pretraining experiments. This learning rate schedule computes the rate as $\frac{1}{\sqrt{\max(n,k)}}$, where n represents the current step, and k is the number of warm-up steps. Raffel et al. [34] used $k = 10,000$, which sets the learning rate of 0.01 for the initial 10k steps, subsequently decreasing exponentially. The learning rate at the end of the pretraining, which consisted of around 1 million steps, was close to 0.001, the same one used during the fine-tuning. Figure 3 illustrates the difference between these optimization strategies.

Attempting to closely mirror the T5 pretraining recipe, we applied this identical schedule in our preliminary experiments. However, we observed a rapid overshoot in the training losses for `t5-large` `t5-3B` within hours; adjusting n only delayed the overshoot without averting it. Transitioning to a constant learning rate of 0.001 solved the overshoot issue, leading to stable pretraining loss across all model sizes and simplifying our experimental setup. Because overshoot was observed only in the larger models, and also given the computational

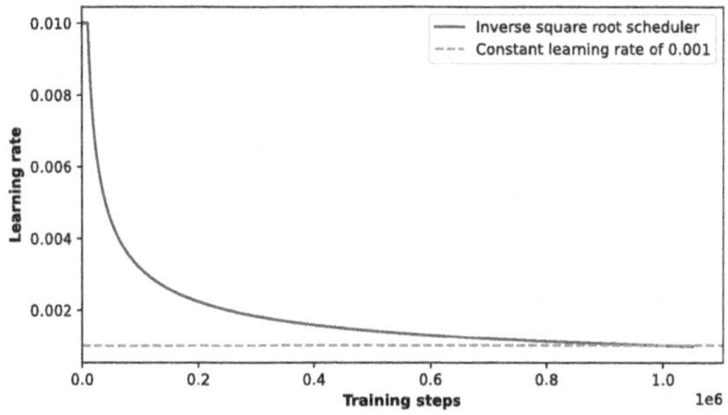

Fig. 3. Learning rate schedules: Constant vs. inverse square root scheduler as a function of training steps.

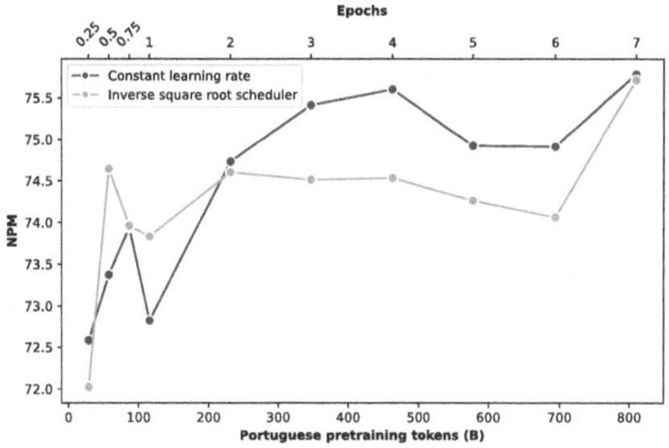

Fig. 4. Effect of scheduler used during pretraining. Epochs are relative to mC4-pt.

costs associated with pretraining the larger models, we performed additional pretraining experiments with "inverse square root" scheduler for `t5-small` and `t5-base` models only.

Number of Pretraining Epochs: In the primary set of experiments, the mC4-pt dataset was fully utilized for pretraining over one epoch. To explore the influence of the number of pretraining epochs on downstream task performance, we conducted experiments over various epochs, including partial epochs (0.25, 0.5, and 0.75 of an epoch). Considering the significant time and computational resources required for extended pretraining, especially with larger models, we limited the pretraining of the `t5-large` model to two epochs and the `t5-3B` model to a single epoch. The reduced epoch duration for the `t5-small` and `t5-base` models enabled more extensive pretraining periods for these configurations.

In Fig. 4, the NPM values for `t5-base` are shown with the constant and the inverse square root scheduler, across a varying number of pretraining epochs. It is observed that there is a difference between these two optimization strategies: the inverse square root scheduler has the advantage for up to two epochs; afterwards, the constant learning rate takes the upper hand, and by the last epoch considered, they reach the same value. An increasing trend in the values of NPM is also noted for more epochs.

5.2 MonoPTT5 Rerankers

The information retrieval tasks reported in 3 represents the performance of our MonoPTT5 experiments, developed with the methodology describe in Sect. 3.3; in this section we also report the results for other approaches, including BM25, and dense retrieval using `multilingual-e5` [47] models. The dense models are used as a single stage retrieval system without reranking; dense indexing and retrieval is performed with A100 and V100 GPUs on Google Colab, leveraging the Pyserini framework. For mRobust-pt, which contain longer documents, we mitigate document truncation by using the same splitting strategy described in the Sect. 3.3, using the maximum score among the document segments as the document score.

Figures 5 and 6 are used to illustrate the discussion presented in this section. Results in Table 3 only shows the effectiveness in each retrieval task of our MonoPTT5 models and the SOTA competitors. For the in-domain retrieval task, the mMarco-pt dataset, we note that BM25 is easily surpassed by all alternatives considered, and the effectiveness figures for MonoPTT5 rerankers and `multilingual-e5` models are similar when consider the size range in common, and MonoPTT5 rerankers effectiveness is above SOTA starting from models of size `t5-base`. For mRobust-pt, representing a zero-shot setting, BM25 is only surpassed by the mT5 reranker from Jeronymo et al. [21], and MonoPTT5 models starting from `t5-base` size.

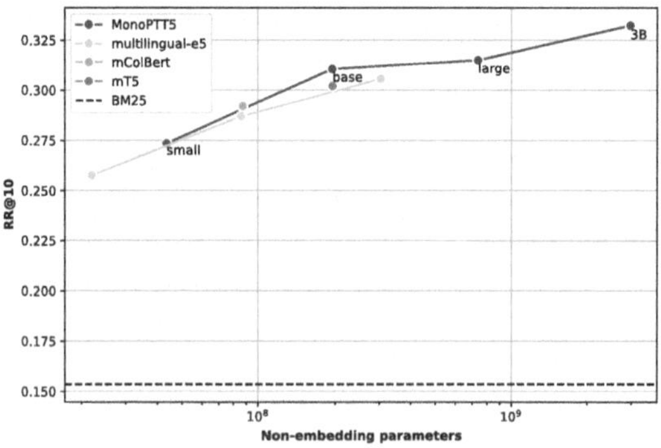

Fig. 5. Retrieval results on mMarco-pt. mColbert and mT5 values are from Bonifacio et al. [3]. Total size excludes embedding parameters.

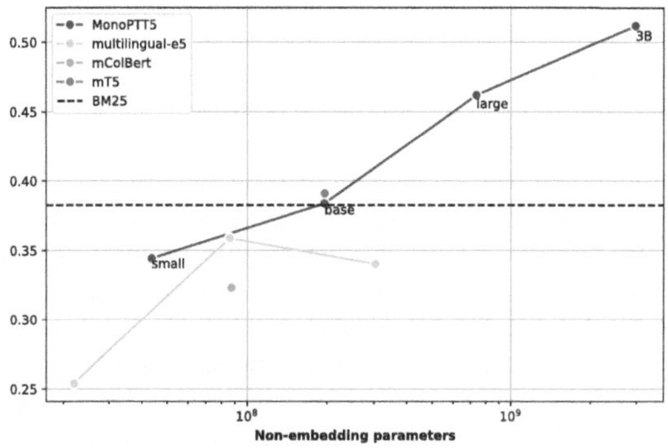

Fig. 6. Retrieval results on mRobust-pt. mColbert and mT5 values are from Jeronymo et al. [21]. Total size excludes embedding parameters.

6 Conclusion

In this study, we introduced `ptt5-v2`, exploring the continued pretraining of T5 models for the Portuguese language. We pretrained T5 models using a Portuguese language tokenizer, over a Portuguese language corpus. The fine-tuned models achieved SOTA on ASSIN2 RTE and TweetSentBr datasets, two of the three downstream tasks considered. Additionally, we applied these pre-trained checkpoints to develop MonoT5 rerankers customized for the Portuguese language, achieving top performance on the mMARCO-pt and mRobust-pt datasets. Our main results supports the evidence of a performance gap favoring monolingual models over English-focused and multilingual models, a gap that narrows as model capacity increases. This underscores the importance of language-specific pretraining, and our analysis of pretraining settings suggests that while data filtering, optimization strategies, and pretraining duration can offer incremental improvements, the overall effects were limited in comparison to our baseline settings and the core pretraining recipe remained robust.

Acknowledgements. We thank Google for the TPU grant through the TRC program.

R Lotufo is partially supported by CNPq (The Brazilian National Council for Scientific and Technological Development) under grant 313047/2022-7.

References

1. Bajaj, P., et al.: MS MARCO: a human generated machine reading comprehension dataset (2018)
2. de Barros, T.M., Pedrini, H., Dias, Z.: Leveraging emoji to improve sentiment classification of tweets. In: Proceedings of the 36th Annual ACM Symposium on Applied Computing, SAC 2021, pp. 845–852. Association for Computing Machinery, New York, NY, USA (2021). https://doi.org/10.1145/3412841.3441960

3. Bonifacio, L., et al.: mMARCO: a multilingual version of the MS MARCO passage ranking dataset (2022)
4. Brum, H., Volpe Nunes, M.d.G.: Building a sentiment corpus of tweets in Brazilian Portuguese. In: Calzolari, N., et al. (eds.) Proceedings of the Eleventh International Conference on Language Resources and Evaluation, LREC 2018, May 2018. European Language Resources Association (ELRA), Miyazaki, Japan (2018)
5. Campiotti, I., Rodrigues, M., Albuquerque, Y., Azevedo, R., Andrade, A.: DeBERTinha: a multistep approach to adapt DebertaV3 XSmall for Brazilian Portuguese natural language processing task (2023)
6. Carmo, D., Piau, M., Campiotti, I., Nogueira, R., Lotufo, R.: PTT5: pretraining and validating the T5 model on Brazilian Portuguese data (2020)
7. Chrabrowa, A., et al.: Evaluation of transfer learning for Polish with a text-to-text model (2022)
8. Chung, H.W., et al.: Scaling instruction-finetuned language models (2022)
9. Conneau, A., et al.: Unsupervised cross-lingual representation learning at scale (2020)
10. Costa, P.B., Pavan, M.C., Santos, W.R., Silva, S.C., Paraboni, I.: BERTabaporu: assessing a genre-specific language model for Portuguese NLP. In: Mitkov, R., Angelova, G. (eds.) Proceedings of the 14th International Conference on Recent Advances in Natural Language Processing, September 2023, pp. 217–223. INCOMA Ltd., Shoumen, Bulgaria, Varna, Bulgaria (2023)
11. Cui, Y., Yang, Z., Yao, X.: Efficient and effective text encoding for Chinese LLaMA and Alpaca (2024)
12. Devlin, J., Chang, M.W., Lee, K., Toutanova, K.: BERT: pre-training of deep bidirectional transformers for language understanding (2019)
13. Finardi, P., Viegas, J.D., Ferreira, G.T., Mansano, A.F., Caridá, V.F.: Bertaú: Itaú bert for digital customer service (2021)
14. Gadre, S.Y., et al.: Language models scale reliably with over-training and on downstream tasks (2024)
15. Garcia, G.L., et al.: Introducing bode: a fine-tuned large language model for Portuguese prompt-based task (2024)
16. Gomes, J.R.S., et al.: Deep learning Brasil at ABSAPT 2022: Portuguese transformer ensemble approaches (2023)
17. Hermann, K.M., et al.: Teaching machines to read and comprehend (2015)
18. Hoffmann, J., et al.: Training compute-optimal large language models (2022)
19. Honnibal, M., Montani, I., Landeghem, S.V., Boyd, A.: spaCy: industrial-strength natural language processing in Python (2020, to appear). https://doi.org/10.5281/zenodo.1212303
20. Hu, J., Ruder, S., Siddhant, A., Neubig, G., Firat, O., Johnson, M.: XTREME: a massively multilingual multi-task benchmark for evaluating cross-lingual generalization (2020)
21. Jeronymo, V., Nascimento, M., Lotufo, R., Nogueira, R.: mRobust04: a multilingual version of the TREC robust 2004 benchmark (2022)
22. Jude Ogundepo, O., Oladipo, A., Adeyemi, M., Ogueji, K., Lin, J.: AfriTeVA: extending "small data" pretraining approaches to sequence-to-sequence models. In: Cherry, C., (eds.) Proceedings of the Third Workshop on Deep Learning for Low-Resource Natural Language Processing, July 2022, pp. 126–135. Association for Computational Linguistics, Hybrid (2022). https://doi.org/10.18653/v1/2022.deeplo-1.14
23. Kaplan, J., McCandlish, S., et al.: Scaling laws for neural language models (2020)

24. Kudo, T., Richardson, J.: SentencePiece: a simple and language independent subword tokenizer and detokenizer for neural text processing. arXiv preprint arXiv:1808.06226 (2018)
25. Larcher, C., Piau, M., Finardi, P., Gengo, P., Esposito, P., Caridá, V.: Cabrita: closing the gap for foreign languages (2023)
26. Lin, J., Ma, X., Lin, S.C., Yang, J.H., Pradeep, R., Nogueira, R.: Pyserini: a Python toolkit for reproducible information retrieval research with sparse and dense representations. In: Proceedings of the 44th Annual International ACM SIGIR Conference on Research and Development in Information Retrieval, SIGIR 2021, pp. 2356–2362 (2021)
27. Lopes, R., Magalhães, J., Semedo, D.: Glória – a generative and open large language model for Portuguese (2024)
28. de Morais, L.P., da Silva Soares, A., da CM Borges, V., da Silva, N.F.F., Pereira, F.S.: Sub-language sentiment analysis in Whatsapp domain with deep learning approaches. Revista de Sistemas de Informaçao da FSMA **1**(31), 32–47 (2023)
29. Nagoudi, E.M.B., Elmadany, A., Abdul-Mageed, M.: AraT5: text-to-text transformers for Arabic language generation (2022)
30. Nogueira, R., Jiang, Z., Pradeep, R., Lin, J.: Document ranking with a pretrained sequence-to-sequence model. In: Cohn, T., He, Y., Liu, Y. (eds.) Findings of the Association for Computational Linguistics, EMNLP 2020, Online, November 2020, pp. 708–718. Association for Computational Linguistics (2020). https://doi.org/10.18653/v1/2020.findings-emnlp.63
31. Oliveira, H.G., Real, L., Fonseca, E. (eds.): Proceedings of the ASSIN 2 Shared Task: Evaluating Semantic Textual Similarity and Textual Entailment in Portuguese, Extended Semantic Web Conference, No. 2583 in CEUR Workshop Proceedings (2020)
32. Pires, R., Abonizio, H., Almeida, T.S., Nogueira, R.: Sabiá: Portuguese Large Language Models, pp. 226–240. Springer, Heidelberg (2023). https://doi.org/10.1007/978-3-031-45392-2_15
33. Rae, J.W., et al.: Scaling language models: methods, analysis & insights from training Gopher (2022)
34. Raffel, C., et al.: Exploring the limits of transfer learning with a unified text-to-text transformer. arXiv preprint arXiv:1910.10683 (2019)
35. Rajpurkar, P., Zhang, J., Lopyrev, K., Liang, P.: SQuAD: 100,000+ questions for machine comprehension of text (2016)
36. Roberts, A., et al.: Scaling up models and data with `t5x` and `seqio` (2022)
37. Rodrigues, J., et al.: Advancing neural encoding of Portuguese with transformer Albertina PT-* (2023)
38. Rosa, G.M., Bonifacio, L.H., de Souza, L.R., Lotufo, R., Nogueira, R.: A cost-benefit analysis of cross-lingual transfer methods (2021)
39. Santos, R., Silva, J., Gomes, L., Rodrigues, J., Branco, A.: Advancing generative AI for Portuguese with open decoder Gervásio PT* (2024)
40. Sarti, G., Nissim, M.: IT5: large-scale text-to-text pretraining for Italian language understanding and generation (2022)
41. Shazeer, N., Stern, M.: Adafactor: adaptive learning rates with sublinear memory cost (2018)
42. Souza, F., Nogueira, R., Lotufo, R.: BERTimbau: pretrained BERT models for Brazilian Portuguese. In: Cerri, R., Prati, R.C. (eds.) LNCS, pp. 403–417. Springer, Cham (2020). https://doi.org/10.1007/978-3-030-61377-8_28
43. Srivastava, A., et al.: Beyond the imitation game: quantifying and extrapolating the capabilities of language models (2023)

44. Wagner Filho, J.A., Wilkens, R., Idiart, M., Villavicencio, A.: The brWaC corpus: a new open resource for Brazilian Portuguese. In: Proceedings of the Eleventh International Conference on Language Resources and Evaluation, LREC 2018 (2018)
45. Wang, A., et al.: SuperGLUE: a stickier benchmark for general-purpose language understanding systems (2020)
46. Wang, A., Singh, A., Michael, J., Hill, F., Levy, O., Bowman, S.R.: GLUE: a multi-task benchmark and analysis platform for natural language understanding (2019)
47. Wang, L., Yang, N., Huang, X., Yang, L., Majumder, R., Wei, F.: Multilingual E5 text embeddings: a technical report (2024)
48. Xue, L., et al.: mT5: a massively multilingual pre-trained text-to-text transformer (2021)

Question Answering with Texts and Tables Through Deep Reinforcement Learning

Marcos M. José[1(✉)], Flávio N. Cação[1,2], Maria F. Ribeiro[1], Rafael M. Cheang[1], Paulo Pirozelli[1], and Fabio G. Cozman[1]

[1] Escola Politécnica, Universidade de São Paulo, São Paulo, Brazil
marcos.jose@alumni.usp.br
[2] Novelo Data, São Paulo, Brazil

Abstract. This paper proposes a novel architecture to generate multi-hop answers to open domain questions that require information from texts and tables, using the Open Table-and-Text Question Answering dataset for validation and training. One of the most common ways to generate answers in this setting is to retrieve information sequentially, where a selected piece of data helps searching for the next piece. As different models can have distinct behaviors when called in this sequential information search, a challenge is how to select models at each step. Our architecture employs reinforcement learning to choose between different state-of-the-art tools sequentially until, in the end, a desired answer is generated. This system achieved an F1-score of 19.03, comparable to iterative systems in the literature.

Keywords: Reinforcement Learning · Large Language Models · Question Answering · Multi-hop

1 Introduction

Question Answering (QA) is concerned with developing programs that can respond to questions posed in natural language. Those systems are widely used to assist users with their queries without the need for direct human intervention. QA datasets typically contain question-answer pairs written in natural language; e.g., SQuAD [17]. While textual data is the most common source of information, other formats, such as graphs [1] and tabular data [6], may contribute in different ways to the task. Tabular data is particularly valuable for structured information, which can be challenging to retrieve in pure textual databases. A mix of textual and tabular data, thus, offers significant potential for improvement in QA systems.

This work addresses QA problems that demand retrieving and aggregating information from both tabular and textual data. Our testbed is the Open Table-and-Text Question Answering (OTT-QA) dataset [5], an open-domain multi-hop QA dataset. In this dataset, questions are provided without supporting texts, and to answer them, one is typically required to navigate through multiple passages,

A. Paes and F. A. N. Verri (Eds.): BRACIS 2024, LNAI 15413, pp. 339–353, 2025.
https://doi.org/10.1007/978-3-031-79032-4_24

textual and tabular. Performance on this dataset remains limited compared to purely textual datasets like SQuAD, where models already outperform humans.

Currently, the most adopted strategies to QA combine retrieval and reader modules: a retriever module fetches relevant passages, whereas a reader module turns these passages into a meaningful answer [4,7,12]. In multi-hop datasets like OTT-QA, this retrieval process needs to be done sequentially, moving from a piece of information to another, in order to get the required data (textual or tabular). Even though recent non-sequential systems have attained greater performance, a sequential decision-making formulation seems to be essential to coordinate retriever and reader schemes.

In this paper we wish to explore the boundaries of performance obtained with sequential decisions applied to standard modules, where we resort to Reinforcement Learning (RL) to learn strategies. RL is suitable for this task: while supervised learning depends on labeled [3], static data for the action sequences, in RL the sequence of decisions is built through evolving information. We thus propose a novel Deep Reinforcement Learning (DRL) approach for this task, which iteratively decides which actions to take (retrieve text, retrieve table, generate answer), using standard modules to conduct these actions.[1]

The paper is organized as follows. Section 2 offers some necessary background. Section 3 presents our proposed RL architecture. Section 4 describes our experimental setup, while Sect. 5 presents the results and their associated discussion. Finally, Sect. 6 presents our concluding thoughts.

2 Background

2.1 Open Table-and-Text Question Answering

Numerous QA datasets and architectures have been designed to address questions that require understanding textual information, such as SQuAD, or tabular data, as MIMICSQL [20]. However, there are not many resources that handle both texts and tables. To bridge this gap, HybridQA has been introduced as a closed-domain multi-hop multimodal question-answering dataset [6]. In HybridQA, each question carries a table and multiple associated passages, and to answer a question, it is necessary to combine several of these sources together.

OTT-QA [5] is an English dataset with 45841 QA pairs built over HybridQA. Its core aim is to serve as an open-domain dataset that combines text and tabular information while simultaneously increasing the complexity of the task. The dataset contains tables and texts with answers for every question in the dataset. An example of a QA pair from OTT-QA is shown in Fig. 1.

To assess the answers, we employ two well-known evaluation metrics (popularized by the SQuAD dataset [17] and used by OTT-QA): **Exact Match**, the percentage of predicted answers that are exactly equal to the golden answers, and **Macro Average F1-score**, the overlap between predicted and golden answers. In this context, precision is the ratio of shared words between the golden and

[1] Data and code available at: https://github.com/MMenonJ/DRL_QA_TT.

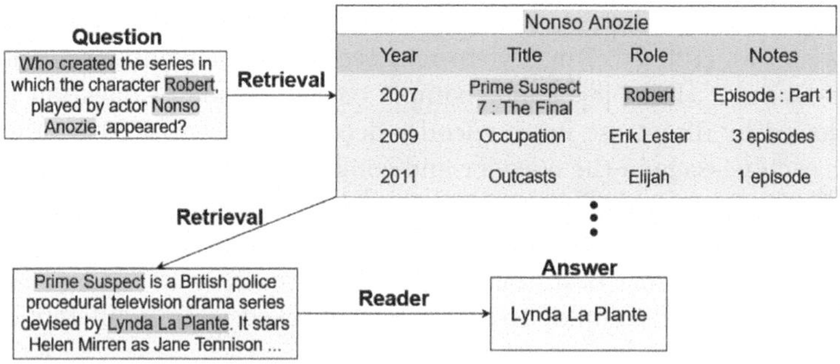

Fig. 1. QA pair from OTT-QA [5]. The question asks who is the author of the series in which Nonso Aznozie played the character Robert. To accurately answer this question, one must locate the relevant table labeled "Nonso Anozie" (highlighted in pink) and identify the cell with "Prime Suspect" (highlighted in yellow), by referencing the role "Robert" (highlighted in green). It is then possible to retrieve the associated text, which reveals that the creator of the movie was "Lynda La Plante" (highlighted in blue). (Color figure online)

prediction answers over the total number of prediction words; and recall is the ratio of shared words over the number of words in the golden answer. The F1-score is the harmonic mean between the precision and the recall.

2.2 Open-Domain and Question Answering

In open-domain QA, the questions might be about any topic and a system must find answers based on a given question and usually within a corpus of information. QA systems typically consist of two main components: a retriever and a reader. The retriever's task is to search the corpus for relevant texts and passages, while the reader processes this retrieved information to generate a suitable answer. Some approaches leverage neural transformers for both retriever and reader tasks [7,12], whereas others use keyword-based models specifically for the retriever [2,4].

Retriever: A widely adopted keyword-based retriever model in QA is BM25 [18]. BM25 relies on TF-IDF (Term Frequency–Inverse Document Frequency) by incorporating document length normalization; hence, shorter documents matching the query are prioritized over longer ones with similar keyword matches. In contrast, Dense Passage Retrieval (DPR) [9] overcomes the limitations of keyword matching by ranking passages based on semantic similarity. DPR utilizes two BERT models trained jointly to produce dense vector representations of questions and passages—similarity between these vectors is estimated by the dot product, enabling more nuanced retrieval based on meaning.

Reader: The best current readers typically are based on transformer neural networks, which take concatenated inputs of questions and retrieved passages to generate answers. For example, BERT-based models, as seen in Guu et al. [7], focus on substring prediction, while encoder-decoder architectures like BART, as in Lewis et al. [12], excel in text-to-text tasks. Another notable reader is the Fusion-in-Decoder (FiD) [8]. Unlike simple concatenation of question and passages followed by direct use in an encoder-decoder network, FiD independently processes each passage in the encoder and combines resulting vectors as input to the decoder. This method efficiently handles a larger number of documents. Similarly, Fusion-in-Encoder (FiE) [10] integrates information from multiple passages within the encoder rather than the decoder. FiE creates a unified representation and incorporates cross-sample attention over all tokens across different samples.

2.3 Architectures Tested on OTT-QA

The authors of OTT-QA [5] presented several solutions, each addressing different retrieval and reading strategies. The first solution, BM25-HYBRIDER, establishes a baseline for OTT-QA. It retrieves 1 to 4 texts or tables from the corpus using the BM25 algorithm. These retrieved passages are then input to a reader, which selects the answer with the highest confidence interval.

The second solution, "Iterative-Retrieval + Cross-Block Reader," employs two iterative retriever variants. **Sparse** uses BM25 to retrieve 10 text passages and 10 table segments. For each text passage, it retrieves 5 table segments, and vice versa. **Dense** utilizes a DPR-like model for retrieval. It starts with 8 text or table passages and iteratively expands the search. Both variants feed the retrieved passages into the Cross-Block Reader, a fine-tuned ETC transformer model, to generate the answer.

The last solution introduced by the authors, the "Fusion Retriever + Cross-Block Reader", was shown to outperform the previous methods. This approach combines tables and text into blocks using BM25 and enhances queries with GPT-2. The Fusion Retriever is a dual-encoder setup, based on DPR, that retrieves k blocks based on their similarity to the question while maintaining the same Cross-Block Reader to generate the answer.

The CORE (Chain Of REasoning) QA system employs a retriever-reader architecture but includes two intermediary components, the Linker and the Chainer [14]. The retriever is built on DPR and is responsible for retrieving 100 tables from the corpus. Subsequently, a Linker system based on BERT models retrieves passages from the corpus based on the rows obtained from the retrieved tables. As the linker may provide an excessive amount of information for the reader to process, the Chainer's role is to select the top 50 chains (table row and text passage), which consist of a table row and a corresponding text passage. The Chainer accomplishes this task by employing a T0 neural network that, in zero-shot manner, assesses the probabilities of the model generating a question based on specific text passages and table rows. Finally, to answer the question, the system uses a FiD model, which relies on the information curated by the Chainer.

The Chain-of-Skills (COS) architecture [15] currently delivers the highest performance on OTT-QA. The authors introduced a (non-sequential) modularization approach that facilitates multi-task training across various open-domain QA datasets. The tasks tackled include single retrieval, expanded query retrieval, entity span proposal, entity linking, and reranking. In the inference phase, the system first identifies 100 tables through single retrieval. These tables are subsequently divided into rows, and the reranking process selects the top 200 rows. For each row, the system collects 10 passages from the expanded query retrieval and one passage from linked entities. The system then leverages the same Chainer used in CORE to select the top 100 passages, which are subsequently input into a fine-tuned FiE reader tailored for OTT-QA.

Another significant contribution is the "Multi-modal Retrieval of Tables and Texts Using Tri-encoder Models" [11], which extends DPR's capabilities by incorporating three small-BERT networks for encoding questions, texts, and tables individually.

2.4 Reinforcement Learning in Question Answering

One of the pioneering approaches to integrate Deep Reinforcement Learning (DRL) into QA is the Reinforced Ranker-Reader (R^3) architecture [21] introduced in 2017. This framework focuses on joint training of the retriever (referred to as the ranker) and the reader, using the REINFORCE algorithm for the ranker and a supervised approach for the reader. A key advantage of this approach is the unsupervised nature of retriever training, which gives rewards based solely on the similarity of the reader's output to the golden answer.

Similarly, MSCQA [22] employs DRL to train an action selector that decides among three components: invoking the reader, selecting additional passages via the retriever, and removing potentially incorrect answers from memory. In [3], a DRL agent is designed with actions for retriever, reader, and cleaner functions, in order to tackle a complex multi-hop QA dataset, where the retriever iteratively uses previously retrieved passages to find new relevant information.

Reinforcement Learning with Human Feedback (RLHF) is an approach that has gained traction in the development of natural language models. RLHF involves training models through a process that combines Reinforcement Learning, where models learn from their actions and consequences, with Human Feedback, which involves human reviewers providing feedback and rankings on model-generated content. One example is ChatGPT (or GPT3.5), a conversational agent built upon the foundation of GPT3. While the specifics of ChatGPT's training process are not fully disclosed, available information indicates that the RLHF training was executed through the PPO algorithm to enhance ChatGPT's alignment with users and minimizing issues related to misinformation, toxicity, and harmful sentiments.

3 Proposed Architecture

The multi-hop and multi-model nature of OTT-QA presents a significant challenge due to the sequential retrieval process it requires. To address these challenges within an explicitly sequential formulation, we propose a novel architecture. Our approach involves training a Deep Reinforcement Learning (DRL) agent that selects from a set of already trained modules in the existing literature.

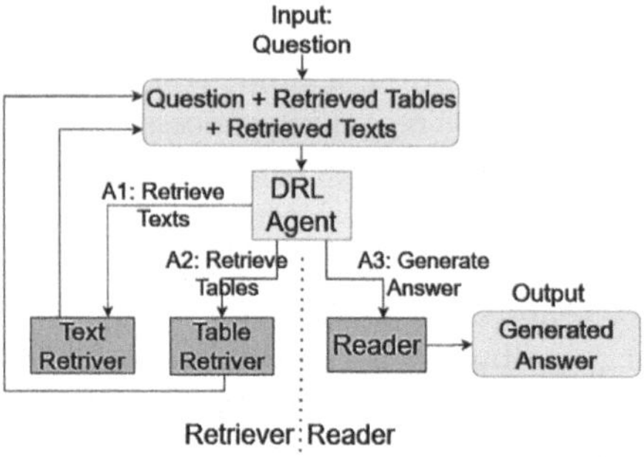

Fig. 2. Proposed architecture. At each time step, the agent selects one of three actions: Retrieve Texts, Retrieve Tables, or Generate the Answer, based on the question and the information gathered so far.

As done for MSCQA [22], we employ a DRL agent as an action selector to determine which pre-trained component should be activated. As depicted in Fig. 2, we have the following actions: Retrieve Texts, Retrieve Tables, and Generate the Answer.

Given that OTT-QA does not specify the correct path for information retrieval, including the sequence of passages to retrieve, we have adopted a Reinforcement Learning approach. We incorporate a delayed reward mechanism, which compares the predicted answers to the gold standard answers at the end of each episode.

One notable advantage of our proposed architecture is its flexibility. All components, including the reader and retrievers, can be easily replaced with newer and superior models as they become available for public use. Furthermore, the system can be enhanced by incorporating additional components, such as a graph retrieval model, or employing multiple readers optimized for specific scenarios. The RL agent learns when to use each component in different situations through experiential training.

3.1 Initial Concepts

Our approach draws inspiration from the Iterative-Retrieval + Cross-Block Reader architecture 5 described in Sect. 2.3, being also an iterative system. For a given question, the initial retrieval step involves searching for 10 passages, which can be either textual or tabular in nature. These passages are used to create 10 different blocks. Should the agent opt to perform an additional retrieval step, it concatenates the items within each block to retrieve four new passages, be they textual or tabular.

The agent may undergo one more retrieval step, leading to the selection of 4 new passages for each block. To maintain the quality of retrieved information and prevent excessive noise, we limit the maximum number of retrieval steps to three. This limitation is intended to avoid an accumulation of uncertainty, as additional passages introduce more information, making it more challenging for the retrieval process to distinguish between valuable information and noise.

Suppose the agent decides to generate the answer or has reached the maximum number of steps. In that case, the reader is presented with the question and all non-repeated passages (note that different blocks may contain the same passage). The reader utilizes the Reader Input to generate a predicted answer a^p, which is subsequently compared to the golden answer a^g. In cases where the reader is invoked before any passage has been retrieved, a preliminary Table Retrieval is run to ensure that a minimum of 10 documents are available for processing.

3.2 Reward

Since we lack prior knowledge of the correct sequence of actions for the agent, we cannot use supervised learning and must rely on the comparison between the expected (golden) answer and the agent's predicted answer to construct the reward function.

We started with the reward function R^3, which trains the retriever indirectly, by using the final answer. However, due to the complexity of the problem, the -1 penalty for a generated answer with no intersection with the golden answer seemed excessive, especially considering that the Iterative-Retrieval + Cross-Block Reader, our inspiration, achieves an F1-score of only 20.9 on the test set. Therefore, we adjusted the penalty to -0.5. The following equation represents the reward at the end of the episode:

$$R(a^g, a^{rc}) = \begin{cases} 2, \text{if } a^g = a^{rc}, \\ F_1(a^g, a^{rc}), \text{ else if } a^g \cap a^{rc} \neq \emptyset, \\ -0.5, \text{else}, \end{cases} \tag{1}$$

where the reward is assigned a value of 2 when the golden answer a^g precisely matches the predicted answer a^p. In cases of partial match, the reward is determined by the F1-score. Lastly, when no match occurs, the reward defaults to -0.5, as mentioned before. We also introduced a minor penalty of -0.02 for each retrieval action taken. This penalty encourages the agent to minimize retrieving unnecessary texts and tables.

3.3 Actions

As previously mentioned, our approach involves three distinct actions: **Retrieve Texts** ($A1$), where a retriever model searches for relevant textual information; **Retrieve Tables** ($A2$), where a retriever model searches for relevant tables; and **Generate Answer** ($A3$), where a reader model produces the answer using extracted textual passages and tables.

Next, we discuss our choices for the reader and retriever models. We opted not to retrain any models from the literature, limiting ourselves to those publicly available.

Retriever: The retriever module gathers information for the reader. We evaluated two types: BM25 [18] and Tri-encoder [11]. This approach allowed us to assess classical sparse retrieval and neural dense retrieval methods, akin to the Iterative-Retrieval + Cross-Block Reader architecture.

Reader: A critical component of any QA system, the reader generates answers. For OTT-QA, the publicly available Fusion-in-Encoder (FiE) model from the COS system [15] demonstrated superior performance. Notably, this model handles passages individually, making it versatile for handling multiple inputs. Following the authors' recommendations, we capped the number of input passages at 50, which yielded optimal results in our validation experiments.

3.4 State/Observation

The observation is the information that the agent uses for the decision-making process. In our case, this information comprises the question and the passages that have been retrieved up to that point. Rather than working directly with pure text, we first convert these elements into numerical data using embeddings.

We employ two distinct encoders to perform this transformation into embeddings. When the retriever is the Tri-encoder, we utilize its embeddings to represent the question, retrieved passages, and retrieved tables. However, in the case of BM25, we leverage the representations provided by a fine-tuned MPNET model trained for semantic search in QA[2]. Each embedding is a vector with length of 512 for Tri-encoder (same as small-BERT) and 768 for the MPNET model.

This information is then translated into a sequence of 11 vectors, with one vector for the question and one for each of the 10 blocks. For each block, we calculate the average of the representation values for each passage, considering the question in isolation, generating the embedding sequence.

3.5 Agent

In DRL applications, the agent's primary function is to process observations and generate corresponding actions. Our specific observations consist of a sequence comprising 11 vectors. In our exploration, we experimented with four distinct neural network architectures: MultiLayer Perceptron (MLP), Long Short-Term Memory (LSTM), Gated Recurrent Unit (GRU), and Transformer.

For the MLP, we flattened the entire sequence into a single vector, which meant adding zeros in the first step to account for the lack of retrieved passages at that point. In contrast, the other networks process this information as a sequence.

To train the agent, we explored two widely-used algorithms: Deep Q Network (DQN) [16] and Proximal Policy Optimization (PPO) [19]. Our implementation is based on the Stable-Baselines3 Python library, selected for its robustness and reliability. The training workflow of the agent is depicted in Fig. 3.

[2] The MPNET model can be at https://huggingface.co/sentence-transformers/multi-qa-mpnet-base-dot-v1.

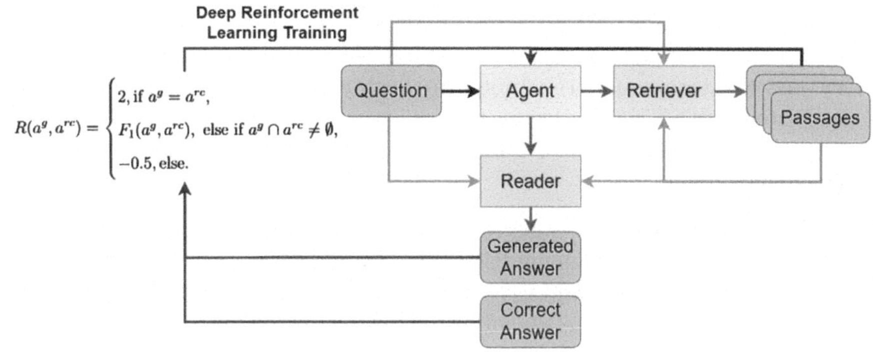

Fig. 3. Training workflow of the proposed architecture.

4 Experiments

4.1 Baselines

Before training the DRL agent, we explored baseline scenarios using our architecture. These baselines involved fixed sets of actions chosen independently of the question and retrieved information. Each conceivable combination of actions was systematically evaluated across a validation dataset. For example, one baseline involved the agent consistently choosing to retrieve texts (Action $A1$) twice and then invoking the reader to generate the answer. In total, we explored 14 different configurations of actions, encompassing two retriever types—BM25 and Tri-encoder—resulting in 28 baselines. The primary purpose of these baseline tests was to establish a benchmark for comparison, determining if the trained DRL agent could outperform or at least match the performance of fixed action paths without contextual understanding.

4.2 Training

The training of the DRL agent involved sampling a random question from the OTT-QA training set for each episode. We evaluated the impact of training both retriever solutions and compared their performance. For the Tri-encoder, we trained for a total of one million timesteps, a reasonable number considering the dataset size of 41,469 training questions. Each question could take up to four timesteps (between retriever and reader), allowing for revisiting the same question multiple times. Due to the higher inference time of BM25 (up to 20 times more than Tri-encoder), we opted to train for only 100 thousand timesteps.

We implemented the DQN and PPO algorithms, employing various neural network architectures, including MLP, LSTM, GRU, and transformer. Subsequent subsections provide detailed explanations of these implementations.

DQN: For training DQN, we made a few adjustments to the dedefault hyperparameters of Stable-Baselines3,[3] primarily to account for the number of training

[3] https://stable-baselines3.readthedocs.io.

steps. We set the buffer size to 500K (maximum number of transitions stored in the experience replay) and learning starts (number of steps taking random actions just to store some transitions in the buffer before training) to 50,000 for the system with Tri-encoder as the retriever. For BM25, we decreased the learning starts by a factor of 5, taking into account that the training has only 100,000 steps.

PPO: Similarly to DQN, we mostly relied on the default hyperparameters for training using the PPO implementation from Stable-Baselines3. We set the number of steps to run before training as 128, with a batch size of 32, and 60 as the number of epochs to optimize the surrogate loss.

As PPO is an Actor-Critic algorithm, it employs two different networks: the actor and the critic. Instead of using entirely separate networks, a common approach is to use a feature extractor to avoid redundant computations. This involves using shared layers between the networks for pre-processing the input, with these shared layers being trained jointly.

Networks: Below are the configurations for the feature extractors we used, maintaining consistency for both training algorithms. For PPO, we employed two hidden layers with 64 and 32 neurons for both the actor and critic networks after the feature extractor:

- **MLP:** for DQN, the MLP network is a simple network with three hidden layers having 512, 128, and 64 neurons. For PPO, the feature extractor is two-layered network with hidden sizes of 512 and 128.
- **LSTM:** we used 2 LSTM layers with a hidden size of 512 for the Tri-encoder and 768 for the BM25 architecture to match the size of the vectors of the embedding sequence E, with a dropout of 0.1. Additionally, there is one hidden layer with 128 nodes.
- **GRU:** similar to LSTM, but with the LSTM layers replaced by GRU layers.
- **Transformer:** for the PPO feature extractor, we opted for two encoder transformer layers with two attention heads, a feature dimension matching the input embedding size, and the same 128 hidden layer. For DQN, we used the same architecture with an extra 64 neurons last layer.

5 Results and Discussion

5.1 Baselines

The complete results of our baselines are provided in Table 1. For the BM25 retriever, it is evident that retrieving tables is less effective than retrieving texts. This imbalance results in the best sequence of actions always being the retrieval of texts and never tables, yielding an F1-score of 19.03. For the Tri-encoder, the results are more balanced. The optimal sequence of actions involves retrieving texts, followed by tables, and then retrieving texts again, resulting in a total F1-score of 8.24. It is notable, however, that the Tri-encoder performance when retrieving texts is considerably lower compared to BM25.

Table 1. Baseline results for the OTT-QA validation dataset using our framework with BM25 and Tri-encoder as the retriever. In this experiment, we assumed that the agent always takes the same sequence of actions, regardless of the question. *A*1 corresponds to retrieving texts, *A*2 is retrieving tables, and *A*3 calls the reader to generate the answer.

Retriever	Action 1	Action 2	Action 3	Action 4	EM (%)	F1-score (%)
BM25	*A*1	*A*3	-	-	12.92	16.80
BM25	*A*2	*A*3	-	-	2.17	4.20
BM25	*A*1	*A*1	*A*3	-	14.60	18.79
BM25	*A*1	*A*2	*A*3	-	13.15	16.88
BM25	*A*2	*A*1	*A*3	-	5.01	7.97
BM25	*A*2	*A*2	*A*3	-	2.12	4.07
BM25	*A*1	*A*1	*A*1	*A*3	**14.81**	**19.03**
BM25	*A*1	*A*1	*A*2	*A*3	14.23	18.21
BM25	*A*1	*A*2	*A*1	*A*3	13.32	17.36
BM25	*A*1	*A*2	*A*2	*A*3	7.59	10.52
BM25	*A*2	*A*1	*A*1	*A*3	5.15	8.08
BM25	*A*2	*A*1	*A*2	*A*3	3.61	6.17
BM25	*A*2	*A*2	*A*1	*A*3	3.21	5.45
BM25	*A*2	*A*2	*A*2	*A*3	2.08	4.15
Tri-encoder	*A*1	*A*3	-	-	3.52	6.32
Tri-encoder	*A*2	*A*3	-	-	2.94	4.41
Tri-encoder	*A*1	*A*1	*A*3	-	4.34	7.11
Tri-encoder	*A*1	*A*2	*A*3	-	4.83	7.54
Tri-encoder	*A*2	*A*1	*A*3	-	5.42	7.89
Tri-encoder	*A*2	*A*2	*A*3	-	2.66	4.40
Tri-encoder	*A*1	*A*1	*A*1	*A*3	4.65	7.35
Tri-encoder	*A*1	*A*1	*A*2	*A*3	4.79	7.56
Tri-encoder	*A*1	*A*2	*A*1	*A*3	**5.55**	**8.24**
Tri-encoder	*A*1	*A*2	*A*2	*A*3	3.66	5.81
Tri-encoder	*A*2	*A*1	*A*1	*A*3	5.51	8.06
Tri-encoder	*A*2	*A*1	*A*2	*A*3	4.29	6.45
Tri-encoder	*A*2	*A*2	*A*1	*A*3	4.07	6.24
Tri-encoder	*A*2	*A*2	*A*2	*A*3	2.89	4.52

5.2 Results for Deep Reinforcement Learning Agent

We present detailed results for the trained DRL agents:

BM25. The performance of trained algorithms using the BM25 retriever is detailed in Table 2. Among the various training algorithms, the PPO agent equipped with a transformer neural network demonstrated remarkable results. Surpassing other configurations, this agent achieved the same level of performance as the best baseline, which involved consistently selecting the *A*1 action regardless of the question or context.

Tri-encoder. Results by algorithms utilizing the Tri-encoder retriever are presented in Table 2. Notably, the best-performing agent emerged from the PPO algorithm, employing the MLP network, demonstrating a F1-score of 6.44%.

Table 2. Results for networks and training algorithms on the OTT-QA dataset.

Retriever	Algorithm	Network	EM (%)	F1-score (%)
BM25	DQN	MLP	11.70	15.44
Bm25	DQN	LSTM	12.33	16.42
BM25	DQN	GRU	13.23	17.16
BM25	DQN	Transformer	11.65	15.12
BM25	PPO	MLP	8.54	11.63
BM25	PPO	LSTM	6.78	9.85
BM25	PPO	GRU	11.25	14.90
BM25	**PPO**	**Transformer**	**14.81**	**19.03**
Tri-encoder	DQN	MLP	3.61	6.01
Tri-encoder	DQN	LSTM	3.52	5.68
Tri-encoder	DQN	GRU	3.79	6.13
Tri-encoder	DQN	Transformer	3.34	5.09
Tri-encoder	PPO	MLP	3.38	5.47
Tri-encoder	PPO	LSTM	3.75	5.53
Tri-encoder	PPO	GRU	2.94	4.93
Tri-encoder	PPO	Transformer	2.80	4.34

5.3 Discussion

Baseline results show a significant advantage for text retrieval over table search with BM25, even though some questions require information that is strictly only in tables. This may be due to BM25's difficulty handling structured tables, where column presence and data types are crucial. Despite its popularity in OTT-QA solutions, the Tri-encoder achieved F1-scores below 10%, potentially due to the lack of multi-hop training and reliance on text retrieval data from other datasets, as OTT-QA lacks this information.

The DRL agents showed performance comparable to or slightly below the best baseline, particularly for BM25, often favoring text retrieval, which aligns with the optimal baseline strategy. Table retrieval proved noisy, especially in iterative search, as tables often contain excessive data with relevant information typically in a single row. This issue may be mitigated by breaking tables into rows or smaller content sets, as suggested by other researchers [14].

The PPO algorithm generally showed slightly lower performance, with one exception: the PPO agent with a Transformer network for BM25, which achieved the best performance by converging to the trivial solution of only picking the best action on average (retrieving texts). The choice of neural network architecture had minimal impact, indicating the need for further experiments.

Our best-performing DRL agent achieved an F1-score of 19.03 using BM25, approaching the 20.7 score of the Iterative-Retrieval (sparse) + Cross-Block solution, which inspired our approach. However, our dense approach reached an F1-score of 8.24, compared to their 18.5, suggesting that the Tri-encoder setting requires further refinement.

Our solution has a few limitations. The exploratory nature of Reinforcement Learning (RL) and the stochastic behavior of neural networks benefit from averaged multiple test runs. However, due to time constraints—each training execution took over a week—we were not able to perform multiple runs. Additionally, we only used 100k training steps with the BM25 retriever. This number of steps is potentially insufficient given the dataset's 41,469 training questions, each potentially requiring up to three steps. Finally, on the Tri-encoder's performance, it is important to stress that this model was not trained for iterative retrieval, likely affecting its efficacy.

6 Conclusion and Future Work

In this work, we introduced a novel system designed to tackle multi-hop questions across texts and tables. Our approach relies on a DRL decision-maker system that iteratively selects among three distinct modules: Text Retriever, Table Retriever and Reader. Our best system achieved an F1-score of 19.03, closely competing with the non-DRL Iterative-Retrieval (sparse) + Cross-Block system, which scored 20.7. We emphasize that our goal here is to explore sequential strategies; other (non-sequential) strategies have been explored with success—for instance, COS, which achieved an F1-score of 63.2 on the dev set. Despite the comparatively lower performance, we believe there is significant value in solutions that explore sequential decisions as they have potential for modular improvement by refining various system modules.

Solutions that combine sequential decisions with non-sequential modules seem to be the most profitable avenue to explore in future work; here we have explored the performance boundary of sequential decisions using reinforcement learning. Future work should also include implementing the Joint-Reranking approach [13], which uses a BERT-based architecture to score and select top passages, enhancing the system's ability to clear excess passages and tables from memory. We are also exploring alternative methods for generating observations for the agent. Currently, we average the embeddings of each block, but we are considering encoding all distinct retrieved passages as a sequence. Additionally, integrating our DRL agent with the COS architecture, which excels in OTT-QA by employing diverse retrieval strategies, could enhance performance. The DRL agent would determine optimal retrieval paths based on the question and retrieved passages.

Acknowledgments. This research has been carried out with support by *Itaú Unibanco S.A.* through the scholarship program *Programa de Bolsas Itaú* (PBI); it is also supported in part by the Coordenação de Aperfeiçoamento de Pessoal de Nível Superior (CAPES), Finance Code 001, Brazil. The authors would like to thank the Center for Artificial Intelligence (C4AI-USP), with support by the São Paulo Research Foundation (FAPESP) under grant number 2019/07665-4 and by the IBM Corporation. Paulo Pirozelli is supported by the FAPESP grant 2019/26762-0.

Any opinions, findings, and conclusions expressed in this manuscript are those of the authors and do not necessarily reflect the views, official policy or position of the Itaú-Unibanco, FAPESP, IBM, and CAPES.

References

1. Berant, J., Chou, A.K., Frostig, R., Liang, P.: Semantic parsing on Freebase from question-answer pairs. In: Conference on Empirical Methods in Natural Language Processing (2013)
2. Cação, F.N., José, M.M., Oliveira, A.S., Spindola, S., Costa, A.H.R., Cozman, F.G.: DEEPAGÉ: answering questions in Portuguese about the Brazilian environment. In: Britto, A., Valdivia Delgado, K. (eds.) Intelligent Systems, pp. 419–433. Springer, Cham (2021)
3. Cação, F.N., Reali Costa, A.H.: A deep reinforcement learning approach to complex open-domain question answering. Master's thesis, Univ. de São Paulo (2023)
4. Chen, D., Fisch, A., Weston, J., Bordes, A.: Reading Wikipedia to answer open-domain questions. In: Annual Meeting of the Association for Computational Linguistics, pp. 1870–1879. Association for Computational Linguistics, Vancouver, Canada (2017). https://doi.org/10.18653/v1/P17-1171
5. Chen, W., Chang, M.W., Schlinger, E., Wang, W.Y., Cohen, W.W.: Open question answering over tables and text. In: International Conference on Learning Representations (2021)
6. Chen, W., Zha, H., Chen, Z., Xiong, W., Wang, H., Wang, W.Y.: HybridQA: a dataset of multi-hop question answering over tabular and textual data. In: Findings of the Association for Computational Linguistics, pp. 1026–1036. Association for Computational Linguistics (2020). https://doi.org/10.18653/v1/2020.findings-emnlp.91
7. Guu, K., Lee, K., Tung, Z., Pasupat, P., Chang, M.W.: REALM: retrieval-augmented language model pre-training. In: International Conference on Machine Learning, PartF168147-6, pp. 3887–3896 (2020)
8. Izacard, G., Grave, E.: Leveraging passage retrieval with generative models for open domain question answering. In: Conference of the European Chapter of the Association for Computational Linguistics, Online, pp. 874–880. Association for Computational Linguistics (2021). https://doi.org/10.18653/v1/2021.eacl-main.74
9. Karpukhin, V., et al.: Dense passage retrieval for open-domain question answering. In: Conference on Empirical Methods in Natural Language Processing, pp. 6769–6781. Association for Computational Linguistics (2020). https://doi.org/10.18653/v1/2020.emnlp-main.550
10. Kedia, A., Zaidi, M.A., Lee, H.: FiE: building a global probability space by leveraging early fusion in encoder for open-domain question answering. In: Conference on Empirical Methods in Natural Language Processing, Abu Dhabi, United Arab Emirates, pp. 4246–4260. Association for Computational Linguistics (2022)

11. Kostić, B., Risch, J., Möller, T.: Multi-modal retrieval of tables and texts using tri-encoder models. In: Fisch, A., et al. (eds.) 3rd Workshop on Machine Reading for Question Answering, Punta Cana, Dominican Republic, November 2021, pp. 82–91. Association for Computational Linguistics (2021). https://doi.org/10.18653/v1/2021.mrqa-1.8
12. Lewis, P., et al.: Retrieval-augmented generation for knowledge-intensive NLP tasks. In: Advances in Neural Information Processing Systems, vol. 33, pp. 9459–9474. Curran Associates, Inc. (2020)
13. Li, A.H., Ng, P., Xu, P., Zhu, H., Wang, Z., Xiang, B.: Dual reader-parser on hybrid textual and tabular evidence for open domain question answering. In: Annual Meeting of the Association for Computational Linguistics and the Joint Conference on Natural Language Processing (Volume 1), pp. 4078–4088. Association for Computational Linguistics (2021). https://doi.org/10.18653/v1/2021.acl-long.315
14. Ma, K., Cheng, H., Liu, X., Nyberg, E., Gao, J.: Open-domain question answering via chain of reasoning over heterogeneous knowledge. In: Conference on Empirical Methods in Natural Language Processing, Abu Dhabi, United Arab Emirates, pp. 5360–5374. Association for Computational Linguistics (2022)
15. Ma, K., Cheng, H., Zhang, Y., Liu, X., Nyberg, E., Gao, J.: Chain-of-skills: a configurable model for open-domain question answering. In: Rogers, A., Boyd-Graber, J., Okazaki, N. (eds.) Annual Meeting of the Association for Computational Linguistics (Volume 1: Long Papers), Toronto, Canada, pp. 1599–1618. Association for Computational Linguistics (2023). https://doi.org/10.18653/v1/2023.acl-long.89
16. Mnih, V., et al.: Playing Atari with deep reinforcement learning. CoRR **abs/1312.5602** (2013)
17. Rajpurkar, P., Zhang, J., Lopyrev, K., Liang, P.: SQuAD: 100,000+ questions for machine comprehension of text. In: Conference on Empirical Methods in Natural Language Processing (ii), pp. 2383–2392 (2016). https://doi.org/10.18653/v1/d16-1264
18. Robertson, S., Zaragoza, H.: The probabilistic relevance framework: BM25 and beyond, vol. 3 (2009). https://doi.org/10.1561/1500000019
19. Schulman, J., Wolski, F., Dhariwal, P., Radford, A., Klimov, O.: Proximal policy optimization algorithms. CoRR **abs/1707.06347** (2017)
20. Wang, P., Shi, T., Reddy, C.K.: Text-to-SQL generation for question answering on electronic medical records. In: Web Conference, pp. 350–361. Association for Computing Machinery, New York (2020). https://doi.org/10.1145/3366423.3380120
21. Wang, S., et al.: R3: reinforced ranker-reader for open-domain question answering. In: AAAI Conference on Artificial Intelligence. AAAI Press (2018)
22. Wang, Y., Jin, H.: A deep reinforcement learning based multi-step coarse to fine question answering (MSCQA) system. AAAI Conf. Artif. Intell. **33**(01), 7224–7232 (2019). https://doi.org/10.1609/aaai.v33i01.33017224

Reinforcement Learning with Utility-Based Semantic for Goals

Christian Delgado Polar[1,2](✉), Karina Valdivia Delgado[1], and Valdinei Freire[1]

[1] University of São Paulo, Cidade Universitária, São Paulo, SP, Brazil
cjdelgadopolar@ime.usp.br, kvd@usp.br, valdinei.freire@usp.br
[2] School of Computer Science, San Pablo Catholic University, Arequipa, Peru
cjdelgado@ucsp.edu.pe

Abstract. Stochastic shortest path problems (SSPs) are Markov decision processes with goal states and the problem is to find policies to achieve the goal with the lowest possible expected cost. Something common in this type of problem is the existence of states from which it is not possible to reach the goal, these states are called dead-end. In this case, it is important to have methods that consider the importance, not only of cost, but also of the probability of achieving the goal. In Reinforcement Learning (RL), the problem of making this trade-off between cost to goal and probability to goal has been little studied and the common strategy to deal with SSPs with dead-ends is using discounts. In some works, penalties are used on exploration in SSPs with dead-ends. However, using discounts and penalties can lead to errors in finding a policy with a desired trade-off. The GUBS criterion (*Goals with Utility-Based Semantic*) considers this type of trade-off without using discounts or penalties, has good semantics based on the expected utility theory and has been used to solve SSPs with dead-end states in the planning area. Thus, in this work, the GUBS criterion is used to propose the first two algorithms for RL to do a trade-off between probability to goal and cost to goal, without using discounts or penalties: the Q-learning-GUBS and Q-learning-eGUBS+Cmax. Theoretical and experimental results show that the proposed algorithms make this trade-off according to the configuration of the GUBS' parameters (Code enabled on https://github.com/QlearningGubs/code).

Keywords: Reinforcement Learning · dead-end · Q-learning · GUBS

1 Introduction

Stochastic shortest path problems (SSPs) are Markov decision processes (MDPs) with a set of goal states. In these problems, the objective is to find a policy with the lowest expected cumulative cost to achieve these goals. In many cases, these problems have states from which it is not possible to reach the goal, these states are called dead-ends. In problems with dead-ends the criterion of the lowest expected cost becomes ill-defined when there is no a proper policy (one that reaches the goal with probability 1) [6].

Some works try to solve SSPs with dead-end, some of them seek to find the policy with the maximum probability of reaching the goal (MAXPROB) [7] or give greater preference to probability and consider cost as a secondary aspect (lexicographic criteria)

A. Paes and F. A. N. Verri (Eds.): BRACIS 2024, LNAI 15413, pp. 354–369, 2025.
https://doi.org/10.1007/978-3-031-79032-4_25

[6,14]. However, it is not always recommended to find policies with the highest probability of achieving the goal as they may have high costs. Sometimes it is better to make a trade-off between probability and cost, for example, decreasing the probability a little to get a large decrease in cost or increasing the cost a little to get a large increase in probability, this characteristic is called *infinite-infinitesimal trade-off* [1]. An example of this type of infinite-infinitesimal trade-off is illustrated in the River problem.

The River problem [4] considers a $N_x \times N_y$ grid. The agent must cross a river from one river bank starting from an initial position s_0 to a destination position on the other bank. There is a bridge at position $y = N_y$. The agent's actions are to move in the four cardinal directions. The actions on the river's bank ($x = 1$ or $x = N_x$) and the bridge are deterministic. The actions are probabilistic within the river. In the river, there is a probability P of following down the river, with the risk of reaching the waterfall located at $y = 1$, where, the agent becomes trapped or killed (dead-end).

The optimal solution for the lexicographical criteria is to walk toward the bridge and traverse it since this policy has a probability 1 to reach the goal, even if the cost is very high. In this problem, for example, a solution with a probability 0.9999 should be accepted if we get a large decrease in cost.

In Reinforcement Learning (RL), most of the theoretical works focus on finite-horizon and infinite-horizon or on loop-free SSP [12]. Also, the problem of making trade-off has been little studied for SSPs with dead-ends. Additionally, a common strategy to follow with SSPs with dead-end states in RL is using discounts. Geibel and Wysotzki [5] consider an MDP with goal and error states (undesirable or dangerous states). The proposed RL solution is based on weighting the cost and the probability of entering error states. This work uses discount factor. Park [8] considers an MDP with goal and fail states. The proposed Q-learning-like algorithm records past failure experiences and uses them to guide the exploration, this work [8] also uses a discount factor. Using a discount factor could have an undesirable effect of biasing importance towards short-term behavior and thus diminishing the incentive to eventually achieve the goal state [1,12]. In some works [12], penalties are used in the exploration problem for SSPs with dead-ends. However, using penalties can lead to errors in finding a policy with a desired trade-off too, as was formally demonstrated by [1].

The GUBS criterion (*Goals with Utility-Based Semantic*) [4] was proposed to make a correct *infinite-infinitesimal trade-off* in planning problems. This criterion allows us to find solutions for SSPs with dead-end, with a parameterization with meaning based on the expected utility theory. Thus, in this work, the GUBS criterion is used to propose the first two algorithms for RL to do a trade-off between probability to goal and cost to goal, without using discounts or penalties, that are: the Q-learning-GUBS and the Q-learning-eGUBS+Cmax. Additionally, to the best of our knowledge, the proposed algorithms are the first to use the expected utility theory to make this trade-off in RL, which allows more precise value criteria to be designed.

2 Problem Formalism and Background

An *Stochastic Shortest Path* (SSP) is defined by a tuple $M = \langle S, A, P, c, \mathcal{G} \rangle$, where: S is a discrete and finite set of completely observed states that model the world, A is a finite set of actions, $P : S \times A \times S \rightarrow [0, 1]$ is a probabilistic transition function, where, $P(s, a, s') = Pr(s_{t+1} = s'|s_t = s, a_t = a)$ is the probability of transition from a state $s \in S$ to a state $s' \in S$ after executing an action $a \in A$, $c : S \times A \rightarrow \mathcal{R}$ is a cost function for each action $a \in A$ in a state $s \in S$, $\mathcal{G} \subset S$ is the set of goal states, i.e. $g \in \mathcal{G}$, such that $c(g, a) = 0$, $P(g, a, g) = 1, \forall a \in A$ and $c(s, a) > 0, \forall s \in S \setminus \mathcal{G}$.

In a state $s_t \in S$ of the environment at time t, the agent executes an action $a_t \in A$ that has a cost $c_t = c(s_t, a_t)$ and probabilistic effects generating a future state s_{t+1} as a function of $P(s_t, a_t, s_{t+1})$. Every interaction that occurs can be summarized in a history of length T represented by $h_T = \{s_1, a_1, c_1, s_2, a_2, c_2, s_3, ..., s_{T-1}, a_{T-1}, c_{T-1}, s_T\}$. The set of all historical possibilities is $\mathcal{H} = (S \times A \times \mathcal{R}_{>0})^* \times S$.

In an SSP, the process follows a history h_T that ends in a goal state $g \in \mathcal{G}$. The solution for an SSP is a *policy* π that can be Stationary ($\pi : S \rightarrow A$) where, each state s is mapped on an action a or not Stationary ($\pi : \mathcal{H} \rightarrow A$) where, a historical h_t is mapped at time t in an action a_t. This policy must be optimal π^* under some criteria. A policy is called *proper* if the process achieves a goal state with probability 1 when it is followed, formally, $\lim_{t \rightarrow \infty} Pr(s_t \in \mathcal{G}| \pi, s_0) = 1$ where, s_0 is the initial state of the process. The optimal value function V^* for an SSP is defined by the Bellman equation $V^*(s) = \min_{a \in A}\{c(s, a) + \sum_{s' \in S} P(s, a, s')V^*(s')\}$. The optimal policy $\pi^*(s)$ for state s is $\pi^*(s) = \arg\min_{a \in A}\{c(s, a) + \sum_{s' \in S} P(s, a, s')V^*(s')\}$.

A state s is called dead-end state when the probability of reaching a goal state from s is 0. Formally, $\lim_{t \rightarrow \infty} Pr(s_t \in \mathcal{G}|\pi, s_0 \in \mathcal{D}) = 0$. $\mathcal{D} \subset S$ is the set of dead-end states. Conventional SSPs do not consider dead-end.

2.1 Criteria for SSPs with Dead-Ends

In environments with dead-end, the agent must decide between finding a policy with a low expected total cost $\overline{C}_G(s)$ from an initial state s until the goal state or a policy with the maximum probability of achieving the goal $P_G(s)$. Several criteria have been proposed to solve SSPs with dead-end some of which prioritize the probability $P_G(s)$ [7,13–15]. Instead of prioritizing probability, an alternative is to make a trade-off between cost to goal or probability to goal. The criterion *fSSPUDE* [6] extends the original SSP with an artificial action q and a penalty D. When q is used, the process transitions to a goal state and the agent pays the cost D, therefore $P(s, q, g) = 1$, $\forall s \in S$ and for a goal state g; and $c(s, q) = D$, $\forall s \in S$. Then, the objective is to find a policy that minimizes the expected accumulated cost $\overline{C}_G(s) = \lim_{T \rightarrow \infty} E[\sum_{t=0}^{T-1} c(s_t, a_t)|\pi, s_0]$. Thus, *fSSPUDE* can give more or less importance to avoid dead-end by choosing an appropriate D. The *discounted cost criterion* uses the discount factor to allow policies with a probability of reaching the goal less than 1, to converge [13]. The objective is to minimize the expected accumulated cost $\overline{C}_G(s) = \lim_{T \rightarrow \infty} E[\sum_{t=0}^{T-1} \gamma^t c(s_t, a_t)|\pi, s_0]$.

The penalty D used in *fSSPUDE* and the discount factor used in *discounted cost criterion* depend on the SSP. Because it is necessary to know the structure of the SSP, both values are difficult to determine to make the desired trade-off between probability and cost [1]. Besides, these criteria do not have the property $\alpha-$*strong probability-to-goal* for $\alpha > 0$ [1].

Definition 1 ($\alpha-$***strong probability-to-goal*** [1]). *For* $0 \leq \alpha \leq 1$, *a criterion has* $\alpha-$*strong probability-to-goal if for every SSP and for any pair of policies* $\pi, \pi' \in \Pi$, *the following condition is true:* $\pi \succ \pi' \implies \frac{P_G^{\pi}(s_0)}{P_G^{\pi'}(s_0)} \geq \alpha$.

Where $\pi \succ \pi'$ means that the decision-maker prefers π to π'.

2.2 Goals with Utility-Based Semantic-GUBS

A criterion that solves SSPs with dead-end with a clear trade-off between cost and probability of achieving the goal is the GUBS [4] criterion. GUBS is based on utility theory and evaluates policies based on:

$$\begin{aligned} V^{\pi}(s) &= \lim_{T \to \infty} E[U(C_T, \beta_T) | \pi, s_0 = s], \\ U(C_T, \beta_T) &= u(C_T) + K_g \beta_T, \end{aligned} \tag{1}$$

where, $u(.)$ is a utility function, $C_T = \sum_{t=0}^{T-1} c(s_t, a_t)$ is the accumulated cost in time T, K_g is the reward for reaching a goal state, and β_T is 1 if a goal is reached.

Unlike the *discounted cost criterion* and *fSSPUDE*, in which the choice of penalty and discount factor depend on the SSP, the value K_g do not depend on the SSP. Additionally, GUBS has the $\alpha-$*strong probability-to-goal* property [1]. Some planning algorithms have been proposed to calculate the GUBS criterion, they are: *GUBS-VI* [4], *eGUBS-VI* [3] and *eGUBS-AO** [1]. The *GUBS-VI* algorithm, which is based on the *Value Iteration* (VI) algorithm solves GUBS for discrete costs and considers as a parameter the maximum accumulated cost C_{max} for every state, including dead-end states, which will have a cost of C_{max}. As the algorithm needs to calculate the utility based on the accumulated cost, the policy found is non-stationary. To be able to use VI, extended states must be used with the cost accumulated so far. The formulation with discrete cost and extended states for GUBS is presented next.

Definition 2 (Discrete Cost [4]**).** *An SSP with dead-end can be restated into an MDP with bounded discrete costs* $M = \langle X, A, P', c', R'_g \rangle$, *where,* $X = S \times \mathbb{N}$ *is an extended state space, which are the states with the accumulated cost so far,* $P'(x, a, x') = P(s, a, s')$ *where,* $x = (s, c)$ *and* $x' = (s', c + c(s, a))$, $x, x' \in X$ *and c is the accumulated cost,* $c'(x, a) = u(c + c(s, a)) - u(c)$[1] *for a utility function* $u(.)$, *and* $R'_g = K_g$ *is the terminal reward for the goal states with value* $K_g \in \mathbb{R}$.

The *GUBS-VI* algorithm uses Definition 2 to calculate the Q-value:

$$Q(s, c, a) \leftarrow c'(s, a) + \sum_{s' \in S \cup \mathcal{G}} P(s, a, s') V(s', c + c(s, a)). \tag{2}$$

A special case of the GUBS criterion is the *Exponential GUBS* (eGUBS) criterion. In this case, the utility function is an exponential function used in risk sensitive SSPs (*Risk Sensitive SSP* - RS-SSP) [9], in which the risk factor λ, that models the agent's risk attitude, is negative. The eGUBS is defined in terms of the lexicographic criterion.

[1] Notice that c' is an utility over costs and the optimal policy must maximize the utility.

Definition 3 (Risk-Sensitive Lexicographic criterion [1,3]**).** *The Risk-Sensitive Lexicographic criterion considers:*

- *If* $P_G^{\pi}(s) > P_G^{\pi'}(s)$*, then* $\pi \succ \pi'$*; or*
- *If* $P_G^{\pi}(s) = P_G^{\pi'}(s)$ *and* $V_{\lambda}^{\pi} > V_{\lambda}^{\pi'}$*, then* $\pi \succ \pi'$*.*

The *eGUBS-VI* algorithm [3] solves the eGUBS criterion exactly. *eGUBS-VI* calls the *Risk-Sensitive-Lexicographic-VI* algorithm [3] which receives an SSP and a λ and calculates the optimal policy π_l^*, the optimal value V_λ and the maximum probability for the goal P_G for the lexicographic criterion described in Definition 3. We will use *Risk-Sensitive-Lexicographic-VI* as part of one of the algorithms proposed in this work.

The *eGUBS-AO** algorithm optimizes the search for the solution considering a finite, acyclic hypergraph, whose nodes are augmented states (s, c) and hyperedges leading (s, c) to its successor state s', c', where s' is the successor state in the original MDP and $c' = c + c(s, a)$ for the action performed a. *eGUBS-AO** solves eGUBS by means of a heuristic search with a variant of the *AO** algorithm.

2.3 Reinforcement Learning

In RL, it is common to divide the process into episodes (represented by e). An episode can end in several ways, for example, when a goal state, a dead-end or a pre-defined time are reached. The Q-Learning algorithm [16] is a well-known algorithm in RL. The Q-value for all $s_t \in S$ in time t when performing action $a_t \in A$ in time t and following the optimal policy π^* is:

$$\begin{aligned} Q(s_t, a_t) &\leftarrow Q(s_t, a_t) + \alpha_t \delta, \\ \delta &= c(s_t, a_t) + \gamma \min_{a' \in A} Q(s_{t+1}, a') - Q(s_t, a_t), \end{aligned} \tag{3}$$

where, α_t is the learning rate ($0 \leq \alpha_t \leq 1$) that determines the learning speed, γ is the discount factor on future values and s_{t+1} is the next state after executing action a_t.

Theorem 1. *(Convergence of the Q-learning* [16]*) Let the costs in each time t be* $c_t \leq C$*, learning rates* $0 \leq \alpha_t < 1$*, and* $\sum_{t=1}^{\infty} \alpha_t(s, a) = \infty, \sum_{t=1}^{\infty}(\alpha_t(s, a))^2 < \infty$ *then* $Q_t(s, a) \rightarrow Q^*(s, a)$ *as* $t \rightarrow \infty$*, for every state s and action a, with probability* 1.

The algorithm proposed by Geibel and Wysotzki [5] is one of the few methods for RL designed specifically to perform a trade-off between cost and probability of reaching an error state. Although this algorithm uses the concept of probability of reaching an error state, this concept is closely related to the concept of probability to reaching the goal used in this work.

Let Φ be the set of error states. The probability of reaching an error state (also called risk) from an initial state s is defined as $\rho^{\pi}(s) = E\left[\sum_{t=0}^{\infty} \bar{\gamma}^t \bar{r}_t | s_0 = s\right]$, where $\bar{\gamma}^t$ is the discount factor for the risk at time t and $\bar{r}_{s,a}(s') = \begin{cases} 1, & \text{If } s \in \Phi \text{ and } s' = \eta \\ 0, & else \end{cases}$, where η is a fictitious absorbing state where the agent transits after reaching an error state. With these concepts, a new value for the risk is defined as $\bar{Q}^{\pi}(s, a) = E\left[\bar{r}_0 + \bar{\gamma}\rho^{\pi}(s')\right]$, where $s_0 = s, a_0 = a$ and s' is the next state. Then a new general value is defined as $Q_{\xi}^{\pi}(s, a) = \xi Q^{\pi}(s, a) - \bar{Q}^{\pi}(s, a)$. The trade-off between the probability of reaching an error state and the cost to goal is determined by varying the weight ξ.

3 RL Under the GUBS Criterion

As shown in the previous section, the GUBS criterion is the only known criterion that guarantees α-strong probability-to-goal priority. In this work, RL algorithms are introduced to solve the GUBS criterion. Some challenges to face in the development of RL algorithms that use the GUBS criterion are: (1) use extended states with cost since GUBS finds non-stationary policies, something that is not the default behavior of RL algorithms; (2) define the type of the initial and final policies used in the algorithms; (3) the equations of the *GUBS-VI* algorithm [4] must be adapted based on temporal difference without using the transition function a priori; and (4) define or compute the maximum accumulated cost C_{max} that is used by the GUBS algorithms. The algorithms introduced are: *Q-learning-GUBS* and *Q-learning-eGUBS+*C_{max}. These algorithms do not need to know transition probabilities or costs and do not use discounts or penalties to find the solution considering the GUBS criterion. They are inspired by two well-known algorithms based on temporal difference, that are: *Q-Learning* and *Dyna-Q* (a model-based algorithm).

3.1 Q-Learning-GUBS

The *Q-learning-GUBS* algorithm is based on the *Q-learning* [16] and *GUBS-VI* [4] algorithms. Like *GUBS-VI*, *Q-learning-GUBS* solves GUBS for discrete costs, can use several utility functions (for example the exponential function and others that hold the properties defined in [1]), and is an approximate algorithm that has as a parameter the maximum accumulated cost C_{max} for every state.

Based on Eqs. 3 and 2, the following is the modified Bellman equation to solve the discrete cost problem (Definition 2).

$$\begin{aligned} Q(x_t, a_t) &\leftarrow Q(x_t, a_t) + \alpha_t \delta, \\ \delta &\leftarrow c'_{t+1}(x_t, a_t) + \max_{a' \in \mathcal{A}} Q(x_{t+1}, a') - Q(x_t, a_t), \end{aligned} \tag{4}$$

where, the Q-value is calculated for each extended state $x_t = (s_t, c_t)$ considering the utility c'_{t+1} of Definition 2. Notice that Eq. 4 does not consider a discount factor. Furthermore, *Q-learning-GUBS* uses a random policy as an initial policy and calculates a non-stationary final policy.

The *Q-learning-GUBS* algorithm (Algorithm 1) first initializes the Q table considering as value the differences between the minimum utility and the utility for each accumulated cost c for states that are not goal states. For the goal states, the algorithm sets the Q-value as K_g (line 1). Then, the algorithm iterates over a set of episodes in which it computes α_e and ϵ_e in line 4 and gets the initial extended state x_0 in line 5. At each step, an action a is chosen for the current state s following an ϵ_e-greedy policy, it acts on the environment, observes the cost of the action $c(s, a)$ and the next state s' (lines 11 and 12). For goal states $s \in \mathcal{G}$, $c(s, a) = 0$ (line 14). Then, with this data, the Q-value is calculated (line 18) for each discrete accumulated cost $c'' \in \{0, ..., C_{max}\}$ based on Eq. 4. The algorithm moves to another episode (i) when a goal state is reached, (ii) when the accumulated cost is greater than or equal to the value C_{max} or (iii) when a dead-end is found (line 9). Finally, the algorithm calculates the GUBS policy based on the Q table, choosing the actions with the highest value for each extended state (line 25).

***Q-learning-GUBS* convergence.** The algorithm finds an approximate policy for the problem regardless of the utility function used. The proof of convergence is based on the following definition and theorems.

Definition 4 (SSPs with discrete cost and maximum accumulated cost C_{max}). *The SSP of Definition 2 can be restated into a new SSP with the same elements but with a finite set of states $X' \subseteq X$ where, $X' = \{x|x = (s, c), c \leq C_{max}\}$ and c is the accumulated cost.*

Theorem 2 (*Q-learning-GUBS* convergence). *Let M be the SSP of Definition 4 where, in the algorithm, $0 \leq \alpha_t < 1$, $\sum_{i=1}^{n} \alpha_t = \infty$ and $\sum_{i=1}^{n} \alpha_t^2 < \infty$, and as the number of episodes $n \rightarrow \infty$, Q-learning-GUBS converges to the optimal value $Q'^*(x, a)$ for M.*

Proof. Since the SSP of Definition 4 has a finite number of states, it is possible to use the same proof of convergence as the *Q-Learning* algorithm (Theorem 1). Thus, the Algorithm 1 converges to the optimal value $Q'^*(x, a)$ for the SSP of the Definition 4.

Theorem 3 (The*Q-learning-GUBS* algorithm is approximate). *Let M be the SSP of Definition 2 where, $0 \leq \alpha_t < 1$, $\sum_{t=1}^{n} \alpha_t = \infty$ and $\sum_{t=1}^{n} \alpha_t^2 < \infty$, and $n \rightarrow \infty$ is the number of episodes, so:*

1. *Q-learning-GUBS converges to the value $Q'^*(x, a)$ where, $Q^*(x, a) - Q'^*(x, a) \geq 0$ for every extended state x and action a, and $Q^*(x, a)$ is the optimal value for the problem in Definition 2;*
2. $\lim_{C_{max} \rightarrow \infty} Q'^*(x, a) = Q^*(x, a)$.

Proof. The *Q-learning-GUBS* algorithm restates the original SSP M (Definition 2) into an SSP M' with maximum cost C_{max} (Definition 4). Since M' has a set of states $X' \subseteq X$, then $Q^*(x, a) - Q'^*(x, a) \geq 0$. Let $Q^{\pi*}$ be the optimal policy value of the original SSP in the SSP of Definition 4, then $Q^*(x, a) \geq Q'^*(x, a) \geq Q^{\pi*}(x, a)$. The difference between these values is given by the traces that do not reach the goal before the cost C_{max}. The greater the value C_{max}, the lower the probability of having these traces, therefore $Q^*(x, a) \rightarrow Q^{\pi*}(x, a)$ and $Q^{\pi*}(x, a) \rightarrow Q'^*(x, a)$. Thus $\lim_{C_{max} \rightarrow \infty} Q'^*(x, a) = Q^*(x, a)$.

3.2 Q-Learning-eGUBS+C_{max}

Q-learning-eGUBS+C_{max} is based on *GUBS-VI*, *eGUBS-VI*, *Q-learning* and *Dyna-Q* [11] algorithms. As *eGUBS-VI*, *Q-learning-eGUBS+C_{max}* computes a lexicographic policy as an initial policy and calculates a non-stationary final policy. Although *Q-learning-eGUBS+C_{max}* is based on the *GUBS-VI* algorithm that solves the GUBS criterion, it calculates the lexicographic policy with the *Risk-Sensitive Lexicographic-VI* algorithm ([3]) which is based on the exponential utility function. Thus, *Q-learning-eGUBS+C_{max}* solves the eGUBS criterion.

Algorithm 1 Q-LEARNING-GUBS

Input: SSP $\mathcal{M}$, maximum accumulated cost C_{max}, utility function u, utility for the goal K_g, number of episodes nepisodes, α_{min} and α_{max} to compute the learning rate α_e, ϵ_{min} and ϵ_{max} to compute ϵ_e.
Output: π^*

Initialize: $Q(s,c,a) \leftarrow u(\infty) - u(c), \forall s \in S, a \in A, c \in \{0,1,...,C_{max} + max_{s,a}c(s,a)\}$; $Q(s,c,a) \leftarrow K_g, \forall s \in \mathcal{G}, c \in \{0,1,...,C_{max} + max_{s,a}c(s,a)\}$.
$e \leftarrow 0$
while $e < nepisodes$ **do**
  compute α_e and ϵ_e
  get initial state s and cost c, $x_0 = (s,c)$
  $endEpisode \leftarrow False$
  while not $endEpisode$ **do**
    if $c \geq C_{max}$ or $s \in \mathcal{G}$ or $s \in \mathcal{D}$ **then**
      $endEpisode \leftarrow True$
    end if
    choose a with ϵ_e-greedy policy derived from Q and take action a
    observe action cost $c(s,a)$ and next state s'
    if $s \in \mathcal{G}$ **then**
      $c(s,a) \leftarrow 0$
    end if
    for each $c'' \in \{0,...,C_{max}\}$ **do**
      $c'(s,a) \leftarrow u(c'' + c(s,a)) - u(c'')$
      $Q(s,c'',a) \leftarrow Q(s,c'',a) + \alpha_e[c'(s,a) + \max_{a' \in \mathcal{A}} Q(s', c'' + c(s,a), a') - Q(s,c'',a)]$
    end for
    $c \leftarrow c + c(s,a)$
    $s \leftarrow s'$
  end while
  $e \leftarrow e + 1$
end while
$\pi^*(s,c') \leftarrow \arg\max_{a' \in \mathcal{A}} Q(s,c',a'), \forall s \in \mathcal{S}$ and cost c'
return π^*

Different from *Q-learning-GUBS*, *Q-learning-eGUBS+*C_{max} allows to calculate the C_{max} value automatically. In the *Q-learning-eGUBS+*C_{max} algorithm, a lexicographic policy π_l is found based on a model of the transition probability $\hat{P}$ and the costs of actions $\hat{c}$ in the SSP, which is calculated at each f episodes with the *Risk-Sensitive Lexicographic-VI* algorithm [3]. As in [11] and [2] the model of probabilities is calculated by $\hat{P}(s,a,s') = \frac{\sum N(s,a,s')}{\sum_{s''} N(s,a,s'')}$, and the model of cost is calculated by $\hat{C}(s,a) = \frac{\sum c(s,a)}{\sum_{s''} N(s,a,s'')}$ where, $N(s,a,s')$ is the number of times a state s is visited, the action a is performed and passed to the state s'.

To compute the C_{max} value, after calculating the lexicographic policy π_l, the algorithm *Q-learning-eGUBS+*C_{max} evaluates the expected cost to goal $\overline{C}_G^{\pi_l}(s)$ of the policy π_l iteratively for each state s based on Eq. 5, without considering the infinite costs of the dead-end states.

$$\overline{C}_G^{\pi_l}(s) = \begin{cases} 0 & , s \in \mathcal{G} \cup \mathcal{D} \\ \hat{c}(s, \pi_l(s)) + \sum_{s' \in S} \hat{P}(s, \pi_l(s), s') \overline{C}_G^{\pi_l}(s') & , otherwise \end{cases} \quad (5)$$

The C_{max} value is equal to the cost of the lexicographic policy from the initial state s_0 to the goal:

$$C_{max} = \overline{C}_G^{\pi_l}(s_0). \quad (6)$$

Since the steps for computing the lexicographic policy and C_{max} are expensive in terms of processing time cost, and this cost depends on the number of states and the type of problem, these steps are only performed in some episodes.

Q-learning-eGUBS+C_{max} (Algorithm 2), in addition to the inputs of the *Q-learning-GUBS* algorithm, receives as input the minimum error ε_l used in the *Risk-Sensitive Lexicographic-VI* algorithm and the amount f of episodes needed before recalculating the lexicographic policy. After initializing the values of Q, $N(s, a)$, $N(s, a, s')$, $\hat{c}(s, a)$, $\hat{C}(s, a)$ and $\hat{P}(s, a, s')$; Algorithm 2 performs a set of episodes. In each episode, it performs the same operations to interact with the environment and to calculate the value Q as in the *Q-learning-GUBS* algorithm. Furthermore, each f episodes, Algorithm 2 recalculates the model of probabilities $\hat{P}$ and costs $\hat{c}$ based on the counts of the visits to states $N(s, a)$ and $N(s, a, s')$ and the accumulated cost of the visits to states $\hat{C}(s, a)$ (lines 8 and 9). It uses these models to calculate the lexicographic policy (line 11) using the *Risk-Sensitive Lexicographic-VI* algorithm [3]. The value C_{max} is calculated in line 12 based on Eq. 6. Algorithm 2 calculates the GUBS value in line 25 based on Eq. 4. Finally, based on the GUBS value the optimal GUBS policy π^* is calculated in line 26. The counts $N(s, a)$ and $N(s, a, s')$ and the accumulated cost $\hat{C}(s, a)$ used in the model calculation, are updated in each episode (line 27).

4 Experiments

The experiments were carried out in two domains: the River [4] (see Sect. 1) and the Navigation domains. In the Navigation domain [10], a robot is walking in a grid with dimensions $N_x \times N_y$. In this domain, the robot starts from an initial position s_0 and tries to go to a destination position. In the first and last lines ($y = 1$ and $y = N_y$) the actions are deterministic and in the intermediate lines, the agent has a probability $P(x, y)$ of disappearing, going to a dead-end state. The probability of disappearing is greater as the agent is closer to the goal.

Although these domains are simple, they have characteristics that make them difficult. These problems have uncertainty, risk and complex decision making. They were modeled as SSPs with extended states, which include the position and accumulated cost. Actions correspond to decisions to move and have cost 1. In the River domain the experiments were carried out in an environment of size $N_y = 20$ and $N_x = 5$, the initial state is on a bank of the river $x_0 = 1, y_0 = 2$, the goal on the other bank at position $x_g = N_x, y_g = 1$ and two probability values were used $P = 0.4$ and $P = 0.6$.

Algorithm 2 Q-LEARNING-EGUBS+C_{max}

Input: SSP $\mathcal{M}$, maximum cost C_{max}, utility function u, utility for the goal K_g, number of episodes nepisodes, α_{min} and α_{max} to compute α_e, ϵ_{min} and ϵ_{max} to compute ϵ_e, minimum error ε_l for Lexicographic-VI, minimum error ε for Iterative-Policy-Evaluation, number of episodes f to calculate lexicographic policy.
Output: π^*

Initialize: $Q(s,c,a) \leftarrow u(\infty) - u(c), \forall s \in S, a \in A, c \in \{0,1,...,C_{max} + max_{s,a}c(s,a)\}$; $Q(s,c,a) \leftarrow K_g, \forall s \in \mathcal{G}, c \in \{0,1,...,C_{max} + max_{s,a}c(s,a)\}$; $N(s,a) \leftarrow 0, \forall s,a$; $N(s,a,s') \leftarrow 0, \forall s,a,s'$; $\hat{c}(s,a) \leftarrow 0, \forall s,a$; $\hat{P}(s,a,s') \leftarrow 0, \forall s,a,s'$; $\hat{C}(s,a) \leftarrow 0, \forall s,a$.
$e \leftarrow 0$
while $e < nepisodes$ **do**
  compute α_e and ϵ_e
  get initial state s and cost c for $x_0 = (s,c)$
  if $mod(e,f) == 0$ **then**
    for each $s'' \in S, a'' \in A$ e $s''' \in S$ **do**
      $\hat{c}(s'',a'') \leftarrow \hat{C}(s'',a'')/N(s'',a'')$
      $\hat{P}(s'',a'',s''') \leftarrow N(s'',a'',s''')/N(s'',a'')$
    end for
    $\pi_l \leftarrow$ Risk-SensitiveLexicographic-VI($\mathcal{M} = \langle \mathcal{S}, \mathcal{A}, \hat{P}, \hat{c}, \mathcal{G} \rangle, \lambda, \varepsilon_l$)
    $C_{max} \leftarrow$ Iterative-Policy-Evaluation($\pi_l, \mathcal{G}, \mathcal{D}, \varepsilon$)
  end if
  $endEpisode \leftarrow False$
  while not endEpisode **do**
    if $c \geqslant C_{max}$ or $s \in \mathcal{G}$ or $s \in \mathcal{D}$ **then**
      $endEpisode \leftarrow True$
    end if
    choose a with ϵ_e-greedy policy derived from Q and take action a
    Observe action cost $c(s,a)$ and next state s'
    if $s \in \mathcal{G}$ **then**
      $c(s,a) \leftarrow 0$
    end if
    $c'(s,a) \leftarrow u(c + c(s,a)) - u(c)$
    $Q(s,c,a) \leftarrow Q(s,c,a) + \alpha_e[c'(s,a) + \max_{a' \in \mathcal{A}} Q(s', c + c(s,a), a') - Q(s,c,a)]$
    $\pi^*(s,c) \leftarrow \arg\max_{a \in \mathcal{A}} Q(s,c,a)$
    $N(s,a) \leftarrow N(s,a) + 1$; $N(s,a,s') \leftarrow N(s,a,s') + 1$; $\hat{C}(s,a) \leftarrow \hat{C}(s,a) + c(s,a)$;
    $c \leftarrow c + c(s,a)$
    $s \leftarrow s'$
  end while
  $e \leftarrow e + 1$
end while
return π^*

These are the same values of P used by [4] and they show different preferences for proper policies. In the Navigation domain, the initial state is at $x_0 = N_x, y_0 = N_y$, the goal is at position $x_g = N_x, y_g = 1$, and two instances were used. An instance of size

Table 1. Parameters for the algorithms.

Parameter	Definition	Value
u	Utility Function	$e^{-\lambda c}$
λ	Risk factor	-0.2
ϵ_{max}	Maximum exploration rate	1.0
ϵ_{min}	Minimal exploration rate	0.1
ε_l	Minimal error for *Risk-Sensitive Lexicographic-VI*	1×10^{-10}
ε	Minimal error for *Iterative-Policy-Evaluation*	1×10^{-3}
Initial C_{max}	Maximum initial cost *Q-learning-eGUBS+*C_{max}	7

Table 2. Parameters for the River and Navigation problems.

Parame-ter	Definition	Value River	Value Navigation
K_g	Goal reward for GUBS	0.01 to 16	0.01 to 1
α_{max}	Maximum learning rate	0.5	0.04
α_{min}	Minimal learning rate	0.001	1×10^{-6}
episodes	Number of episodes used in the algorithms	4×10^5	1×10^6
f	Number of episodes for C_{max} recalculation	4×10^3	1×10^4
C_{max}	Maximal Cost for *Q-learning-GUBS* River	100	
C_{max}	Maximal Cost for *Q-learning-GUBS* Navigation $N_x = 15 \times N_y = 3$		55
C_{max}	Maximal Cost for *Q-learning-GUBS* Navigation $N_x = 10 \times N_y = 4$		45

$N_x = 15 \times N_y = 3$, with a minimum probability of disappearing of 0.2 and another instance of size $N_x = 10 \times N_y = 4$ with a minimum probability of disappearing of 0.1. Table 1 shows the parameters used by the algorithms. A function with linear decay was used to vary α and ϵ from their respective maximum values to the minimum values of both parameters. The initial extended state was chosen randomly for each episode. Table 2 shows the parameters used for the River and Navigation problems.

In Sects. 4.1 and 4.2, the proposed algorithms are compared with the *GUBS-VI* algorithm. This choice is because the proposed algorithms are based on *GUBS-VI*. The purpose of this comparison is to verify if the proposed algorithms found the same results obtained by *GUBS-VI*. Theoretically, the proposed algorithms should obtain the same result, however, since they use sampling, they could obtain a very close solution within the number of episodes used. In the figures shown in Sects. 4.1 and 4.2, the x axis has the values of K_g expressed in logarithm base 2 and the y axis shows the probability and cost of reaching the goal from the initial state. Each point in the figures is the average of 5 executions of algorithms. Additionally, the standard deviation is shown. We also record the learning curves of these experiments. These results show that the algorithm that converges the fastest is Q-learning-GUBS algorithm. We do not include these curves for space reasons.

In Sect. 4.3, we compare the proposed algorithms with the Geibel and Wysotzki criterion [5]. Notice that, the GUBS criterion and other criteria (fSSPUDE and discount cost) were compared by [1].

4.1 Results of the River Domain

Figures 1a) and 1b) show the probability to goal obtained by the *GUBS-VI* algorithm and the proposed algorithms *Q-learning-GUBS* and *Q-learning-eGUBS+C_{max}* in the River problems with $P = 0.4$ and $P = 0.6$, respectively. The probabilities obtained by the algorithms are very close to the results obtained by the algorithm *GUBS-VI* for both instances of the domain. The proper policy for the $P = 0.4$ instance is reached when $K_g > 16$ ($\log(K_g) = 4$), and the proposed algorithms find policies with probabilities increasingly closer to 1. For the instance with $P = 0.6$, the optimal policy when $K_g \geq 0.5$ ($\log(K_g) = -1$) is a proper policy (i.e. crossing the bridge). In this case, the proposed algorithms find the proper policy. Figures 1c) and 1d) compare the costs reached from the initial state by the three algorithms for the same instances of the River domain. The results are very close between the algorithms for both instances.

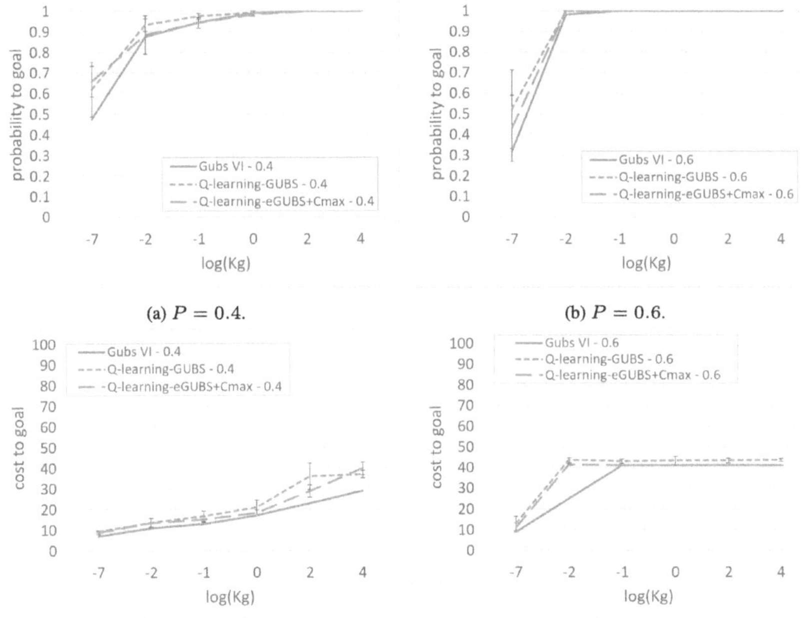

(a) $P = 0.4$. (b) $P = 0.6$. (c) $P = 0.4$. (d) $P = 0.6$.

Fig. 1. Probability to goal and cost to goal found by *Q-learning-GUBS*, *Q-learning-eGUBS+C_{max}* and *GUBS-VI* algorithms for different values of K_g for instance $N_y = 20 \times N_x = 5$ of the River domain.

4.2 Results of the Navigation Domain

Figures 2a) and 2b) show the probabilities to goal obtained by the three algorithms in the two described instances of the Navigation domain. The probabilities found by the proposed algorithms are very close to the probabilities found by the *GUBS-VI* algorithm for both instances of the problem. However, there is a larger standard deviation for values of K_g that have policies close to MAXPROB and correspond to the furthest route from the goal (the probability for the goal is 0.8 for the instance $N_x = 15 \times N_y = 3$, and 0.81 for the instance $N_x = 10 \times N_y = 4$).

The algorithms find a policy according to K_g. For values $K_g \geq 0.3$ ($\log(K_g) = -1.7$) in instance $N_x = 15 \times N_y = 3$ and $K_g \geq 0.05$ ($\log(K_g) = -4.3$) in the instance $N_x = 10 \times N_y = 4$, the algorithm *Q-learning-GUBS* can find the MAXPROB. However, due to the parameters used, the algorithm *Q-learning-eGUBS+*C_{max} can not find the MAXPROB policy in the instance $N_x = 15 \times N_y = 3$ in any of the 5 executions.

Figures 2c) and 2d) compare the costs reached from the initial state by the three algorithms for the same two instances of the Navigation domain. The policy costs calculated by the *Q-learning-eGUBS+*C_{max} algorithm are very close to the policy costs calculated by the *a*lgorithm GUBS-VI.

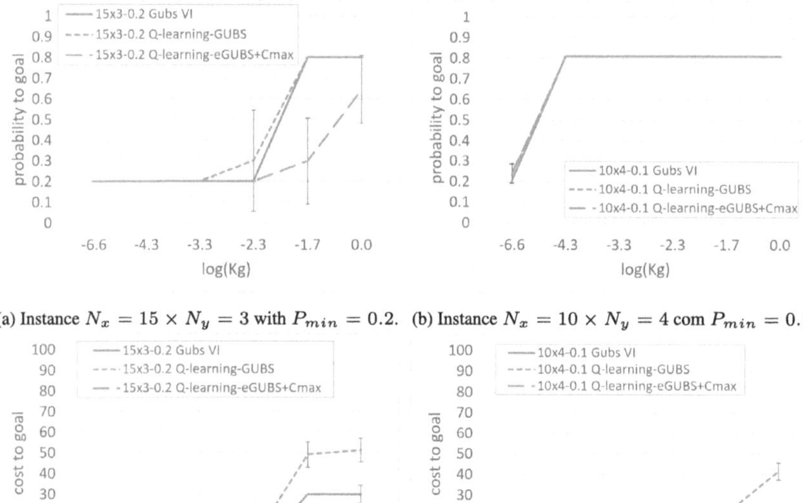

(a) Instance $N_x = 15 \times N_y = 3$ with $P_{min} = 0.2$. (b) Instance $N_x = 10 \times N_y = 4$ com $P_{min} = 0.1$.

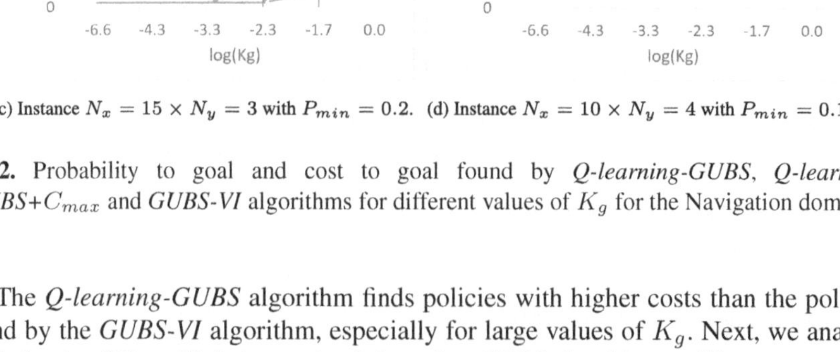

(c) Instance $N_x = 15 \times N_y = 3$ with $P_{min} = 0.2$. (d) Instance $N_x = 10 \times N_y = 4$ with $P_{min} = 0.1$.

Fig. 2. Probability to goal and cost to goal found by *Q-learning-GUBS*, *Q-learning-eGUBS+*C_{max} and *GUBS-VI* algorithms for different values of K_g for the Navigation domain.

The *Q-learning-GUBS* algorithm finds policies with higher costs than the policies found by the *GUBS-VI* algorithm, especially for large values of K_g. Next, we analyze this behavior. When K_g is large, the *Q-learning-GUBS* algorithm oscillates more especially during the learning of probabilities in states with uncertainty, it is difficult to reach the exact value of the actions since, in addition, the difference between action values are very small in this problem. Then, the found policy uses actions with values close to the optimal values for some extended states, but not the optimal one. It causes the formation of cycles, in which the policy can return to the same real state but with a higher cost. Thus, the total cost to the goal is increased. As far as we have analyzed, by changing the parameters, it is possible to reach the right cost, for example, by increasing the number of episodes.

4.3 Comparison with the Geibel and Wysotzki Criterion

We compare the proposed algorithms with the Geibel and Wysotzki criterion [5]. For this comparison, we varied the parameter ξ between the values $[0.1, 1 \times 10^7]$ to evaluate if Geibel and Wysotzki criterion can compute the optimal policies for eGUBS only by choosing the parameters appropriately and, if not, how close it can get to the optimal value. The discounts $\gamma = 0.9$ and $\bar{\gamma} = 1$ were used.

Figure 3 shows the $log_2(\xi)$ on the x axis and, on the y-axis, the values corresponding to the optimal policy found by the *GUBS-VI*, *Q-learning-GUBS* and *Q-learning-eGUBS+Cmax* algorithms with $K_g = 5$ and $C_{max} = 100$, and the values of the optimal policies found by the Geibel and Wysotzki criterion evaluated with the GUBS criterion for the River domain of size $N_y = 20$ and $N_x = 5$ with $P = 0.4$. Each point in the figure is the average of 5 executions of the algorithms.

The values found by the algorithms *Q-learning-eGUBS+Cmax*, *Q-learning-GUBS* and *GUBS-VI* are close. When $log_2(\xi) = 1$, the values of the policies found by the Geibel and Wysotzki criterion are close to the ones found by GUBS-VI, *Q-learning-eGUBS+Cmax* and *Q-learning-GUBS*.

Notice that, to carry out this experiment, an indefinite set of values for ξ had to be varied, this is because there is no clear procedure to choose a certain value of ξ for this criterion, to have a guarantee of finding the resulting policy under GUBS. In certain situations, it may not be feasible to carry out this type of procedure to guarantee finding certain policies.

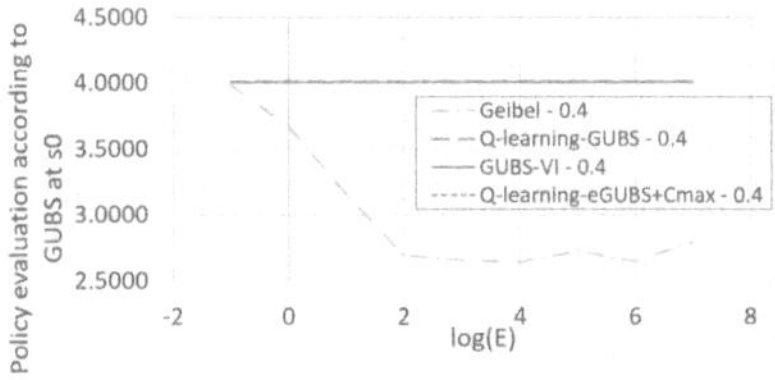

Fig. 3. Comparison of the value of the optimal policy for GUBS-VI, Q-learning-GUBS and Q-learning-eGUBS+Cmax with the value of optimal policies for the Geibel and Wysotzki criterion for different values of ξ for the River domain $N_y = 20$ and $N_x = 5$ with $P = 0.4$.

5 Conclusion

In this work, we propose the first two algorithms for RL that use the GUBS criterion: *Q-learning-GUBS* and *Q-learning-eGUBS+*C_{max} to make a trade-off between probability to goal and cost to goal. *Q-learning-GUBS* is an approximate algorithm and can use several utility functions. *Q-learning-eGUBS+*C_{max} is a model-based algorithm that calculates the C_{max} value for the particular case of the exponential utility function.

Experiments were carried out with the proposed algorithms with several instances corresponding to two domains. The results show that these algorithms can find policies that make a trade-off between probability and cost according to the parameter K_g and according to the utility function used. We also compare the proposed algorithms with Geibel and Wysotzki criterion [5]. The values of the policies found by Geibel and Wysotzki criterion for some values of parameters are close to those found by the proposed algorithms. Different from Geibel and Wysotzki criterion that is based on ξ and discount factors, where the choice of parameters depends on the problem; the parameters of GUBS (K_g, U_{max} and U_{min}) do not depend on the problem, so, the GUBS parameters of the proposed algorithms can be rationally set.

The proposed algorithms could be the basis for designing more sophisticated RL algorithms that can work in more complex domains.

Acknowledgments. This study was supported in part by the *Conselho Nacional de Desenvolvimento Científico e Tecnológico* (CNPq) – Finance Code 001, by the São Paulo Research Foundation (FAPESP) grant #2018/11236-9 and the Center for Artificial Intelligence (C4AI-USP), with support by FAPESP (grant #2019/07665-4) and by the IBM Corporation.

References

1. Crispino, G.N., Freire, V., Delgado, K.V.: GUBS criterion: arbitrary trade-offs between cost and probability-to-goal in stochastic planning based on expected utility theory. Artif. Intell. **316**(C) (2023)
2. Faycal, T., Zito, C.: Dyna-t: dyna-q and upper confidence bounds applied to trees. arXiv abs/2201.04502 (2022)
3. Freire, V., Delgado, K.V., Reis, W.A.S.: An exact algorithm to make a trade-off between cost and probability in SSPs. In: International Conference on Automated Planning and Scheduling 2019, vol. 29, pp. 146–154 (2019)
4. Freire, V., Delgado, K.V.: GUBS: a utility-based semantic for goal-directed Markov decision processes. In: International Conference on Autonomous Agents and Multiagent Systems, pp. 741–749 (2017)
5. Geibel, P., Wysotzki, F.: Risk-sensitive reinforcement learning applied to control under constraints. J. Artif. Intell. Res. **24**, 81–108 (2005)
6. Kolobov, A., Mausam, Weld, D.S.: A theory of goal-oriented MDPs with dead ends. In: Conference on Uncertainty in Artificial Intelligence, pp. 438–447. AUAI Press, Arlington, Virginia, USA (2012)
7. Kolobov, A., Mausam, M., Weld, D., Geffner, H.: Heuristic search for generalized stochastic shortest path MDPs. In: International Conference on Automated Planning and Scheduling 2011, vol. 21, pp. 130–137 (2011)
8. Park, I.W., Kim, J.H., Park, K.H.: Accelerated Q-learning for fail state and action spaces. In: 2008 IEEE International Conference on Systems, Man and Cybernetics, pp. 763–767 (2008)

9. Patek, S.D.: On terminating Markov decision processes with a risk-averse objective function. Automatica **37**(9), 1379–1386 (2001)
10. Sanner, S., Yoon, S.: IPPC results presentation In: International Conference on Automated Planning and Scheduling (2011). http://users.cecs.anu.edu.au/ssanner/IPPC_2011/IPPC_2011_Presentation.pdf. Accessed Aug 2024
11. Sutton, R.S.: Integrated architectures for learning, planning, and reacting based on approximating dynamic programming. In: International Conference on Machine Learning, pp. 216–224. Morgan Kaufmann (1990)
12. Tarbouriech, J.: Goal-oriented exploration for reinforcement learning. Ph.D. thesis, Université de Lille (2022)
13. Teichteil-Konigsbuch, F., Vidal, V., Infantes, G.: Extending classical planning heuristics to probabilistic planning with dead-ends. In: AAAI Conference on Artificial Intelligence, pp. 1017–1022. AAAI Press (2011)
14. Teichteil-Königsbuch, F.: Stochastic safest and shortest path problems. In: AAAI Conference on Artificial Intelligence, vol. 26, pp. 1825–1831 (2012)
15. Trevizan, F.W., Teichteil-Königsbuch, F., Thiébaux, S.: Efficient solutions for stochastic shortest path problems with dead ends. In: Conference on Uncertainty in Artificial Intelligence (2017)
16. Watkins, C.J.C.H., Dayan, P.: Technical note. Q-Learning. Mach. Learn. **8**(7), 279–292 (1992)

SARA - A Generative AI for Legal Process Summarization Based on Chain of Density Prompt Engineering

Francisco das Chagas Jucá Bomfim(✉), Joao Araujo Monteiro Neto, Gilson Bezerra Filho, Vasco Furtado, and Vládia Pinheiro

Universidade de Fortaleza, Fortaleza, CE, Brazil
{fjuca,joaoneto,vasco,vladiacelia}@unifor.br,
gilsonbezerram@edu.unifor.br
http://www.unifor.br

Abstract. Generative AI, particularly Large Language Models (LLMs), holds significant promise for enhancing judicial tasks, especially in automating the generation of legal cases during the sentencing phase. This paper introduces SARA (System for Analysis and summaRization of legal Actions), an innovative method for abstractive and multi-document summarization of legal proceedings. SARA employs GPT-4o, trained exclusively through in-context learning, utilizing Chain of Density (CoD) and CO-STAR prompt engineering techniques. These methods, adapted from judicial procedural knowledge, significantly improve the quality of generated summaries. Traditional evaluation metrics reveal effective training strategies but highlight their limitations in assessing summary quality. Therefore, we propose a qualitative evaluation methodology based on expert-generated questionnaires, focusing on essential content inclusion and proper report structuring. Inspired by this methodology, we trained another GPT model for large-scale summary evaluation. Evaluations of summaries of fifteen first-degree court cases from the Court of Justice of the State of Ceará show a significant advantage of in-context learning with CoD, emphasizing the role of domain knowledge and report style.

Keywords: Text Summarization · Legal Documents · Generative AI · Prompt Engineering

1 Introduction

Generative Artificial Intelligence (GAI), particularly Large Language Models (LLMs), exhibits remarkable language generation capabilities shows great potential for various judicial tasks. Recently, the Federal Superior Court issued public call 01/2023 to develop AIs tailored for these purposes [17], aligning with the goals set by the CNJ (National Council Of Justice) to enhance procedural efficiency and adhere to global judicial innovation practices [12]. A key objective is

A. Paes and F. A. N. Verri (Eds.): BRACIS 2024, LNAI 15413, pp. 370–383, 2025.
https://doi.org/10.1007/978-3-031-79032-4_26

the automatic generation of the process report, crucial for ensuring judges are well-informed about essential case issues, thereby justifying their final decisions adequately.

A process report must accurately narrate the facts and allegations presented, documenting the parties involved, their arguments, summaries of requests and defenses, and major events during the process. Producing such a report is a complex, time-consuming task that requires domain knowledge and synthesis ability, making it well-suited for abstractive summarization. Unlike extractive summarization, which merely selects phrases from the text, abstractive summarization generates a coherent synthesis that captures the essence of the case.

Research on using LLMs for abstractive summarization of legal documents has progressed, but it faces challenges, such as handling the length and volume of legal texts and evaluating summary quality, which often relies on syntactic metrics that inadequately capture semantic context [11]. Notable State-of-the-Art (SOTA) models include SimCLS [15] for English, which employs Contrastive Learning (CL), LegalSumm for Portuguese legal documents [13], and CLSJUR.BR [14], which combines CL with free-reference evaluation techniques and a language model for legal documents in Brazilian Portuguese - Legal-BERT.PT [19]. These models, however, have limitations related to the original document size and the nature of the generated summaries, typically producing only syllabuses. Traditional metrics like ROUGE and BLEU, though useful for comparison, fall short of accurately reflecting how well a summary captures the essence of the document.

In this paper, we introduce SARA (System for Analysis and summaRization of legal Actions), a novel method for abstractive and multi-document summarization of legal proceedings aimed at generating judicial reports. SARA's architecture is based on an LLM (GPT-4o [1]) trained exclusively through in-context learning. This training employs two prompt engineering techniques: Chain of Density (CoD) [2] for summarization and CO-STAR [3] for structuring prompt content. We demonstrate that these techniques, adapted from procedural knowledge provided by judges and legal academics, significantly enhance the informative quality of the generated summaries.

Results using traditional model quality assessment metrics indicate which training strategies are most effective, but they also highlight the limitations of these metrics in evaluating summary quality. Consequently, we propose a qualitative evaluation methodology based on expert-generated questionnaires. This methodology focuses on essential content inclusion and the appropriate structuring of process reports.

Beyond guiding prompt engineering and enabling human-quality analysis of summaries, this methodology inspired the training of ChatGPT [6] to evaluate generated summaries, facilitating large-scale assessments. Evaluations of summaries from fifteen first-degree judicial cases in a large Brazilian State Court, conducted by both machines and humans, reveal a significant advantage of summaries produced using in-context learning with CoD, enhanced by domain knowledge and report style familiarity.

2 Related Work

Automatic summarization can be classified into two types: abstractive and extractive. Abstractive summarization generates new text that captures the original content's essence, often rephrasing it concisely. Extractive summarization selects direct excerpts from the original text. Summarization can also be categorized as single-document or multi-document, with the latter integrating information from multiple sources into a cohesive summary. For first-degree judicial processes, the most suitable automatic summarization is abstractive and multi-document.

Several works before the GPT era proposed various approaches to legal text summarization. For example, [15] employed Contrastive Learning for abstractive summarization, while [7] used Deep Reinforcement Learning, combining extractive and abstractive techniques. "DELSumm: Incorporating Domain Knowledge for Extractive Summarization of Legal Case Documents" [10] utilized domain-specific knowledge for extractive summarization, achieving high ROUGE metrics (ROUGE-1: 51.2) but requiring manual annotations. A similar approach was attempted by [9], who enhanced summarization by incorporating markers like <Issue> and <Argument>, leading to improved ROUGE scores (ROUGE-1: 50.73). However, manual tagging presents a scalability limitation.

For Brazilian Portuguese, LegalSumm [13] employs Transformer models [16] and Contrastive Learning to summarize judicial decisions from Brazil's Supreme Federal Court, generating concise summaries. Using the Ruling.BR [13] corpus for training and testing, results on ROUGE metrics demonstrated that LegalSumm improved the quality of generated summaries compared to traditional methods. In [14], the authors proposed the CLSJUR.BR model, an evolution of LegalSumm, which uses the CL approach along with a new technique for generating and evaluating candidate summaries. CLSJUR.BR is based on the MBart neural model [18] and the LegalBERT.PT language model [19]. The results, using the same LegalSumm test set, showed an improvement in the generation of syllabuses with ROUGE-L from 0.35 (LegalSumm) to 0.4789 (CLSJUR.BR). The limitation of CLSJUR.BR for generating court reports is the limitation for mono documents and the size of the input text (1024 tokens).

In summary, while research on abstractive and multi-document automatic legal document summarization (ALDS) has made advances with SimCLS, LegalSumm, and CLSJUR.BR models, significant challenges remain due to the complexity and volume of legal texts. Recent work consistently highlights the need to explore the potential of Generative AI (GAI) and Large Language Models (LLMs) to improve the efficiency and accuracy of judicial report generation, contributing to a more agile and well-founded justice system.

3 SARA Generating the Report Process

In this section, we introduce the architecture of SARA, a GAI designed for the analysis and summarization of legal cases, and the prompt engineering process that led to the final prompt for multi-document abstractive summarization of legal cases. SARA specifically generates the section known as the Report in the Brazilian judicial system, whose main objective is to summarize the key factual and legal elements of the procedural relationship, including the names of the parties, the identification of the case with a summary of the request and objection, and a documentation of the main events that occurred during the process.

A critical aspect of court reporting is the ability to summarize from multiple documents in a cohesive and accurate manner. Therefore, the core of SARA is a multi-document summary prompt capable of integrating information from diverse sources, including initial pleadings, responses, replies, terms of hearing, closing briefs, and other documents within the judicial process that are essential for understanding the legal demand and reaching a decision in the case. This efficient summarization process is essential for procedural speed, allowing an integrated and holistic view of cases.

3.1 SARA Architecture

Figure 1 presents SARA's basic architecture [https://l1nk.dev/HPiPv] with the following modules:

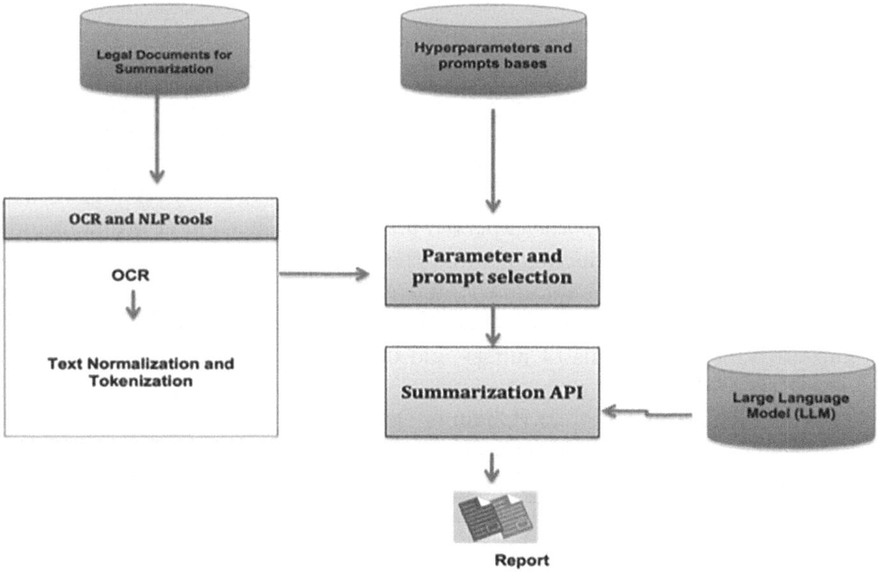

Fig. 1. SARA's General Architecture

- **OCR and NLP Tools** - responsável por realizar o reconhecimento óptico de caracteres (OCR) e verificações como validade do documento em relação

ao tamanho, linguagem válida e normalização dos múltiplos documentos de entrada;
- **Parameter and Prompt Selection** - responsável por selecionar o prompt a ser usado na sumarização e configurá-lo com base nos seguintes hiperparâmetros: lista de documentos a serem resumidos `<DOCS LIST>`; modelo padrão do relatório judicial a ser gerado `<REPORT TEMPLATE>`; tamanho inicial do relatório `<T>`; número `<N>` de etapas no processo de densificação (estratégia CoD); número `<M>` de entidades ausentes; lista de informações obrigatórias que devem ser incluídas no resumo `<REQUIRED-INFO>`;
- **Summarization API** - rotina que envia a instrução de sumarização (prompt) para o LLM junto com os documentos dos casos a serem resumidos.

3.2 Versions of SARA's Prompts

Prompt engineering is an essential skill, focusing on creating clear and precise instructions to guide LLMs in generating the desired outcomes [5]. To instruct SARA in generating high-quality judicial process reports, an incremental approach to prompt engineering was used, focusing on enhancing the following qualitative aspects of judicial reports: form, clarity, precision, and density. In total, four versions of prompts were developed following the CO-STAR framework [3] and the CoD strategy [2].

The CO-STAR framework offers an organized and consistent structure for prompts, dividing them into layers of form and content to facilitate the understanding and analysis of information. The form layers include CONTEXT, which provides an overview of the scenario; OBJECTIVE, which defines the task's purpose; STYLE, which specifies the type of language; TONE, which indicates the desired attitude or emotion; AUDIENCE, which identifies the recipients; and RESPONSE, which presents the expected final format. He content layers include specific details about the documents and the legal scenario, explanations of the mandatory elements in the summary, and guidelines on appropriate language and register, ensuring that the legal summaries are accurate, coherent, and well-organized to meet the needs of judges and other legal professionals involved in the case.

On the other hand, the CoD (Chain of Density) prompt engineering strategy ensures that every word in the text significantly contributes to the overall content, removing redundancies and superfluous information. This technique has proven effective in increasing the informational density of summaries, improving clarity and conciseness. The generic densification process is detailed below:

Step 1: Identification of Missing Entities Identify from one to <M> missing informative entities that are absent in the previously generated summary. Repeat step 2 for <N> times.
Step 2: Write a new, denser summary of similar length that covers all entities and details from the previous summary, as well as the missing entities. The final summary must contain only the essential text of the

story, without any instructions, steps, or guidelines. A missing entity is: relevant to the main story; specific - descriptive but concise (6 words or less); new - not in the previous summary; faithful - present in the original documents; located anywhere in the original documents. Avoid specific terms that are not generalizable; use standardized language according to court thesauri;

This section outlines the evolution of the SARA prompts, with each version improving accuracy, clarity, and density in generated summaries. Reports from each prompt version were qualitatively evaluated by three judges from the Court of Justice of Ceará. The evaluation revealed that the initial use of CoD and CO-STAR techniques failed to meet judicial report requirements, such as maintaining chronological order and adhering to legal guidelines. To address this, a template was introduced in the prompt, ensuring facts were presented in the correct order and format, incorporating domain knowledge and legal style.

The *Prompt-CoD-baseline* represents the starting point, using the CoD strategy proposed in [2] to generate dense summaries, but it lacked the necessary nuances and precision to capture the complexity of legal texts [9].

The next version, called *Prompt-CoD-Summ-zeroShot*, applied zero-shot learning techniques, allowing the model to generalize its knowledge from other domains to the specific task of legal summarization without additional training on the legal corpus. The hyperparameters for this version were defined by empirical experiments from the STF 01/2023 public call [17]. The CONTEXT and OBJECTIVE sections specify which documents will be processed by SARA and what the AI should do, outlining the mandatory information to include in the summary, such as the nature of the action, urgent relief, involved parties, legal arguments, relevant citations, and requests made by the parties.

The GUIDELINES section provides detailed instructions for executing the summary based on the Chain of Density (CoD) strategy. The essential segments of this section are:

1) They provide guidance on maintaining the most relevant information, ensuring that it is presented clearly and comprehensibly for anyone.
2) They define the summary length (T = 1000 words), with an emphasis on clarity and informational density.
3) To ensure informational density, it is instructed that the summary be rewritten N = 5 times to improve the flow and make room for additional entities. Less informative sentences should be merged, compressed, or removed to make the text more compact and meaningful. The following excerpt from the prompt, (GUIDELINES section), is responsible for this densification task:
"Follow these steps:

Sample Heading (Fourth Level) Step 1: Identify 1 to 5 missing informative entities (delimited by ";") that are absent in the previously generated summary.
Repeat Step 2 five times.

Step 2: Write a new, denser summary of similar length that covers all entities and details from the previous summary, as well as the missing entities. The final summary must contain only the essential text of the story, without any instructions, steps, or guidelines."

These guidelines aim to ensure that the summaries are clear, concise, densely informative, and faithful to the original content. Continuous rewriting of the summary, according to CoD instructions, is essential to maintain the density and quality of the information, facilitating the understanding and use of the summarized documents.

The other sections, STYLE, TONE, AUDIENCE, and RESPONSE, aim to guide the language model in generating formal and professional texts with an assertive and informative tone, suitable for legal professionals, and with a logical narrative free of value judgments.

The qualitative evaluation of the reports generated by SARA using the Prompt-CoD-Summ-zeroShot version indicated that, despite their overall quality, they exhibited the following issues: (1) Form: They do not meet legal formalities, as they do not follow the chronological order of events in the process; (2) Clarity: Important information is omitted, such as the minutes of the evidentiary hearing and the Public Prosecutor's opinion; (3) Accuracy: Only one of the parties' requests is mentioned, whereas there were multiple requests in the process.

These issues are critical and can cause the nullity of the judgment. If the order of events is unclear or if there are relevant omissions, it can result in a violation of the adversarial principle. If the facts are not presented in an organized and complete manner, the decision may be considered insufficiently substantiated. Additionally, if the narration of the facts is confusing and disorganized, the defense may be unable to present arguments or adequately contest the points raised.

The Prompt-CoD-Summ-DomainKnowledge version was designed to enhance report quality by incorporating legal domain knowledge, ensuring compliance with judicial standards, emphasizing clear citations, organized information, and the use of standardized legal language. The *Prompt-CoD-Summ-DK-Styles* version further advanced this by integrating domain knowledge with specific writing styles, introducing command and content layers, and using a REPORT-TEMPLATE developed by magistrates. This version focused on accuracy and presentation, receiving the highest evaluations from magistrates for its clarity, form, and content accuracy.

Table 1 presents the final structure of SARA's prompt, indicating the hyperparameters to be instantiated.

Table 1. General Structure of the SARA Abstractive Summarization Prompt Prompt-CoD-Summ-DK-Styles

Section	Instructions and Hyperparameters
Version	Prompt-CoD-Summ-DK-Styles
CONTEXT	You will receive a set of legal documents related to a case. These documents include `<DOCS_LIST>`.
OBJECTIVE	Summarize legal documents using the `<REPORT_TEMPLATE>`, write the report conservatively, maintaining fidelity to the information in the original documents, act with precision and clarity, ensuring that all important information is preserved, generating a report for a judicial decision. The summary must include the following mandatory information: `<REQUIRED_INFO>`. All information about the plaintiff must come before the defendant's information. If any of this information is not available in the documents, note its absence in the summary.
STYLE	The writing style should be formal and professional, using appropriate legal terminology. Summaries and the report should be clear, direct, and objective, allowing quick understanding of legal issues. In narrative form.
TONE	The tone should be assertive and informative, reflecting the seriousness and importance of the legal content. Accuracy is essential, ensuring that all information presented is precise and relevant to the case.
AUDIENCE	The target audience includes judges and other legal professionals involved in the case, including members of the legal team or other court staff. The report should meet their need for concise but comprehensive information to support legal and strategic decision-making.
RESPONSE	The final report should be in narrative form, without making value judgments or opinions about the case, strictly adhering to the summary, and under no circumstances structured in topics, maintaining logic, starting with an introduction to the case, followed by document summaries.
GUIDELINES	The summary should ensure clarity in writing, allowing people outside the legal field to understand it; be concise, focusing on the essential without omitting significant details; be faithful to the content of the original vote, ensuring complete coverage of essential elements. The first summary should be long (approximately `<T>` words), mainly covering all important entities and containing little information beyond the defined entities. Use overly detailed language and fillers (e.g., "this text contains") to reach about `<T>` words. Make every word count: rewrite the previous summary to improve flow and make room for additional entities. It is essential to indicate the legislation cited in the reference text in the summary, with appropriate reference to the original text. Follow these steps: Step 1: Identify 1 to `<M>` missing informative entities (delimited by ";") that are absent in the previously generated summary. Repeat Step 2 for `<N>` times. Step 2: Write a new, denser summary of similar length that covers all entities and details from the previous summary, as well as the missing entities. Avoid specific terms that are not generalizable; use standardized language according to court thesauri.

4 Empirical Evaluation

The experimental evaluation consisted of generating reports for 15 first-degree cases from the Court of Justice of the state of Ceará by SARA, using the following versions of prompts: V0 - Prompt-CoD-baseline; V1 - Prompt-CoD-Summ-zeroShot; V2 - Prompt-CoD-Summ-DomainKnowledge; V3 - Prompt-CoD-Summ-DK-Styles. The reports were then compared with the reference report using traditional ROUGE [20] and BLEU [21] metrics, as the available metrics are not yet sufficiently reliable for this purpose [4,6]. Finally, a qualitative evaluation was performed by the human experts and by LLM GPT-4o using an evaluation prompt designed in this study and a questionnaire formulated by expert magistrates from the Court of Justice of the state of Ceará.

4.1 Quantitative Evaluation

Table 2 presents the average results of the ROUGE (F1-Score) and BLEU metrics for the reports of the processes generated by SARA with prompt versions V0 to V3, compared to the reference report produced by the magistrates. Individual results for each report are available at https://l1nk.dev/HPiPv.

In summary, version V2 performed best in ROUGE-1 F1, showing better word accuracy. Version V3 excelled in ROUGE-2 F1 and ROUGE-L F1, capturing more bigrams and preserving sentence structure and fluency. While V2 had the highest BLEU score, it may have been penalized for incorrect sequences. Version V0 had the lowest scores, struggling with both word accuracy and structure. Despite a lower BLEU score, V3 demonstrated superior structural quality and coherence.

Table 2. Average ROUGE (F1-Score) and BLEU Metrics for the different prompt versions.

Prompt Version	ROUGE-1 F1	ROUGE-2 F1	ROUGE-L F1	BLEU Score
V0	0.230088	0.088335	0.123917	0.043988
V1	0.421370	0.173248	0.178351	0.049367
V2	0.436677	0.180113	0.180793	0.055854
V3	0.380506	0.182838	0.192885	0.039577

4.2 Qualitative Evaluation

The quantitative metrics (see Table 2) demonstrated weak correlations between the reports generated by SARA and those generated by the magistrates, being insufficient to decide on one prompt strategy over the other. In this work, we conducted two qualitative evaluation strategies to capture quality criteria for the generated reports.

In the first strategy, three magistrates evaluated all reports generated by the prompt versions V0 - Prompt-CoD-baseline and V1 - Prompt-CoD-Summ-zeroShot, based on the following criteria: form, accuracy, clarity, and density. All reports were rejected on all criteria, being considered incorrect and insufficient to support judicial decisions. Interestingly, version V1 achieved one of the best results in the ROUGE-1 F1 metric but was rejected by the experts. To address the quality issues, domain knowledge and style information (template) were incorporated into the subsequent versions V2-Prompt-CoD-Summ-DomainKnowledge and V3- Prompt-CoD-Summ-DK-Styles. Reports were generated by these versions and evaluated by the magistrates. Version V2 fully met the criteria of form, accuracy, and density, and partially met the clarity criterion due to the absence of some legal reasoning. Version V3 fully met all the evaluated criteria.

In a second strategy, a qualitative questionnaire consisting of 13 questions was developed (Table 3) by the magistrate experts, with the responses ranging from 1 (report does not meet) to 5 (report fully meets). In order to systematize and automate a qualitative evaluation of reports on a large scale, focusing on form, clarity, accuracy, and utility, the following prompt was submitted to ChatGPT 4o along with the questionnaire and five versions of reports for each case in the sample: the reference report and four summaries generated by SARA using the four prompt versions V0 to V3:

> "you are a judicial analyst who will receive 5 versions of judicial process summaries and should evaluate them according to a provided <questionnaire>. At the end, you should indicate the summary that best answers the questions in the evaluation questionnaire and justify your response. <summary1> <summary2> <summary3> <summary4> <summary5> <questionnaire>"

Versions V1 and V2 were fully reproved by the human experts, even though V1 was not so bad in quantitative metrics. Among the versions V2 and V3, version V3 stood out in the human evaluation on all the criteria established by the evaluation questionnaire, as well as by the GPT-4o language model in ratings, with the language model choosing version V3 based on criteria defined in the evaluation questionnaire. Among the findings, we can highlight that, although all versions demonstrated clarity in identifying the involved parties, version V3 stood out by maintaining this clarity throughout the entire document. This aspect significantly facilitates the reader's understanding, as evidenced by the consistent high ratings in all evaluations of "Identification of the Parties" in the analyzed versions.

Version V3 provided a detailed and accurate summary of both the claim and the defense, covering all essential aspects without significant omissions. For example, while versions V1 and V2 received high evaluation for clarity, both presented omissions in important specific details. Version V3, on the other hand, ensured the complete inclusion of all essential information, as indicated by the consistent five-star ratings in all evaluated criteria, including "Summary of the Claim" and "Summary of the Defense."

Table 3. Qualitative Evaluation Questionnaire and average of answers given by Chat-GPT 4.0 for the questionnaire, for each report generated by SARA's prompt versions V0 to V3.

Item	Topic	Question	V0	V1	V2	V3
1	Identification of the Parties	Were the parties involved in the judicial action clearly identified?	4.200	4.900	4.900	5.000
2	Summary of the Claim	Was the claim or claims made by the plaintiff in the initial petition clearly summarized?	4.100	4.800	4.900	5.000
3	Summary of the Defense	Was the defense presented by the defendant clearly summarized?	4.000	4.700	4.800	5.000
4	Contextualization of the Facts	Were the main facts alleged by the parties clearly contextualized?	4.300	4.900	4.900	5.000
5	Legal Background of the Initial Petition	Were the legal background of the initial petition properly summarized?	3.900	4.400	4.500	5.000
6	Legal Background of the Contestation	Were the legal background of the refutation properly summarized?	4.000	4.500	4.600	5.000
7	Relevant Procedural Occurrences	Were the main occurrences in the process clearly described?	4.200	4.800	4.800	5.000
8	Produced Evidence	Was the evidence produced by both parties clearly described?	4.100	4.900	4.900	5.000
9	Cited Legal Norms	Were the legal norms cited by the parties clearly identified?	4.200	4.900	4.900	5.000
10	Legal Reasoning	Was the application of legal norms to the facts of the case accurately presented?	4.300	4.800	4.900	5.000
11	Coherence with Arguments and Evidence	Is the report consistent with the arguments and evidence presented by the parties?	4.200	4.900	4.900	5.000
12	Chronological Order	Was the report written in chronological order?	4.100	4.800	4.800	5.000
13	Preparation for Decision	Did the report adequately prepare the ground for the reasoning and dispositive part of the judgment?	4.300	4.900	4.900	5.000

Version V3 provided a comprehensive and well-structured contextualization of facts, offering a cohesive narrative with clear and detailed accounts of procedural occurrences and evidence. While other versions performed well, they lacked the same level of detail and cohesion. Version V3 also excelled in detailing the legal backgrounds in both the initial petition and the contestation, which is crucial for preparing the magistrate's decision. Additionally, V3 followed a clear and precise chronological order, ensuring fluidity and understanding. In summary, V3 was the most complete and well-structured version, meeting the evaluation criteria with the highest clarity, detail, and comprehensiveness. Despite being tested in a specific context of civil courts in Ceará, V3 was recognized for its superior performance. No hallucinations were detected, but human review remains essential for such tasks.

5 Conclusion

The integration of prompt engineering principles, such as the Chain of Density and the CO-STAR framework, along with domain knowledge, has led to the creation of more comprehensible and useful summaries for legal experts. These methods emphasize the importance of essential elements in the process report and the manner, especially the order, in which the report should be generated, effectively meeting the needs of specialists in the legal field.

The combination of generic in-context learning techniques with domain-specific knowledge has proven essential for automatically generating the process report. This application has significant social implications, as it reliably accelerates a crucial stage of the judicial process. Although we focused on a specific type of summary, the generic nature of the process report, which involves various elements, suggests that our approach is broadly applicable to other types of legal document summaries. Further investigation in this area is warranted.

Quantitative metrics for evaluating legal summaries remain limited. Expert-created questionnaires and large-scale validations using GPT models present a promising alternative for the qualitative and quantitative evaluation of different approaches.

Finally, it is worth highlighting the practical viability of SARA, which is already in use by magistrates in a major Brazilian court, significantly contributing to the streamlining of National Justice. The deployment of SARA in a real-world judicial setting underscores its robustness and adaptability to the complex requirements of legal processes. By automating the generation of process reports, SARA not only reduces the time and effort required by legal professionals but also enhances the consistency and accuracy of the reports, ensuring that essential details are systematically captured and presented.

The successful implementation of SARA demonstrates its potential to be scaled and adapted for use in other courts across different jurisdictions, further amplifying its impact on the judicial system. As courts globally face increasing caseloads and the pressure to expedite legal proceedings, tools like SARA offer a viable solution to these challenges by providing a reliable, efficient, and standardized method for handling procedural documentation.

As SARA continues to be refined and expanded, its role in the judicial process could extend beyond report generation to encompass other aspects of legal document management and analysis. The ongoing development and integration of such AI-driven tools represent a transformative step toward a more modern, efficient, and accessible judicial system, ultimately benefiting legal professionals, litigants, and society as a whole.

References

1. OPENAI. Hello GPT-4o. Disponível em: https://openai.com/index/hello-gpt-4o/. Acesso em: 30 June 2024
2. Adams, G., Fabbri, A., Ladhak, F., Lehman, E., Elhadad, N.: From sparse to dense: GPT-4 summarization with chain of density prompting. In: Proceedings of the 4th New Frontiers in Summarization Workshop. Association for Computational Linguistics (2023). Disponível em: https://aclanthology.org/2023.propor-1.33. Acesso em: 1 set. 2024
3. Gunawan, A.: How I Won Singapore's GPT-4 Prompt Engineering Competition. Disponível em: https://towardsdatascience.com/how-i-won-singapores-gpt-4-prompt-engineering-competition-34c195a93d41. Acesso em: 30 June 2024
4. Dias, P., Mendes, A., Martins, A.F.T.: Improving abstractive summarization with energy-based re-ranking. arXiv (2022). https://doi.org/10.48550/arXiv.2210.15553
5. IBM.: Engenharia de prompts: a arte de fazer perguntas a IA. IBM (2023). Disponível em: https://www.ibm.com/br-pt/topics/prompt-engineering. Acesso em: 24 June 2024
6. Liu, Y., Iter, D., Xu, Y., Wang, S., Xu, R., Zhu, C.: G-Eval: NLG Evaluation using GPT-4 with Better Human Alignment. arXiv (2023). https://doi.org/10.48550/arXiv.2303.16634
7. Heilmann, C.M.: Text summarization of legal documents using reinforcement learning: a study, pp. 403–414 (2021). https://doi.org/10.1007/978-981-19-2894-9_30
8. Kanapala, A., Pal, S., Pamula, R.: Text summarization from legal documents: a survey. Artif. Intell. Rev. **51**(3), 371–402 (2017). https://doi.org/10.1007/s10462-017-9566-2
9. Elaraby, M., Litman, D.J.: ArgLegalSumm: improving abstractive summarization of legal documents with argument mining. arXiv, (2022). http://arxiv.org/abs/2209.0165010.48550/arXiv.2209.01650
10. Bhattacharya, P., Poddar, S., Rudra, K., Ghosh, K., Ghosh, S.: Incorporating domain knowledge for extractive summarization of legal case documents. arXiv (2021). Disponível em: 2209.01650. Acesso em: 30 June 2024
11. Kanapala, A., Jannu, S., Pamula, R.: Summarization of legal judgments using gravitational search algorithm. Neural Comput. Appl. **31** (2019)
12. CONSELHO NACIONAL DE JUSTIÇA (CNJ): Automação processual amplia eficiência de atendimentos em tribunal. Disponível em: https://www.cnj.jus.br/revista-cnj-automacao-processual-amplia-eficiencia-de-atendimentos/break-em-tribunal/. Acesso em: 30 jun. 2024
13. Feijó, D.d.V.: Summarizing Legal Rulings. Universidade Federal do Rio Grande do Sul, Instituto de Informática, Programa de Pós-Graduação em Computação. Porto Alegre (2021)
14. Lins, A.A., Carvalho, C., Bomfim, F.dC.J., Bentes, D.d.C., Pinheiro, V.: CLSJUR.BR - a model for abstractive summarization of legal documents in Portuguese language based on contrastive learning. In: Proceedings of the 16th International Conference on Computational Processing of Portuguese, vol. 1, Santiago de Compostela, Galicia/Espanha, março 2024. Association for Computational Linguistics. Disponível em: https://aclanthology.org/2024.propor-1.33
15. Liu, Y., Liu, P.: SimCLS: A Simple Framework for Contrastive Learning of Abstractive Summarization. arXiv (2021). Disponível em: 2106.01890. Acesso em: 1 jul. 2024

16. Vaswani, A., et al.: Attention is all you need. In: Advances in Neural Information Processing Systems, pp. 5998–6008 (2017)
17. BRASIL. Supremo Tribunal Federal. Chamamento Público para Pesquisa sobre IA no STF. Disponível em: https://www.stf.jus.br/arquivo/cms/noticiaNoticiaStf/anexo/ChamamentoPblicoIASTF.pdf. Acesso em: 1 July 2024
18. Liu, Y., Lapata, M.: mBART: multilingual denoising pre-training for neural machine translation. In: Proceedings of the 58th Annual Meeting of the Association for Computational Linguistics, pp. 7871–7880 (2020)
19. Silveira, R., Ponte, C., Almeida, V., Pinheiro, V., Furtado, V.: LegalBERT-PT: a pretrained language model for the Brazilian Portuguese legal domain. In: Naldi, M.C., Bianchi, R.A.C. (eds) BRACIS 2023. LNCS, vol. 14197, pp. 268–282. Springer, Cham (2023). https://doi.org/10.1007/978-3-031-45392-2_18
20. Lin, C.-Y.: ROUGE: a package for automatic evaluation of summaries. In: Proceedings of the Workshop on Text Summarization Branches Out (WAS 2004), Barcelona, Espanha, pp. 25–26. Association for Computational Linguistics (2004). Disponível em: https://aclanthology.org/W04-1013. Acesso em: 1 jul. 2024
21. Papineni, K., Roukos, S., Ward, T., Zhu, W.-J.: BLEU: a method for automatic evaluation of machine translation. In: Proceedings of the 40th Annual Meeting of the Association for Computational Linguistics (ACL), pp. 311–318. Association for Computational Linguistics, Philadelphia (2002). Disponível em: https://aclanthology.org/P02-1040. Acesso em: 1 jul. 2024

Semi-supervised Predictive Clustering Trees for Multi-label Protein Subcellular Localization

Leonardo U. Alcantara[1], Isaac Triguero[2], and Ricardo Cerri[3(✉)]

[1] Department of Computing, Federal University of São Carlos, São Carlos, SP, Brazil
utidaleonardo@gmail.com

[2] Department of Computer Science and Artificial Intelligence, University of Granada, Granada, Spain
triguero@decsai.ugr.es

[3] Institute of Mathematics and Computer Science, University of São Paulo, São Carlos, SP, Brazil
cerri@icmc.usp.br

Abstract. Protein subcellular localization is an important classification task because the location of proteins in a cell is directly linked to their functions. Since a protein can act at two or more locations simultaneously, multi-label classification algorithms are necessary. The currently used algorithms are usually based on supervised learning, which presents some disadvantages such as (i) a need for a large amount of labeled instances for training; (ii) a waste of valuable information that labeled instances can provide; and (iii) a high cost involved in obtaining labeled instances for training. To overcome these disadvantages, semi-supervised learning can be applied, where classifiers exploit both labeled and unlabeled data. Thus, in this paper, we propose a new semi-supervised algorithm for multi-label protein subcellular localization. Our proposal is based on decision tree classifiers induced using predictive clustering trees. We investigate many semi-supervised protein subcellular localization scenarios to test whether unlabeled instances can improve the multi-label classification process. Our results show that the proposal can achieve competitive or better results when compared to the pure supervised version of the predictive clustering trees.

Keywords: predictive clustering trees · protein subcellular localization · multi-label classification · semi-supervised learning

1 Introduction

Determining the protein subcellular localization (PSL) is a crucial process for deducing and understanding protein functions and for genome annotation [26]. PSL is especially important in the medical field because it is fundamental for discovering and analyzing new kinds of medicines [21]. It can also help to identify some diseases, such as cancer or Alzheimer, due to abnormal PSL being correlated to these diseases [9].

A. Paes and F. A. N. Verri (Eds.): BRACIS 2024, LNAI 15413, pp. 384–399, 2025.
https://doi.org/10.1007/978-3-031-79032-4_27

Although experimental approaches are the most reliable methods to determine PSL, they are very costly and labor-intensive. In recent years, with the completion of many large-scale genomics and proteomics sequencing projects, a vast number of protein sequences have been discovered, making experimental approaches even more unfeasible. Thus, Machine Learning (ML) computational predictors have been used, since they can handle this problem quickly and accurately [1,27].

In addition, proteins can move between two or more locations within a cell or reside at more than one location simultaneously. Thus, multi-label classification methods have also been developed to deal with PSL problems [12,29,33,36]. All these methods use training and test datasets normally acquired only from proteins whose localizations were experimentally defined, referred to as Labeled Proteins (LPs) [6].

Although using only LPs is a well-established method in the literature, it presents some disadvantages: (i) LPs are hard to obtain because they need to be annotated with experimental approaches, which, are labor-intensive and time-consuming; (ii) these experimental protein sequences are just a small part of the overall sequences in the knowledge bases [6,35]. About only 5.5% of all instances in UniProtKB/Swiss-Prot are LPs; and (iii) normally not all LPs retrieved from the knowledge bases are used because some of them are homologous or similar proteins to each other. This may cause overfitting, by the use of models or procedures that include more terms than are necessary or use more complicated approaches than are necessary [14]. This problem is undesirable because it leads to models that adjust themselves really well to a particular set of data but fail to fit additional data or predict future observations reliably, therefore creating weaker classifiers [6,35].

Usually, most works in the literature use supervised ML to deal with PSL. Therefore, these works disregard proteins whose localizations are missing, referred to as Not Labeled Proteins (NLPs). However, the small number of available labeled proteins can lead to datasets with an insufficient number of instances. As an example, this is the case of the virus benchmark dataset used in the works of Wan et al. [33] and Shen and Chou [23], which has only 207 proteins, only 8 of them located in the "viral capsid".

Supervised methods ignore the fact that NLPs can provide valuable information for the classification. For example, they contain information about the joint probability distribution over sequence elements [7,35]. In situations where there is a large amount of unlabeled data, and the economical commitment necessary to label them all is not affordable, Semi-supervised Learning (SSL) [30] provides strategies to overcome these limitations. SSL is a learning paradigm focused on learning from both labeled and unlabeled data, mainly centered on addressing problems where labeled data is hard to obtain, and unlabeled data is abundant.

The main advantage of using learning strategies that take advantage of unlabeled data for PSL prediction is that we can construct reliable classifiers to predict the PSL of organisms or groups of organisms with very few labeled instances. This is the case, for example, of domains such as Archaea and Bacteria, groups of organisms such as Alveolata and Amoebozoa and Viruses. As an

Table 1. Comparison between LPs and NLPs in UniprotKB/Swiss-Prot

	Taxonomy	LPs (%)	NLPs (%)	Total N of proteins
	Archaea	201 (≈1.02%)	19452 (≈98.98%)	19653
	Bacteria	2845 (≈0.85%)	332221 (≈99.15%)	335066
Eukaryota	**Total**	33534 (≈17.33%)	159987 (≈82.67%)	193521
	Alveolata	115 (≈9.93%)	1043 (≈90.07%)	1158
	Amoebozoa	219 (≈4.97%)	4184 (≈95.03%)	4403
	Euglenozoa	104 (≈11.93%)	768 (≈88.07%)	872
	Fungi	6244 (≈17.66%)	29116 (≈82.34%)	35360
	Metamonada	11 (≈11%)	89 (≈89%)	100
	Viridiplantae	5803 (≈14.18%)	35122 (≈85.82%)	40925
	Viruses	1151 (≈6.77%)	15863 (≈93.23%)	17014
	Total	37731 (≈10.86%)	309704 (≈89.14%)	347435

example, Table 1 presents the number of LP and NLP proteins of some domains and groups of organisms in the UniprotKB/Swiss-Prot in the 2021_03 release[1].

In this paper, we propose a semi-supervised multi-label method based on the concept of Predictive Clustering Trees (PCT), a decision tree induction algorithm well-established in the multi-label classification literature [32]. We investigate many semi-supervised protein subcellular localization scenarios in datasets related to fungus, viruses, and plants in order to test whether unlabeled instances can improve the multi-label prediction of protein subcellular localization. Our results show that the proposal can achieve competitive or better results when compared to the original pure supervised version of the predictive clustering trees algorithm.

The remainder of this paper is organized as follows. Section 2 presents the main theoretical concepts needed for the development of our proposal. Section 3 presents our proposed semi-supervised PCT. Our methodology is presented in Sect. 4, while our experiments are presented and discussed in Sect. 5. Finally, Sect. 6 presents our conclusions and future work.

2 Theoretical Foundation

This section presents the main concepts necessary to develop our proposal.

2.1 Multi-label Classification

Traditional classification algorithms associate a single label/class with each instance in a dataset. However, there are many problems where we need to

[1] http://ftp.uniprot.org/pub/databases/uniprot/previous_releases/release-2021_03/.

associate an instance with two or more labels simultaneously. In multi-label classification, an instance can be classified into two or more labels simultaneously. Documents, images, videos, music, emotions, and proteins are examples of multi-label data Given $L = \{l_j : j = 1, ..., q\}$ a finite label set, multi-label algorithms associate to each instance X_i of a dataset, a set of labels Y_i such that $Y_i \subseteq L$ [31].

Multi-label classification methods can be characterized into three approaches [18]: i) Problem Transformation, which transforms multi-label problems into single-label problems and then employs traditional classification algorithms; ii) Algorithm Adaptation, which creates algorithms to handle multiple labels simultaneously; and iii) Ensemble, which combines a set of multi-label classifiers to solve the problem.

In this work, our proposed semi-supervised predictive clustering tree is based on the algorithm adaptation approach since the literature has already shown the advantages of algorithm adaptation over problem transformation, such as considering label correlations [4,32].

2.2 Semi-supervised Classification

The main goal of Semi-Supervised Classification (SSC) is to construct a model that uses labeled and unlabeled data, obtaining a better prediction performance than the one trained only with labeled data. Thus, SSC learning can be seen as a midway between supervised learning, in which we only learn from labeled instances, and unsupervised learning, in which we only learn from unlabeled instances [22].

SSC can be helpful in situations where (i) the labels of the instances are hard to obtain, requiring human specialists, special devices, slow experiments, etc., and (ii) there is a large amount of unlabeled data available. In these cases, SSL requires less human effort and can potentially improve the accuracy of the classifiers. Many domains struggle with the abovementioned problems, such as speech recognition, bioinformatics, text processing, and image categorization [19, 22].

SSC can be divided into two categories: transductive learning and inductive learning. While inductive learning focuses on building a good classifier capable of predicting unseen data accurately, transductive learning aims to predict the labels from the unlabeled instances given in the training set [30].

In this work, our proposed semi-supervised predictive clustering tree is based on inductive learning.

2.3 Predictive Clustering Trees

Predictive clustering trees (PCTs) [3] are a generalization of decision trees that can be used for clustering and classification. In regular decision trees, the leaves contain the classes, and the branches from the root to the leaves contain conditions for the classification. PCTs view a decision tree as a hierarchy of clusters in which the root of a PCT represents a cluster containing all the data and is recursively partitioned into smaller clusters using a criterion to select the best split. The leaves of a PCT correspond to the lowest level of the cluster hierarchy, and each leaf is labeled with a cluster prototype (multi-label prediction) [16].

PCTs can be constructed using a conventional top-down induction of decision trees heuristic (TDIDT) [5]. The algorithm receives a set of examples (E) as input and outputs the final PCT. To do this, it uses a heuristic (h) to select the best test (t^*) between the tests (t), also called possible splits. The heuristic used in PCT is the variance reduction caused by partitioning the instances from a node into many data partitions (child nodes) according to the best test (t^*).

To find the best test at each node, the BestTest procedure searches for the best acceptable test by calculating the variance reduction for each possible test. The procedure returns the best test found, the heuristic value, and the new partitions of the node data that will be used as child nodes. This process is repeated for every child node until no acceptable test can be found, which means that no test significantly reduces the variance of the node. Then, the algorithm creates a leaf and computes the prototype of the instances belonging to that leaf to create the predicted labels. This whole PCT induction process is shown in Algorithm 1.

Algorithm 1. Top-down induction for decision trees construction used in Clus

procedure InduceTree
Input: A dataset E
Output: A predictive clustering tree
$(t^*, h^*\ P^*) = BestTest(E)$
if $t^* \neq none$ **then**
  for each $E_i \in P^*$ **do**
    $tree_i = InduceTree(E_i)$
  return $node(t^*, \cup_i\{tree_i\})$
else
  return $leaf(Prototype(E))$

procedure BestTest
Input: A dataset E
Output: The best test (t^*), its heuristic score (h^*) and the partition (P^*) it induces on the dataset (E)
$(t^*, h^*\ P^*) = (none, 0, \emptyset)$
for each possible test t **do**
  $P =$ partition induced by t on E
  $h = |E| Impurity(E) - \sum_{E_i \in P} |E_i| Impurity(E_i)$
  if $(h > h^*) \wedge Acceptable(t, P)$ **then**
    $(t^*, h^*, P^*) = (t, h, P)$
return (t^*, h^*, P^*)

Maximizing the variance reduction when splitting each node ensures that the corresponding clusters' homogeneity increases with the depth of the tree, improving the prediction performance in general [3]. The main difference between PCTs and regular decision trees is that PCTs treat the variance and prototype functions as hyperparameters that are instantiated based on the learning task. Thus, PCTs can be used in many tasks, such as clustering [3,25], multi-output classification and regression [2,24], time series data analysis [10], interaction prediction [20], and hierarchical multi-label classification [32]. In our experiments, we used the Clus implementation of the PCTs [16,32].

3 Semi-supervised Predictive Clustering Trees

Our proposal consists of adapting the Clus induction procedure to create a Semi-Supervised Predictive Clustering Tree (SSL-PCT) able to induce a clustering decision tree using labeled and unlabeled data simultaneously. Our method preserves the appealing characteristics of using a decision tree, such as being fast to learn and being readily interpretable (unlike the models learned by most semi-supervised algorithms), while making use of unlabeled data.

Some essential modifications are needed to adapt the Clus algorithm to make it semi-supervised. The first one is related to the initial set of instances (E) since now we have to modify Clus to accept labeled and unlabeled instances as input. Our dataset is now represented as $E = E_l \cup E_u$, where E_l is the labeled part of the training set and E_u is the unlabeled part. The second modification is related to the impurity calculation in the BestTest procedure. We must create a new heuristic that considers labeled and unlabeled data in a node. Our proposed impurity comprises two parts, a supervised and an unsupervised. A ω hyperparameter controls the influence of each part. The proposed impurity is presented in Eq. 1.

$$\text{Impurity}_{\text{SSL}}(E) = \underbrace{\frac{\omega}{T} \cdot \sum_{i=1}^{T} \text{Impurity}(E_l, Y_i)}_{\text{Supervised part}} + \underbrace{\frac{1-\omega}{D} \cdot \sum_{i=1}^{D} \text{Impurity}(E, X_i)}_{\text{Unsupervised part}} \quad (1)$$

In Eq. 1, E is the set of instances, Y_i represents each of the T target features (labels), X_i represents each of the D descriptive features, and ω is a value between 0 and 1 that weights the supervision and unsupervision used in the method. If $\omega = 1$, the method works using only the labeled part of the dataset, behaving as a supervised learning algorithm. On the other hand, if $\omega = 0$, the algorithm behaves as an unsupervised learning algorithm, not using the labels to define the splits. Any value between 0 and 1 means the algorithm uses labeled and unlabeled data. Recall that in Eq. 1, the whole dataset E is considered an unsupervised dataset since, regardless of the instances being labeled or not, the unsupervised part of the equation uses only descriptive features to search for the best split. What changes is the weight of the supervision employed during induction.

One of our method's main advantages is the controlled level of supervision. When performing semi-supervised learning, the unlabeled data can always negatively affect prediction performance. By adapting the level of supervision, we can guarantee that our SSL-PCT achieves better or equal prediction performance compared to its supervised counterpart.

The main impurity measure used in the original PCT algorithm is the Gini Impurity (GI). GI is a measure used in decision trees to calculate how much a set of instances in a node is pure to find the best split. GI can be calculated for a set of data E and a specific label Y according to Eq. 2.

$$\text{Gini}(E,Y) = 1 - \sum_{i=1}^{C} p_i^2 \tag{2}$$

In Eq. 2, C is the number of possible values for the label Y. As an example, in a binary classification, $C = 2$. Also, p_i is the a priori probability of class c_i.

Since we now have labeled and unlabeled instances, we need to define the impurity calculation in each case. The impurity over the target features (labels) is calculated using only the labeled instances (E_l). It is referred to as a labeled impurity. In contrast, the impurity over the descriptive features is calculated using all the labeled and unlabeled instances (E) and is referred to as an unlabeled impurity. Equation 3 presents the labeled impurity calculation for a node of our SSL-PCT. In the equation, Y_i is a label, E_l^{train} represents the whole labeled set of instances before the root node split, and E_l is a subset of instances resulting after the root node split.

$$\text{Impurity}(E_l, Y_i) = \frac{\text{Gini}(E_l, Y_i)}{\text{Gini}(E_l^{train}, Y_i)} \tag{3}$$

The unlabeled impurity is calculated over the descriptive features, which, differently from the labels, can assume numeric or categorical values. Thus, we need to define an impurity for each case. The unlabeled impurities for a dataset E over categorical and numeric descriptive feature X_i are calculated according to Eqs. 4 and 5.

$$\underbrace{\text{Impurity}(E, X_i) = \frac{\text{Gini}(E, X_i)}{\text{Gini}(E^{train}, X_i)}}_{\text{For categorical features}} \tag{4}$$

$$\underbrace{\text{Impurity}(E, X_i) = \frac{\text{Var}(E, X_i)}{\text{Var}(E^{train}, X_i)}}_{\text{For numeric features}} \tag{5}$$

The variance of the ith feature in the set E over the descriptive feature X_i is calculated according to Eq. 6.

$$\text{Var}(E, X_i) = \frac{\sum_{j=1}^{N}(X_i^j)^2 - \frac{1}{N}\cdot(\sum_{j=1}^{N} X_i^j)^2}{N} \tag{6}$$

After modifying the impurity calculations, our proposed SSL-PCT can now handle datasets containing labeled and unlabeled instances, exploiting them according to the ω weight hyperparameter.

4 Methodology

This section presents experiments, software, datasets, and evaluation measures.

4.1 Clus

Clus is an open-source decision tree and rules learning system that implements the predictive clustering framework [3]. It generalizes the decision tree concept by learning trees interpreted as cluster hierarchies, called PCTs. To create a cluster during induction, Clus can be optimized using different heuristics to choose the best split in each internal node, improving prediction performance.

In this work, we used the implementation provided by Vens et al. [32], modifying it to use our proposed semi-supervised impurity function. All implementations and experiments were performed using the Java program language.

4.2 Datasets

UniProt Knowledge Base (UniProtKB) [28] is an expertly curated database, acting as a central access point for integrated protein information with cross-references to multiple sources. The knowledge base consists of UniProtKB/SwissProt and UniProtKB/TrEMBL. The former contains manual and high-quality annotations with information extracted from literature and curator-evaluated computational analysis. The latter contains computationally analyzed records enriched with automatic annotation and classification.

We extracted three UniProtKB datasets: viruses, plants, and fungi. We used data extracted from the UniProtKB/Swiss-Prot 2021_03 version. We used Gene Ontology (GO) [13] terms as features for each protein. We extracted GO terms associated with each protein from the UniProt-GOA [15] database.

The GO is a set of defined vocabularies and terms representing gene functions and products across different species. Two strategies were used to construct feature vectors: 1 and 0 values to indicate the presence or absence of GO terms and the frequency (TF) of each GO term for each protein. Therefore, we constructed six protein datasets, three with categorical descriptive features (1/0) and three with numeric descriptive features (TF).

Table 2 describes the characteristics of the datasets used in our experiments. Card refers to cardinality, the average number of labels per instance in a multi-label dataset. Dens refers to density, which is a dimensionless level of the cardinality. Equations 7 and 8 show how these two measures are calculated. In the equations, Y_i represents a label of the dataset D.

Table 2. Characteristics of the datasets used in the experiments

Name	Num. Instances	Type of feature	Num. Labels	Card	Dens	MeanIR
Virus categorical	645	Categorical	6	1.356	0.226	4.079
Virus numeric	645	Numeric	6	1.356	0.226	4.079
Plants categorical	5283	Categorical	12	1.413	0.118	8.639
Plants numeric	5283	Numeric	12	1.413	0.118	8.639
Fungus categorical	5506	Categorical	10	1.287	0.129	29.168
Fungus numeric	5506	Numeric	10	1.287	0.129	29.168

$$Card(D) = \sum_{i=1}^{|D|} \frac{|Y_i|}{|D|} \quad (7) \qquad Dens(D) = \frac{Card(D)}{|Y|} \quad (8)$$

In Table 2, MeanIR refers to the mean imbalance ratio of the datasets, which is a value that represents the average level of imbalance in a multi-label dataset [8]. This measure is essential because most of the multi-label datasets suffer from a high level of imbalance, making the learning task more complicated.

4.3 Evaluation Measures

To evaluate the performance of our proposed SSL-PCT, we used the multi-label evaluation measures defined in the work of Godbole and Sarawagi [11]. They are presented in Eqs. 9, 10, 11 and 12.

$$Accuracy = \frac{1}{m} \sum_{i=1}^{m} \frac{|Y_i \cap Z_i|}{|Y_i \cup Z_i|} \quad (9) \qquad Precision = \frac{1}{m} \sum_{i=1}^{m} \frac{|Y_i \cap Z_i|}{|Z_i|} \quad (10)$$

$$Recall = \frac{1}{m} \sum_{i=1}^{m} \frac{|Y_i \cap Z_i|}{|Y_i|} \quad (11) \qquad F1 = \frac{1}{m} \sum_{i=1}^{m} \frac{2|Y_i \cap Z_i|}{|Z_i| + |Y_i|} \quad (12)$$

Consider that each multi-label instance is represented by $(\mathbf{x}_i, Y_i)$ in which $i = 1 \ldots m$ (with m the total number of instances). In the equations, $Y_i \subseteq L$ is the set of true labels of instance $\mathbf{x}_i$, and $L = \{\lambda_j : j = 1 \ldots q\}$ is the set of all labels in the problem. Given a specific instance $\mathbf{x}_i$, the set of labels predicted by a classifier is represented by Z_i. The F1 measure in Eq. 12 is the harmonic mean of precision and recall.

5 Experiments and Discussion

We executed our experiments using a 5-fold cross-validation procedure, splitting the data into five partitions containing 20% of the original data. For each fold, the algorithm is trained using the remaining four folds, referred to as a training set, and then used to label the instances in the fold, referred to as the test set. The test set is kept aside from the training data to evaluate the inductive performance of our method. The transductive performance is also evaluated, labeling the instances of the four folds used as the training set. The final result is the average of the evaluation measures calculated for each fold.

Each training set is then divided into labeled and unlabeled portions of instances. Following the recommendation of Wang et al. [34], we do not keep class proportion in both labeled and unlabeled sets since the objective of semi-supervised learning is to exploit the use of unlabeled data to improve prediction performance. The selection of instances to be labeled is performed randomly until a specific criterion is met, and the remaining instances are used as unlabeled. In this case, the labels of the instances are removed and replaced with the symbol "?" so that our algorithm recognizes these instances as unlabeled. We ensure that every labeled portion of the training set has at least one representative instance for each class.

We used different ratios when dividing the training set to test many semi-supervised learning scenarios for each dataset. We performed as follows. Consider C as the number of classes in a dataset E. For every dataset, the labeled portion of the training set was constructed with $C \times \alpha$ instances. The α hyperparameter is a value that controls the number of instances in the labeled portion. We used three values for α: 5, 10 and 20. This strategy was performed for every dataset in Table 2.

Considering the ω hyperparameter, which defines the contribution of supervised and unsupervised parts in the calculation of the impurity value (Eq. 1), we tested six different values: 0.0, 0.2, 0.4, 0.6, 0.8 and 1.0. Thus, we can evaluate if the unlabeled instances improved the prediction performance.

It is worth noting that our proposed SSL-PCT can present different performance on different datasets. This happens because semi-supervised methods are known to be domain dependent [17].

Table 3 shows the Accuracy and F1 values for all the experiments. Acc I and F1 I stand for Accuracy and F1 Inductive performances, while Acc T and F1 T stand for Accuracy and F1 Transductive performances. The best results for each dataset are highlighted in boldface to see the best performance among the variations of the ω hyperparameter. Each value is the Accuracy and F1 average values across 5-fold cross-validation. The rows containing the symbol "-" represent cases in which our heuristic could not find any acceptable test to put in the nodes of the decision trees. Thus, no tree was created in these cases. This only happened for some experiments with the numeric versions of the datasets when the ω hyperparameter was set to 0.0, i.e., when only unlabeled instances were used to calculate the impurity.

As can be seen from our results, in most cases, the best results were obtained when the maximum supervision level was used ($\omega = 1$), indicating that the algorithm could not improve the classification with the unlabeled data. It is possible to see that for numeric datasets, semi-supervision slightly improved the result for some cases compared to the purely supervised version. Even in these cases, the increase was not very significant. In some configurations where $\omega = 0.8$, a semi-supervised strategy could increase the performance by more than 1%.

An unusual point of these results is that the evaluation metrics returned very high values even with very few labeled instances. This is even more unusual when we see that in the experiments with $\alpha = 5$, we have approximately 30 Virus labeled instances (4.65% of the total), 60 Plant labeled instances (1.14% of the total), and 50 Fungus labeled instances (0.91% of the total).

A hypothesis raised to explain this point is that we have several types of GO terms, each one representing a concept[2]. As an example, we have the type ***involved_in***, which tells if a protein is involved in some biological process, or the type ***enables***, which tells whether a protein allows for some biological process to occur. A possible problem may be related to our features since in the cellular

[2] http://geneontology.org/docs/go-annotation-file-gaf-format-2.2/.

Table 3. Results in the Virus, Plant and Fungus Datasets.

	Virus categorical - $\alpha = 5$				Plants categorical - $\alpha = 5$				Fungus categorical - $\alpha = 5$			
ω	Acc I	F1 I	Acc T	F1 T	Acc I	F1 I	Acc T	F1 T	Acc I	F1 I	Acc T	F1 T
1.0	**75.11%**	**79.32%**	**77.48%**	**81.39%**	**74.63%**	**78.06%**	**75.25%**	**78.68%**	**77.55%**	**81.64%**	**77.72%**	**81.70%**
0.8	69.38%	74.08%	72.51%	77.12%	70.61%	74.46%	70.78%	74.69%	69.65%	74.90%	70.42%	75.49%
0.6	69.86%	74.74%	72.94%	77.72%	69.70%	73.63%	69.55%	73.57%	69.60%	74.87%	70.41%	75.49%
0.4	68.58%	73.76%	71.56%	76.60%	69.07%	73.21%	68.96%	73.17%	68.08%	73.70%	68.86%	74.31%
0.2	68.68%	73.46%	71.71%	76.37%	69.09%	73.23%	69.00%	73.19%	68.08%	73.70%	68.86%	74.31%
0.0	70.20%	75.44%	73.34%	78.32%	69.09%	73.23%	69.00%	73.19%	67.42%	73.20%	68.27%	73.86%
	Virus categorical - $\alpha = 10$				Plants categorical - $\alpha = 10$				Fungus categorical - $\alpha = 10$			
ω	Acc I	F1 I	Acc T	F1 T	Acc I	F1 I	Acc T	F1 T	Acc I	F1 I	Acc T	F1 T
1.0	**83.14%**	**86.43%**	**86.91%**	**89.69%**	**79.58%**	**83.14%**	**79.93%**	**83.39%**	**81.27%**	**84.61%**	**81.59%**	**84.82%**
0.8	80.56%	84.20%	83.61%	86.96%	75.26%	78.99%	75.33%	79.15%	77.21%	81.20%	77.70%	81.63%
0.6	79.52%	83.30%	83.89%	87.22%	75.00%	78.67%	75.27%	78.94%	76.71%	80.28%	77.33%	80.84%
0.4	78.40%	82.57%	82.35%	86.01%	75.16%	78.79%	75.41%	79.05%	75.70%	79.42%	76.22%	79.94%
0.2	77.46%	81.73%	81.65%	85.26%	74.88%	78.56%	75.19%	78.89%	75.36%	78.95%	75.76%	79.38%
0.0	75.67%	79.97%	80.10%	83.89%	74.93%	78.56%	75.21%	78.88%	74.11%	77.78%	74.76%	78.46%
	Virus categorical - $\alpha = 20$				Plants categorical - $\alpha = 20$				Fungus categorical - $\alpha = 20$			
ω	Acc I	F1 I	Acc T	F1 T	Acc I	F1 I	Acc T	F1 T	Acc I	F1 I	Acc T	F1 T
1.0	**88.12%**	**90.94%**	**89.30%**	**91.82%**	**83.28%**	**86.26%**	**83.64%**	**86.55%**	**82.86%**	**86.16%**	**83.24%**	**86.50%**
0.8	85.02%	88.15%	86.93%	89.62%	79.94%	83.31%	80.44%	83.89%	82.27%	85.42%	82.88%	86.01%
0.6	83.49%	86.86%	87.00%	89.99%	78.58%	82.01%	79.05%	82.51%	81.67%	84.82%	81.99%	85.11%
0.4	82.81%	85.98%	86.26%	89.12%	78.46%	81.94%	78.66%	82.16%	80.72%	84.10%	80.94%	84.30%
0.2	82.71%	85.82%	86.57%	89.39%	77.91%	81.46%	78.18%	81.74%	78.65%	82.23%	79.04%	82.55%
0.0	79.38%	83.06%	82.21%	85.57%	78.36%	81.78%	78.48%	81.96%	77.99%	81.45%	78.41%	81.81%
	Virus numeric - $\alpha = 5$				Plants numeric - $\alpha = 5$				Fungus numeric - $\alpha = 5$			
ω	Acc I	F1 I	Acc T	F1 T	Acc I	F1 I	Acc T	F1 T	Acc I	F1 I	Acc T	F1 T
1.0	**75.70%**	**80.64%**	**77.54%**	**82.11%**	**72.24%**	**76.07%**	**73.33%**	**76.97%**	75.33%	79.22%	75.59%	79.50%
0.8	75.26%	80.42%	77.24%	82.03%	71.69%	75.63%	72.34%	76.10%	**79.30%**	**82.98%**	**79.32%**	**82.91%**
0.6	69.27%	73.51%	72.35%	76.59%	68.49%	72.42%	69.44%	73.35%	77.01%	81.36%	76.99%	81.21%
0.4	68.25%	72.59%	71.21%	75.63%	62.85%	67.99%	63.60%	68.70%	72.84%	77.20%	73.44%	77.61%
0.2	56.88%	63.04%	60.61%	66.57%	59.13%	64.62%	60.28%	65.76%	57.18%	63.14%	57.19%	63.03%
0.0	45.09%	53.95%	46.41%	55.49%	-	-	-	-	20.60%	22.42%	21.44%	23.29%
	Virus numeric - $\alpha = 10$				Plants numeric - $\alpha = 10$				Fungus numeric - $\alpha = 10$			
ω	Acc I	F1 I	Acc T	F1 T	Acc I	F1 I	Acc T	F1 T	Acc I	F1 I	Acc T	F1 T
1.0	**82.97%**	**86.82%**	**84.92%**	**88.18%**	80.03%	83.18%	80.83%	83.80%	**81.59%**	**84.82%**	**81.51%**	**84.66%**
0.8	80.68%	84.50%	81.24%	84.78%	**81.11%**	**84.26%**	**81.88%**	**84.90%**	80.58%	83.48%	80.21%	83.10%
0.6	82.76%	86.51%	83.49%	86.89%	80.30%	83.74%	81.18%	84.45%	80.59%	83.84%	80.52%	83.73%
0.4	76.77%	81.59%	78.28%	82.67%	78.26%	82.43%	78.81%	82.86%	77.24%	81.41%	76.84%	80.96%
0.2	69.18%	74.97%	71.52%	77.21%	70.87%	75.06%	71.80%	76.03%	72.08%	77.09%	72.40%	77.42%
0.0	38.18%	42.04%	38.69%	42.42%	31.64%	34.82%	32.34%	35.42%	-	-	-	-
	Virus numeric - $\alpha = 20$				Plants numeric - $\alpha = 20$				Fungus numeric - $\alpha = 20$			
ω	Acc I	F1 I	Acc T	F1 T	Acc I	F1 I	Acc T	F1 T	Acc I	F1 I	Acc T	F1 T
1.0	87.39%	90.12%	**89.03%**	**91.46%**	**85.70%**	**88.43%**	**86.10%**	**88.60%**	**86.50%**	**89.09%**	**87.34%**	**89.81%**
0.8	**87.90%**	**90.79%**	88.34%	91.18%	84.79%	87.72%	85.34%	88.07%	85.65%	88.45%	86.50%	89.12%
0.6	87.48%	90.48%	87.99%	90.67%	83.89%	87.02%	84.49%	87.44%	84.99%	87.95%	85.87%	88.63%
0.4	85.96%	89.36%	87.04%	89.99%	83.54%	86.70%	84.34%	87.33%	82.56%	86.01%	84.02%	87.15%
0.2	82.67%	86.53%	84.51%	88.01%	80.85%	84.48%	81.95%	85.33%	78.98%	83.51%	79.68%	84.02%
0.0	-	-	-	-	43.17%	46.78%	44.43%	48.10%	10.97%	12.54%	11.02%	12.47%

Table 4. Results in the Virus, Plant and Fungus Datasets without *located_in* Terms

	Virus categorical - $\alpha = 5$				Plants categorical - $\alpha = 5$				Fungus categorical - $\alpha = 5$			
ω	Acc I	F1 I	Acc T	F1 T	Acc I	F1 I	Acc T	F1 T	Acc I	F1 I	Acc T	F1 T
1.0	**36.10%**	42.99%	**39.71%**	**46.33%**	**27.33%**	**30.33%**	**28.68%**	**31.68%**	**43.80%**	**48.92%**	**43.49%**	48.64%
0.8	33.65%	41.20%	36.68%	44.04%	25.00%	27.68%	25.57%	28.18%	39.98%	43.91%	40.01%	**43.84%**
0.6	33.27%	41.29%	36.60%	44.16%	25.11%	27.87%	25.70%	28.37%	39.99%	43.92%	40.01%	**43.84%**
0.4	34.13%	41.91%	37.02%	44.42%	25.11%	27.87%	25.71%	28.38%	39.99%	43.92%	40.01%	**43.84%**
0.2	35.91%	**43.90%**	38.50%	46.18%	25.11%	27.87%	25.71%	28.38%	39.99%	43.92%	40.01%	**43.84%**
0.0	35.31%	43.31%	37.91%	45.67%	25.13%	27.90%	25.71%	28.38%	39.88%	43.86%	39.91%	43.80%
	Virus categorical - $\alpha = 10$				Plants categorical - $\alpha = 10$				Fungus categorical - $\alpha = 10$			
ω	Acc I	F1 I	Acc T	F1 T	Acc I	F1 I	Acc T	F1 T	Acc I	F1 I	Acc T	F1 T
1.0	**44.16%**	**49.80%**	**44.04%**	**50.24%**	**34.02%**	**38.15%**	**33.68%**	**37.65%**	**52.65%**	**58.02%**	**53.26%**	**58.52%**
0.8	35.64%	41.10%	38.73%	44.84%	25.05%	28.08%	25.11%	28.12%	39.37%	44.09%	40.66%	45.39%
0.6	36.67%	43.14%	39.29%	46.20%	24.58%	27.69%	25.06%	28.11%	39.83%	44.54%	41.12%	45.82%
0.4	37.40%	44.09%	39.50%	46.54%	24.09%	27.00%	24.66%	27.58%	40.04%	44.79%	41.21%	45.94%
0.2	37.10%	43.86%	39.87%	47.08%	24.06%	26.89%	24.52%	27.36%	40.26%	45.03%	41.23%	45.97%
0.0	39.27%	46.19%	41.14%	48.40%	24.07%	26.93%	24.48%	27.33%	40.19%	44.96%	41.20%	45.94%
	Virus categorical - $\alpha = 20$				Plants categorical - $\alpha = 20$				Fungus categorical - $\alpha = 20$			
ω	Acc I	F1 I	Acc T	F1 T	Acc I	F1 I	Acc T	F1 T	Acc I	F1 I	Acc T	F1 T
1.0	**45.00%**	**51.76%**	**53.35%**	**59.79%**	**36.75%**	**40.16%**	**39.42%**	**42.93%**	**55.00%**	**59.43%**	**55.67%**	**60.15%**
0.8	43.46%	50.45%	51.61%	57.86%	28.85%	32.09%	31.16%	34.59%	42.13%	46.19%	43.40%	47.45%
0.6	40.91%	47.43%	47.98%	54.01%	29.16%	32.61%	31.25%	34.80%	41.79%	45.63%	42.78%	46.69%
0.4	40.51%	46.98%	48.69%	54.64%	28.96%	32.51%	31.00%	34.68%	41.76%	45.60%	42.84%	46.77%
0.2	39.25%	45.40%	48.53%	54.43%	28.91%	32.40%	30.84%	34.51%	41.92%	45.70%	42.68%	46.58%
0.0	40.25%	46.58%	47.76%	53.90%	29.11%	32.61%	30.81%	34.46%	41.66%	45.39%	42.42%	46.28%
	Virus numeric - $\alpha = 5$				Plants numeric - $\alpha = 5$				Fungus numeric - $\alpha = 5$			
ω	Acc I	F1 I	Acc T	F1 T	Acc I	F1 I	Acc T	F1 T	Acc I	F1 I	Acc T	F1 T
1.0	**44.15%**	**51.45%**	**46.38%**	**53.03%**	26.32%	30.70%	27.53%	31.94%	**44.14%**	**48.42%**	**44.57%**	**48.84%**
0.8	42.41%	50.47%	44.14%	51.68%	26.23%	30.09%	27.06%	31.12%	41.47%	47.53%	42.49%	48.65%
0.6	41.18%	47.03%	42.30%	47.59%	26.55%	31.14%	27.27%	31.92%	41.65%	47.27%	42.16%	47.82%
0.4	39.61%	46.27%	40.29%	45.94%	28.02%	32.70%	28.20%	33.00%	36.63%	42.21%	37.13%	42.69%
0.2	38.07%	44.43%	38.87%	44.49%	**28.84%**	**33.01%**	**28.97%**	**33.21%**	33.46%	40.17%	33.98%	40.51%
0.0	40.29%	44.17%	40.06%	43.72%	11.31%	13.63%	11.54%	13.93%	12.39%	15.33%	12.61%	15.52%
	Virus numeric - $\alpha = 10$				Plants numeric - $\alpha = 10$				Fungus numeric - $\alpha = 10$			
ω	Acc I	F1 I	Acc T	F1 T	Acc I	F1 I	Acc T	F1 T	Acc I	F1 I	Acc T	F1 T
1.0	44.22%	50.92%	46.66%	53.53%	31.58%	35.21%	32.34%	36.06%	**52.69%**	**57.74%**	**53.34%**	**58.49%**
0.8	44.10%	51.00%	**47.73%**	**54.47%**	30.96%	34.97%	31.42%	35.49%	50.34%	54.90%	50.77%	55.55%
0.6	**45.13%**	**52.15%**	47.42%	54.42%	**32.14%**	**35.71%**	**32.57%**	**36.18%**	49.54%	54.22%	49.98%	54.75%
0.4	40.55%	47.25%	43.60%	50.47%	30.53%	34.52%	31.42%	35.49%	47.06%	52.33%	47.61%	52.86%
0.2	44.44%	51.34%	45.91%	53.24%	30.37%	34.28%	31.03%	35.01%	42.55%	48.43%	42.54%	48.43%
0.0	39.13%	44.56%	40.18%	45.62%	10.10%	11.67%	10.89%	12.48%	-	-	-	-
	Virus numeric - $\alpha = 20$				Plants numeric - $\alpha = 20$				Fungus numeric - $\alpha = 20$			
ω	Acc I	F1 I	Acc T	F1 T	Acc I	F1 I	Acc T	F1 T	Acc I	F1 I	Acc T	F1 T
1.0	44.37%	50.67%	53.06%	59.04%	35.33%	38.85%	36.77%	40.45%	**54.74%**	**59.27%**	**55.97%**	**60.45%**
0.8	45.99%	52.85%	54.31%	60.66%	35.31%	39.17%	37.40%	41.23%	54.27%	58.71%	55.07%	59.45%
0.6	**46.18%**	**52.91%**	52.69%	58.92%	35.80%	39.50%	37.91%	41.63%	53.75%	58.00%	54.69%	59.01%
0.4	43.91%	51.15%	**54.53%**	**60.94%**	**37.22%**	**41.20%**	**38.84%**	**42.89%**	52.80%	57.59%	54.40%	59.16%
0.2	40.92%	47.76%	49.77%	56.20%	32.74%	37.13%	34.76%	39.29%	46.98%	52.02%	47.10%	52.07%
0.0	42.08%	48.24%	49.53%	55.57%	20.07%	22.60%	21.29%	23.93%	17.58%	19.97%	18.14%	20.49%

components' ontology, we have annotations with the term ***located_in***. This term appears in a protein when we have evidence that it is present in any cellular component, i.e., in this case, we have a descriptive feature representing almost the same concept of the target feature: subcellular localization. This can explain why the evaluation measures achieved high results, even with very few proteins, given that these ***located_in*** features will always be great candidates to be the best features to partition the dataset. New experiments were executed to confirm this hypothesis, removing all GO terms of the type ***located_in***. Table 4 shows the results of these new experiments. Again, the best results are highlighted in boldface: Acc I and F1 I stand for Accuracy and F1 Inductive performances, and Acc T and F1 T stand for Accuracy and F1 Transductive performances.

From these new results, we can see that our hypothesis seems to be correct since the decision trees trained in the datasets without the ***located_in*** terms obtained much lower values for the evaluation measures. However, these values better represent the reality of the taxonomies with few proteins with annotated subcellular localizations.

6 Conclusions and Future Work

In this paper, we proposed a Semi-Supervised Predictive Clustering Tree (SSL-PCT) algorithm capable of classifying multi-label instances exploiting labeled and unlabeled data. Our proposal modifies the Clus algorithm [32], introducing a new impurity measure for the decision tree heuristic. Our new impurity uses labeled and unlabeled instances to find the best split in a decision tree node.

Our proposed algorithm is also flexible, as we have an ω hyperparameter that controls the tree's supervision level. If $\omega = 1$, the tree acts precisely like a supervised PCT. If $\omega = 0$, the algorithm only uses the descriptive features from the labeled and unlabeled instances to construct the tree, without the information about the labels.

We applied our method to different protein subcellular localization datasets. We also compared many values of the ω hyperparameter, varying its value from 0.0 to 1.0 in steps of 0.2. We also investigated the performance of SSL-PCT when used with different amounts of labeled instances. The results show that our new method can exploit unlabeled data to help the classification in many scenarios.

Our experiments showed that the best results typically relied only on the labeled data. However, the experiments also showed that the unsupervised part of the algorithm can help improve the classification results.

In future work, we plan to expand the experiments in various ways, such as investigating more well-known multi-label datasets, better investigating the variation in the ω hyperparameter to analyze the influence of the level of supervision in the datasets, evaluating the prediction performance with more multi-label evaluation measures, and perform experiments with more variations in the proportions of labeled and unlabeled data in our datasets.

We will also study the possibility of creating semi-supervised learning random forests within our SSL-PCT. Random Forests are one of the best classification algorithms, and their ability to generate a feature ranking can help define the best features in multi-label datasets. We also plan to apply some pre-processing to the protein datasets to improve the results using unlabeled data. Since, in general, only a few features are used to construct the trees, there is an indication that we may remove some less important features in the used protein datasets.

Acknowledgment. This study was financed by the São Paulo Research Foundation (FAPESP) grants #2016/25220-1, #2017/24807-1 and #2022/02981-8. I. Triguero is funded by a Maria Zambrano Senior Fellowship at the University of Granada. I. Triguero's work is also supported by PID2023-149128NB-I00.

References

1. Almagro Armenteros, J.J., Sønderby, C.K., Sønderby, S.K., Nielsen, H., Winther, O.: DeepLoc: prediction of protein subcellular localization using deep learning. Bioinformatics **33**(21), 3387–3395 (2017)
2. Basgalupp, M., Cerri, R., Schietgat, L., Triguero, I., Vens, C.: Beyond global and local multi-target learning. Inf. Sci. **579**, 508–524 (2021)
3. Blockeel, H., Raedt, L.D., Ramon, J.: Top-down induction of clustering trees. In: Proceedings of the Fifteenth International Conference on Machine Learning, ICML 1998, pp. 55–63. Morgan Kaufmann Publishers Inc. (1998)
4. Bogatinovski, J., Todorovski, L., Džeroski, S., Kocev, D.: Comprehensive comparative study of multi-label classification methods. Expert Syst. Appl. **203**, 117215 (2022)
5. Breiman, L., Friedman, J., Olshen, R., Stone, C.: Classification and Regression Trees. Wadsworth and Brooks, Monterey (1984)
6. Cao, J., Liu, W., He, J., Gu, H.: Mining proteins with non-experimental annotations based on an active sample selection strategy for predicting protein subcellular localization. PLOS One **8**, e67343 (2013)
7. Caragea, C., Caragea, D., Silvescu, A., Honavar, V.: Semi-supervised prediction of protein subcellular localization using abstraction augmented Markov models. BMC Bioinform. **11**(Suppl 8), S6 (2010)
8. Charte, F., Rivera, A.J., Del Jesus, M.J., Herrera, F.: Addressing imbalance in multilabel classification: measures and random resampling algorithms. Neurocomputing **163**, 3–16 (2015)
9. Cui, Q., JiangEmail, T., Liu, B., Ma, S.: Esub8: a novel tool to predict protein subcellular localizations in eukaryotic organisms. BMC Bioinform. **5**, 66 (2004)
10. Džeroski, S., Gjorgjioski, V., Slavkov, I., Struyf, J.: Analysis of time series data with predictive clustering trees. In: Džeroski, S., Struyf, J. (eds.) KDID 2006. LNCS, vol. 4747, pp. 63–80. Springer, Heidelberg (2007). https://doi.org/10.1007/978-3-540-75549-4_5
11. Godbole, S., Sarawagi, S.: Discriminative methods for multi-labeled classification. In: Dai, H., Srikant, R., Zhang, C. (eds.) PAKDD 2004. LNCS (LNAI), vol. 3056, pp. 22–30. Springer, Heidelberg (2004). https://doi.org/10.1007/978-3-540-24775-3_5
12. Guo, X., Liu, F., Ju, Y., Wang, Z., Wang, C.: Human protein subcellular localization with integrated source and multi-label ensemble classifier. Sci. Rep. **6**, 28087 (2016)

13. Harris, M.A., Clark, J., Ireland, A.: The gene ontology (go) database and informatics resource. Nucl. Acids Res **32**, D258-61 (2004)
14. Hawkins, D.M.: The Problem of Overfitting. J. Chem. Inf. Comput. Sci. **44**, 1–12 (2004)
15. Huntley, R.P., et al.: The GOA database: gene ontology annotation updates for 2015. Nucl. Acids Res. **43**, D1057–D1063 (2015)
16. Kocev, D., Slavkov, I., Dzeroski, S.: Feature ranking for multi-label classification using predictive clustering trees. In: Proceedings of Companion Publication of the European Conference on Machine Learning and Knowledge Discovery in Databases (2013)
17. Levatić, J., Ceci, M., Kocev, D., DźEroski, S.: Semi-supervised classification trees. J. Intell. Inf. Syst. **49**(3), 461–486 (2017)
18. Madjarov, G., Kocev, D., Gjorgjevikj, D., Džeroski, S.: An extensive experimental comparison of methods for multi-label learning. Pattern Recogn. **45**(9), 3084–3104 (2012)
19. Pise, N.N., Kulkarni, P.: A survey of semi-supervised learning methods. In: 2008 International Conference on Computational Intelligence and Security, vol. 2, pp. 30–34 (2008)
20. Pliakos, K., Vens, C.: Drug-target interaction prediction with tree-ensemble learning and output space reconstruction. BMC Bioinform. **21** (2020)
21. Rey, S., Gardy, J.L., Brinkman, F.S.: Assessing the precision of high-throughput computational and laboratory approaches for the genome-wide identification of protein subcellular localization in bacteria. BMC Genom. **6**, 162 (2005)
22. Sadarangani, A., Jivani, A.: A survey of semi-supervised learning. Int. J. Eng. Sci. Res. Technol. **5**(10), 138–143 (2016)
23. Shen, H., Chou, K.: Virus-mPLoc: a fusion classifier for viral protein subcellular location prediction by incorporating multiple sites. J. Biomol. Struct. Dyn. **28**(2), 175–186 (2010)
24. Struyf, J., Džeroski, S.: Constraint based induction of multi-objective regression trees. In: Bonchi, F., Boulicaut, J.-F. (eds.) KDID 2005. LNCS, vol. 3933, pp. 222–233. Springer, Heidelberg (2006). https://doi.org/10.1007/11733492_13
25. Struyf, J., Džeroski, S.: Clustering trees with instance level constraints. In: Kok, J.N., Koronacki, J., Mantaras, R.L., Matwin, S., Mladenič, D., Skowron, A. (eds.) ECML 2007. LNCS (LNAI), vol. 4701, pp. 359–370. Springer, Heidelberg (2007). https://doi.org/10.1007/978-3-540-74958-5_34
26. Su, E., Chiu, H., Lo, A., Hwang, J., Sung, T., Hsu, W.: Protein subcellular localization prediction based on compartment-specific features and structure conservation. BMC Bioinform. **8**, 330 (2007)
27. Su, R., He, L., Liu, T., Liu, X., Wei, L.: Protein subcellular localization based on deep image features and criterion learning strategy. Brief. Bioinform. **22**(4), bbaa313 (2020)
28. The UniProt Consortium: UniProt: the universal protein knowledgebase. Nucl. Acids Res. **45**(Issue D1), D158–D169 (2017)
29. Thumuluri, V., Almagro Armenteros, J.J., Johansen, A., Nielsen, H., Winther, O.: DeepLoc 2.0: multi-label subcellular localization prediction using protein language models. Nucl. Acids Res. **50**(W1), W228–W234 (2022)
30. Triguero, I., Garcia, S., Herrera, F.: Self-labeled techniques for semi-supervised learning: taxonomy, software and empirical study. Knowl. Inf. Syst. **42**(2), 245–284 (2015)

31. Tsoumakas, G., Katakis, I., Vlahavas, I.: Mining multi-label data. In: Maimon, O., Rokach, L. (eds.) Data Mining and Knowledge Discovery Handbook, pp. 667–685. Springer, Boston (2009). https://doi.org/10.1007/978-0-387-09823-4_34
32. Vens, C., Struyf, J., Schietgat, L., Džeroski, S., Blockeel, H.: Decision trees for hierarchical multi-label classification. Mach. Learn. **73**, 185–214 (2008)
33. Wan, S., Mak, M.W., Kung, S.Y.: mGOASVM: multi-label protein subcellular localization based on gene ontology and support vector machines. BMC Bioinform. **13**, 290 (2012)
34. Wang, Y., Xu, X., Zhao, H., Hua, Z.: Semi-supervised learning based on nearest neighbor rule and cut edges. Knowl.-Based Syst. **23**(6), 547–554 (2010)
35. Xu, Q., Hu, D., Xue, H., Yu, W., Yang, Q.: Semi-supervised protein subcellular localization. BMC Bioinform. **10**(Suppl 1), S47 (2009)
36. Zhang, Q., et al.: Accurate prediction of multi-label protein subcellular localization through multi-view feature learning with RBRL classifier. Brief. Bioinform. **22**(5) (2021)

Siamese Network-Based Prioritization for Enhanced Multi-document Summarization

Klaifer Garcia and Lilian Berton(✉)

Universidade Federal de Sao Paulo, Avenida Cesare Mansueto Giulio Lattes, nř 1201, Eugênio de Mello, SP, Brazil
lberton@unifesp.br

Abstract. Methods for document summarization hold significance in numerous applications, particularly in scenarios involving extensive content, such as news and social media monitoring. Nevertheless, enhancing the task of summarizing multiple documents remains an area for improvement. Systems that generate summaries from multiple documents must take into account additional challenges such as document redundancies and inconsistencies. On the other hand, as multi-document databases contain multiple descriptions of the same content, they offer benefits that go beyond the availability of more content and we believe that this benefit has not been explored in the literature. Different authors may write different descriptions of the same event, some more objective, others more detailed, or even with different terms or writing styles. As a way to take advantage of this availability of different descriptions, we present a new approach to evaluating documents in order to identify those that can contribute most effectively to the generation of automatic summaries. For this, we employed a Siamese network, which was trained using the ROUGE metric observed in individual documents. Additionally, we demonstrated how to apply the outcomes of document evaluation to different summarization techniques. Our experiments included a comparison with SOTA approaches (LeadSum, TextRank, PacSum, BertSum) in multiple datasets encompassing news, events, Wikipedia texts, and scientific publications (Multi-News, WCEP, WikiSum, arXiv, Multi-XScience) and the results indicated our approach had a relevant improvement in the production of summaries, with statistical significance in the evaluation with Wilcoxon rank sum for a confidence factor of 95%.

Keywords: Multi-document summarization · Siamese network · NLP

1 Introduction

Automatic text summarization is the process that applies computational methods to produce a condensed version of one or more input documents while preserving the most relevant information [7]. Particularly in the case of summarizing

A. Paes and F. A. N. Verri (Eds.): BRACIS 2024, LNAI 15413, pp. 400–415, 2025.
https://doi.org/10.1007/978-3-031-79032-4_28

multiple documents, in addition to selecting the most relevant content, there is the challenge of avoiding redundancies and inconsistencies. News summarization is an example where an ongoing event can be reported by different channels, resulting in different descriptions with different updates.

According to [22], many of the multi-document summarization models combine inputs as a flat sequence and do not consider relationships between documents. We believe the combination of input documents can help provide a more accurate description of the content, and our goal is to improve summarization by leveraging this variety of input documents.

More specifically, we will apply a Siamese network [13] to evaluate different input documents and identify those capable of producing better summaries. The difference between documents arises, for example, from the different volumes of information and objectivity. To emphasize these differences, we will consider different extractive summarization techniques. Extractive summarization aims to produce a summary selecting a representative subset of the sentences or paragraphs from the source text without making any changes in them, opposed to abstractive summarization which aims to represent the input content with new words and sentences. Despite the promising results with abstractive methods, it is not possible to guarantee that the summaries produced will be consistent with the facts [32], which can be limiting in some usage scenarios.

The Siamese Neural Network is a natural choice for our approach, as these networks are efficient in comparison learning. We use datasets composed of multiple input documents and at least one reference summary. Every input document will be evaluated using the ROUGE metric and, after that, it will be reorganized into pairs indicating which one obtained the best evaluation. This arrangement is compatible with the Siamese network application, which needs to learn to recognize the qualities that enable one document to be evaluated more highly than another. We used a training dataset and a separate test dataset, demonstrating generalization to unseen data. As an additional benefit, reorganizing documents into pairs produces a larger volume of training data.

Our main contributions are as follows:

- Proposal of a document comparison method, based on Siamese networks, applied to the automatic summarization of multiple documents;
- Description of changes made to a variety of summarization methods to take advantage of document prioritization;
- Experiments validated with statistical tests pointing to relevant differences in our results;
- Comparison of some popular summarization techniques using the same datasets and testing methodology.
- Our document prioritizing code is made publicly available[1].

The remainder of this article is organized as follows. We review the related work in Sect. 2. Section 3 presents the proposed model description. Section 4 describes the datasets, the baseline methods, the changes made to the methods to

[1] https://github.com/Klaifer/DocScore.

include document prioritization and information about the testing methodology. In Sect. 5 the results and discussions are presented. Finally, Sect. 6 brings the conclusion and future works.

2 Related Work

Cross-document relations should be taken into account while summarizing multiple documents. These relationships are typically used by summarizing techniques to reduce repetition and identify the most important details. Applying clustering techniques, such as in [8,33], is a way to reduce redundancy.

Graphs are often used to represent the relationship between pieces of text, as in [31]. This work used heterogeneous vertices to represent information of larger granularity, such as sentences, or smaller granularity, such as words. The vertices contained embeddings to represent information, which were updated with graph attention networks [29]. In [2] GRAPHSUM is presented, which analyzes text fragments to select sentences. Text fragments are represented as nodes and a variation of the Pagerank method is proposed to favor positive correlations.

The encoder-decoder framework is a widely used method in abstractive summarization approaches, including multi-document scenarios. With this method, the encoder layer aims to create a representation of the input texts, and the decoder generates new sentences using this representation. The methodology has shown good results, although there are still some challenges, especially with long documents. To mitigate this difficulty, [15] proposed the use of graphs to guide abstractive summarization.

A popular datasets for evaluating multi-document approaches is presented in [9], which also proposes a summarization technique. The suggested approach is a change to the Point-Generator summarizer [26], applying an attention layer that includes the importance of sentences. To determine the importance of sentences, a bidirectional LSTM network is applied to produce embeddings. Using the sentence representations as input, a similar process is applied to produce the document representation. This outputs are used to identify the most important sentences, which occurs with a variation of the MMR [3] technique.

In the works retrieved, the main motivation for extracting cross-document relations is the reduction of redundancy and identification of the most important sentences. In contrast to these approaches, our goal is to evaluate entire documents. Beyond just finding documents with the most pertinent information, our method seeks to identify those that are capable of producing the best summaries. A possible application, which will be demonstrated in our experiments, is the use of an additional layer in the evaluation of sentences, prioritizing the sentences that are present in the best documents.

3 Model Description

The suggested approach seeks to determine which articles produce the best summaries, allowing documents to be ranked in order of importance for information extraction. The hypothesis is that better summaries can be produced when more sentences are extracted from the best documents. In the datasets we consider, which are presented in Sect. 4, we observe that each sample is composed of a variable number of input documents. For simplicity, we suggest a technique for evaluating pairs of documents, which allows us to rank documents by applying multiple comparisons.

In this way, the model needs to be able to determine which of the two input texts has the highest likelihood of yielding a good summary, and Siamese networks are useful for learning by comparison [16]. In its simplest configuration, a Siamese network is composed of two attribute extraction networks, one for each input, and a comparison head. This architecture is flexible enough to allow deep-learning techniques to be applied to the feature extraction layer, which is a configuration that has achieved state-of-the-art results in different tasks [1,14,30]. Additionally, the two feature extraction networks have identical connection weights, ensuring a similar processing method for both input data.

In our model, as depicted in Fig. 1, each input document is processed by a neural network and produces a representation as document embeddings. These embeddings are used in a classifier to identify the best document. During training, the error observed in the classification is used to adjust the entire network, including the two networks that produce the document embeddings. With this approach, the reduction of errors in the classifier occurs through the optimization of document representations, which requires the network to learn how to identify the most useful attributes for evaluating document quality.

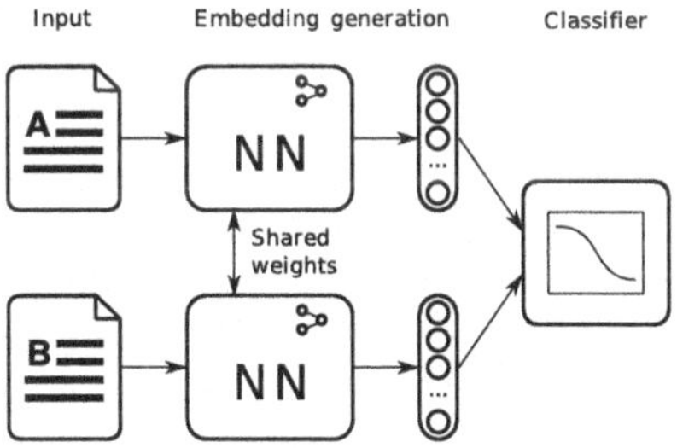

Fig. 1. Document comparison using Siamese network. Two documents are evaluated by neural networks with the same connection values. The output is embeddings that are input to a classifier.

Figure 2 shows the network in more detail up until the point where the document representation stage is completed. In this figure, the input documents are used to produce word embeddings that are processed by a convolutional neural network (CNN) [17]. CNN provide adjustable configurations that enable evaluating the precision increase as a function of computing cost in addition to the possibility of applying parallel processing. Furthermore, this type of network has been successfully applied to a variety of problems such as news classification [10], medical image classification [5] and energy demand prediction [25].

Our network is composed of one-dimensional filters, with multiple planes and different window sizes. This means that the filters consider each word embedding as a multidimensional point, that the filters are moved along the text, and that at each iteration several tokens are evaluated simultaneously, depending on the size of the window. Filter results are compiled in a max-pooling layer. During the training, we included a dropout procedure, which in our experiments helped to increase generalization. At the end, a fully connected layer is included to performs dimensionality reduction and produce the document representation. The step after producing document embeddings, not represented in this figure, is the union of two document representations in a fully connected layer and a classifier activated with a Sigmoid function.

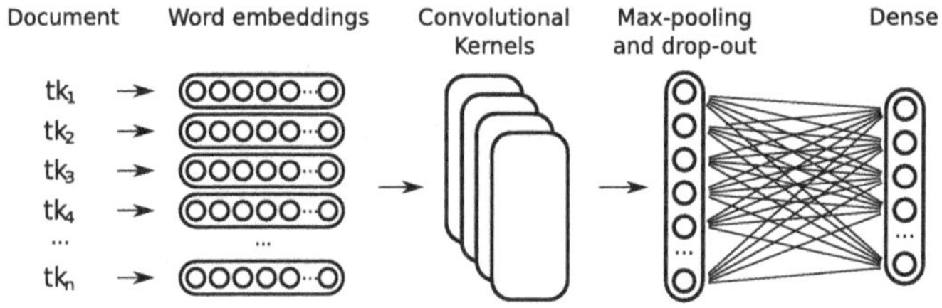

Fig. 2. Document embedding generation. Tokens are converted to chains of word-embeddings, which are evaluated by convolutional filters. Outputs are reduced with max-pooling and document embeddings are produced with a dense layer.

To train the network, we need labeled pairs of documents that we construct using the training data. We start by evaluating all documents with the ROUGE metric. Using this evaluation, we choose two documents, d_1 and d_2, and assign the label 1 if the ROUGE evaluation of d_1 is better than d_2 and 0 otherwise. We take care during this process to maintain class balance and avoid duplicating documents in order to prevent overfitting. Furthermore, as pointed out by [28], differences in the lengths can interfere with the evaluation with ROUGE, so we are using summaries of the input documents truncated with the same number of tokens. For simplicity, summaries are produced by extracting the initial tokens from documents and we leave other approaches for future work.

To apply the trained network to order sets of documents, we need a strategy using pairwise comparisons. Since each sample we considered was composed of a few input documents, for the sake of simplicity, we are applying the comparison to all possible pairs. In this way, considering all possible pairs of documents, we count the number of times that a document d_x obtained a score higher than d_y, according to Eq. 1. In this equation, D is the set of input documents, and $S(d_i)$ is the output of the Siamese network.

$$score(d_x) = \mid \{d_y \in D \mid d_y \neq d_x,\ S(d_x) > S(d_y)\} \mid \tag{1}$$

Equation 1 is repeated for all documents allowing the identification of the best ones. To simplify integration with the other methods that will be presented in Sect. 4, we normalize the document scores to sum to 1.

4 Materials and Methods

This section presents the materials used in the experiments, including the datasets and baseline methods. It also describes the changes we made to include document prioritization and some implementation details.

4.1 Datasets

Few datasets are available for multi-document summarization, mainly because producing reference summaries is labor-intensive. Some of the datasets we use are composed of news, where it is possible to find different sources describing the same event.

Multi-News [9]**:** Contains articles extracted from the news aggregation Newser[2]. This site contains human-written news summaries that include citations to external sources of information, gathering information from over 1,500 news channels. To compose this dataset, summaries are used as references and the news is cited as sources. There are a few versions of this data and we are using the pre-processed version without truncation.

WCEP [11]**:** Contains documents extracted from the Wikipedia Current Events Portal[3]. Following Wikipedia guidelines, summaries are short, approximately 35 words, written in the present tense and avoiding opinions and sensationalism. Each summary averages only 1.2 cited sources, but this number has been incremented through documents extracted from the Common Crawl News dataset[4]. We are using the version distributed by the authors, which was truncated at a maximum of 100 documents per sample.

WikiSum [18]**:** Wikipedia was used to create this dataset by providing reference summaries. The task is to create the lead, or first part, of the Wikipedia page using the references listed in the article and ten Google search engine results. However, due to its enormous size, some works only use a part of this data. In this work we are using the partition of [20], which contains the first 40 paragraphs selected with a logistic regression method that was trained using ROUGE-2-Recall from the comparison of paragraphs with the reference summary. Since the input data is made up of separate paragraphs that are not identified with the document from which it was taken, we are treating each paragraph as a separate document.

[2] www.newser.com.

[3] https://en.wikipedia.org/wiki/Wikipedia:How_the_Current_events_page_works.

[4] https://commoncrawl.org/2016/10/news-dataset-available/.

arXiv [4]**:** Scientific publications taken from arXiv[5] are included in this dataset. Since many of the datasets now in use are composed of news articles or other shorter content, the objective was to develop a dataset composed of longer documents. The extracted content contains the abstracts, which are used as reference summaries, and the article sections which are used as input. Since we are interested in multi-document datasets, we are considering each section as an input document.

Multi-XScience [21]**:** It is a dataset of scientific articles, where the objective is to produce the related works section using the abstract of the article itself and a collection of reference documents. The articles were extracted from arXiv, and related documents were identified by applying a set of heuristics in Microsoft Academic Graph [27].

Table 1 presents a summary of the evaluated datasets. The "Train Test Val" corresponds to the number of samples in each data partition. "Doc" is the average number of input documents in each sample. "Doc/Ref Len" is the average number of tokens per document and reference, calculated using the nltk[6] library.

Table 1. Dataset attributes

Data	Train Test Val	Doc	Doc/Ref Len
Multi-News	44,972 5,622 5,622	2.8	760.7 264
WCEP	8,158 1,022 1,020	63	483 31.8
WikiSum	1,578,588 38,205 34,353	39.5	57.1 107.6
arXiv	203,037 6,440 6,436	5.5	1,202.6 286.4
XScience	30,369 5,093 5,066	4.4	176.3 116.1

4.2 Baseline and Integration Methods

In this section, we describe the methods we used to compare the results and the procedure we applied to integrate with document prioritization.

LeadSum: Since a part of the datasets is composed of news, we are considering extracting content from the beginning of articles. This strategy is justified by the fact that in journalistic content the first sentences are usually the most important [12]. In this way, all content is merged into a single large document, and the initial content is extracted until the length defined for the summary is reached.

DocScore+LeadSum: To use document prioritization, we are reordering the input documents, moving the most relevant ones to the beginning. This way, if

[5] https://arxiv.org/.
[6] www.nltk.org.

the most relevant document is larger than the size of the summary, all content will be extracted from this source. Otherwise, content from other documents will be used following the same order of relevance.

TextRank [23]: Is an unsupervised extractive summarization method based on graphs. In this method, the sentences are the vertices and the similarity between the sentences is used to define connection weights. Using this graph, PageRank is applied to assign a score to each sentence, which is used to rank the sentences. Summaries are produced by selecting the sentences with the highest score. In our implementation, the similarity between sentences was calculated with Tf-Idf (term frequency–inverse document frequency).

DocScore+TextRank: The result of processing with TextRank is a score value related to each sentence. To integrate with document prioritization, we are adding the document score to the paragraph score, resulting in a greater probability of selecting the paragraphs from the best documents. In this sum, we are applying a weighting factor, according to Eq. 2, where $T(s_{d,i})$ is the score obtained with TextRank for sentence i of document d, $S(d)$ is the document score obtained with our model and α is an adjustable parameter.

$$score(s_{d,i}) = T(s_{d,i}) + \alpha * S(d) \tag{2}$$

PacSum [34]**:** This is a summarization method that represents the input contents in the form of digraphs. The authors argue that when selecting sentences to construct summaries, there are those that contain the most relevant information and those that are complementary. A simple approach is proposed to identify the most relevant sentences considering their position of occurrence in the text. In the proposed digraph, the vertices are sentence embeddings, and the direction of the connections is defined by the order of occurrence of the paragraphs in the text. The most central sentences are selected, and centrality is defined with Eq. 3. In this equation, i and j are positions of sentences in the input documents, s_i is a sentence, $e_{i,j}$ is a measure of similarity between sentences s_i and s_j and λ_1 and λ_2 are adjustable parameters that determine the importance of the initial and final sentences, respectively.

$$centrality(s_i) = \lambda_1 \sum_{j<i} e_{i,j} + \lambda_2 \sum_{j>i} e_{i,j} \tag{3}$$

DocScore+PacSum: PacSum's adjustable parameters are calibrated through a supervised process. In particular, the parameters λ_1 and λ_2 are configured to prioritize the initial and final sentences. This way, the method already has a mechanism in addition to sentence similarity that we will reuse to prioritize the most important documents. To achieve this, our approach was to reorder the input documents, keeping the most relevant ones first, which allows the parameters λ_1 and λ_2 to be adjusted more effectively.

BertSum [19]**:** It is a supervised method for producing extractive summaries using the BERT [6] network. The network receives the content to be summarized, which has been changed to include sentence separation and produce contextualized sentence embeddings. To select sentences, some approaches are evaluated, including a simple classifier, an inter-sentence Transformer, and an LSTM network. As an additional resource to avoid redundancy, summaries are produced avoiding inserting sentences with repeated trigrams.

DocScore+BertSum: It is challenging to use BertSum when there is a large amount of input data because it processes all input sentences at once. In the BERT network, input documents are truncated into 512 tokens, which in our test case can lead to the deletion of complete documents. To minimize this limitation, we reorganize the input documents in a manner similar to our approach using PacSum, allowing the best documents to participate in the summarization.

Oracle: Inserted as an upper bound for extractive methods. We are using a greedy approach similar to the one presented in [19]. Starting with an empty set, in each iteration, we include a sentence, and this sentence must be the one that increases the ROUGE metric the most of all sentences in all documents. Furthermore, repeated sentences should not be included. The process is repeated until the size defined for the summary is reached or there is no sentence that increases the summary score.

Oracle Lead: An additional upper bound that exclusively takes document ordering into account. Input documents were evaluated with ROUGE and ordered best first. Using the data in this sequence we apply LeadSum, which is the simplest classifier.

4.3 Implementation Details

Our goal when running the experiments is to find out whether adding document prioritization improved the quality of summaries. In order to achieve this, we conducted experiments on the datasets and evaluated how well the baseline and modified versions of the algorithms performed. All training of the document prioritization model was carried out with the training data and followed up with the validation data. We terminated training when no reductions in classification error were observed in the validation set for three consecutive iterations.

To determine the model settings, we performed tests on the Multi-News dataset, applying the same configuration to all datasets. In our evaluation, we only considered the quality of the results, disregarding the computational cost, although in our studies the larger network did not always produce better results. In this way, 128 convolutional kernels were used, and we produced document embeddings with a length of 32 values. The input data was converted to lowercase and we are using a vocabulary with the most frequent 10,000 lemmatized tokens. The word embeddings were generated with fastText [24], which has a subword mechanism capable of avoiding out-of-vocabulary occurrences. We used the distribution trained with Common Crawl and allowed these representations to be optimized in the training. As mentioned in Sect. 3, the input documents are truncated to the same size as expected for the summary.

Some of the baseline methods also require defining attributes. In the modified version of TextRank, we include the α variable that needs to be defined. In order to achieve this, we generate summaries for various values of α using the validation data. We used the first 1,000 samples from the validation set to evaluate ten α values between 0 and 1.0, using the best-performing configuration to produce the summaries in the test set. For PacSum we use the optimizer provided by the authors, which selects values using the grid-search method. In this process, in the same way as in the original article, we use the validation set. We also use the version of BERT that was fine-tuned by the method's authors.

We configured BertSum to select sentences with a classifier in the output layer and maintained the default settings, which include the block-trigram strategy. We train the model with the training dataset, evaluating with the validation set every 1,000 training iterations. The training was stopped after three consecutive iterations with no decrease in error on the validation set. Because it is a very large dataset, when training BertSum with WikiSum, we reduced the validation set to the initial 3,200 samples. We also reduced the training set to 20% of the total, which is enough to complete training in the first epoch, since in our experiments training ended approximately after using 5% of the data.

According to [28], evaluation using ROUGE can be influenced by the length of the summaries. This way, when generating summaries, enough sentences are chosen and the final sentence is truncated to reach the number of tokens specified by the summary, producing summaries that are precisely the same length across all methods. Summary lengths were chosen through a literature search. This way we are using 300 tokens for Multi-News, 40 for WCEP, 60 for WikiSum, 220 for arXiv, and 120 for Multi-XScience.

All experiments were repeated 10 times, always with different seeds of random numbers. Using Wilcoxon rank-sum statistics, we compared the statistical significance of the results from the prioritized version with the results from the original methods. For statistical tests, it is also necessary to repeat the tests with the original versions of the methods. Since in the LeadSum, PacSum, and BertSum methods, the order of documents is important, we are randomly changing the order of the input documents in each experiment. With this approach, we are comparing the random reordering of input documents with the reordering using our prioritization method.

The original version of TextRank does not use random numbers and is not affected by the order of documents, so we use another comparison approach. In its original implementation, the ROUGE metric is presented along with confidence intervals, which are calculated with bootstrap for a 95% confidence factor. In this way, we performed ten repetitions of our modified version and compared the median with this confidence interval.

5 Results and Discussion

5.1 Classification Scores

In this section, the classification results using the Siamese network are presented. The input data are the document pairs produced as indicated in Sect. 3, so these results refer to binary classification to identify the best document. Document pairs were produced for the training, testing, and validation data, and we are presenting the results with the testing data after the training is complete. The results are presented in Table 2.

Table 2. Siamese Network classification results

	f1-Score	Acc.	Prec.	Recall	Corr.
Multi-News	0.72	0.71	0.71	0.73	0.986
WCEP	0.64	0.65	0.67	0.62	0.974
XScience	0.61	0.62	0.62	0.60	0.947
arXiv	0.78	0.78	0.78	0.77	0.994
WikiSum	0.75	0.75	0.74	0.75	0.986

We conclude from the results that the suggested network helps identify the best documents. Despite this, especially in the Multi-XScience dataset, the accuracy was not very high. We therefore conducted experiments to determine the situations in which the model made the most mistakes. In particular, we were interested in identifying the relationship between the difference between ROUGE values and classification performance.

Since each training sample is composed of a pair of documents, we consider the difference in ROUGE evaluations between documents, reusing the evaluation applied to construct the training data. Ordering the documents by this difference, we divide the test data into sets of 500 samples. Thus, the first group corresponds to documents with similar evaluations and the last group contains documents with greater differences in quality.

For each group produced, we calculate the f1-Score metric and the mean difference of ROUGE. We calculated the Spearman correlation between these values and the results are in the last column of Table 2, where it is possible to verify a very strong correlation. This indicates that the proposed Siamese network works better when the input documents have a large difference in the evaluation and not so well when the evaluations are very similar.

These results are also represented in Fig. 3. To simplify visualization, this figure represents a simplified version of the correlation experiment. In this experiment, we applied the same ordering procedure by ROUGE difference, with the change that the results were separated into just 10 groups. This figure shows that classifier performance is not particularly significant while the difference in evaluations with ROUGE is close to zero, but it advances to substantially greater performance when the difference with ROUGE increases.

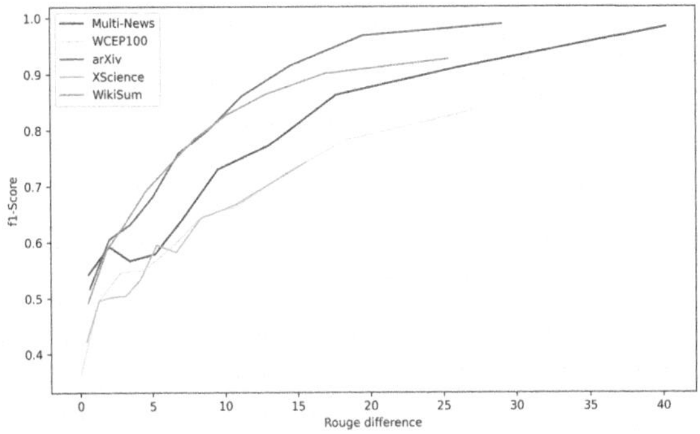

Fig. 3. F1-score as a function of the difference in quality of documents evaluated with ROUGE.

5.2 Summarization Scores

We evaluate the results with the ROUGE 1, 2, and L metrics. In Table 3 are the results with ROUGE-1, and correspond to the medians after 10 repetitions of the experiments. The results with the other metrics were equivalent and are available with our code. We are highlighting in bold the best results that have statistical significance in the evaluation with Wilcoxon rank-sum for a confidence factor of 95%. At TextRank we highlight the best results that exceed confidence margins calculated with bootstrap and a confidence factor of 95%. When the results are not different enough to achieve statistical significance, we highlight both values in bold.

Evaluating these results, we can see that document prioritization had a significant impact on the results, which is evident in Oracle-Lead. Even applying the simplest summarizer, the perfect prioritization of input documents brought the best results. Although without achieving perfect ordering, the proposed model for prioritizing documents allowed important improvements. In these results presented, the median of ROUGE-1 increments was 2.7.

The prioritization of documents brought benefits in most results, and in the worst case, it obtained results equivalent to the non-prioritized version. Equivalent results were obtained with TextRank on the WCEP and Multi-XScience datasets. The original version of TextRank is not influenced by document order and prioritization gives an advantage to the best documents. In the case of WCEP, considering that it is the dataset with the largest volume of input documents, it might be necessary to increase the advantage of the best documents, which could be achieved by applying higher values of α in Eq. 2. The Multi-XScience dataset had the smallest increase in the evaluation and was also the dataset with the lowest precision presented in Sect. 5.1.

Table 3. Multi-News: results F1-Score (%)

	Multi-News	WCEP	XScience	arXiv	WikiSum
Oracle	51.58	56.27	39.23	55.98	28.80
Oracle-Lead	48.19	49.05	34.08	46.03	40.82
LeadSum	44.00	30.55	29.22	35.39	20.73
LeadSum_d	**45.84**	**35.13**	**30.43**	**44.12**	**29.87**
TextRank	42.56	**32.28**	**31.46**	39.76	28.02
TextRank_d	**44.89**	**32.98**	**31.33**	**43.09**	**31.11**
PacSum	44.29	30.54	30.33	38.36	21.72
PacSum_d	**45.87**	**32.74**	**30.56**	**44.07**	**28.96**
BertSum	44.32	29.97	31.61	35.74	26.73
BertSum_d	**45.60**	**34.83**	**32.04**	**42.93**	**32.96**

6 Conclusions

We present a document comparison approach using a Siamese network that is useful for identifying the best documents in multi-document summarization tasks. As shown in the results, document prioritization can help increase the accuracy of summaries produced with different summarization techniques.

The integration process with several methods from the literature was presented, and a possible future work is integration with other methods.

Since each experiment was conducted with identical tools and procedures, in addition to evaluating document prioritization, it is possible to compare different summarization techniques.

In pairs, the original and prioritized versions were compared with statistical tests, indicating a significant difference in the results.

Acknowledgments. This work has been supported by the Brazilian research agencies Coordenação de Aperfeiçoamento de Pessoal de Nível Superior (CAPES) and Conselho Nacional de Desenvolvimento Científico e Tecnológico (CNPq).

References

1. Bandara, W.G.C., Patel, V.M.: A transformer-based Siamese network for change detection. In: IGARSS 2022 - 2022 IEEE International Geoscience and Remote Sensing Symposium, pp. 207–210 (2022)
2. Baralis, E., Cagliero, L., Mahoto, N., Fiori, A.: GraphSum: discovering correlations among multiple terms for graph-based summarization. Inf. Sci. **249**, 96–109 (2013)
3. Carbonell, J., Goldstein, J.: The use of MMR, diversity-based reranking for reordering documents and producing summaries. In: Proceedings of the 21st Annual International ACM SIGIR Conference on Research and Development in Information Retrieval, SIGIR 1998, New York, NY, USA, pp. 335–336. Association for Computing Machinery (1998)

4. Cohan, A., et al.: A discourse-aware attention model for abstractive summarization of long documents. In: Walker, M., Ji, H., Stent, A. (eds.) Proceedings of the 2018 Conference of the North American Chapter of the Association for Computational Linguistics: Human Language Technologies, Volume 2 (Short Papers), New Orleans, Louisiana, pp. 615–621. Association for Computational Linguistics, June 2018
5. De Castro Santos, M.A., Berton, L.: An enhanced framework for overcoming pitfalls and enabling model interpretation in pneumonia and COVID-19 classification. IEEE Access **11**, 115330–115347 (2023)
6. Devlin, J., Chang, M.W., Lee, K., Toutanova, K.: BERT: pre-training of deep bidirectional transformers for language understanding. In: Burstein, J., Doran, C., Solorio, T. (eds.) Proceedings of the 2019 Conference of the North American Chapter of the Association for Computational Linguistics: Human Language Technologies, Volume 1 (Long and Short Papers), Minneapolis, Minnesota, pp. 4171–4186. Association for Computational Linguistics, June 2019
7. El-Kassas, W.S., Salama, C.R., Rafea, A.A., Mohamed, H.K.: Automatic text summarization: a comprehensive survey. Expert Syst. Appl. **165**, 113679 (2021)
8. Ernst, O., et al.: Proposition-level clustering for multi-document summarization. arXiv e-prints arXiv:2112.08770, December 2021
9. Fabbri, A.R., Li, I., She, T., Li, S., Radev, D.R.: Multi-news: a large-scale multi-document summarization dataset and abstractive hierarchical model. In: ACL 2019 - 57th Annual Meeting of the Association for Computational Linguistics, Proceedings of the Conference, pp. 1074–1084 (2020)
10. Garcia, K., Shiguihara, P., Berton, L.: Breaking news: unveiling a new dataset for Portuguese news classification and comparative analysis of approaches. PLOS ONE **19**(1), 1–15 (2024)
11. Ghalandari, D.G., Hokamp, C., Pham, N.T., Glover, J., Ifrim, G.: A large-scale multi-document summarization dataset from the wikipedia current events portal. arXiv preprint arXiv:2005.10070 (2020)
12. Grenander, M., Dong, Y., Cheung, J.C.K., Louis, A.: Countering the effects of lead bias in news summarization via multi-stage training and auxiliary losses. In: Inui, K., Jiang, J., Ng, V., Wan, X. (eds.) Proceedings of the 2019 Conference on Empirical Methods in Natural Language Processing and the 9th International Joint Conference on Natural Language Processing (EMNLP-IJCNLP), Hong Kong, China, pp. 6019–6024. Association for Computational Linguistics, November 2019
13. Koch, G., Zemel, R., Salakhutdinov, R., et al.: Siamese neural networks for one-shot image recognition. In: ICML Deep Learning Workshop, Lille, vol. 2 (2015)
14. Lan, M., Zhang, J., He, F., Zhang, L.: Siamese network with interactive transformer for video object segmentation. In: Proceedings of the AAAI Conference on Artificial Intelligence, vol. 36, no. 2, pp. 1228–1236 (2022)
15. Li, W., Xiao, X., Liu, J., Wu, H., Wang, H., Du, J.: Leveraging graph to improve abstractive multi-document summarization. In: Jurafsky, D., Chai, J., Schluter, N., Tetreault, J. (eds.) Proceedings of the 58th Annual Meeting of the Association for Computational Linguistics, pp. 6232–6243. Association for Computational Linguistics, Online, July 2020
16. Li, Y., Chen, C.L.P., Zhang, T.: A survey on Siamese network: methodologies, applications, and opportunities. IEEE Trans. Artif. Intell. **3**(6), 994–1014 (2022)
17. Li, Z., Liu, F., Yang, W., Peng, S., Zhou, J.: A survey of convolutional neural networks: analysis, applications, and prospects. IEEE Trans. Neural Netw. Learn. Syst. **33**(12), 6999–7019 (2022)

18. Liu, P.J., et al.: Generating Wikipedia by summarizing long sequences. arXiv e-prints arXiv:1801.10198, January 2018
19. Liu, Y.: Fine-tune BERT for extractive summarization. arXiv e-prints arXiv:1903.10318, March 2019
20. Liu, Y., Lapata, M.: Hierarchical transformers for multi-document summarization. In: Korhonen, A., Traum, D., Màrquez, L. (eds.) Proceedings of the 57th Annual Meeting of the Association for Computational Linguistics, Florence, Italy, pp. 5070–5081. Association for Computational Linguistics, July 2019
21. Lu, Y., Dong, Y., Charlin, L.: Multi-XScience: a large-scale dataset for extreme multi-document summarization of scientific articles. In: Webber, B., Cohn, T., He, Y., Liu, Y. (eds.) Proceedings of the 2020 Conference on Empirical Methods in Natural Language Processing (EMNLP), pp. 8068–8074. Association for Computational Linguistics, Online, November 2020
22. Ma, C., Zhang, W.E., Guo, M., Wang, H., Sheng, Q.Z.: Multi-document summarization via deep learning techniques: a survey. ACM Comput. Surv. **55**(5) (2022)
23. Mihalcea, R.: Graph-based ranking algorithms for sentence extraction, applied to text summarization. In: Proceedings of the ACL 2004 on Interactive Poster and Demonstration Sessions. ACLdemo 2004, USA, p. 20–es. Association for Computational Linguistics (2004)
24. Mikolov, T., Grave, E., Bojanowski, P., Puhrsch, C., Joulin, A.: Advances in pre-training distributed word representations. In: Proceedings of the International Conference on Language Resources and Evaluation (LREC 2018) (2018)
25. Rick, R., Berton, L.: Energy forecasting model based on CNN-LSTM-AE for many time series with unequal lengths. Eng. Appl. Artif. Intell. **113**, 104998 (2022)
26. See, A., Liu, P.J., Manning, C.D.: Get to the point: summarization with pointer-generator networks. In: Barzilay, R., Kan, M.Y. (eds.) Proceedings of the 55th Annual Meeting of the Association for Computational Linguistics (Volume 1: Long Papers), Vancouver, Canada, pp. 1073–1083. Association for Computational Linguistics, July 2017
27. Sinha, A., et al.: An overview of microsoft academic service (MAS) and applications. In: Proceedings of the 24th International Conference on World Wide Web, WWW 2015 Companion, New York, NY, USA, pp. 243–246. Association for Computing Machinery (2015)
28. Sun, S., Shapira, O., Dagan, I., Nenkova, A.: How to compare summarizers without target length? Pitfalls, solutions and re-examination of the neural summarization literature. In: Bosselut, A., et al. (eds.) Proceedings of the Workshop on Methods for Optimizing and Evaluating Neural Language Generation, Minneapolis, Minnesota, pp. 21–29. Association for Computational Linguistics, June 2019
29. Veličković, P., Cucurull, G., Casanova, A., Romero, A., Liò, P., Bengio, Y.: Graph attention networks. arXiv e-prints arXiv:1710.10903, October 2017
30. Viji, D., Revathy, S.: A hybrid approach of weighted fine-tuned BERT extraction with deep Siamese Bi - LSTM model for semantic text similarity identification. Multimedia Tools Appl. **81**(5), 6131–6157 (2022)
31. Wang, D., Liu, P., Zheng, Y., Qiu, X., Huang, X.: Heterogeneous graph neural networks for extractive document summarization. In: Jurafsky, D., Chai, J., Schluter, N., Tetreault, J. (eds.) Proceedings of the 58th Annual Meeting of the Association for Computational Linguistics, pp. 6209–6219. Association for Computational Linguistics, Online, July 2020

32. Xu, S., Zhang, X., Wu, Y., Wei, F., Zhou, M.: Unsupervised extractive summarization by pre-training hierarchical transformers. In: Proceedings of the 2020 Conference on Empirical Methods in Natural Language Processing: Findings, pp. 1784–1795. Association for Computational Linguistics, Online, November 2020
33. Zhao, J., et al.: SummPip: unsupervised multi-document summarization with sentence graph compression. In: Proceedings of the 43rd International ACM SIGIR Conference on Research and Development in Information Retrieval, SIGIR 2020, New York, NY, USA, pp. 1949–1952. Association for Computing Machinery (2020)
34. Zheng, H., Lapata, M.: Sentence centrality revisited for unsupervised summarization. In: Proceedings of the 57th Annual Meeting of the Association for Computational Linguistics, Florence, Italy, pp. 6236–6247. Association for Computational Linguistics, July 2019

Standing on the Shoulders of Giants

Lucas F. F. Cardoso[1](✉), José de Sousa Ribeiro Filho[2], Vitor C. A. Santos[3], Regiane S. Kawasaki Francês[1], and Ronnie C. O. Alves[3]

[1] Federal University of Pará - UFPA, Belém, Brazil
lucas.cardoso@icen.ufpa.br, kawasaki@ufpa.br
[2] Federal Institute of Pará - IFPA, Belém, Brazil
jose.ribeiro@ifpa.edu.br
[3] Vale Technological Institute - ITV, Belém, Brazil
{vitor.cirilo.santos,ronnie.alves}@itv.org

Abstract. Although fundamental to the advancement of Machine Learning, the classic evaluation metrics extracted from the confusion matrix, such as precision and F1, are limited. Such metrics only offer a quantitative view of the models' performance, without considering the complexity of the data or the quality of the hit. To overcome these limitations, recent research has introduced the use of psychometric metrics such as Item Response Theory (IRT), which allows an assessment at the level of latent characteristics of instances. This work investigates how IRT concepts can enrich a confusion matrix in order to identify which model is the most appropriate among options with similar performance. In the study carried out, IRT does not replace, but complements classical metrics by offering a new layer of evaluation and observation of the fine behavior of models in specific instances. It was also observed that there is 97% confidence that the score from the IRT has different contributions from 66% of the classical metrics analyzed.

Keywords: Machine Learning · Classification · Item Response Theory

1 Introduction

Artificial Intelligence (AI) systems are present in all spheres of society. Despite the different types of AI that exist, Machine Learning (ML) techniques are the most commonly used, especially for more objective tasks such as classification problems [7]. The increasing advancement of its applications requires models with increasingly higher success rates and a low chance of error [8].

The binary classification task is a classic ML task, in which the model is challenged with a dataset that has only two classes, normally translated into class 1 (positive) and 0 (negative). This format is used because in binary problems a class tends to be the objective of classification, e.g., in a dataset that has information on patients in a hospital who do or do not have a certain disease,

A. Paes and F. A. N. Verri (Eds.): BRACIS 2024, LNAI 15413, pp. 416–430, 2025.
https://doi.org/10.1007/978-3-031-79032-4_29

the objective of a classifier would be to identify whether a new patient has the disease (positive) or does not (negative) [18].

With this standard, it is possible to evaluate the model's performance based on its confusion matrix and extract the most classic evaluation metrics used in ML, such as Accuracy, F1 Score and Recall [18]. Despite being widely used, classical metrics can only evaluate in a quantitative way, as they only count errors and successes and then evaluate the model over the entire test set, without differentiating the instances that are classified [11].

Recent work aims to overcome this limitation by applying other forms of evaluation within the ML context. An interesting approach that has been gaining ground is the use of concepts from psychometrics to evaluate models. Works such as [1,6,19] use Item Response Theory (IRT), to more accurately measure the ability of models, due to its main characteristic of being instance-level. With IRT it is possible to evaluate the performance of a model against a specific instance of the test set, in a way that allows the model and data relationship to be further explored. The application of IRT in the context of ML is not to replace classical metrics, but rather to add more information to prior knowledge.

Given the above, this work investigates how IRT can be used to open the box of the classical confusion matrix. The confusion matrix typically delivers its evaluation in a batch of instances format, therefore, it does not allow observing small fluctuations that the instances cause in the generalization of the models, in their learning. It is observed that IRT is capable of bringing an extra evaluative layer by standing on the shoulders of the confusion matrix and thus reinforcing the assessment of the models' ability, enabling a more assertive decision regarding the comparison between models that eventually perform in a similar way. A new performance metric was also introduced that evaluates intrinsic aspects of learning, not yet explicitly covered by classical performance metrics.

The remainder of the paper is organized as follows: Sect. 2 presents the necessary theoretical framework of the article, and explains the concepts of using IRT applied to ML; Sect. 3 deals with the methodology used to carry out the experiments proposed in the article; Sect. 4 presents the results obtained by the experiments and discusses what was observed; Sect. 5 concludes the article and presents future work.

2 Background

2.1 Item Response Theory

The objective of this section is to explain the psychometric concepts of IRT, how it works and how it can be used in Machine Learning. As already mentioned in Sect. 1, IRT emerged within psychometrics as a new way of evaluating individuals' performance on the same test. Just like in classic ML metrics, the classic way of evaluating people is done through a test where performance is measured based on the individual's mistakes or successes in the test, so that 1 is added for each correct answer and 0 for each error. Although it is still widely used, this format is not capable of measuring an individual's real ability, as it does not consider the complexity of the different items that exist in the test [3].

The IRT then emerges as a new assessment approach that is centered on the items and not on the test itself. Its objective is to calculate the probability of success of a given respondent j on a specific item i considering the ability θ that this individual has and the parameters that describe item i. There are 3 main parameters: (a_i) discrimination, whose value represents how well a given item is able to distinguish respondents with high ability from respondents with low ability, the higher its value, the more discriminating the item is; (b_i) difficulty, restates how difficult an item is to answer correctly, the higher its value, the more difficult an item is; (c_i) guessing, the parameter that indicates the minimum chance that an item has of being answered correctly or the chance of a low-skill respondent getting the item right at random [4].

IRT has different logistic models that are used to calculate the probability of a correct answer, determined by the number of item parameters they use. The logistic model that uses the three item parameters is called the Three-Parameter Logistic model (3PL) for dichotomous items, i.e., items where only whether it was answered correctly or not is considered. The 3PL logistic model is defined by Eq. 1, where $P(U_{ij} = 1|\theta_j)$ is the probability of the respondent θ_j correctly answering the dichotomous item U_{ij} [3,4].

$$P(U_{ij} = 1|\theta_j) = c_i + (1 - c_i)\frac{1}{1 + e^{-a_i(\theta_j - b_i)}} \tag{1}$$

Despite being item-centered, IRT is also capable of generating a final score on an individual's performance. To do this, you can use the concept of True Score [9], which generates a score $TrueS_j$ for each respondent based on the sum of the probabilities of a correct answer $P(U_{ij} = 1|\theta_j)$ calculated for each test item, the True Score is defined by Eq. 2.

$$TrueS_j = \sum_{i=1}^{N} P(U_{ij} = 1|\theta_j) \tag{2}$$

Item Characteristic Curve. One of the main products of the IRT is the Item Characteristic Curve (ICC), which allows analyzing the relationship between item parameters and the probability of a correct answer given the variation in ability. Figure 1 presents an example of the ICC. It is possible to visualize the behavior of the probability of a correct answer increasing as the ability θ also increases. In the ICC the guessing parameter c determines the lowest probability of a correct answer that the curve will have, while the discrimination a determines the slope of the curve, so that the higher the value of a the steeper the curve becomes. The difficulty b and the ability θ are calculated on the same scale, so that if $\theta = b$ the chance of a correct answer is equal to 50%. The difficulty determines the point on the curve where this happens [3].

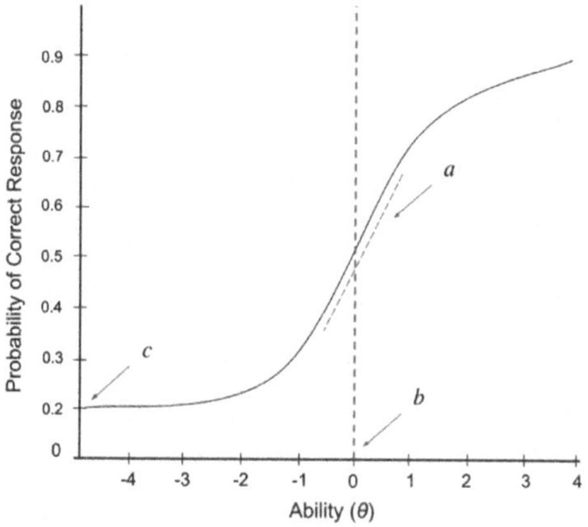

Fig. 1. Example of Item Characteristic Curve.

3 Methodology

In this section, the methodology used to carry out the experiments with the application of IRT in ML will be presented. As mentioned in Sect. 1, this work aims to explore how IRT can be useful for evaluating ML models considering the confusion matrix obtained in a classical classification problem[1].

Although its use is aimed at evaluating individuals, as explained in Sect. 2.1, a simple analogy is enough to apply IRT in the context of ML when considering respondents as models while items are test instances. To calculate the probability of a correct answer, the 3PL logistic model was used, due to the fact that it considers the probability of a random hit, as it is possible for the models to get the instances right by guessing. To estimate the item parameters of the dataset instances and the ability of the models, the Birnbaum method [10] was used, where the item parameters are first estimated by considering only the responses of all models and subsequently the ability is estimated from the item parameters obtained and the response vector of each model.

As a case study, the Heart-Statlog[2] dataset was chosen. The Heart-Statlog is a heart disease dataset, which is composed of 270 instances with 13 features. Figure 2 illustrates the methodology used to calculate the IRT estimators, which is separated into 6 main steps.

1. The Heart-Satolog dataset is separated into training and testing. The necessary pre-processing of the data is also carried out at this stage, such as coding categorical features.

[1] Link to the source code of the methodology: https://github.com/LucasFerraroCardoso/IRT_Confusion_Matrix.

[2] Heart-Statlog Link: https://www.openml.org/search?type=data&status=active&id=53.

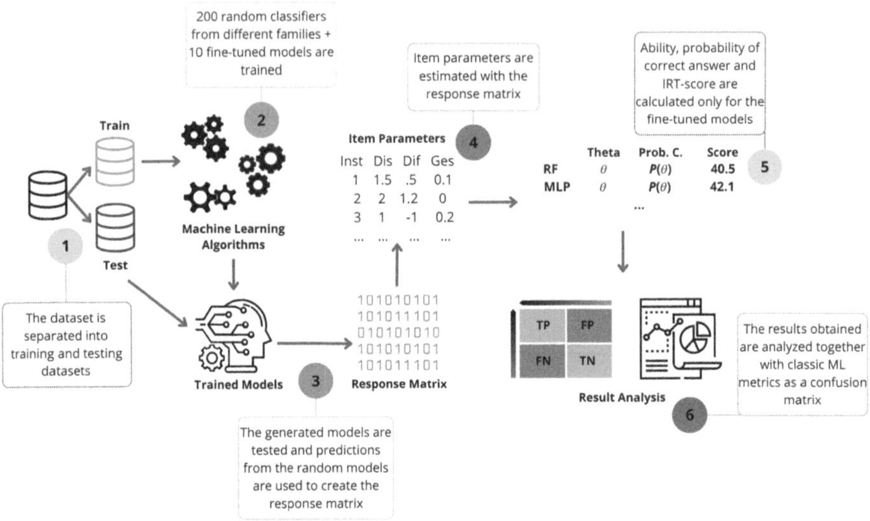

Fig. 2. Flowchart of the methodology.

2. Next, 200 random models are generated from different classification algorithms that will be trained with the training dataset. At this stage, 10 fine-tuned models are also trained, whose performance will be evaluated later.
3. The trained random models must then classify the instances of the test dataset to generate the response matrix from their predictions. The fitted models also classify the test dataset. The response matrix is a matrix where the columns are the instances and the rows are the dichotomous response vectors for each instance, with 1 being correct and 0 being incorrect.
4. The response matrix that was obtained is used to estimate the item parameters of the test dataset.
5. With the item parameters calculated, it is possible to estimate the ability of the fine-tuned models and then calculate the probability of a correct answer and the IRT scores.
6. The results obtained are then divided according to the confusion matrix to analyze the performance of the models.

To estimate accurate values of item parameters, 200 random classifiers from 10 different learning algorithms were used Decision Tree (DT), Random Forest (RF), Ada Boost (ADA), Gradient Boosting (GB), Bagging (BAG), Multilayer Perceptron (MLP), k-Nearest Neighbors (KNN), Support Vector Machine (SVM), Linear Support Vector Machine (LSVM) and Linear Discriminant Analysis (LDA). To achieve this, 20 models were generated with randomly chosen hyperparameters for each of the 10 algorithms used. This way, it is possible to generate a diversity of answers, as if there were different students of different levels answering a test. The Scikit-Learn library [14] was used to generate all the classification algorithms.

For performance analysis, 10 models were generated, one for each algorithm used. These 10 models are trained and hyper-parameterized using the grid search strategy with cross-validation provided by Scikit Learn, with 5-folds for cross-validation. The R package Ltm [16] was used to estimate the IRT item parameters from the response matrix. Subsequently, the ability of the fitted models was estimated using the Catsim [12] Python package, this same package was used to calculate the probability of a correct answer and the ICC's of the test instances.

Although IRT already has the True-Score, presented in Sect. 2.1, it is understood that it is important to also consider the error for a fairer evaluation of the models. For this, the concept of Total-Score, presented in [19], was used, where the probability of error is subtracted from the score for instances that were incorrectly classified. The Total-Score is defined by the Eq. 3, where i' corresponds to items that were answered correctly while i'' corresponds to items answered incorrectly.

$$TotalS_j = \sum_{i=1}^{i'} P(U_{ij} = 1|\theta_j) - \sum_{i=1}^{i''} 1 - P(U_{ij} = 1|\theta_j) \tag{3}$$

4 Results and Discussion

In this section, the results obtained from the experiments carried out in this research are presented and discussed[3]. As presented in Sect. 3, the Heart-Statlog dataset was chosen as a case study because it is a known and easy-to-understand dataset. The Heart-Statlog is a dataset that contains data from patients who may or may not have heart disease. It has 270 instances with 13 patient diagnostic features, with 120 instances of patients who have heart disease (positive class) and 150 instances of patients who do not have heart disease (negative class), so the dataset has a balance of 55.55% for the negative class (majority) and 44.45% for the positive class (minority). In this work, the division 70% was used for training and 30% for testing.

To understand the performance of models, it is important to first understand the complexity of the dataset. Figure 3 generally shows the item parameters estimated for the Heart-Statlog, in which the instances were organized into a histogram for each item parameter separating the majority and minority class instances. According to IRT concepts, the dataset can be considered not very difficult due to the negative average of difficulty with −2.67 and suitable for comparing respondents due to the positive average of discrimination with 3.13.

However, despite the positive average discrimination, it is important to highlight that the dataset presents around 24.7% of instances with negative discrimination. While instances with high discrimination values mean that they are good at certifying the models' ability, instances with negative values mean that they are not well formed or even defined, so if a model hits many instances with these characteristics it means that this model may not be reliable.

[3] All results generated are available at the link: https://github.com/LucasFerraroCardoso/IRT_Confusion_Matrix/tree/main/Results.

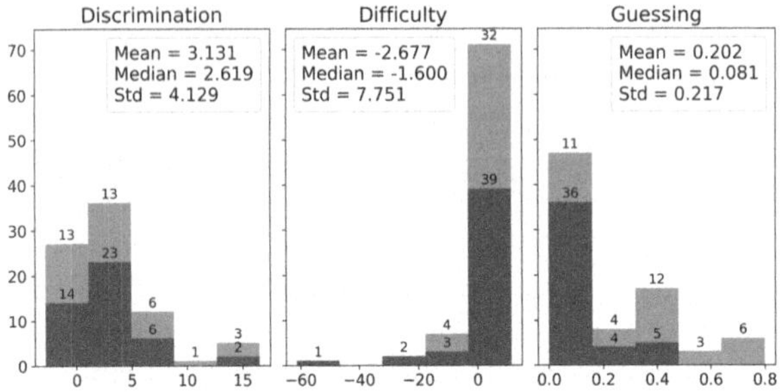

Fig. 3. Heart-Statlog item parameter histograms. The orange bars are the instances of the minority class while the blue ones are the majority class. (Color figure online)

It is also noted that there is a balance between the frequency of majority and minority instances present in the 5 bins for the difficulty and discrimination histograms, this was already expected given the balance of 55%–45% between the classes. For the guessing parameter, the minority class presented the highest frequency of instances with a high probability of random success, despite balancing. Previous research has already highlighted that models tend to hit instances of the minority class by chance more frequently [5].

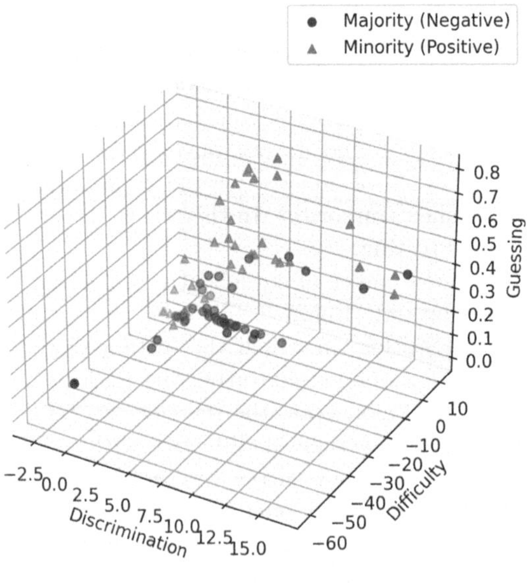

Fig. 4. Heart-Statlog instances arranged by item parameters.

This distribution of instances by the guessing parameter becomes even more evident when analyzing Fig. 4. Although the largest number of instances of the majority class have 0 guessing, which means that chance does not tend to influence the correct answer, there are still 9 instances (11%) that have a greater chance of being classified by guessing. From the figure it is also evident that some instances of the minority class have difficulty outliers, where there are three instances that are considered much easier than the others because they have very negative values. It is understood that in a dataset it is expected that there will be determining examples of a class distribution.

Once you understand the behavior of the data, you can then look at the classifiers. Table 1 presents the results of classic ML metrics for the 10 fine-tuned models with their respective ranking positions. When observing only the classic metrics, the GB model presents the highest score achieved in 5 of the 6 metrics considered. If the objective of this experiment was to choose the best model, with a brief analysis it is possible to indicate that the GB would be ideal. However, GB only has the 6th best score if only the recall is considered and in this case this is very pertinent information. Simply put, the recall metric represents how well a model was able to correctly classify True Positives instances, and in the case of a context-sensitive dataset that aims to correctly identify patients who have heart disease, the recall metric has great relevance. If only recall is considered, the model that presents the best performance is LSVM.

Table 1. Scores obtained by each model for classic ML metrics.

	accuracy	F1 score	precision	recall	auc score	specificity
RF	0.802 (2)	0.778 (2)	0.778 (5)	0.778 (2)	0.800 (2)	0.822 (5)
GB	**0.827 (1)**	**0.781 (1)**	**0.893 (1)**	0.694 (6)	**0.814 (1)**	**0.933 (1)**
BAG	0.790 (5)	0.761 (5)	0.771 (6)	0.750 (3)	0.786 (6)	0.822 (5)
ADA	0.765 (8)	0.725 (8)	0.758 (7)	0.694 (6)	0.758 (8)	0.822 (5)
KNN	0.617 (10)	0.523 (10)	0.586 (10)	0.472 (10)	0.603 (10)	0.733 (10)
DT	0.728 (9)	0.686 (9)	0.706 (9)	0.667 (9)	0.722 (9)	0.778 (8)
SVM	0.790 (5)	0.754 (7)	0.788 (4)	0.722 (5)	0.783 (7)	0.844 (3)
MLP	0.802 (2)	0.771 (4)	0.794 (3)	0.750 (3)	0.797 (3)	0.844 (3)
LDA	0.802 (2)	0.758 (6)	0.833 (2)	0.694 (6)	0.792 (4)	0.889 (2)
LSVM	0.790 (5)	0.773 (3)	0.744 (8)	**0.806 (1)**	0.792 (4)	0.778 (8)

Although LSVM has the best recall score, the model only has the 5th best average performance among all classic metrics. RF has the second best average performance, behind only GB, and has the second best recall score, a condition that would qualify it as a candidate to be chosen as the best model. The precision and recall metrics tend to have inverse performances, i.e. to increase one the other tends to decrease, and this occurs in this example where the RF only presents the 5th precision score. If you also consider the high variation in the ranking as an elimination criterion, it is the MLP that has the 3rd best performance on average and also has the lowest variation in position among the rankings, with high values in both precision and recall. *In this scenario, what would be the best model to choose? How else can models be evaluated appropriately?*

To try to answer these questions, the performance of the models will be analyzed from the perspective of IRT. The first step was to calculate the IRT scores proposed in Sect. 3. For the True Score, which only considers the respondents' probability of being correct, it was the BAG model that appears first, and it is important to highlight that the difference between True Scores is only 0.0002 for the second (GB). For the Total Score that considers whether there was an error or a correct answer, GB appears again with the highest score, this helps to reaffirm its position as the best model in this scenario (see Table 2).

Table 2. True and Total Scores for the models.

	True Score	TrueS Rank	Total Score	TotalS Rank
RF	0.7849	3	0.5874	2
GB	0.7851	2	**0.6122**	**1**
BAG	**0.7853**	**1**	0.5754	4
ADA	0.7603	9	0.5258	8
KNN	0.6692	10	0.2865	10
DT	0.7659	6	0.4943	9
SVM	0.7790	5	0.5691	5
MLP	0.7651	7	0.5676	6
LDA	0.7840	4	0.5865	3
LSVM	0.7651	7	0.5552	7

To assess whether there is a gain in using IRT scores in conjunction with classical metrics, the Friedman [15] and Nemenyi [13] tests were performed on the results of all metrics to confirm whether there are different distributions. The Friedman test is a non-parametric test used to check whether there are different distributions among sampled data. Its result is a p-value, which if it is less than a certain limit, then it is considered that there are different distributions in the data. The Nemenyi test is complementary and is used to check which distributions are different when performing pairwise comparisons. The Friedman test resulted in a *p-value* of approximately 1.9×10^{-9}, when considering a *p-value* threshold of 0.05 it can be stated that there are different distributions between scores. Next, a Nemenyi test was performed to verify which evaluation metrics are different.

Figure 5 presents the result of the Nemenyi test, where it is possible to see that the *p-value* values of the True Score do not show any difference from the classic metrics. On the other hand, interesting results are observed for the Total Score, where the f1 and recall metrics are the only ones that do not present a *p-value* smaller than 0.05. Thus, applying the confidence measure, there is 52.25% confidence that f1 is statistically different from the Total Score, while there is

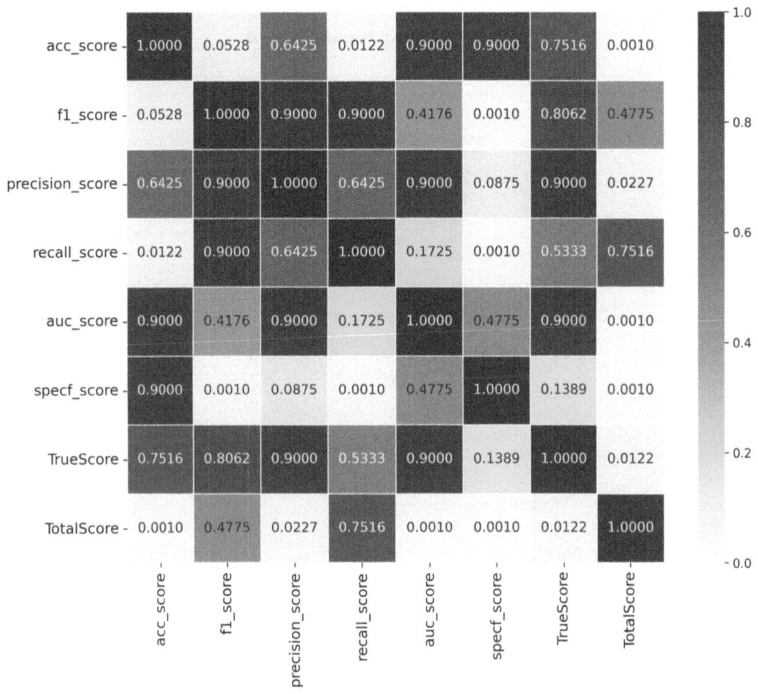

Fig. 5. Nemenyi test heatmap for evaluation metrics.

24.84% confidence that the recall is statistically different, for the other metrics the lowest confidence is 97.73%. Such results point to the Total Score as a metric capable of giving visibility to new nuances in the models' performance.

To confirm the result obtained, the instances were then opened using the concepts of the IRT Item Characteristic Curves. An ideal confusion matrix should present values only on its main diagonal. With this in mind, the ICC's of each instance were separated according to the models' confusion matrix, called the Item Characteristic Confusion Matrix Curve (ICCMC).

Figure 6 shows the ICCMC of the GB model, in which it is possible to observe that the model performs well for True Positives and True Negatives instances. Therefore, all TP instances have high discrimination values and are considered good instances to attest to the model's ability. Furthermore, only one well-formed instance appears among the FP, it is expected that only instances with negative discrimination appear between the FP and FN, as can be observed in the set of FN instances of the GB model. The point of greatest attention is the TN set, where 7 (16.67%) are instances that are not suitable for evaluating the model's performance (red lines). This can also be observed by the average information value of the instances, where the set of TPs presents a high average information value with 0.320, while the TNs present a low average information value with 0.043.

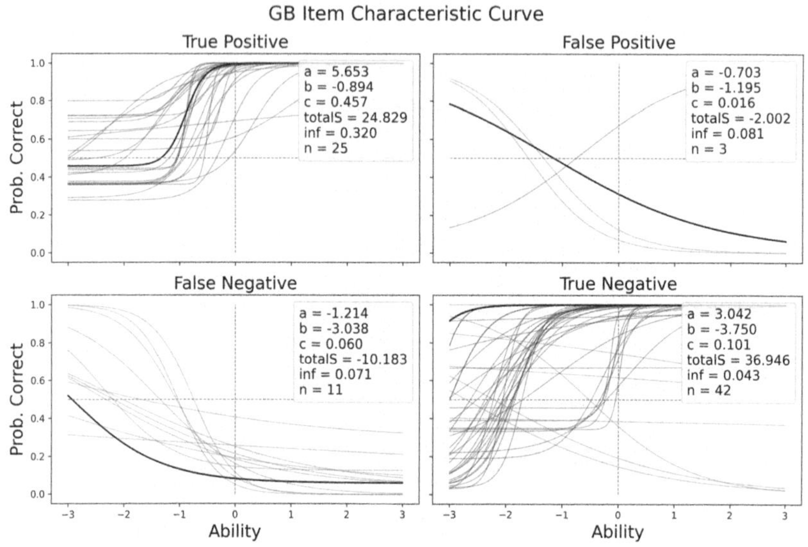

Fig. 6. ICCMC of the GB model, where a is the average discrimination value, b is the average difficulty, c is the average guessing, $totalS$ is the Total Score value, inf is the average information of the instances and n is the number of instances. The blue lines represent instances with positive discrimination, the red lines are instances with negative discrimination, and the black line is the average ICC of the instances. (Color figure online)

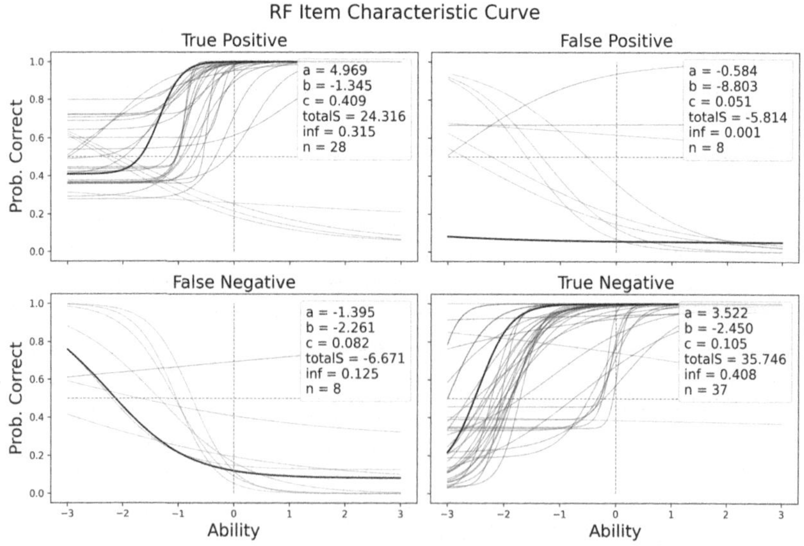

Fig. 7. ICCMC of the RF model.

In direct comparison with the ICCMC of the RF, it is highlighted that the RF has a lower performance compared to the GB (see Fig. 7). This can be confirmed by analyzing the sets of TP and TN instances, where the GB Total Score was higher in both cases. It is also noted that, although RF hits more TP instances (28) than GB (25), 4 instances classified by RF present negative discrimination. In addition, 3 well-formed instances appear in the FP and FN sets. Although the Recall of RF is greater than GB, the set of TP instances of GB is composed only of well-formed items while RF presents instances that are not suitable for evaluating the model.

Another interesting detail is the high guessing value for the instances of the TP set seen in all ICCMC presented, as previously observed in Fig. 6, the minority class has a high chance of getting it right by chance. This raises the question of whether the models have actually learned how to classify this class or are just getting lucky. A possible future work would be to investigate more incisively the impact of the guessing parameter and whether these instances are really well formed.

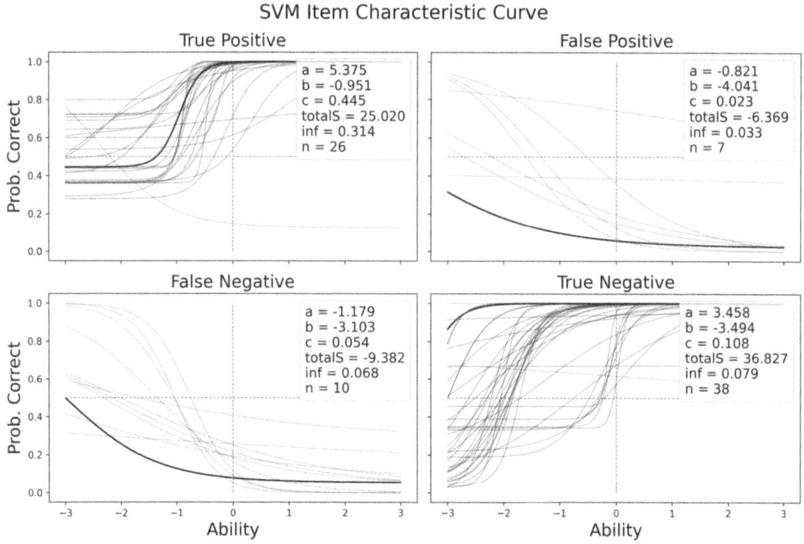

Fig. 8. ICCMC of the SVM model. (Color figure online)

Other models with interesting results are SVM and LDA, where the best configuration of ICC curves of the instances arranged in the confusion matrix sets is noted. As can be seen in Fig. 8 for the SVM there is only one instance considered not suitable for evaluation in the TP set and only instances with negative discrimination in the FP and FN sets as expected in an ideal scenario. For the TN set, there are only 2 malformed instances, but it is interesting that among them there is an instance that is repeated as TN in all models. For LDA, the TP set is made up of instances suitable for evaluation, while there are three instances with negative discrimination (red lines) in the TN set.

In the work [17] the authors indicate that not all instances of a dataset are equally good for learning, it is common knowledge that models need quality data to be trained. If the same thinking is applied to the evaluation, where instances with negative discrimination are removed from the analysis because they are considered bad by the IRT, the result would be that LDA and SVM would present the best performance in all classical ML metrics (see Table 3). Furthermore, when analyzing the ICCs within the confusion matrix, LDA would be preferable because it presents the most reliable TP set for the Heart-Statlog scenario. Furthermore, another point that makes LDA preferable is its explanatory nature [2]. In a context-sensitive scenario such as the one analyzed, a simpler model to explain its predictions is also more suitable.

Which model is ideal for a given dataset will depend on the context in which it is inserted and determining the ideal model should only be done after a more thorough analysis of performance considering the complexity of the data. A deeper analysis of the relationship between instances at the feature level would be interesting to further explore the use of ICC's in the confusion matrix.

Table 3. Ranking of scores obtained for classic metrics without negative discrimination.

	accuracy	F1 score	precision	recall	auc score	specificity
RF	0.951	0.941	0.923	0.96	0.952	0.944
GB	0.984	0.980	0.962	1.00	0.986	0.972
BAG	0.918	0.902	0.885	0.92	0.918	0.917
ADA	0.918	0.902	0.885	0.92	0.918	0.917
KNN	0.738	0.667	0.696	0.64	0.723	0.806
DT	0.869	0.846	0.815	0.88	0.871	0.861
SVM	**1.000**	**1.000**	**1.000**	**1.00**	**1.000**	**1.000**
MLP	0.967	0.962	0.926	1.00	0.972	0.944
LDA	**1.000**	**1.000**	**1.000**	**1.00**	**1.000**	**1.000**
LSVM	0.934	0.926	0.862	1.00	0.944	0.889

5 Final Considerations

The empirical evaluation of ML models remains the most common way to analyze the performance of a classifier on a dataset. This work was developed with the aim of deepening the classic form of evaluation considering the performance of models at the instance level using IRT. A case study was presented where the aim was to choose the best model for a given heart disease classification problem with the Heart-Satlog dataset. IRT allowed a deeper understanding of the relationship between model and data, by combining the traditional way of evaluating the confusion matrix with the psychometric concepts of IRT, represented by the idea of the Item Characteristic Confusion Matrix Curve (ICCMC). Furthermore, the direct relationship between IRT scores and classic metrics was explored and it was shown that the Total Score has statistically significant differences in 66% of the evaluated metrics, with a confidence of 97%. An experiment was carried out with 10 models from 10 different classification algorithms, while the classic metrics pointed to GB as the most suitable model, the IRT estimators allowed us to understand the data and how the models behaved in each specific instance. In a scenario where the entire test dataset is used for evaluation, IRT allowed us to point out with greater confidence that the GB would be the most suitable, but in another scenario with a filter of the evaluated instances, the LDA model would be the most suitable. If it was possible to see further in models' abilities it is by standing on the shoulders of confusion matrix.

As future work, it is interesting to explore more datasets of different types and contexts, and evaluate whether it would be possible to derive a new metric that would allow evaluating models directly considering the complexity of the data in relation to its features. Another aspect to be observed is the limitations of the methodology used. It is noted that the results obtained by the IRT are conditioned by the population of respondents and items selected for evaluation. Thus, the results may be different if another set of models and items is selected. The IRT itself already has tools to perform the evaluation considering different populations; it is interesting to explore this in future work.

Disclosure of Interests. The authors have no competing interests to declare that are relevant to the content of this article.

References

1. Araujo Santos, V.C., Cardoso, L., Alves, R.: The quest for the reliability of machine learning models in binary classification on tabular data. Sci. Rep. **13**(1), 18464 (2023)
2. Arrieta, A.B., et al.: Explainable artificial intelligence (XAI): concepts, taxonomies, opportunities and challenges toward responsible AI. Inf. Fusion **58**, 82–115 (2020)
3. Baker, F.B.: The Basics of Item Response Theory. ERIC, College Park (2001)
4. Cai, L., Choi, K., Hansen, M., Harrell, L.: Item response theory. Ann. Rev. Stat. Appl. **3**, 297–321 (2016)
5. Cardoso, L.F., et al.: Explanation-by-example based on item response theory. In: Xavier-Junior, J.C., Rios, R.A. (eds.) BRACIS 2022. LNCS, vol. 13653, pp. 283–297. Springer, Cham (2022). https://doi.org/10.1007/978-3-031-21686-2_20
6. Cardoso, L.F.F., Santos, V.C.A., Francês, R.S.K., Prudêncio, R.B.C., Alves, R.C.O.: Decoding machine learning benchmarks. In: Cerri, R., Prati, R.C. (eds.) BRACIS 2020. LNCS (LNAI), vol. 12320, pp. 412–425. Springer, Cham (2020). https://doi.org/10.1007/978-3-030-61380-8_28
7. Lee, R.S.: Artificial Intelligence in Daily Life. Springer, Singapore (2020). https://doi.org/10.1007/978-981-15-7695-9

8. Linardatos, P., Papastefanopoulos, V., Kotsiantis, S.: Explainable AI: a review of machine learning interpretability methods. Entropy **23**(1), 18 (2020)
9. Lord, F.M., Wingersky, M.S.: Comparison of IRT true-score and equipercentile observed-score "equatings". Appl. Psychol. Meas. **8**(4), 453–461 (1984)
10. Martínez-Plumed, F., Prudêncio, R.B., Martínez-Usó, A., Hernández-Orallo, J.: Making sense of item response theory in machine learning. In: Proceedings of the Twenty-second European Conference on Artificial Intelligence, pp. 1140–1148 (2016)
11. Martínez-Plumed, F., Prudêncio, R.B., Martínez-Usó, A., Hernández-Orallo, J.: Item response theory in AI: analysing machine learning classifiers at the instance level. Artif. Intell. **271**, 18–42 (2019)
12. Meneghetti, D.D.R., Junior, P.T.A.: Application and simulation of computerized adaptive tests through the package CATSim. arXiv preprint arXiv:1707.03012 (2017)
13. Nemenyi, P.: Distribution-free multiple comparisons. In: Biometrics, vol. 18, p. 263. International Biometric Soc 1441 I ST, NW, SUITE 700, WASHINGTON, DC 20005-2210 (1962)
14. Pedregosa, F., et al.: Scikit-learn: machine learning in python. J. Mach. Learn. Res. **12**, 2825–2830 (2011)
15. Pereira, D.G., Afonso, A., Medeiros, F.M.: Overview of Friedmans-test and post-hoc analysis. Commun. Stat.-Simul. Comput. **44**(10), 2636–2653 (2015)
16. Rizopoulos, D.: LTM: an R package for latent variable modeling and item response theory analyses. J. Stat. Softw. **17**(5), 1–25 (2006)
17. Smith, M.R., Martinez, T.: Reducing the effects of detrimental instances. In: 2014 13th International Conference on Machine Learning and Applications, pp. 183–188. IEEE (2014)
18. Sokolova, M., Lapalme, G.: A systematic analysis of performance measures for classification tasks. Inf. Process. Manag. **45**(4), 427–437 (2009)
19. de Sousa Ribeiro Filho, J., Cardoso, L.F.F., da Silva, R.L.S., Carneiro, N.J.S., Santos, V.C.A., de Oliveira Alves, R.C.: Explanations based on item response theory (eXirt): a model-specific method to explain tree-ensemble model in trust perspective. Expert Syst. Appl. **244**, 122986 (2024)

SurveySum: A Dataset for Summarizing Multiple Scientific Articles into a Survey Section

Leandro Carísio Fernandes[1](✉), Gustavo Bartz Guedes[2], Thiago Soares Laitz[2,3], Thales Sales Almeida[2,3], Rodrigo Nogueira[2,3], Roberto Lotufo[2,4], and Jayr Pereira[2,5]

[1] Instituto de Pesquisa para Economia Digital, Brasília, DF, Brazil
carisio@gmail.com
[2] UNICAMP, Campinas, SP, Brazil
jayr.pereira@ufca.edu.br
[3] Maritaca AI, Campinas, SP, Brazil
[4] NeuralMind.ai, Campinas, SP, Brazil
[5] Universidade Federal do Cariri, Juazeiro do Norte, CE, Brazil

Abstract. Document summarization is a task to shorten texts into concise and informative summaries. This paper introduces a novel dataset designed for summarizing multiple scientific articles into a section of a survey. Our contributions are: (1) SurveySum, a new dataset addressing the gap in domain-specific summarization tools; (2) two specific pipelines to summarize scientific articles into a section of a survey; and (3) the evaluation of these pipelines using multiple metrics to compare their performance. Our results highlight the importance of high-quality retrieval stages and the impact of different configurations on the quality of generated summaries.

Keywords: Dataset · Multi-document summarization · Document summarization · Text summarization · Scientific writing

1 Introduction

Document summarization is a field of artificial intelligence aimed at creating concise and informative summaries from extensive texts [13]. This process allows users to quickly grasp the main information of a document without needing to read the entire original content. Summarization can be performed in two main ways: extractive and abstractive [13]. Extractive summarization selects and copies key parts of the original text to compose the summary, preserving the verbatim of the source text. On the other hand, abstractive summarization rewrites the content, generating a summary that may include paraphrasing and synthesizing information more creatively and flexibly.

L. C. Fernandes, G. B. Guedes, T. S. Laitz and T. S. Almeida—Equal contribution.

A. Paes and F. A. N. Verri (Eds.): BRACIS 2024, LNAI 15413, pp. 431–444, 2025.
https://doi.org/10.1007/978-3-031-79032-4_30

Multi-document summarization (MDS) extends these concepts to handle multiple texts simultaneously [13]. The goal is to generate a single summary that incorporates information from a set of documents, eliminating redundancies and highlighting essential cross-referenced information. This technique is useful in situations where information on a specific topic is spread across various documents.

MDS datasets advance MDS research by providing essential data for training, testing, and validating models. These datasets support the development of algorithms capable of handling the complexities of summarizing multiple documents, enabling researchers to compare different summarization methods effectively [8]. Essentially, an MDS dataset is composed of summaries generated from multiple source documents. The composition of this dataset can be done in different ways, such as by selecting a set of documents and asking humans to write summaries, or by using existing summaries and their source documents. The latter is the case for the MULTI-NEWS dataset [4], which uses news articles and their summaries to create an MDS dataset.

A comprehensive survey is a review that provides a broad overview of a specific topic, covering the most relevant and recent research in the field. These surveys are essential for researchers to understand the state-of-the-art and identify gaps in the literature. The automatic generation of a survey's text can be considered a MDS task, where a set of scientific articles is summarized to produce a cohesive text. This task is challenging due to the complexity of scientific literature and the need to maintain the technical accuracy of the original content. However, most available summarization datasets are designed with a general-purpose approach, covering a broad range of topics and document types. There is a lack of datasets dedicated to the task of summarizing multiple scientific articles to generate a survey's text, which poses a substantial barrier to advancements in this specialized area of research. This type of dataset assists the development and benchmarking of summarization models focusing on the generation of scientific surveys.

This paper introduces SURVEYSUM[1], a dataset designed for the task of summarizing multiple scientific articles into survey sections. Our contribution includes: (1) the SURVEYSUM dataset and the methodology to create it, which aims to address the lack of in domain-specific summarization tools; (2) two specific pipelines to summarize scientific articles into a section of a survey; and (3) evaluation of these pipelines in respect to consistency, quality, and relevance of the generated summaries.

2 Related Work

Multi-document summarization (MDS) addresses the problem of extracting information that is spread across multiple documents, making it more challenging than single-document summarization. It is still an evolving field, with no single approach to solve this problem. The development of this field relies on datasets

[1] https://huggingface.co/datasets/unicamp-dl/SurveySum.

that serve as benchmarks for evaluating and comparing different summarization methods. The scientific literature includes some examples of such datasets, each varying in domain, structure, size, and summarization objective. These datasets are important for advancing research and improving summarization models.

Outside the scientific domain, MULTI-NEWS [4] presents a large volume of news articles and summaries, focusing on the abstractive summarization of multiple documents in the journalistic context. Also in this context, Ghalandari et al. [5] proposed a large-scale dataset for multi-document summarization that contains concise human-written summaries of news events. In the legal field, MULTI-LEXSUM tackles the summarization of legal cases, presenting a set of civil rights litigation summaries with multiple granularities. The summarization of instructional content is the focus of the HOWSUMM dataset, derived from WikiHow articles [1]. The FINDSUM dataset focuses on the challenge of summarizing long text and multiple tables [10].

In the scientific domain, the need to manage the vast amount of literature has generated interest in datasets specific to the summarization of scientific articles. MULTI-XSCIENCE [12] is a dataset focused on the generation of "related work" sections by summarizing multiple scientific articles. BIGSURVEY [9] was designed to create structured summaries of academic articles, focusing on consolidating literature reviews. SUMPUBMED [6] and SCISUMMNET [19] aim the summarization of single scientific articles.

Several other datasets can be used for multi-document summarization. Koh et al. [8] work is an extensive survey on long document summarization, including ten other MDS datasets.

However, to the best of our knowledge, there is no dataset designed to summarize a set of scientific articles in order to generate a section of a scientific survey. The closest dataset, MULTI-XSCIENCE [12], focuses on writing the "related work" section of a paper based on the abstract of the paper and the abstract of the articles it references. It does not consider the text of the referenced articles. Our dataset, on the other hand, adresses this gap: it considers the full text of the referenced articles to write a specific section on a subject. Consequently, our proposed dataset is truly a multi-document summarization dataset, capable of handling lengthy documents.

3 Method

In this section, we describe the methodology to create SURVEYSUM. First, we describe our approach in selecting existing surveys from the scientific literature. Later, we detail the text and citations extraction process.

3.1 Survey Selection

The selection of surveys was based on the authors' expertise in the field of artificial intelligence, natural language processing, and machine learning. The surveys were chosen to cover a broad range of topics within these areas, ensuring the dataset's diversity and applicability. For surveys to be eligible for inclusion in the dataset, they had to attend to the following criteria: (1) be a comprehensive survey in the field of artificial intelligence, natural language processing, or machine learning; (2) be divided into sections, each with citations of scientific articles; (3) be freely available online; and (4) be written in English.

First, we conducted a search on Google Scholar using the keywords *comprehensive*, *survey*, *artificial intelligence*, *natural language processing*, and *machine learning*. We then filtered the results to include only surveys published in the last five years and that are freely available. The surveys were manually reviewed to ensure they met the eligibility criteria, resulting in the selection of six surveys.

Usually, surveys are structured into sections to organize the content and facilitate reading and understanding. This makes each section a self-contained summary of a specific aspect of the survey, while the cited papers are the source documents. Thus, each section is a summary of the cited papers, making it an ideal target for the MDS task.

Finally, SurveySum samples are comprised by the sections texts of surveys with the corresponding citation papers. This ensures that the dataset can be used to evaluate the summarization of specific topic areas within the surveys.

3.2 Dataset Creation

In this section, we describe the methodology to create the dataset, illustrated in Fig. 1. This process is applied to all the surveys selected in the method described in the previous section.

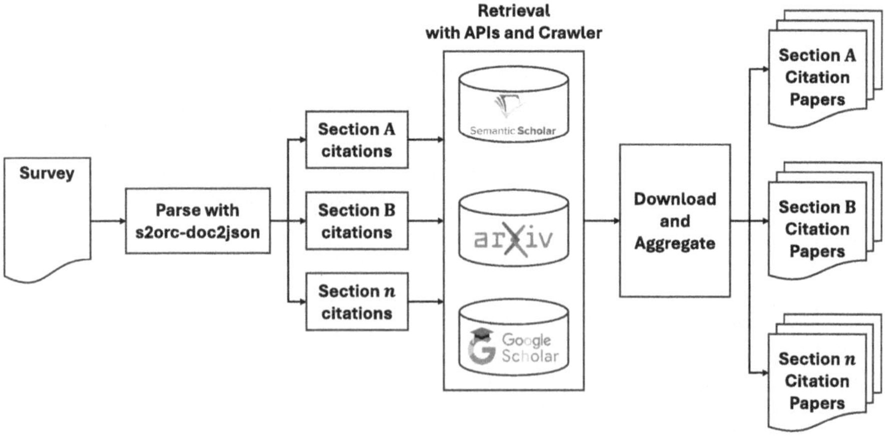

Fig. 1. Dataset Creation Process.

Initially, we parsed each survey with s2orc-doc2json[2], extracting the sections along with the corresponding citations. s2orc-doc2json converts sci-

[2] https://github.com/allenai/s2orc-doc2json.

entific articles from various formats, such as PDFs and LaTeX, to JavaScript Object Notation (JSON) format. Additionally, S2ORC-DOC2JSON extracts the citations from scientific papers, indicating both the position of the citation within the text and the referenced paper.

The next step is to retrieve the full text of the cited papers. For this, we conducted searches, utilizing available APIs (e.g., arXiv search API), across online repositories including arXiv[3] and Semantic Scholar[4] supplemented by a crawler for Google Scholar[5].

Finally, we manually verified each extracted citation against the original surveys. This step was necessary because S2ORC-DOC2JSON is a machine learning based process, thus susceptible to inconsistencies. Also, some corresponding papers were not retrieved using the APIs or the crawler, so a manual search and retrieval was necessary.

Using this approach, the SURVEYSUM was created. Data was extracted from 79 sections across 6 surveys. Each section is related to an average of 7.38 articles, with a standard deviation of 4.95 articles (median is 7 articles). Thus, the task is to generate a survey's section text given its title and a collection of papers.

4 Pipelines

4.1 General Pipeline

The text generation of a survey's section is a multi-document summarization problem. Starting from an input topic, the goal is to find papers related to the topic and summarize them into a single cohesive, coherent, and technically correct text. We present a general pipeline, divided in three stage, to automatically generate a survey's section text from reference papers, as shown in Fig. 2 and detailed as follows:

1. Stage 1: Definition of the title of the survey, the title of the sections, and related papers;
2. Stage 2: Split the relevant papers texts into chunks and use a search algorithm to retrieve the relevant chunks according to the title of the survey's section;
3. Stage 3: Generate the text for the section with a Large Language Model (LLM).

The first stage can be done manually, with the researcher choosing a topic/title for the subject and providing a set of pre-selected articles. Or, it can be automated: starting from a topic, one can use a scientific article search engine (e.g. Semantic Scholar [18]) to find related papers and then refine the results.

The effectiveness of the next stages of the pipeline relies on the quality of the retrieved papers. If the selection is manual, it is assumed that the researcher is

[3] https://info.arxiv.org/help/api/index.html.
[4] https://www.semanticscholar.org/product/api.
[5] https://scholar.google.com.br.

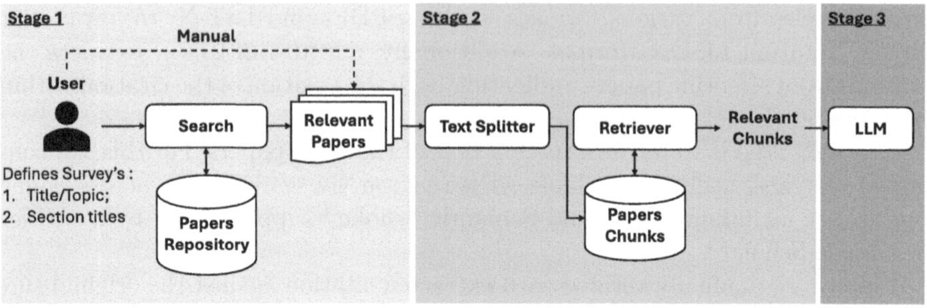

Fig. 2. Three stages to generate a section of a survey.

aware of which papers are relevant. If the selection is automatic, the quality of the result will be closely tied to the retriever.

The second stage receives the data from the previous stage and selects excerpts from the texts to be consolidated into a single text. This typically includes splitting the text into smaller units, called chunks, storing them in a database, and retrieving the most relevant ones.

Creating the chunks can be done with different strategies. The text can be split by characters, sentences, or paragraphs. It can involve some degree of overlap or not. The database where the chunks will be stored can be either textual or vector-based. If it's a textual database, the chunks can be enriched using some method (e.g. doc2query [15]). As for a vector database, it's necessary to choose a model to generate the chunk embeddings. Then, for retrieval, one can use a variety of methods – for example, a lexical search using BM25 [17] followed by one or more reranking steps.

Finally, the third stage takes as input the titles of the survey, the section (from stage 1) and the relevant chunks (from stage 2) to summarize using a LLM. To achieve this, it's essential to select the model[6] that will be used.

4.2 Pipeline 1

In this pipeline, it is assumed that the subject and title of the section are given to the system, chosen by the researcher. The researcher has the option to provide the papers for each section or let the system search online for relevant papers, in which case the selected articles are downloaded in PDF format and then have their text extracted.

In the second stage, the text from each article are segmented in chunks of 150 words, with an overlap of 50%. Each chunk is then judged by a neural ranker model, monoT5-3B, which demonstrated a strong zero-shot performance in numerous works [14]. With the score monoT5-3B gives to each chunk, we select the top n chunks and provide them to the LLM. Figure 3 shows the prompt template used for generating the content of a section.

[6] At this stage, there are no restrictions on which LLM to use, but for the purposes of this paper, we experimented only with those from OpenAI.

[System]
You are an expert in the scientific literature review, your job is, given a series of papers and their summaries, write a paragraph with a given title citing relevant information in the traditional format (e.g [1,2,3] [1]) from the provided papers.

[User]
Paper 1
Title: Evaluating GPT-3.5 and GPT-4 Models on Brazilian - University Admission Exams
Summary: the present study aims to explore the capabilities of Language Models (LMs) in tackling high-stakes multiple-choice tests (...)

(...)

Paper 5
Title: BLUEX: a benchmark based on Brazilian Leading Universities Entrance eXams
Summary: To enable future comparisons, we evaluated our dataset using several language models, ranging from 6B to 66B parameters (...)

Paragraph Subject: The capacity and impact of ChatGPT in brazilian education

[AI]
Since its introduction, ChatGPT proved that it was capable of handling numerous tasks, [1,2] showed that GPT4 in particular is quite skilled at solving some of the most challenging high school level standardized tests in Brazil, namely ENEM [1], Unicamp and USP [5]...

[User]
Paper 1
Title: {TITLE_ARTICLE_1}
Summary: {SUMMARY_ARTICLE_1}

Paper {N}
Title: {TITLE_ARTICLE_N}
Summary: {SUMMARY_ARTICLE_N}

Paragraph Subject: {PARAGRAPH_SUBJECT}

Fig. 3. Prompt used to write the section of the survey in pipeline 1.

4.3 Pipeline 2

In this pipeline, the task is to generate the text for a survey section given the corresponding cited papers associated with it in SurveySum.

In the second stage of the general pipeline, each paper's contents are segmented into chunks of 7 sentences with an overlap of 2 sentences to preserve the surrounding context. Then, the embeddings for each chunk are generated using the SPECTER2 model [3] and stored in a FAISS vector database [7]. SPECTER2 was trained to generate embeddings for scientific articles considering the title of the paper and its abstract. In this pipeline, each chunk was encoded considering the title and the chunk.

The retriever selects the 20 most relevant chunks using the section title as the query term. Each chunk is sent to GPT [2], which provides a score between 0 and 5 using the prompt shown in Fig. 4.

[System]
You are a renowned scientist who is writing the section '{TITLE_SECTION}' of a survey entitled '{TITLE_SURVEY}'.

[User]
I've found a text excerpt from a scientific article that might be useful for the section '{SECTION_TITLE}' of your survey about '{SURVEY_TOPIC}'.
Your task is to generate a score for it ranging from 0 to 5 indicating its importance to the section that you are writing.
The score of a text written in a language other than English must be 0.
If the text contains only metadata of an article (title, authors, venue, etc), i.e. if the text belongs to the references session, the score should be 0.
You should also explain why you chose this score.
Your answer MUST be enclosed in an RFC8259-compliant JSON object with two properties, "score" and "reasoning", containing the score and the reasoning for it. Remember to enclose the value of the reasoning property in quotes.
Do not answer with anything besides the JSON object. Do not insert any text before or after the RFC8259-compliant JSON object.
Use the following format to answer (write the reasoning property first and than write the score):

```
{
  "score": {SCORE},
  "reasoning": {REASONING}
}
```

[AI]
Sure, send me the text I will give you what you need. I will answer with only a JSON object containg the score and the reasoning.

[User]
{CHUNK_OF_TEXT}

Fig. 4. Prompt used to rerank the chunks in pipeline 2.

[System]
You are a renowned scientist who is writing a survey entitled '{TITLE}'.

[User]
Your task is to write the text of the section '{SECTION_TITLE}' of the survey. To complete this task, I will give you a list of documents that should be used as references. Each document has a text and an alphanumeric ID.
When writing the section, you MUST follow this rules:
- be aware of plagiarism, i.e., you should not copy the text, but use them as inspiration.
- when using some reference, you must cite it right after its use. You should use the IEEE citing style (write the id of the text between square brackets).
- you are writing the paragraphs of the section. You MUST write only this section.
- you MUST NOT split the section in subsections, nor create introduction and conclusion for it.
- DO NOT write any conclusion in any form for the subsection.
- DO NOT write a references section.
- DO NOT begin the text writing that the context is 'context', as this is obvious from the title of the survey.
Do you understand your task?

[AI]
Sure, send me a list of text and I will write a section about '{SECTION_TITLE}' using them as references. I am aware that I should use the IEEE citing style.

[User]
ID: {REF0}
Text: {CHUNK_0}
...
ID: {REFN}
Text: {CHUNK_N}

Fig. 5. Prompt used to write the section of the survey in pipeline 2.

After this step, all chunks that received a score of 0 are discarded. The remaining ones are reordered by the score, and only the top n best-scored are selected and sent to the third stage. These chunks are inserted into the prompt shown in Fig. 5 and sent to GPT to generate the section text.

4.4 Evaluation Metrics

References F1 Score
When using LLMs, particularly in the context of Retrieval-Augmented Generation (RAG) systems, it's important to assess the quality of the generated content

and the model's effectiveness in using the provided references to generate text. In scientific literature, it is common for each claim to be supported by multiple references. Therefore, one approach is to request the model to produce a summary along with the references it used. This enables the calculation of the F1 score for the references used by the model against the references provided in the input, serving as the ground truth. This method allows for an evaluation of which model or prompt more effectively uses the given information.

G-eval

G-eval is a metric for evaluation content generated by LLMs [11]. It measures the quality by comparing the generated text to a reference text. The final score is calculated by sampling a list of scores provided by an LLM with decoding temperature set to 1, top-p sampling set to 1 and number of samples set to 20. It is defined as follows:

$$score = \frac{1}{N} \sum_{i=1}^{N} p(s_i) \times s_i \tag{1}$$

where N is the number of samples, $p(s_i)$ is the probability of the i-th sample, and s_i is the score of the i-th sample. The score is generated using the prompt shown in Fig. 6.

Check-Eval

We also used the Check-Eval metric to evaluate the generated text. This metric is based on the methodology proposed by Pereira et al. [16]. The Check-Eval metric evaluates the consistency between the candidate text and the reference text using an evaluation checklist. The checklist is generated by a LLM model, which is then used to evaluate the candidate text. The presence of the checklist elements in the candidate text is quantified to determine its information quality. The metric calculates a score reflecting the consistency between the candidate and reference text. For example, if the evaluation checklist consists of ten elements and the candidate text includes eight of these, the Check-Eval score will be 0.8, i.e., the metric suggests that the candidate text maintains an 80% consistency level with the reference text.

```
[System]
You will be given a text written for a survey section and a ground truth section. Your task is to rate the content of the survey section on one metric comparing this text with the ground truth which has the maximum score. Please make sure you read and understand the instructions carefully. Please keep the document open while reviewing, and refer to it as needed.

[User]
Evaluation Steps
1 - Carefully read the content to identify the main topic and key points.
2 - Evaluate whether the content adequately addresses the main topic stated in the title and provides a comprehensive technical description of it.
3 - Assign a score to the text on a scale of 1 to 5, where 1 represents the lowest score and 5 represents the highest score, according to the Evaluation Criteria.
# - #
Survey Name: {SURVEY_NAME}
Survey Section: {SURVEY_SECTION}
Content: {GENERATED_CONTENT}
# - #
Ground Truth Text: {GROUND_TRUTH}
# - #
Evaluation Form (scores ONLY - numbers from 1.0 to 5.0)
Score (1.0 - 5.0):
```

Fig. 6. Prompt used for G-Eval.

5 Results

This section presents the results of generating the texts of each section of SURVEYSUM using the two pipelines described in Sect. 4 with different setups.

Pipeline 1 was executed using the monoT5-3B[7] model to retrieve the chunks and the gpt-3.5-turbo-0125 model to generate the final text. We have evaluated different setups for this pipeline as follows:

- 1.1: Section generation with 5 chunks, with the corresponding cited articles from SURVEYSUM;
- 1.2: Section generation with 10 chunks, with the corresponding cited articles from SURVEYSUM;
- 1.3: Section generation with 10 chunks, with articles retrieved from the Semantic Scholar API.

Pipeline 2 was executed with six different setups after the chunks retriever stage. After selecting the top 20 chunks, a GPT model was used to rerank the chunks and to write the section. Each configuration differs in the GPT model

[7] https://huggingface.co/castorini/monot5-3b-msmarco.

used and the number of chunks considered to summarize the documents. In all of them, only the articles from each section of the dataset were considered as input for the models:

- 2.1: gpt-3.5-turbo-0125 model. Section written with a single chunk;
- 2.2: gpt-3.5-turbo-0125 model. Section written with up to 5 chunks;
- 2.3: gpt-3.5-turbo-0125 model. Section written with up to 10 chunks;
- 2.4: gpt-4-0125-preview model. Section written with a single chunk;
- 2.5: gpt-4-0125-preview model. Section written with up to 5 chunks;
- 2.6: gpt-4-0125-preview model. Section written with up to 10 chunks;

Table 1 shows the metrics for nine configurations evaluated. Regarding the metrics used (G-eval and CHECK-EVAL), both metrics generally agreed with each other. However, unlike the evaluation checklist, G-eval tends to concentrate the results around certain numbers.

Pipeline 1.3 had the worst results in both CHECK-EVAL and G-eval scores among all pipelines. This was the only configuration tested that used articles retrieved from a search engine instead of articles from SURVEYSUM. This result shows a strong dependence of text summarization on the retriever stage.

For pipelines of the same type ([1.1 and 1.2], [2.1, 2.2, and 2.3], [2.4, 2.5, and 2.6]), the more chunks used, the higher the F1 score tends to be, as there is a higher probability of selecting chunks from different articles. However, it cannot be inferred that increasing the number of chunks will necessarily improve text quality. Although the experiments indicated this average behavior for this number of chunks, there are samples where increasing the number of chunks worsened the result. The quality of the chunk—directly influenced by the retriever stage—is very important.

Regardless of the metric considered in Table 1, the results for pipelines 2.4, 2.5, and 2.6 (using the gpt-4-0125-preview model) were better than the results for pipelines 2.1, 2.2, and 2.3 (equivalent pipelines using gpt-3.5-turbo-0125). This suggests that the LLM used is relevant to the quality of result.

Table 1. Results of pipelines. All checklist evaluations were performed using gpt-3-5-turbo and all G-Eval scores were generated by gpt-4.

Pipeline	LLM	Number of Chunks	Source	IR Pipeline	CHECK-EVAL Score	F1 Score	G-Eval Score
1.1	gpt-3.5-turbo-0125	5	SURVEYSUM	MonoT5	**0.415**	0.64	3.8
1.2	gpt-3.5-turbo-0125	10	SURVEYSUM	MonoT5	0.244	0.59	3.74
1.3	gpt-3.5-turbo-0125	10	Semantic Scholar	MonoT5	0.141	–	2.84
2.1	gpt-3.5-turbo-0125	1	SURVEYSUM	Embeddings	0.188	0.36	3.42
2.2	gpt-3.5-turbo-0125	5	SURVEYSUM	Embeddings	0.257	0.66	3.9
2.3	gpt-3.5-turbo-0125	10	SURVEYSUM	Embeddings	0.284	0.7	3.9
2.4	gpt-4-0125-preview	1	SURVEYSUM	Embeddings	0.271	0.36	3.82
2.5	gpt-4-0125-preview	5	SURVEYSUM	Embeddings	0.323	0.67	3.99
2.6	gpt-4-0125-preview	10	SURVEYSUM	Embeddings	0.352	**0.72**	**4.01**

The G-Eval and Check-Eval scores are high correlated (76%), indicating consistency in their evaluation of summary quality. However, G-Eval tends to produce results with lower variability compared to the Check-Eval score. The Check-Eval score, being more granular and item-specific, captures a broader range of evaluation aspects. These insights underscore the importance of employing multiple evaluation metrics to obtain a comprehensive understanding of the performance of summarization models.

6 Conclusion

In this paper, we introduced SurveySum, a dataset designed for summarizing sections within scientific surveys in artificial intelligence, natural language processing, and machine learning. Our contributions include the creation of the dataset, two pipelines to summarize scientific articles, and their evaluation using various metrics.

As with various other summarization challenges, the summarization of articles for the generation of a survey section does not yield a single definitive answer. Different researchers may use the same set of papers to produce a section on a given topic, each covering different aspects of that topic. Nevertheless, it is important to establish parameters for comparison. SurveySum addresses a gap in document summarization by providing a dataset that assists the development and benchmarking of models focused on scientific surveys.

Our results highlight the dependence on the quality of the retriever, as the effectiveness of the summarization pipelines was influenced by the accuracy and relevance of the documents retrieved during the initial stages. High-quality retrieval may ensure that the most relevant information was available for summarization.

References

1. Boni, O., Feigenblat, G., Lev, G., Shmueli-Scheuer, M., Sznajder, B., Konopnicki, D.: HowSumm: a multi-document summarization dataset derived from wikihow articles (2021)
2. Brown, T.B., et al.: Language Models are Few-Shot Learners (2020)
3. Cohan, A., Feldman, S., Beltagy, I., Downey, D., Weld, D.S.: SPECTER: document-level representation learning using citation-informed transformers. In: ACL (2020)
4. Fabbri, A.R., Li, I., She, T., Li, S., Radevm, D.R.: Multi-news: a large-scale multi-document summarization dataset and abstractive hierarchical model (2019)
5. Ghalandari, D.G., Hokamp, C., Pham, N.T., Glover, J., Ifrim, G.: A large-scale multi-document summarization dataset from the wikipedia current events portal (2020)
6. Gupta, V., Bharti, P., Nokhiz, P., Karnick, H.: SumPubMed: summarization dataset of PubMed scientific articles. In: Kabbara, J., Lin, H., Paullada, A., Vamvas, J. (eds.) Proceedings of the 59th Annual Meeting of the Association for Computational Linguistics and the 11th International Joint Conference on Natural Language Processing: Student Research Workshop, Online, August 2021, pp. 292–303. Association for Computational Linguistics (2021)

7. Johnson, J., Douze, M., Jégou, H.: Billion-scale similarity search with GPUs. IEEE Trans. Big Data **7**(3), 535–547 (2019)
8. Koh, H.Y., Ju, J., Liu, M., Pan, S.: An empirical survey on long document summarization: datasets, models, and metrics. ACM Comput. Surv. **55**(8), 1–35 (2022)
9. Liu, S., Cao, J., Yang, R., Wen, Z.: Generating a structured summary of numerous academic papers: Dataset and method. In: Proceedings of the Thirty-First International Joint Conference on Artificial Intelligence, IJCAI-2022. International Joint Conferences on Artificial Intelligence Organization, July 2022
10. Liu, S., Cao, J., Yang, R., Wen, Z.: Long text and multi-table summarization: dataset and method (2023)
11. Liu, Y., Iter, D., Xu, Y., Wang, S., Xu, R., Zhu, C.: G-eval: NLG evaluation using GPT-4 with better human alignment. In: Conference on Empirical Methods in Natural Language Processing (2023)
12. Yao, L., Dong, Y., Charlin, L.: Multi-xscience: a large-scale dataset for extreme multi-document summarization of scientific articles (2020)
13. Nenkova, A., McKeown, K.: Automatic summarization. Found. Trendső Inf. Retr. **5**(2–3), 103–233 (2011)
14. Nogueira, R., Jiang, Z., Pradeep, R., Lin, J.: Document ranking with a pretrained sequence-to-sequence model. In: Cohn, T., He, Y., Liu, Y. (eds.) Findings of the Association for Computational Linguistics: EMNLP 2020, pp. 708–718. Association for Computational Linguistics, Online, November 2020
15. Nogueira, R., Yang, W., Lin, J., Cho, K.: Document expansion by query prediction (2019)
16. Pereira, J., Lotufo, R.: Check-Eval: a checklist-based approach for evaluating text quality (2024)
17. Robertson, S.E., Walker, S., Jones, S., Hancock-Beaulieu, M., Gatford, M.: Okapi at TREC-3. In: Harman, D.K. (ed.) Proceedings of The Third Text REtrieval Conference, TREC 1994, Gaithersburg, Maryland, USA, 2–4 November 1994, volume 500-225 of *NIST Special Publication*, pp. 109–126. National Institute of Standards and Technology (NIST) (1994)
18. Semantic Scholar. https://www.semanticscholar.org/
19. Yasunaga, M., et al.: ScisummNet: a large annotated corpus and content-impact models for scientific paper summarization with citation networks. In: Proceedings of AAAI 2019 (2019)

Traffic Forecasting Using Federated Randomized High-Order Fuzzy Cognitive Maps

Omid Orang[3], Felipe A. R. da Silva[1], Petrônio C. L. Silva[2,3](✉), Pedro H. S. S. Barros[3], Heitor S. Ramos[3], and Frederico G. Guimarães[3]

[1] Graduate Program in Electrical Engineering, Federal University of Minas Gerais, Belo Horizonte, Brazil
fars@ufmg.br

[2] Federal Institute of Northern Minas Gerais, Januaria, Brazil
petronio.candido@ifnmg.edu.br

[3] Department of Computer Science, Federal University of Minas Gerais, Belo Horizonte, Brazil
{omid.orang,pedro.barros,ramosh}@dcc.ufmg.br, fredericoguimaraes@ufmg.br

Abstract. Numerous machine learning (ML) and deep learning (DL) forecasting techniques have emerged for predicting traffic flow and traffic speed. Despite their notable achievements, they encounter challenges related to preserving in-vehicle user privacy. Alternatively, Federated Learning (FL) offers a decentralized ML strategy to strike a balance between prediction accuracy and privacy preservation. This paper proposes a novel fuzzy-based method called FL-RHFCM, which integrates the principles of Randomized High-order Fuzzy Cognitive Maps (R-HFCM) with FL. R-HFCM is akin to an echo state network (ESN), where the only trainable component is the output layer using least squares (LS) minimization. The reservoir layer's weights are initialized randomly and remain unchanged during training. Thus, the LS coefficients represent the only parameters shared with the server for aggregation in our FL-RHFCM approach. FL-RHFCM introduces an efficient distributed forecasting method using FCMs, significantly reducing communication costs. The effectiveness of our proposed model is assessed using two datasets, demonstrating its promise compared to some existing baseline methods.

Keywords: Traffic Prediction · Federated Learning · Randomized High-order Fuzzy Cognitive Maps · Echo State Network · Least Squares

1 Introduction

Intelligent Transportation Systems (ITS) have recently gained attention due to rising road safety and efficiency concerns. Traffic prediction is crucial for route planning, optimizing vehicle dispatching, controlling traffic congestion, etc. [1,2]. Traffic prediction (travel time, flows, speeds, occupancy, and demand) utilizes a

A. Paes and F. A. N. Verri (Eds.): BRACIS 2024, LNAI 15413, pp. 445–459, 2025.
https://doi.org/10.1007/978-3-031-79032-4_31

trainable function to analyze past traffic data and forecast future traffic conditions. It relies on two primary types of data: traffic flow, indicating total detected vehicles over a period, and traffic speed, representing average vehicle velocity in the same area during the same time frame [3]. This study will refer to traffic flow and speed as traffic.

Forecasting traffic presents a formidable challenge given the non-linear, time-dependent, and spatio-temporal properties of traffic time series [1,4]. Consequently, developing an accurate traffic prediction model is imperative. Accordingly, a wide range of traffic forecasting approaches have been presented in the literature, which can be broadly categorized into conventional parametric statistical models and non-parametric machine learning (ML) and deep learning (DL)-based methods [1,5]. AutoRegressive Integrated Moving Average (ARIMA) models and their variants, including Seasonal ARIMA (SARIMA) and Vector ARIMA (VARIMA), are commonly used statistical techniques in traffic prediction, as discussed in [5,6]. However, these models encounter challenges due to the complexity, non-stationarity, and non-linearity of traffic data. Additionally, ARIMA models generally necessitate a substantial volume of historical data to perform effectively, which may not always be accessible [5,7].

To tackle these challenges, ML and DL methods are extensively utilized in traffic prediction tasks owing to their capacity to automatically extract crucial features from historical traffic data, thus eliminating the need for complex mathematical model design. Convolutional Neural Networks (CNN), Recurrent Neural Networks (RNN), and their variants, such as Long Short-Term Memory (LSTM) and Gated Recurrent Unit (GRU), have shown remarkable performance [5,8]. In addition, Graph Neural Networks (GNNs) [4] are considered state-of-the-art methods particularly well-suited for traffic forecasting challenges due to their capacity to capture spatial dependencies. As discussed in [4], several types of GNNs have been developed, including graph recurrent neural networks (GRNN) [9], graph-structured recurrent neural networks (GSRNN) [10], and graph LSTM (GLSTM) [11], among others. Attention mechanisms and transformers represent another category of DL techniques that have proven effective in the field of traffic prediction [5,12,13].

Despite the considerable success and frequent usage of DL methods, they suffer from limitations such as interpretability, parsimony, and high computational costs in terms of time and resources. Fuzzy Time Series (FTS) forecasting methods offer an alternative due to their simplicity, interpretability, updatability, scalability, and capacity to handle uncertainty and complex systems [14]. FTS is a methodology for time series forecasting (TSF) that involves converting the numerical time series into a linguistic time series representation using fuzzy sets. Then, the transitions between the fuzzy sets in the historical data are used to find rules for forecasting. Fuzzy Cognitive Maps (FCMs), a subset of weighted FTS methods, represent a specialized category in fuzzy modeling and forecasting techniques [15]. FCMs, as interpretable ML models, have a good capacity for dealing with uncertainty and effectively simulating the dynamic behavior of non-linear and complex systems [15,16]. Few studies among the proposed

FTS methods have utilized FCM for traffic prediction. Examples include interpretable deep attention FCM in [17], interpretable deep FCM (DFCM) in [18], FCM learned with an evolutionary algorithm (FCMEVOL) in [19].

From another perspective, existing centralized ML and DL traffic prediction methods necessitate collecting raw data for model training, leading to significant privacy risks. DL algorithms typically require vehicles to transmit raw data, including sensitive information like location, to a central server for training the proposed models in a centralized manner [20]. If the central server is compromised, the entire forecasting system is vulnerable to a single point-of-failure attack, risking severe privacy breaches for vehicles. Additionally, heavy reliance on extensive data for centralized training increases communication overhead and the risk of data leakage and pollution [21]. To address these problems, federated learning (FL) [22], which shares model updates without exchanging raw data, has recently been introduced as an efficient solution [23]. FL performs local learning model calculations and then sends the local learning model parameters to the central server for global model aggregation. Accordingly, it can avoid direct interaction with original data and achieve the trade-off between model performance, communication overhead, and data privacy.

To the best of our knowledge, only two investigations have explored the use of FL and FCMs. The authors in [24] implemented FL methods for FCMs in medical applications, specifically for classifying dengue in Colombian cities. A blind federated learning approach without an initial model was introduced in [25] proposing two innovative methodologies for PSO-based FL of FCMs to classify breast cancer and demographic features related to adult income. Consequently, there are no references regarding federated FCMs in forecasting applications. Thus, the focal contribution of this study is to fill this gap by presenting a novel univariate forecasting method termed FL-RHFCM for the first time in the literature. FL-RHFCM is a hybrid method integrating FL and randomized high-order FCM (R-HFCM) to predict traffic flow and traffic speed. R-HFCM [26] is a new class of FCMs combining FTS, FCM, and echo-state network (ESN), trained via least squares (LS). It functions as an ESN with input, reservoir, and output layers. The reservoir contains parallel sub-reservoirs (L) with unalterable weights, a unique feature that sets R-HFCM apart from typical FCM-based methods. LS is then applied to train the output layer and determine the LS coefficients (λ_i). Therefore, in FL-RHFCM, each user is trained locally to obtain the optimal values of λ_i. Afterward, these values are shared with the server to be aggregated using the average method to update the global model. Finally, clients are updated locally, receiving updated global model parameters from the aggregator. This process is executed for 15 rounds using the Flower platform.

The rest of this paper is structured as follows: Sect. 2 shares a brief review of traffic prediction; Sect. 3 provides a comprehensive presentation of the proposed approach; Sect. 4 details the experimental results and discussion; and lastly, Sect. 6 encapsulates the paper's findings and delineates potential directions for future research.

2 Literature Review

Plenty of univariate and multivariate traffic forecasting techniques have been developed in the literature [3,27], which can be grouped into three main classes, including statistical methods, traditional ML, and DL methods. According to [27,28], statistical methods, particularly ARIMA and its variants such as SARIMA, SARIMAX, Kohenen ARIMA (KARIMA), and Vector ARIMA, have been frequently employed to predict traffic. These methods are generally suitable for simpler and less complex datasets but require a substantial volume of historical data to perform effectively. This dual requirement means they face limitations when dealing with non-stationary, non-linear, spatio-temporal, and complex datasets. ML models excel in generalization and adaptability to changing traffic network conditions compared to statistical methods [27,28]. They are typically divided into three categories: feature-based methods, Gaussian process models, and state-space models, capable of handling non-linear and complex time series data [29]. They have pros and cons, as explained in [27]. Artificial neural networks (ANN) [30], k-nearest neighbor (KNN) [31], and support vector regression (SVR) are some examples of this category [32].

DL methods are extensively used in traffic prediction tasks because of their strong capacity to capture stochastic and nonlinear relationships in traffic data [27]. Based on this reference, Multi-layer Perceptrons (MLP), Autoencoder (AE), CNN, RNN, LSTM, GNNs, Restricted Boltzmann Machines (RBMs), Deep Belief Networks (DBN), Graph Convolution Network (GCN), Wavelet Neural Network (WNN) and Attention-based models have been employed to predict traffic. Also, the authors in [3] reviewed 37 DL traffic forecasting models to predict spatial and/or temporal traffic flow and speed, confirming various types of DNN methods. For instance, in [33], DBN, k-means clustering, and Dempster-Shafer theory are utilized for traffic flow prediction, while in [34], CNN with Pearson correlation-based theory is employed to predict traffic speed. Graph Neural Networks (GNNs) with DL are another traffic forecasting method elaborated in [4], including models like graph LSTM (GLSTM), graph multi-attention network (GMAN), graph attention temporal convolutional network (GATCN), among others. The next sub-group of DNN forecasting methods are hybrid methods such as LSTM and bidirectional LSTM [35], encoder-decoder LSTM and FNN-based attention module [36], encoder-decoder GRU, and graph diffusion [37], encoder-decoder GRU, FNN and graph attention network [38], among others.

Fuzzy-based forecasting models are the last category of traffic forecasting techniques that can solve some limitations regarding DL methods regarding interpretability, training time, scalability, and complexity. A new fuzzy-based CNN method in [39], an evolving fuzzy neural network (EFNN) in [40], and an adaptive hybrid fuzzy rule-based system in [41] are some studies in this group. Also, some researchers combined fuzzy and DL such as fuzzy deep convolution network (FDCN) in [42], spatiotemporal fuzzy-graph convolutional network model in [43], and a combination of fuzzy logic, LSTM, and decision trees (DTs) in [44].

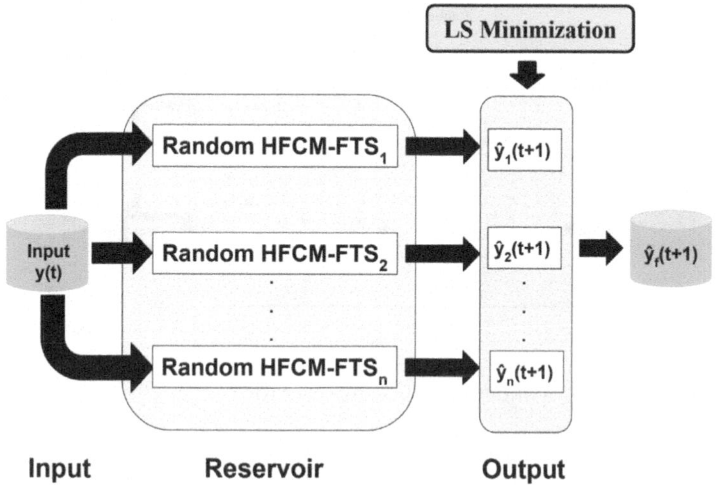

Fig. 1. Generic structure of the proposed R-HFCM technique

Despite accurately forecasting centralized methods, data privacy poses significant challenges. To address this, some researchers have utilized decentralized FL methods to make a tradeoff between prediction accuracy and privacy preservation. An integration of FL and GRU in [45], a combination of FL and attention-based spatial-temporal GNN (ASTGNN) in [46], graph attention networks (GAT), LSTM, and FL in [21], clustering-based hierarchical and two-step-optimized FL in [47], federated community GCN (FCGCN) in [48], and FL with asynchronous GCN in [49] are some examples of FL-based models for traffic flow and speed forecasting.

FCMs, as a neuro-fuzzy method, have shown considerable success in capturing the dynamics of various complex systems and effectively handling uncertainties [26]. Despite this, there is a notable gap in the literature regarding using decentralized FCMs for predictive applications, such as traffic forecasting. To address this, our research introduces a novel federated FCM forecasting model in this field.

3 Proposed FL-RHFCM Method

As explained in [15], FCMs are composed of a set of concepts and signed directed connections, known as weights. Therefore, weight matrices form the core element of each FCM, and various training methods have been developed to optimize both the weights and structure.

This research introduces a novel hybrid univariate forecasting method called FL-RHFCM, a fusion of R-HFCM and FL, to predict traffic flow and speed. R-HFCM [26] is a centralized FCM-based forecasting method with a different

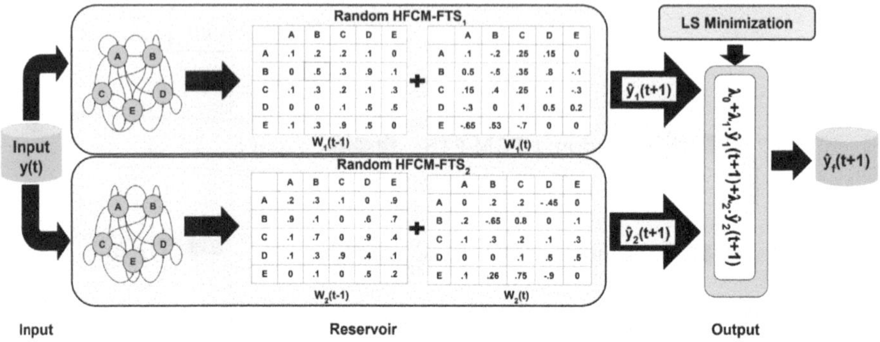

Fig. 2. A simple example R-HFCM model with L = 2, Ω = 2 and k = 5.

structure from regular FCMs. Accordingly, this section is divided into two subsections. Section 3.1 details R-HFCM, and Sect. 3.2 provides information regarding our proposed FL-RHFM technique.

3.1 Centralized R-HFCM Method

R-HFCM is a hybrid method combining the concepts of FTS, FCMs, and echo state networks (ESN) [26]. More precisely, as shown in Fig. 1, R-HFCM consists of three layers: the input layer, reservoir, and output layer. The reservoir layer is composed of a specific number of sub-reservoirs (L). Each sub-reservoir employs the HFCM-FTS method [50], with weights randomly initialized following the ESN approach and kept constant throughout training. Then, output from each sub-reservoir, generated from the defuzzification step, is fed to the output layer. Finally, LS is employed to train the output layer and identify the optimal values of LS coefficients (λ_i).

From another point of view, R-HFCM can be seen as an ESN (or reservoir computing) method such that only the output layer is trainable. In R-HFCM, the LS coefficients (λ_i) serve as the sole trainable parameters.

Thus, R-HFCM is not focused on training weights among the concepts. This property distinguishes R-HFCM from other FCM-based methods and makes it much faster than methods trained via population-based techniques such as genetic algorithms (GA) or particle swarm optimization (PSO). Figure 2 a simple topology of the R-HFCM method with $L = 2$, $k = 5$ (concepts), and $\Omega = 2$ (order). In this case, LS trains the output layer to find LS coefficients. The number of coefficients is three (λ_0, λ_1 and λ_2) which means that the number of LS coefficients directly depends on the number of sub-reservoirs. As such, the number of LS coefficients is equal to $L+1$. The details of training and forecasting procedures are described in detail as follows:

A. Training Process

1) Weight and bias initialization: The weights among the concepts C_i and C_j are randomly chosen from a uniform distribution in [−1,1] and scaled inspiring by ESN weight initialization using:

$$\mathbf{W}^t = \mathbf{W}^{\mathbf{rand}} \cdot \left(\frac{\epsilon}{\rho_{\mathbf{max}}(\mathbf{W}^{\mathbf{rand}})} \right) \tag{1}$$

where $\rho_{\mathbf{max}}(\mathbf{W}^{\mathbf{rand}})$ is the largest absolute eigenvalue of $\mathbf{W}^{\mathbf{rand}}$, and $\epsilon = 0.5$. The bias vector $\mathbf{w}^{\mathbf{0}}$ is initialized similarly:

$$\mathbf{w}^0 = \mathbf{w}^{\mathbf{0}}_{\mathbf{rand}} \cdot \left(\frac{\epsilon}{\mathbf{S}} \right) \tag{2}$$

where $\mathbf{S}$ is the maximum singular value of $\mathbf{w}^{\mathbf{0}}_{\mathbf{rand}}$.

2) Partitioning: Firstly, the Universe of Discourse (UoD) is determined using the formula $UoD = [\min(Y_i) - D_1, \max(Y_i) + D_2]$, where $D_1 = \min(Y_i) \times 0.2$ and $D_2 = \max(Y_i) \times 0.2$. Then, UoD is partitioned into k even-length intervals (representing concepts of FCMs) using grid partitioning and the triangular membership function.

3) Fuzzification: In this step, the activation state $a_i(t)$ of each concept $C_i \in C$, $\forall y(t) \in Y$, is computed to convert the crisp time series Y into a fuzzy series A. Each fuzzified sample $a(t) \in A$ is given by $a_i(t) = \mu_{C_i}(y(t))$, $\forall i \in \{1, \ldots, k\}$.

4) Activation: The state value of each concept within each sub-reservoir at time t+1 is updated based on the following formula:

$$\mathbf{a}_j(t+1) = f\left(\mathbf{w}^0 + \sum_{l=1}^{\Omega} \mathbf{W}^l \cdot \mathbf{a}(t-l+1) \right) \tag{3}$$

5) Defuzzification: The output from each sub-reservoir is generated through this process using the below equation:

$$\hat{y}_j(t+1) = \frac{\sum_{i=1}^{k} a_{ji}(t+1) \cdot mp_i}{\sum_{i=1}^{k} a_{ji}(t+1)} \tag{4}$$

where $a_j(t+1)$ is the activation state of each concept at time $t+1$ and mp_i represents the center of each concept C_i.

6) Calculating LS coefficients: From the outputs $y_j(t+1)$ $\forall j \in \{1, \ldots, L\}$, and for each input sample $y(t) \in Y$, $t = 1 \ldots T$, a design matrix X is formed for the linear system $Y = \lambda X$. The LS method is then used to find the coefficient vector λ to minimize the mean squared error.

B. Forecasting Process

1) Fuzzification: Same as the third stage of the training process.

2) Activation: Same as the fourth stage of the training process.

3) Defuzzification: First, the defuzzified value of each sub-reservoir is calculated using Eq. 4. Then, the linear combination of $\hat{y}_j(t+1)$ and λ_j is considered to generate the final predicted value as follows:

$$\hat{y}_f(t+1) = \lambda_0 + \sum_{j=1}^{L} \lambda_j \cdot \hat{y}_j(t+1) \tag{5}$$

3.2 Decentralized R-HFCM Method

This section introduces FL-RHFCM as a decentralized adaptation of the R-HFCM technique, wherein FL is incorporated into R-HFCM to enhance data privacy. Thus, in FL-RHFCM, data is not shared among the users, and each user is trained using the local dataset. Figure 3 exhibits our proposed FL-RHFCM approach. Given m users, denoted as $\{U_1, U_2, \ldots, U_m\}$, each associated with its local data denoted by $\{Y_1, Y_2, \ldots, Y_m\}$. In FL-RHFCM, the UoD for each client is calculated using $UoD = [\min(Y_i) - D_1, \max(Y_i) + D_2]$, as expressed earlier.

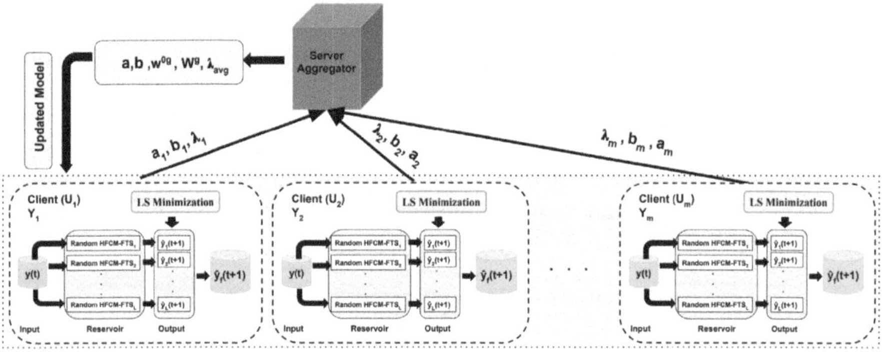

Fig. 3. The generic architecture of FL-RHFCM, considering m clients with m different datasets

In Fig. 3, a_i and b_i respectively represent $\min(Y_i)$ and $\max(Y_i)$, $\forall i \in m$. Furthermore, w_0, W, and λ_{avg} represent bias, weight, and average LS coefficients, respectively. As mentioned earlier in Sect. 3.1, the only trainable parameters in the R-HFCM method are LS coefficients. Accordingly, each local node is in charge of training a local model and sending the local model to the server node for aggregation. More specifically, the steps of our proposed federated method are expressed as follows:

- Step 1 The server initializes w_0, W, a, b, and λ, then transmits them to the clients. Besides, w_0 and W remain fixed and are the same during all executions.
- Step 2 Each client (U_i) is trained using local time series (Y_i), calculating a_i, b_i and $\lambda_i \in \{\lambda_{0i}, \lambda_{1i}, \ldots, \lambda_{li}\}$, $\forall l \in L$.
- Step 3 The obtained a_i, b_i and λ_i are transferred to the server for aggregation. The server updates a, b, and λ as follows: $a = \min_{i=1}^{m} a_i$, $b = \max_{i=1}^{m} b_i$ and $\lambda = \frac{1}{m} \sum_{i=1}^{m} \lambda_i$.

Step 4 The updated values are shared with the users. The performance of the proposed model for each user is calculated in terms of accuracy metrics.
Step 5 This process is repeated in 15 rounds.

It is worth noting that we deployed FL-RHFCM with three client nodes, each trained using distinct time series datasets, as explained in the following section.

4 Computational Experiments

4.1 Case Studies

Two different traffic datasets are exploited to validate our approach:

1. Traffic flow (hourly)[1]: This dataset collected from sensors contains 48,120 observations of the number of vehicles in four *junctions*;
2. Traffic speed (minutely)[2]: contains 6 road *segments* of the Xueyuan Road in Beijing, China.

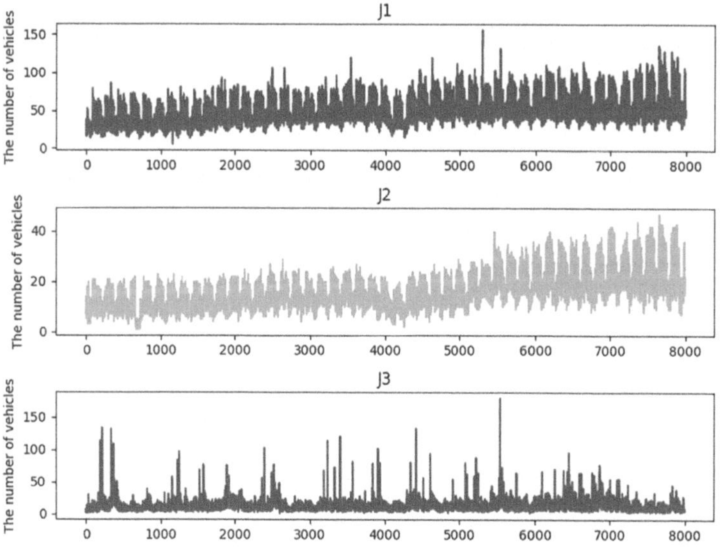

Fig. 4. Plotting of 8,000 samples from three junctions (J1, J2, and J3)

This study employs datasets from three junctions (J1, J2, and J3) and three segments (R3, R4, and R5). Detailed information about these time series is reported in Table 1. In addition, Figs. 4 and 5 display traffic flow and traffic speed time series, respectively.

[1] https://www.kaggle.com/datasets/fedesoriano/traffic-prediction-dataset/.
[2] https://github.com/BuaaPercy/Traffic-DataSet-of-Beijing-Road.

Table 1. Summary statistics of the 6 datasets used in the experiments

Time series	Min	Max	Avg	Std	samples
J1	5.000	156.000	55.149	20.855	8000
J2	1.000	47.000	16.485	7.270	8000
J3	1.000	180.000	16.038	11.049	8000
R3	5.090	70.0360	35.990	8.794	8000
R4	5.370	84.600	48.286	15.387	8000
R5	6.330	69.600	47.609	14.357	8000

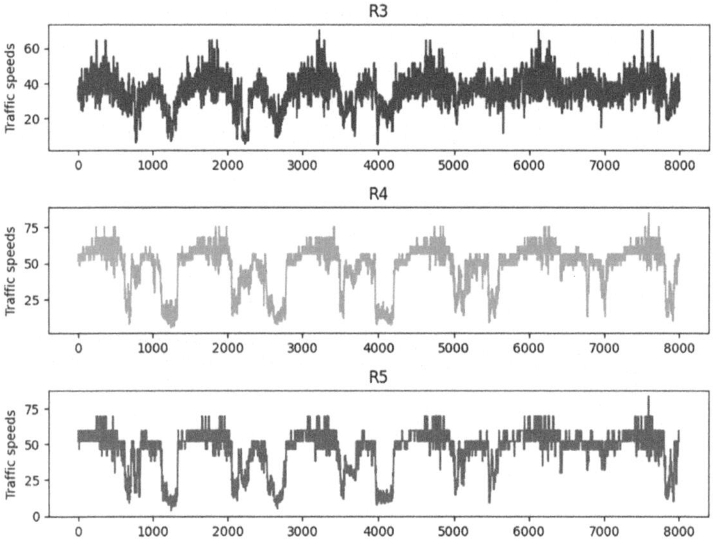

Fig. 5. Plotting of 8,000 samples from three road segments (R3, R4, and R5)

4.2 Experimental Methodology

Root mean squared error (RMSE) and normalized RMSE (NRMSE) are employed to evaluate the model's accuracy. RMSE is calculated using

$$RMSE = \sqrt{\frac{1}{n}\sum_{i=1}^{n}(y_i - \hat{y}_i)^2},$$

where y_i represents the actual values, $\hat{y}_i$ represents the predicted values, and n is the total number of samples. NRMSE is then obtained by normalizing RMSE as follows:

$$NRMSE = \frac{RMSE}{y_{max} - y_{min}}.$$

These metrics provide insights into the model's performance. It is noteworthy that 80% of each dataset is designated for training, while the remaining 20% is for testing. The model's Python code is publicly available for replication via the provided link: https://github.com/OMIDUFMG2019/FL-RHFCM-model.

5 Results and Discussion

This section analyzes the accuracy of FL-RHFCM in comparison to various centralized methods, including R-HFCM, PWFTS, LSTM, CNN, CNN-LSTM, and ARIMA. As we mentioned, FL-RHFCM includes three client nodes. Two scenarios are considered: (i) in the first scenario, each traffic speed time series (R3, R4, and R5) is directed to each node, and (ii) in the second scenario, each node receives its traffic flow time series (J1, J2, and J3).

To obtain the optimized performance of the FL-HFCM technique, multiple experiments are conducted, exploring various combinations of hyperparameters (HPs) including $k \in \{3, 4, \ldots, 10, 20, 30\}$, $L \in \{2, 3, \ldots, 10, 20, 40\}$, $\Omega \in \{2, 3, \ldots, 10\}$, and $f \in \{sigmoid, tanh, ReLU\}$).

In Scenario 1, the best performance is achieved with $L = 20$, $k = 3$, $\Omega = 5$, and $f = tanh$, whereas $L = 8$, $k = 3$, $\Omega = 5$, and $f = tanh$ yield the best result in Scenario 2. It is worth observing that the randomized grid search is utilized to find the best HPs for competing methods.

Table 2 showcases the experimental results of all methods to compare the performance of FL-RHFCM with other centralized techniques. The evaluation metrics used are RMSE and NRMSE, where the top result for each dataset is highlighted in bold and the second-best is underscored.

Table 2. Comparison of the FL-RHFCM method with other centralized models

Models	FL-RHFCM		R-HFCM		PWFTS		LSTM		CNN		CNN-LSTM		ARIMA	
Datasets	RMSE	NRMSE	RMSE	NRMSE	RMSE	NRMSE	RMSE	NRMSE	RMSE	NRMSE	RMSE	NRMSE	RMSE	NRMSE
R3	**3.860**	**0.069**	3.933	0.071	4.066	0.073	4.641	0.083	5.706	0.102	4.582	0.082	9.247	0.167
R4	3.260	0.043	**3.125**	**0.041**	3.261	0.043	3.609	0.048	4.258	0.057	3.923	0.052	15.492	0.207
R5	3.740	0.050	**3.479**	**0.047**	3.968	0.053	4.094	0.055	4.710	0.064	3.950	0.053	14.013	0.190
J1	10.820	0.095	6.627	0.058	6.584	0.058	5.836	0.051	6.371	0.056	**5.817**	**0.051**	22.451	0.198
J2	3.500	0.087	**2.955**	**0.073**	3.046	0.076	3.318	0.083	3.850	0.096	3.256	0.081	6.487	0.162
J3	**5.580**	**0.060**	5.815	0.062	6.582	0.070	6.244	0.067	7.943	0.085	6.104	0.065	11.303	0.121
Average NRMSE	-	0.067	-	0.058	-	0.062	-	0.065	-	0.077	-	0.064	-	0.174
Rank	-	5	-	1	-	2	-	4	-	6	-	3	-	7

The recorded results in Table 2 suggest that FL-RHFCM performs better predicting traffic speed than forecasting traffic flow. In more detail, FL-RHFCM outperforms other competing methods with R3 and J3 datasets, while centralized R-HFCM is superior for R4, R5, and J2. To this end, employing the average NRMSE facilitates a meticulous comparison of the accuracy of the methods. According to the table, R-HFCM demonstrates the most accurate predictor. Although PWFTS, CNN-LSTM, and LSTM outperform FL-RHFCM, their performances only marginally exceed that of FL-RHFCM. Furthermore, FL-RHFCM maintains data privacy and security, as data is not shared among clients, unlike centralized methods. Moreover, this method is simpler, has faster training speed, and incurs reduced communication costs compared to deep learning

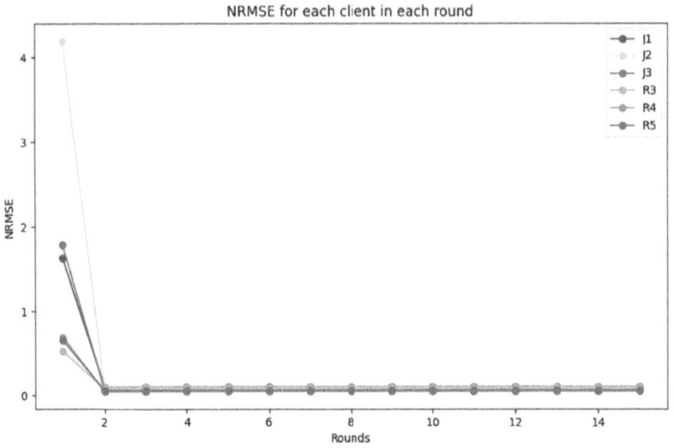

Fig. 6. The NRMSE accuracy of the FL-RHFCM method for each client per round.

models. Also, Fig. 6 indicates that FL-RHFCM converges for each client after the second round, demonstrating its efficiency in reaching stable predictions quickly. Thus, FL-RHFCM's competitive performance and additional benefits make it a valuable method for spatio-temporal forecasting in federated learning environments. The combination of accuracy, privacy, and efficiency positions FL-RHFCM as a noteworthy approach in the landscape of forecasting methodologies.

5.1 Limitations

1. The number of clients involved in the experiments is relatively low.
2. It is not compared with complex state-of-the-art techniques such as GNNs in this area.
3. While federated experiments were conducted to evaluate our proposed approach, there needs to be more comparison with other federated learning approaches in the literature.

6 Conclusion and Future Works

This study introduces the first proposal of a distributed framework for FCM-based forecasting methods called FL-RHFCM. FL-RHFCM integrates FL and centralized R-HFCM to predict both traffic speed and traffic flow. The experimental results illustrate that FL-RHFCM is a robust and efficient forecasting method, especially advantageous in scenarios requiring data privacy and decentralized learning. While centralized R-HFCM demonstrates the highest overall accuracy, FL-RHFCM's competitive performance and additional benefits make it a valuable method in spatio-temporal time series forecasting scenarios. Since traffic datasets are high-dimensional and non-stationary time series, future work will extend FL-RHFCM to predict multivariate time series, considering these aspects.

Acknowledgment. This work was partially funded by grant APQ-00426-22, Fundação de Amparo a Pesquisa do Estado de Minas Gerais (FAPEMIG), grant 2023/00721-1, São Paulo Research Foundation (FAPESP), and grant 312682/2021-2, Conselho Nacional de Desenvolvimento Científico e Tecnológico (CNPq).

References

1. Boukerche, A., Tao, Y., Sun, P.: Artificial intelligence-based vehicular traffic flow prediction methods for supporting intelligent transportation systems. Comput. Netw. **182**, 107484 (2020)
2. Cunha, F., et al.: Vehicular Networks to Intelligent Transportation Systems, pp. 297–315. Springer Singapore, Singapore (2018)
3. Tedjopurnomo, D.A., Bao, Z., Zheng, B., Choudhury, F.M., Qin, A.K.: A survey on modern deep neural network for traffic prediction: Trends, methods and challenges. IEEE Trans. Knowl. Data Eng. **34**(4), 1544–1561 (2020)
4. Jiang, W., Luo, J.: Graph neural network for traffic forecasting: a survey. Expert Syst. Appl. **207**, 117921 (2022)
5. Wang, Z., Sun, P., Hu, Y., Boukerche, A.: A novel hybrid method for achieving accurate and timeliness vehicular traffic flow prediction in road networks. Comput. Commun. (2023)
6. Boukerche, A., Tao, Y., Sun, P.: Artificial intelligence-based vehicular traffic flow prediction methods for supporting intelligent transportation systems. Comput. Netw. **182**, 107484 (2020)
7. Yin, X., Wu, G., Wei, J., Shen, Y., Qi, H., Yin, B.: A comprehensive survey on traffic prediction, ArXiv, abs/ arXiv: 2004.08555 (2020)
8. Almeida, A., Brás, S., Oliveira, I., Sargento, S.: Vehicular traffic flow prediction using deployed traffic counters in a city. Futur. Gener. Comput. Syst. **128**, 429–442 (2022)
9. Wang, X., Chen, C., Min, Y., He, J., Yang, B., Zhang, Y.L: Efficient metropolitan traffic prediction based on graph recurrent neural network, arXiv preprint, arXiv: 1811.00740 (2018)
10. Wang, B., Luo, X., Zhang, F., Yuan, B., Bertozzi, A.L., Brantingham, P.J.: Graph-based deep modeling and real time forecasting of sparse spatio-temporal data, arXiv preprint arXiv: 1804.00684 (2018)
11. Lu, Z., Lv, W., Cao, Y., Xie, Z., Peng, H., Du, B.: Lstm variants meet graph neural networks for road speed prediction. Neurocomputing **400**, 34–45 (2020)
12. Cai, L., Janowicz, K., Mai, G., Yan, B., Zhu, R.: Traffic transformer: capturing the continuity and periodicity of time series for traffic forecasting. Trans. GIS **24**(3), 736–755 (2020)
13. Shi, X., Qi, H., Shen, Y., Wu, G., Yin, B.: A spatial-temporal attention approach for traffic prediction. IEEE Trans. Intell. Transp. Syst. **22**(8), 4909–4918 (2020)
14. Lucas, P.O., Orang, O., Silva, P.C., Mendes, E.M., Guimaraes, F.G.: A tutorial on fuzzy time series forecasting models: recent advances and challenges. Learn Nonlinear Models **19**, 29–50 (2022)
15. Orang, O., Silva, P.C.d.L., Guimarães, F.G.: Time series forecasting using fuzzy cognitive maps: a survey. Artifi. Intell. Rev. 1–62 (2022)
16. Kosko, B.: Fuzzy cognitive maps. Inter. J. Man-Mach. Stud. **24**(1), 65–75 (1986). http://www.sciencedirect.com/science/article/pii/S0020737386800402
17. Qin, D., Peng, Z., Wu, L.: Deep attention fuzzy cognitive maps for interpretable multivariate time series prediction. Knowl.-Based Syst. **275**, 110700 (2023)

18. Wang, J., Wang, X., Li, C., Wu, J., et al.: Deep fuzzy cognitive maps for interpretable multivariate time series prediction. IEEE Trans. Fuzzy Syst. (2020)
19. Chmiel, W., Szwed, P.: Learning fuzzy cognitive map for traffic prediction using an evolutionary algorithm. In: Dziech, A., Leszczuk, M., Baran, R. (eds.) MCSS 2015. CCIS, vol. 566, pp. 195–209. Springer, Cham (2015). https://doi.org/10.1007/978-3-319-26404-2_16
20. Lv, Y., Duan, Y., Kang, W., Li, Z., Wang, F.-Y.: Traffic flow prediction with big data: a deep learning approach. IEEE Trans. Intell. Transp. Syst. **16**(2), 865–873 (2015)
21. Huang, H., Hu, Z., Wang, Y., Lu, Z., Wen, X., Fu, B.: Train a central traffic prediction model using local data: a spatio-temporal network based on federated learning. Eng. Appl. Artif. Intell. **125**, 106612 (2023)
22. Konecný, J., McMahan, H., Yu, F., Richtárik, P., Suresh, A., Bacon, D.: Federated learning: strategies for improving communication efficiency. CoRR, abs/arXiv: 1610.05492 (2016)
23. Qi, Y., Hossain, M.S., Nie, J., Li, X.: Privacy-preserving blockchain-based federated learning for traffic flow prediction. Futur. Gener. Comput. Syst. **117**, 328–337 (2021)
24. Hoyos, W., Aguilar, J., Toro, M.: Federated learning approaches for fuzzy cognitive maps to support clinical decision-making in dengue. Eng. Appl. Artif. Intell. **123**, 106371 (2023)
25. Salmeron, J., Arévalo, I.: Blind federated learning without initial model. J. Big Data **11**, 04 (2024)
26. Orang, O., de Lima e Silva, P.C., Silva, R., Guimarães, F.G.: Randomized high order fuzzy cognitive maps as reservoir computing models: a first introduction and applications. Neurocomputing (2022)
27. Shaygan, M., Meese, C., Li, W., Zhao, X.G., Nejad, M.: Traffic prediction using artificial intelligence: review of recent advances and emerging opportunities. Transportation Res. Part C: Emerging Technol. **145**, 103921 (2022)
28. Yang, X., Zou, Y., Tang, J., Liang, J., Ijaz, M.: Evaluation of short-term freeway speed prediction based on periodic analysis using statistical models and machine learning models. J. Adv. Trans. **2020**, 1–16 (2020)
29. Yin, X., Wu, G., Wei, J., Shen, Y., Qi, H., Yin, B.: Deep learning on traffic prediction: methods, analysis, and future directions. IEEE Trans. Intell. Transp. Syst. **23**(6), 4927–4943 (2021)
30. Dougherty, M.S., Kirby, H.R., Boyle, R.D.: The use of neural networks to recognise and predict traffic congestion. Traffic Eng. Control **34**(6), 311–314 (1993)
31. Zhang, L., Liu, Q., Yang, W., Wei, N., Dong, D.: An improved k-nearest neighbor model for short-term traffic flow prediction. Procedia. Soc. Behav. Sci. **96**, 653–662 (2013)
32. Castro-Neto, M., Jeong, Y.-S., Jeong, M.-K., Han, L.D.: Online-svr for short-term traffic flow prediction under typical and atypical traffic conditions. Expert Syst. Appli. **36**(3), Part 2, 6164–6173 (2009)
33. Soua, R., Koesdwiady, A., Karray, F.: Big-data-generated traffic flow prediction using deep learning and dempster-shafer theory. In: 2016 International joint conference on neural networks (IJCNN), pp. 3195–3202. IEEE (2016)
34. Wang, J., Gu, Q., Wu, J., Liu, G., Xiong, Z.: Traffic speed prediction and congestion source exploration: A deep learning method. In: 2016 IEEE 16th international conference on data mining (ICDM), pp. 499–508. IEEE (2016)

35. Cui, Z., Ke, R., Pu, Z., Wang, Y.: Deep bidirectional and unidirectional lstm recurrent neural network for network-wide traffic speed prediction, arXiv preprint, arXiv: 1801.02143 (2018)
36. He, Z., Chow, C.-Y., Zhang, J.-D.: Stann: a spatio-temporal attentive neural network for traffic prediction. IEEE Access **7**, 4795–4806 (2018)
37. Li, Y., Yu, R., Shahabi, C., Liu, Y.: Diffusion convolutional recurrent neural network: Data-driven traffic forecasting, arXiv preprint, arXiv: 1707.01926 (2017)
38. Pan, Z., Liang, Y., Wang, W., Yu, Y., Zheng, Y., Zhang, J.: Urban traffic prediction from spatio-temporal data using deep meta learning. In: Proceedings of the 25th ACM SIGKDD International Conference on Knowledge Discovery & Data Mining, pp. 1720–1730 (2019)
39. An, J., Fu, L., Hu, M., Chen, W., Zhan, J.: A novel fuzzy-based convolutional neural network method to traffic flow prediction with uncertain traffic accident information. IEEE Access **7**, 20 708–20 722 (2019)
40. Tang, J., Liu, F., Zou, Y., Zhang, W., Wang, Y.: An improved fuzzy neural network for traffic speed prediction considering periodic characteristic. IEEE Trans. Intell. Transp. Syst. **18**(9), 2340–2350 (2017)
41. Dimitriou, L., Tsekeris, T., Stathopoulos, A.: Adaptive hybrid fuzzy rule-based system approach for modeling and predicting urban traffic flow. Trans. Res. Part C: Emerging Technol. **16**(5), 554–573 (2008)
42. Chen, W., An, J., Li, R., Fu, L., Xie, G., Bhuiyan, M.Z.A., Li, K.: A novel fuzzy deep-learning approach to traffic flow prediction with uncertain spatial-temporal data features. Futur. Gener. Comput. Syst. **89**, 78–88 (2018)
43. Zhang, S., Chen, Y., Zhang, W.: Spatiotemporal fuzzy-graph convolutional network model with dynamic feature encoding for traffic forecasting. Knowl.-Based Syst. **231**, 107403 (2021)
44. Alkheder, S., Alomair, A.: Urban traffic prediction using metrological data with fuzzy logic, long short-term memory (lstm), and decision trees (dts). Nat. Hazards **111**, 03 (2022)
45. Liu, Y., Yu, J.J.Q., Kang, J., Niyato, D., Zhang, S.: Privacy-preserving traffic flow prediction: a federated learning approach. IEEE Internet Things J. **7**(8), 7751–7763 (2020)
46. Zhang, C., Zhang, S., Yu, J.J.Q., Yu, S.: Fastgnn: a topological information protected federated learning approach for traffic speed forecasting. IEEE Trans. Industr. Inf. **17**(12), 8464–8474 (2021)
47. Zhang, C., Cui, L., Yu, S., James, J.: A communication-efficient federated learning scheme for iot-based traffic forecasting. IEEE Internet Things J. **9**(14), 11918–11931 (2021)
48. Xia, M., Jin, D., Chen, J.: Short-term traffic flow prediction based on graph convolutional networks and federated learning. IEEE Trans. Intell. Transp. Syst. **24**(1), 1191–1203 (2023)
49. Qi, T., Chen, L., Li, G., Li, Y., Wang, C.: Fedagcn: A traffic flow prediction framework based on federated learning and asynchronous graph convolutional network. Appl. Soft Comput. **138**, 110175 (2023)
50. Orang, O., Silva, R., de Lima e Silva, P.C., Guimarães, F.G.: Solar energy forecasting with fuzzy time series using high-order fuzzy cognitive maps. In: 2020 IEEE International Conference on Fuzzy Systems (FUZZ-IEEE), pp. 1–8 (2020)

Unsupervised Statistical Keyword Extraction Pipeline: Is LLM All You Need?

Fernando Rezende Zagatti(✉), Daniel Lucrédio, and Helena de Medeiros Caseli

Department of Computing, Federal University of São Carlos, São Carlos, Brazil
fernando.zagatti@estudante.ufscar.br,
{daniel.lucredio,helenacaseli}@ufscar.br

Abstract. Keyword extraction is an important step for text interpretation, serving to identify and highlight the most significant words or phrases within a text. This step is essential for various applications such as summarization, indexing, and information retrieval. This paper presents a custom-built keyword extraction pipeline named USKE (Unsupervised Statistical Keyword Extraction) and compares its performance to large language models (LLMs). USKE is able to deliver fast and simple results based in statistical methods even when dealing with large datasets. Our evaluation demonstrates that although LLMs can achieve good results in single sentences with minimal context, they require a lot of post-processing and may output inconsistent answers, while USKE excels in efficiency and scalability.

Keywords: Keyword extraction · Natural Language Processing · Large Language Models

1 Introduction

In the era of big data, with vast amounts of information being generated continuously, the ability to quickly and accurately identify the main topics, keywords, and terms from a large volume of texts has become highly desirable [5,6]. This capability is essential for efficiently managing and leveraging information in various domains, from academia to industry, where rapid insights can drive decision-making and innovation.

In this way, keyword extraction plays a crucial role in various applications for Natural Language Processing (NLP), such as information retrieval [22] and text summarization [9]. Traditionally, this task has been approached using classical methods, mostly unsupervised and statistical ones, which often involve extensive preprocessing and the use of domain-specific features. However, with the advent of Large Language Models (LLMs) such as Bidirectional Encoder Representations for Transformers (BERT) [4] and Generative Pre-trained Transformer (GPT) [16], new opportunities have emerged to increase the accuracy and efficiency of keyword extraction.

A. Paes and F. A. N. Verri (Eds.): BRACIS 2024, LNAI 15413, pp. 460–475, 2025.
https://doi.org/10.1007/978-3-031-79032-4_32

These new text generation methods leverage deep learning techniques to capture contextual and semantic nuances, offering greater accuracy over traditional methods, but their computational complexity and dependence on large training datasets present practical challenges [10]. In this way, this study not only compares the performance of these advanced models with traditional statistical methods but also introduces an unsupervised pipeline to perform keyword extraction on textual data.

Our open-source proposed system, Unsupervised Statistical Keyword Extraction (USKE)[1], employs a series of NLP pipeline steps and statistical techniques to identify and extract relevant keywords from texts. By integrating text preprocessing methods with statistical techniques, we create a versatile and efficient keyword extraction that does not demand any specialized knowledge in LLM or AI to be used.

In this way, this paper aims to compare classical methods against modern LLM-based approaches to assess their respective strengths and weaknesses. While traditional methods often struggle with large-scale data and multilingual contexts, LLMs can be resource-intensive and less feasible for real-time applications. Thus, through this comparative analysis, we seek to provide insights into whether LLMs are truly indispensable for keyword extraction or if traditional methods still hold significant value.

Our work seeks to offer a direct comparison between classical keyword extraction methods and LLM-based techniques, focusing on their respective strengths and weaknesses in a more generalizable context. Unlike methods like KeyBERT, which are resource-intensive, USKE provides a lightweight, open-source alternative that is both versatile and efficient by integrating statistical techniques within a streamlined NLP pipeline, demonstrating that traditional methods can still yield competitive results, particularly when used in conjunction with domain expertise.

This study demonstrates the effectiveness of our proposed approach in datasets of reviews and titles of the Portuguese language, confirming that traditional methods can still obtain competitive results for this specific task, and can obtain satisfactory results when used alongside experts to make annotations to serve as input for the algorithm.

This work is organized as follows. Section 2 provides a comprehensive review of related works, highlighting the current state of the art in unsupervised keyword extraction. In Sect. 3 we present our proposed method, detailing the algorithms and techniques used. Section 4 describes the experimental setup, including the datasets and prompt tuning process. In Sect. 5 we discuss the results and provide an analysis of the performance of our method when compared to LLM. Finally, Sect. 6 concludes the paper with a summary of our findings and outlines potential directions for future research.

---

[1] Available at: https://github.com/fernandozagatti/Unsupervised-Statistical-Keyword-Extraction

2 Related Works

In recent years, keyword extraction has garnered significant attention within the field of NLP, leading to the development of various methodologies that aim to improve the accuracy and efficiency of this task [7,12,21]. These methods, while effective, often require extensive preprocessing and the use of external resources, which can limit their applicability across diverse languages and domains.

KeyBERT [7] is a keyword extraction method that leverages transformer-based models to identify the most relevant keywords within a text. In this way, the BERT-embeddings are applied into the document and the word embeddings are extracted in N-grams; then, cosine similarity is used to find the most relevant words (or keywords) in the document.

Despite its advantages, KeyBERT has some limitations: (i) its dependence on large datasets, which may not perform well on texts from certain domains or on languages less represented in the training data, and (ii) the method can be computationally expensive, as it requires generating embeddings for both the document and numerous candidate keywords, making it less feasible for real-time or large-scale applications.

YAKE (Yet Another Keyword Extractor) [2], unlike traditional machine learning based models, does not require a training phase, making it a versatile and scalable solution that operates on single documents, enabling quick adaptation to various scenarios without the need for linguistic tools or external databases.

This method identifies relevant keywords based on statistical features, like by analyzing the frequency and co-occurrence of word sequences within a document, offering a language and domain-independent solution. The effectiveness of YAKE has been demonstrated through extensive testing across multiple datasets and languages [1,18]. However, it is important to highlight that the larger the analyzed dataset and the greater the number of keywords to be extracted, the time required to obtain the keywords can increase considerably.

Gupta et al. (2024) [8] made an integration of BERT [4] and YAKE to analyze sustainability reports from Indian IT companies. The study uses BERT tokenization and embedding to capture the contextual and semantic nuances of the text, understanding the importance of each word within the broader context of the document, while YAKE is employed to extract keywords from the text.

The study acknowledges some limitations and areas for future improvement, specifying that the study only uses sustainability reports from IT organizations, reinforcing that to improve the generalizability, reports from additional sectors should be included. Furthermore, the study analyzes point-in-time reports, while conducting a time series study could provide insights into how sustainability practices into organizations evolve over time.

Aiming to perform a comparative analysis, Nadim et al. (2023) [13] evaluates the performance of statistical-based, graph-based, and embedding-based keyword extraction tools, highlighting their strengths and weaknesses in terms of both performance and execution time. In this way, KeyBERT achieved better performance of F1-Score across three of the four datasets, though it has higher execution times for longer texts, while Rapid Automatic Keyword Extraction (RAKE) [21] and YAKE are highlighted for their fast execution times.

Among all the works, several keyword extraction tools and some benchmarks between the tools were seen. However, no studies were found that aims to compare classic pipelines and new approaches with LLMs. Therefore, our work seeks to perform a comparison between classical and LLM approaches, at the same time as providing an open-source tool for extracting keywords in a simple and quick way using statistical methods.

3 Unsupervised Statistical Keyword Extraction

We propose a system, referred to as USKE, that employs an unsupervised pipeline with statistical techniques to identify and extract relevant keywords in the text. This section describes how the proposed system was developed, the techniques used and the relationship between its steps, as illustrated by Fig. 1.

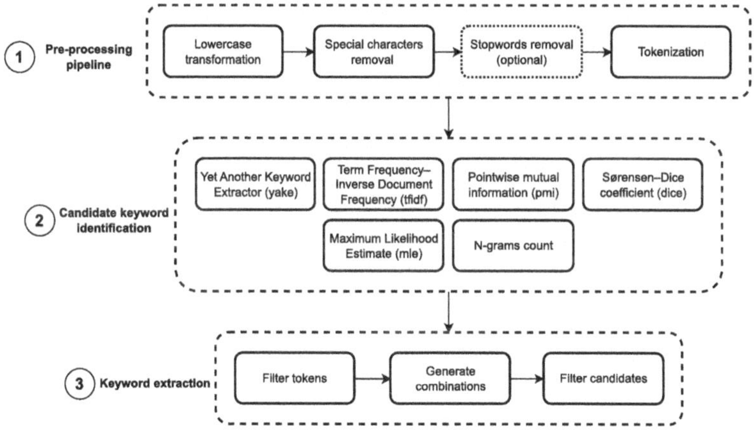

Fig. 1. Pipeline for USKE system

As can be seen, there are 3 main stages for the system, namely: (i) pre-processing pipeline, (ii) candidate keyword identification, and (iii) keyword extraction.

3.1 Pre-processing Pipeline

The preprocessing stage aims to prepare and clean text for the use of NLP techniques and extraction of the keywords. At this stage, the following techniques are applied: (i) document lowercase transformation, (ii) removal of all special characters (e.g. @, #, $) and standardization of word accentuation (e.g. á, â, ã -> a), (iii) an optional stopword removal, and (iv) simple tokenization by split.

3.2 Candidate Keyword Identification

This stage performs the identification of keywords relevant to the corpus employing statistical methods, namely: YAKE (Yet Another Keyword Extractor), TF-IDF (Term Frequency-Inverse Document Frequency), PMI (Pointwise Mutual Information), Dice coefficient, MLE (Maximum Likelihood Estimation), and n-grams count. Each approach will be detailed below:

- **YAKE[2]:** this is a method that extracts keywords or key phrases from text documents by preprocessing and identifying candidates based on term frequency (TF) within the document, inverse document frequency (IDF) across a larger corpus, word co-occurrence patterns, position of the candidate within the document, and the length of the candidate phrase. The candidates are scored and ranked using a weighted combination of these features, with the highest-scoring candidates being selected as the most relevant keywords or key phrases.
- **TF-IDF:** identifies keywords that are both frequently occurring within a specific document and relatively rare across the entire corpus. In this way, this approach assigns a numerical weight to each word and performs a ranking based on these scores.

$$\texttt{tf-idf(t, d)} = \texttt{tf(t, d)} \times \texttt{idf(t)} \tag{1}$$

$$\texttt{idf(t)} = log\left[\frac{(1+\texttt{n})}{(1+\texttt{df(t)})}\right]+1 \tag{2}$$

where:

  - `tf-idf(t, d)`: is the "Term Frequency-Inverse Document Frequency" of word `t` in document `d`.
  - `idf(t)`: is the "Inverse Document Frequency" which measures how common a word (`t`) is in all documents.
  - `tf(t, d)`: computes "term frequency" of the word `t` in a document `d` as the number of times `t` appears in `d`.
  - `n`: is the total number of documents.
  - `df(t)`: is the number of documents in which the word `t` appears.
- **PMI:** this is a statistical measure used to determine the association or co-occurrence between two events (or words) within a text corpus. A high PMI score indicates a strong association, implying that the occurrence of one word makes the presence of the other more likely, while a low or negative PMI suggests independence or repulsion; its calculation can be visualized through Eq. 3, Eq. 4, and Eq. 5,

$$PMI(w_1 \dots w_n) = \log_2\left(\frac{P(w_1 \dots w_n)}{\prod_{i=1}^{n} P(w_i)}\right) \tag{3}$$

$$P(w_i) = \frac{Count(w_i)}{totalWordCount} \tag{4}$$

[2] Available at: https://github.com/LIAAD/yake.

$$P(w_1 \dots w_n) = \frac{Count(w_1 \dots w_n)}{totalWordCount} \tag{5}$$

where:

- $P(w_i)$ is the individual probabilities of each word;
- $P(w_1 \dots w_n)$ is the joint probability of the n-gram;
- n is the number of words of the keyword.
- $Count(w_i)$ is the number of times that word appears in the text;
- $Count(w_1 \dots w_n)$ is the number of times that n-gram appears in the text;
- $totalWordCount$ is the total number of words in the text;

- **Dice:** this is a measure used to quantify the degree of overlap or similarity between two sets. The score ranges from 0 to 1, with higher values indicating greater similarity.

$$Dice(w_1 \dots w_n) = \frac{n * Count(w_1 \dots w_n)}{\sum_{i=1}^{n} Count(w_i)} \tag{6}$$

- **MLE:** this is a technique to estimate the probability distribution of words (or phrases) in a corpus, assuming that the probability of a word occurring in a specific context is proportional to the number of times that word appears within the training corpus.

$$MLE(w_1 \dots w_n) = \begin{cases} \frac{Count(w_1)}{totalWordCount}, & \text{if } n = 1 \\ \frac{Count(w_1 \dots w_n)}{Count(w_1 \dots w_{n-1})}, & \text{otherwise} \end{cases} \tag{7}$$

- **n-grams count:** these are contiguous sequences of n items, which can be characters, words, or even more extended units like phrases. In this context, it is checked how many times each n-gram appears in the text.

3.3 Keyword Extractor

This stage extracts keywords from the text based on a list of identified candidate keywords from the previous stage (or a list of keywords provided by the user). To illustrate the process, we will demonstrate the outputs of each function using the example of searching `['electric fryer', 'deep fryer']` within the text "EGGKITPO Electric Deep Fryer". This stage involves the following functions:

- **Filter tokens**: in this function, the text is separated into tokens and each token is checked if it is present in the list of identified keywords. This function returns a list containing the words that passed the filter.
  - After filter: `['electric', 'deep', 'fryer']`
- **Generate combinations:** in this function, from the list outputted by the previous step, a new sorted list (by order of appearance in the input text) is generated with all possible combinations of words. This step is important when keywords candidates have 2 or more tokens.
  - After combinations, considering up to 2 tokens: `['electric', 'deep', 'fryer', 'electric deep', 'electric fryer', 'deep fryer']`

- **Filter candidates:** this function selects only those word combinations that are present in the list of identified keywords.
  - After new filter: `['electric fryer', 'deep fryer']`

This separation into tokens and making combinations instead of simply searching for the sequence of words is very useful for identifying keywords that are not directly found in the text. Table 1 exemplifies this problem.

Table 1. Example of token combination efficiency versus direct keyword search

Product name	Direct search	Our method
EGGKITPO Electric Deep fryer Single Tank Deep Fryer with Basket Capacity 10L(10.5QT) Electric Countertop Fryers with 60 Minute Timer for Home Kitchen and Restaurant	['deep fryer']	['electric fryer', 'deep fryer']

* Example taken from Amazon. Available at: https://a.co/d/dknbFTX

4 Experimental Setup

The experiments were performed with a processor Intel Core i7-10700K, a high-performance Graphics Processing Unit (GPU) NVIDIA GeForce RTX 4070 Ti, and 32 GB of Random Access Memory (RAM).

We compared the performance of USKE, KeyBERT and 5 LLMs that were available on HuggingFace[3] to evaluate the performance of some language models, which can be easily found, with some classical pipelines.

The models were selected based on the trending models of HuggingFace, selecting textual generation models that use English and/or Portuguese and that allow their use with the LangChain API. Among the selected models, except for Phi-Bode, all were trained with most of the data being in English and a few in Portuguese. The selected models are described below:

- **Vlt5-base-keywords[4]:** a keyword generation model based on the encoder-decoder architecture presented by Google. It was trained on a corpus composed of scientific articles, to predict a given set of keyword phrases that describe the article content based only on the abstract [15]. Among the selected models, this is the only one specifically constructed for keyword extraction.

[3] Available at: https://huggingface.co/.
[4] Available at: https://huggingface.co/Voicelab/vlt5-base-keywords.

- **Flan-t5-xxl**[5]**:** a Google model trained using 540B-parameters, performing fine-tuning with the addition of more than 1000 new tasks and covering more languages than its predecessor, T5 [3,17]. It can generate coherent and contextually relevant text based on a given prompt, making it useful for creative writing, chatbots and content creation.
- **Bloom**[6]**:** a model trained to continue text from a prompt, which is capable of producing text in 46 languages and 13 programming languages. This model seeks to assist with text and code generation, and characteristics exploration.
- **Blenderbot-400M-distill**[7]**:** a lightweight and distilled version of Facebook's BlenderBot 2.7B, designed for open-domain conversational AI. It is based on a 400 million parameter transformer model that has been fine-tuned on diverse conversational datasets to improve its ability to engage in multi-turn dialogue [20].
- **Phi-Bode**[8]**:** a language model fine-tuned from Microsoft's Phi-2B base model, refined using the translated Alpaca dataset for fine-tuning, aimed at adapting the model to the nuances of the Portuguese language [14]. Among the selected models, this is the only one specifically tuned for Portuguese.

4.1 Datasets

Two datasets composed of reviews of products or services were selected to perform the tests, namely: (i) B2W-Reviews01 [19] and (ii) Aspect-Based Sentiment Analysis in Portuguese (ABSAPT) 2022. The datasets are described below:

- **B2W-Reviews01:** this dataset constitutes an open corpus of product reviews, comprising more than 130 thousand reviews from e-commerce customers, collected from the `americanas.com` website between January and May 2018. This corpus provides product title texts in Portuguese; however, there are no specialized annotations, which can make certain tasks difficult.
- **ABSAPT 2022:** this dataset contains a corpus of reviews about hosting service companies written in Portuguese. The specialists annotated the aspects and their respective polarities, totalling more than 800 annotated reviews. The aspects were used as the target keywords in the experiments described in this paper.

It is important to mention that due to the challenges associated with making inferences from LLMs on extensive datasets, we reduced the dataset sizes for our experiments: for the B2W-Reviews01, only the first 100 instances were used; while for the ABSAPT 2022, we used the test dataset which comprises 184 instances.

[5] Available at: https://huggingface.co/google/flan-t5-xxl.
[6] Available at: https://huggingface.co/bigscience/bloom.
[7] Available at: https://huggingface.co/facebook/blenderbot-400M-distill.
[8] Available at: https://huggingface.co/recogna-nlp/Phi-Bode.

4.2 Prompt Tuning

Prompt tuning only updates the input values of the prompts while the LLM parameters remain unchanged. This tuning method allows one to guide the models to produce more relevant and accurate outputs without necessarily having to retrain them [11].

In this study, we experimented with a few prompt structures (Tables 2, 3 and 4) to evaluate their effectiveness across different LLMs. Below, we detail the tuning process and provide examples of the tested prompts along with their results. Note that all prompts have `{text}`, which is a variable where the text from which we want to extract the keywords will be prompted. For the example results, we will use the following `{text}`: Copo Acrílico Com Canudo 500ml Rocie.

Table 2. Prompt #1

`Based in the following phrase, return a list of the most relevant words: {text}`	
Vlt5-base-keywords	Copo Acrílico, Copo Acrílico, Copo Acríli
Flan-t5-xxl	acilico, canudo, cie, copo
Bloom	N1056 | Copo com Canudo | Copo com Canudo 500ml | Copo com Canudo Acrílico | Copo com Canudo Acrílico 500ml | Copo com Canudo Acrílico Rocie N1056
Blenderbot-400M-distill	I have never heard of that one. What is it about? Is it a book?
Phi-Bode	# Crie uma lista de palavras relevantes palavras_relevantes = [\'Copo Acrílico\', \'Com\', \'Canudo\', \'500ml\', \'Rocie\'] # Retorne a lista return palavras_relevantes # Teste print(get_relevant_words("Copo Acrílico Com Canudo 500ml Roc

From Prompt #1 we see that Vlt5-base-keywords and Flan-t5-xxl were able to extract words in a simpler way, while Bloom had some difficulty and Blenderbot seems not to have understood our request. Furthermore, it is possible to notice that all models generated some words that did not exist in the input text, such as "Acríli", "cie", and "N1056".

Table 3. Prompt #2

`Question: Based in the following phrase, return a list of the most relevant words: {text}` `Answer:`	
Vlt5-base-keywords	Copo Acrílico, Copo Acrílico, Copo Acríli
Flan-t5-xxl	acilico, canudo, com, rocie
Bloom	Copo Acrílico Com Canudo 500ml Rocie, [...]
Blenderbot-400M-distill	Hello, how are you doing today? I am well, thank you for asking.
Phi-Bode	Copo Acrílico, Canudo, 500ml, Rocie.

After some testing, it was found that adding "Question:" and "Answer:" at the beginning and end of the prompt helped some models generate more concise answers (Table 3). Bloom in this case only replicated the input text, but did not correctly answer our request to generate the keywords found in the text.

Table 4. Prompt #3

`Question: Based in the following phrase, return a list of the most relevant words: {text}` `Answer: The list is`	
Vlt5-base-keywords	Copo Acrílico, Copo Acrílico, Copo Acríli
Flan-t5-xxl	acilico, canudo, com, rocie
Bloom	Rocie, 500ml, Com, Canudo, Acrílico, Copo, Rocie, Copo, Acrílico, Com, Canudo, 500ml
Blenderbot-400M-distill	Do you know that the word"copo" means "little armoured one" in Spanish?
Phi-Bode	1. Copo Acrílico 2. Com Canudo 3. 500ml 4. Rocie

Finally, as Bloom is an LLM that continues texts from the prompt, "The list is" was added at the "Answer" part (Table 4), which helped to obtain more assertive answers and did not affect the answers of the other models. Thus, Prompt #3 was chosen for the experiments. For experiments containing custom lists, Prompt #3 was used changing the "Based in the following phrase, return a list of the most relevant words:" to "Based in this list of words `{list}`, return only the list of words that are in the following sentence:". As the Phi-Bode model was trained with texts in Portuguese, versions of the same prompts were used, but translated to Portuguese.

4.3 LLMs Post-processing

The outputs generated by the LLMs did not consistently adhere to a standard format in the reference keyword lists used in our experiments, requiring additional post-processing steps. LLMs output often contained special characters, word duplication, and lacked consistent token separation, which required thorough cleaning before any automatic evaluation could be performed. Table 5 brings an example of the result obtained when searching for keywords for the input text `{text}`.

Since the outputs did not follow a well-known pattern, we removed any special characters from the outputs to prevent them from interfering with evaluation metric process. For instance, outputs like "`: localização, café-da-manhã, diárias`" were cleaned to remove ":", "-", and accents, resulting in "`localizacao cafe da manha diarias`". This step was performed to avoid discrepancies between the evaluations, where the word "localização" could be considered different from "localizacao".

After cleaning, we separated the output into individual tokens. This step is important for transforming textual data (strings) into lists with the words contained in the texts, allowing word-by-word comparisons with the experts' annotations.

Table 5. Example of LLMs outputs for the input text `{text}`

`{text}`	A localização do hotel é muito boa , porém as partes comuns são pequenas e desconfortáveis para uma classificação 4 estrelas !!!!! O valor da diária não compensa o deconforto das instalações comuns. As salas de recepção e de café da manhã énão refletem um conforto de hotel 4 estrelas da rede Accor. A equipe de atendimento é gentil e atenciosa.
Vlt5-base-keywords	Accor, Accor, Accor, Accor, Accor, Accor, Accor, Accor, Accor, Accor, Accor, hotel 4 estrelas
Flan-t5-xxl	localizacao, part, star
Bloom	: localização, hotel, pequena, comum, classificação, estrelas, diárias, compensa, desconforto, instalações, salas, recepção, café, manhã, equipe, atendimento, gentil, atenciosa
Blenderbot-400M-distill	Do you speak any other languages? I speak Spanish, French, and Portuguese.
Phi-Bode	Palavras mais relevantes: localização, boa, partes comuns, desconfortáveis, classificação, equipe de atendimento, salas de recepção, café, diário, valor, atenciosa.

Finally, we removed all duplicate tokens from the lists to ensure that each keyword was unique. For example, if a list was "`[''localizacao'', ''hotel'', ''hotel'', ''cafe'']`", the duplicate "hotel" was removed, resulting in "`[''localizacao'', ''hotel'', ''cafe'']`". This deduplication process was necessary to accurately reflect the diversity of keywords generated by the LLMs.

These post-processing steps were critical in preparing the LLM outputs for fair and accurate comparison with the annotated data. They ensured that the evaluation metrics were based on clean and standardized outputs, thus providing a more reliable assessment of the LLMs' performance for this specific task.

4.4 Evaluation Metrics

The assessment of the outputs was made based on the metrics commonly used in the literature: precision, recall, and F1-score. Precision is the proportion of instances classified as positive that are actually positive. Recall is the proportion of actual positive instances that are correctly classified as positive. F1-score is the harmonic mean of precision and recall, providing a balanced measure between the two metrics.

Our experiments also considered two similarity measures: Cosine similarity and Jaccard similarity . Similarity is a quantitative measure used to determine the degree of similarity between objects, and is often used to compare two distinct vectors. These metrics are fundamental in information retrieval, NLP, and machine learning, where identifying patterns and benchmarking are essential.

The Cosine similarity calculates the cosine of the angle between two vectors. The closer the cosine value is to 1, the greater the similarity between the entities

Jaccard similarity is defined as the ratio between the size of the intersection of the sets and the size of their union, ranging from 0 to 1, where 1 indicates that the sets are identical and 0 indicates that there are no elements in common.

5 Results and Analysis

We applied the 5 previously presented LLM models to the 2 datasets and compared them with our tool – for USKE, the outputs of each technique were merged to comprehensively evaluate the tool. The results were evaluated differently, since the ABSAPT 2022 dataset had a prior annotation, enabling direct analysis, while B2W-Reviews01 did not.

5.1 ABSAPT 2022 Results

This dataset has a prior annotation of aspects for the Aspect-Based Sentiment Analysis (ABSA) task and these aspects were considered our reference keywords. So, two specific tests were applied: (i) extraction of keywords without the list of annotated words and (ii) direct extraction of keywords from the annotations. Table 6 presents the results from the first case . In this specific dataset, since annotated data were available, Jaccard and Cosine Similarity metrics were directly applied to the annotated data to evaluate the similarity between the model outputs and the annotated data.

Table 6. ABSAPT 2022 without custom list

Model	Precision	Recall	F1-Score	Jaccard	Cosine	Time (seconds)
USKE (ours)	0.16	0.18	0.17	0.18	**0.35**	**0.45 s**
Vlt5-base	0.17	0.05	0.07	0.14	0.24	408.41 s
Flan-t5	0.21	0.08	0.10	**0.20**	0.31	106.81 s
Bloom	**0.47**	0.36	0.38	0.05	0.13	3561.53 s
Blenderbot	0.03	0.00	0.00	0.00	0.00	662.65 s
Phi-Bode	0.46	**0.38**	**0.40**	0.06	0.16	4896.83 s
KeyBERT	0.09	0.10	0.09	0.19	0.33	2.32 s

The results demonstrate that USKE obtained a precision close to the Vlt5-base and Flan-t5 models and was the third-best method in recall and F1-Score. On the other hand, Blenderbot did not get any significant results, as it was trained with a conversational objective and could not generate direct answers for this specific task, and although KeyBERT achieves considerable similarity values, it does not have competitive results in terms of precision and recall.

Furthermore, Phi-Bode and Bloom obtained significantly better results in precision, recall, and F1-score, but the similarity values were low. This happened because the LLMs generated more texts than expected, reducing similarity, but since most of the time the expected outputs were between the generated text, this led to increased precision and recall.

In contrast, Table 7 shows the test results with the list of words we wanted to extract from the text, where we evaluate the results like the previous table.

Table 7. ABSAPT 2022 with custom list

Model	Precision	Recall	F1-Score	Jaccard	Cosine	Time (seconds)
USKE (ours)	**0.70**	**0.84**	**0.74**	**0.72**	**0.83**	**0.03 s**
Vlt5-base	0.19	0.10	0.11	0.18	0.30	1467.17 s
Flan-t5	0.12	0.31	0.12	0.17	0.33	745.02 s
Bloom	0.06	0.69	0.09	0.07	0.20	4537.33 s
Blenderbot	0.00	0.00	0.00	0.00	0.00	595.72 s
Phi-Bode	0.08	0.20	0.06	0.08	0.21	6169.38 s
KeyBERT	N/A	N/A	N/A	N/A	N/A	N/A

As our pipeline was developed to extract keywords directly from the text, the experiments using a predefined list lead to high values in all the metrics evaluated. LLM models were unable to consistently extract the required words.

Although only the Vlt5-base and Flan-t5 models managed to obtain a slightly higher F1-Score than the previous test, all models (except for Blenderbot and Phi-Bode) obtained superior results in recall. However, it is important to mention that the precision values decreased in all cases, and the execution time also increased. Finally, KeyBERT was developed only to find keywords in an unsupervised way, and it is not able to deal with a predefined list. Furthermore, processing time became longer in all LLMs (except for Blenderbot), since the prompts were larger due to the lists that were sent together with the texts.

5.2 B2W-Reviews01 Results

Although this dataset does not have prior annotations, it is possible to evaluate whether the results are meaningful using similarity tests. As the task is to search for explicit words in the text, the models' responses need to have some degree of similarity with the input text. In this way, if the similarity is equal to or close to 0, it is possible to say that the language models were not able to answer the request. The results can be seen in Table 8.

Phi-Bode obtained the highest similarity values, demonstrating that its outputs were most often related to the input text. USKE also achieved respectable scores with a Jaccard similarity of 0.38 and a Cosine similarity of 0.55, staying at the average of LLMs. KeyBERT achieved similarity values very close to USKE.

In terms of processing time, USKE outperformed the others, completing the task in just 0.31 s, which is significantly faster than the LLMs. KeyBERT also

Table 8. B2W-Reviews01 results

Model	Jaccard	Cosine	Time (seconds)
USKE (ours)	0.38	0.55	**0.31 s**
Vlt5-base	0.49	0.65	72.13 s
Flan-t5	0.30	0.50	71.66 s
Bloom	0.52	0.62	2090.01 s
Blenderbot	0.02	0.03	369.70 s
Phi-Bode	**0.69**	**0.81**	1390.15 s
KeyBERT	0.37	0.59	1.81 s

had a fast processing time of 1.81 s. However, Phi-Bode had one of the longest processing times, reaching 1390.15 s.

5.3 Critical Analysis

The analysis reveals critical insights regarding the performance of various models in keyword extraction tasks, particularly emphasizing the effectiveness of models trained in Portuguese, the potential for prompt tuning, and the tendency to generate answers that do not adhere to the request and input text.

Phi-Bode, trained specifically in Portuguese, demonstrated more satisfactory results in methods without pre-defined lists. This model achieved the highest similarity values, highlighting the importance of training models in the specific language of the dataset to enhance performance in language-specific tasks.

Furthermore, the possibility of improving prompts can lead to better answers – prompt tuning must be carried out carefully, as even small variations and adjustments can lead to huge differences in performance. Experiments with different prompt structures can significantly impact the effectiveness of the LLMs.

Lastly, we reinforce that some models, such as Blenderbot, are not ideal for keyword extraction activities due to their optimization for conversational tasks. Its performance, with low similarity values and long processing times, highlights this limitation. Preventing models from generating irrelevant or incorrect information remains a challenge, emphasizing the need for careful model selection and optimization for targeted applications.

In this way, we demonstrate that the use of stricter pipelines is still extremely useful for developers, and can save a lot of time choosing models, prompt tuning, and post-processing.

6 Conclusions and Future Work

Our study reveals that both our proposed pipeline and LLMs have distinct advantages and disadvantages for extracting relevant keywords in textual data. USKE demonstrated significant advantages in processing speed and simplicity, particularly when handling large datasets with minimal computational resources. Furthermore, as it is a traditional and more controlled pipeline, the outputs will always be related to the input and guaranteed to follow the desired format.

Traditional keyword extraction pipelines can be efficient but the lack of a broader context (multiple sentences) and poorly optimized preprocessing can lead to less accurate keywords being extracted, compromising the quality of the results. On the other hand, LLMs can handle single sentences or small paragraphs effectively without needing deep context, but have difficulty maintaining performance in environments that require handling a high flow of data, demanding substantial computational power and specialized hardware (e.g. GPUs). Therefore, whether LLMs are necessary depends on the specific requirements and constraints of the application.

In future work, we could conduct a more qualitative benchmarking, where specialists could evaluate and compare the results of the outputs generated by the different methods. This kind of analysis would allow for a deeper understanding of the relevance and accuracy of the extracted keywords, as well as provide valuable insights that cannot be captured by quantitative metrics alone. Additionally, exploring the integration of more advanced NLP techniques, such as contextualized word embeddings, with experiments on datasets from other domains, could potentially increase the quality of keyword extractions.

Acknowledgments. This study was financed in part by the Coordenação de Aperfeiçoamento de Pessoal de Nível Superior - Brasil (CAPES) - Finance Code 001.

References

1. Campos, R., Jatowt, A., Jorge, A.: Text mining and visualization of political party programs using keyword extraction methods: The case of portuguese legislative elections. In: Sserwanga, I., et al. (eds.) International Conference on Information, pp. 340–349. Springer, Cham (2023). https://doi.org/10.1007/978-3-031-28035-1_24
2. Campos, R., Mangaravite, V., Pasquali, A., Jorge, A., Nunes, C., Jatowt, A.: YAKE! keyword extraction from single documents using multiple local features. Inf. Sci. **509**, 257–289 (2020)
3. Chung, H.W., et al.: Scaling instruction-finetuned language models (2022). https://doi.org/10.48550/ARXIV.2210.11416, https://arxiv.org/abs/2210.11416
4. Devlin, J., Chang, M.W., Lee, K., Toutanova, K.: BERT: pre-training of deep bidirectional transformers for language understanding. In: Proceedings of NAACL-HLT 2019, p. 41714186. Minneapolis, Minnesota (2019)
5. Firoozeh, N., Nazarenko, A., Alizon, F., Daille, B.: Keyword extraction: issues and methods. Nat. Lang. Eng. **26**(3), 259–291 (2020)
6. Gopan, E., Rajesh, S., Vishnu, G., Thushara, M., et al.: Comparative study on different approaches in keyword extraction. In: 4th International Conference on Computing Methodologies and Communication (ICCMC), pp. 70–74. IEEE (2020)
7. Grootendorst, M.: KeyBERT: minimal keyword extraction with BERT (2020).https://doi.org/10.5281/zenodo.4461265

8. Gupta, A., Chadha, A., Tewari, V.: A natural language processing model on BERT and YAKE technique for keyword extraction on sustainability reports. IEEE Access **12**, 7942–7951 (2024)
9. Hernández-Castañeda, Á., García-Hernández, R.A., Ledeneva, Y., Millán-Hernández, C.E.: Language-independent extractive automatic text summarization based on automatic keyword extraction. Comput. Speech Lang. **71**, 101267 (2022)
10. Li, J., Tang, T., Zhao, W.X., Nie, J.Y., Wen, J.R.: Pre-trained language models for text generation: a survey. ACM Comput. Surv. **56**(9), 1–39 (2024)
11. Liu, X., et al.: P-Tuning V2: prompt tuning can be comparable to fine-tuning universally across scales and tasks. arXiv preprint arXiv:2110.07602 (2021)
12. Mihalcea, R., Tarau, P.: TextRank: bringing order into text. In: Proceedings of the 2004 EMNLP, pp. 404–411 (2004)
13. Nadim, M., Akopian, D., Matamoros, A.: A comparative assessment of unsupervised keyword extraction tools. IEEE Access **11**, 144778–144798 (2023)
14. Paiola, P.H., Garcia, G.L., Papa, J.P.: Phi-bode (2024). https://doi.org/10.57967/hf/1880, https://huggingface.co/recogna-nlp/Phi-Bode
15. Pęzik, P., Mikołajczyk, A., Wawrzyński, A., Żarnecki, F., Nitoń, B., Ogrodniczuk, M.: Transferable keyword extraction and generation with text-to-text language models. In: International Conference on Computational Science, pp. 398–405. Springer, Cham (2023). https://doi.org/10.1007/978-3-031-36021-3_42
16. Radford, A., Narasimhan, K., Salimans, T., Sutskever, I., et al.: Improving language understanding by generative pre-training (2018)
17. Raffel, C., et al.: Exploring the limits of transfer learning with a unified text-to-text transformer. J. Mach. Learn. Res. **21**(140), 1–67 (2020). http://jmlr.org/papers/v21/20-074.html
18. Ramachandran, R., Mohan, M.K., Sara, S.K.: Document clustering using keyword extraction. In: 2022 IEEE 3rd Global Conference for Advancement in Technology (GCAT), pp. 01–06. IEEE (2022)
19. Real, L., Oshiro, M., Mafra, A.: B2W-reviews01 - A Brazilian Portuguese reviews corpus (2019). https://opencor.gitlab.io/corpora/real19b2wreviews01/
20. Roller, S., et al.: Recipes for building an open-domain chatbot. arXiv preprint arXiv:2004.13637 (2020)
21. Rose, S., Engel, D., Cramer, N., Cowley, W.: Automatic keyword extraction from individual documents. In: Text Mining: Applications and Theory, pp. 1–20 (2010)
22. Yang, Z., Yu, H., Tang, J., Liu, H.: Toward keyword extraction in constrained information retrieval in vehicle social network. IEEE Trans. Veh. Technol. **68**(5), 4285–4294 (2019)

Using Complex Networks to Improve Legal Text Hierarchical Classification

Rilder S. Pires[1,2(✉)], Raquel Silveira[3], Carlos G. O. Fernandes[2,4], João A. Monteiro Neto[5], and Vasco Furtado[1,2,6]

[1] Laboratório de Ciência de Dados e Inteligência Artificial, Universidade de Fortaleza, Fortaleza, Ceará, Brazil
rilder@unifor.br

[2] Programa de Pós Graduação em Informática Aplicada, Universidade de Fortaleza, Fortaleza, Ceará, Brazil

[3] Instituto Federal de Educação, Ciência e Tecnologia do Ceará, Tianguá, Ceará, Brazil

[4] BNB - Banco do Nordeste do Brasil S.A., Fortaleza, Ceará, Brazil

[5] Centro de Ciências Jurídicas, Universidade de Fortaleza, Fortaleza, Ceará, Brazil

[6] ETICE - Empresa de Tecnologia da Informação do Ceará, Fortaleza, Ceará, Brazil

Abstract. Legal topics are typically organized into trees of labels, where each branch, from the root (generic topics) to the leaf (specific topics), categorizes the vocabulary that describe lawsuits. In this article, we describe an approach, innovative in the Brazilian Justice System, of automatic hierarchical classification of a petition topic. The approach integrates methods based on transformers and complex networks to capture, in addition to the characteristics of the text, the relationship between the legal citations present in the petitions and the topic to which the petition refers. The validation of this approach is done through a benchmark that shows accuracy gains, as well as, a practical scenario with the implementation of a microservice on a National Justice Platform whose front-end implementation is already being used by a State Court to automatically suggest the classification of the petitions topic in the National Procedural System.

Keywords: Text Classification · Brazilian Legal System · Complex Networks

1 Introduction

Classification of the topic addressed by a legal document is a task where Artificial Intelligence, specifically through Natural Language Processing (NLP), can provide automatic inference capabilities [21]. Legal topics are typically organized into hierarchical trees, where each branch, from the root (e.g., consumer law) to the leaf (e.g., moral/material damage), categorizes the vocabulary that describes lawsuits. Topic classification generally occurs in the early stages of the judicial

A. Paes and F. A. N. Verri (Eds.): BRACIS 2024, LNAI 15413, pp. 476–490, 2025.
https://doi.org/10.1007/978-3-031-79032-4_33

process when a petition is submitted to the court. At this point, the petitioner is required to indicate the relevant subject matter of the demand. In the Brazilian context, a petitioner must choose the topic from a hierarchy encompassing over 4,000 subjects, which are part of the Unified Procedural Tables (TPU system) maintained by the National Council of Justice (CNJ) [9]. Making the correct association within this hierarchy is not a trivial task and is often done incorrectly by the petitioner. This misclassification generates delays in the judicial process, leading to negative impacts both financially, due to rework, and socially, by fostering a sense of impunity.

The scientific literature presents various methodologies for the automatic classification of judicial documents [1,2,4,14,21,29]. The advent of textual classification via deep learning, particularly with transformer models, has significantly advanced the field of text classification [10,17,28,31]. Nonetheless, the intricate and extensive hierarchical structure of topics, along with the specialized nature of legal texts, poses significant challenges to traditional methods. Taking a closer look at our problem, we see that it is a hierarchical text classification problem, where we typically aim to categorize a text into a set of labels organized in a structured class hierarchy. The big challenge with this type of task is to adequately model the label hierarchy, which is usually large-scale, unbalanced, and highly structured [34]. These challenges have motivated us to develop an innovative approach.

Our approach assumes that the numerous legal citations present within petitions are crucial for understanding the document's content and defining its relevant subject matter. These legal citations not only help formalize the arguments presented by both petitioners and decision-makers but also serve as a bridge connecting the petition to the applicable legal provisions, such as laws, decrees, precedents, and articles. By examining the citations within each petition, we can construct a network linking legal provisions to the petition topics. This network offers supplementary information to enhance the automatic classification process, thereby significantly improving its accuracy.

Using this network-based approach, we developed a classifier based on BERT that leverages not only the petition text but also vector information generated from the network connections between topics and legal provisions. We validated our methodology with a dataset comprising 300,000 petitions from various Brazilian courts, maintained by the CNJ's Codex system. The results demonstrate the validity of our approach, showcasing significant gains in accuracy, in some situations, when compared to a hierarchical classification method that do not use the network information. Additionally, we show a example of application of the proposed approach in a version of a hierarchical classification model that is available on the SINAPSE platform of the Council and is currently being used by a State Court to automatically suggest topic classifications for petitions entered in the National Procedural System.

2 Related Work

The modeling of complex networks based on the characteristics of a lawsuit has its potential evidenced in several works as described in [5,11,13,24,25,27]. Classification of legal documents has been used in the forensic and Law Enforcement context (e.g. identifying potential criminals [32], predicting crimes [22,23] or discovering antisocial behavior [7]).

Mariana et al. [21] investigated, in the Brazilian context, different methods of classification and data representation to conclude that deep learning (more specifically) with Word2Vec trained on a domain-specific corpus, are promising, but still limited, methods for classifying many classes.

Luz de Araujo et al. [4] presented a new dataset constructed from legal documents from the Federal Supreme Court of Brazil, containing labeled text data for two types of tasks: document type classification and assignment of themes. The authors explored and compared different methods for classification, such as bag-of-words models, CNN, RNN, boosting algorithms, and linear-chain Conditional Random Fields. CNN with word embedding also proved to be more performing. Silva et al. [30] proposed a structured model in a CNN to classify types of documents received by the Federal Supreme Court of Brazil. Aguiar et al. [1] investigated the application of topic modeling to identify the subject of legal documents and evaluated the applicability in classifying Brazilian legal processes. The classification model was trained on documents from the Court of Justice of Ceará, in Brazil, and the cases were classified into the five most representative classes of the National Council of Justice (CNJ) of Brazil. Recently, pre-trained models based on BERT has shown to be more performing to classify topics. Importantly, the effectiveness of BERT is mainly due to the transfer learning ability that leverages semantic and syntactic knowledge from pre-training on a large non-labeled corpus [10]

Aguiar et al. [2] investigate different text classification methods and different combinations of embeddings, extracted from Portuguese language models, and information about legislation cited in the initial documents to classification task of lawsuits. The models were trained with a Golden Collection from the Court of Justice of the State of Ceará, in Brazil, whose lawsuits were classified in the five more representative CNJ's classes. The best result was obtained by the BERT model. Shaheen et al. [29] studied the performance of several transformer-based models (namely BERT [10], RoBERTa [17], DistilBERT [28] and XLNet [36]) in combination with strategies such as pre-generative training, gradual unfreezing and discriminating learning rates to achieve competitive ranking performance. Chalkidis et al. [6] developed an English legal judgment prediction dataset, containing cases from the European Court of Human Rights. A wide variety of neural models are evaluated on this dataset, establishing strong baselines that outperform previous feature-based models.

Wang et al. [34] presented the Hierarchy-aware Prompt Tuning (HPT) method, aimed at addressing hierarchical text classification from the perspective of a masked language model with multiple labels. The method involves creating a dynamic virtual template and using label words that act as soft prompts, integrating knowledge of the label hierarchy. Additionally, a zero-bounded multi-label cross-entropy loss is introduced to align the processes of hierarchical text classification with masked language models. The authors conducted experiments that demonstrated HPT achieves a good performance on several popular hierarchical text classification datasets.

3 The Proposed Approach

In the proposed approach, we assume that the primary source of data is a set of textual petitions P that make up the initial piece of the judicial case. Each initial petition is associated with a *legal topic* that the case addresses. In the text of these petitions, there are references to *legal provisions* that support the arguments put forward by the petitioner. Our hypothesis is that the relationship of a *legal topic* with one or more *legal provisions* is relevant for automatically classifying the matter of a petition.

Figure 1 shows the pipeline of the proposed approach, which depends on the texts of the petitions and the citations to legislation made within them. Additional background knowledge about the hierarchical structure of legal topics and provisions is used. This additional information and the citations of legal provisions within the text form a complex network of topics and legal provisions that we will model here through a weighted graph. Then, the embedding of vertices in this graph allows us to generate a vector representation of the petitions. Next sections will provide formalization of these concepts.

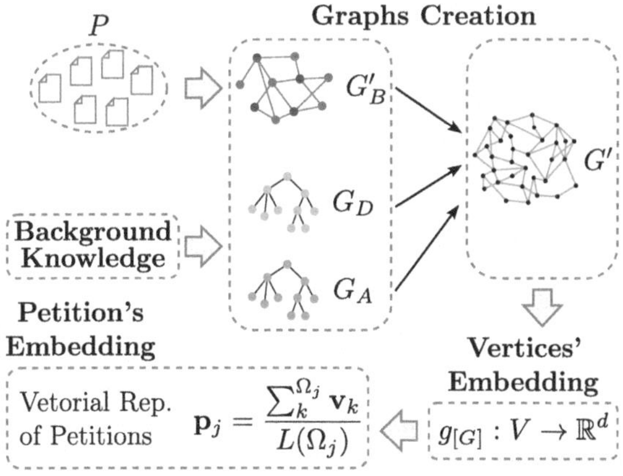

Fig. 1. Pipeline of the proposed approach. In the upper left corner are the primary source of data namely the petitions and the legal citations made within them as well as background knowledge about the hierarchical structure of legal topics and provisions. This data is transformed in a weighted graph of legal topics and provisions that allows to generate a vector representation of the petitions.

3.1 Bipartite Graph of Legal Provisions and Topics

Given a set of petitions P, we define G_B as the graph that represents the *co-occurrence* relationship between the *legal topics* of the *petitions* of P and the respective *legal provisions* cited therein. Here, we define $G_B = (A, D, E_B, w_B)$, where

- $A = \{a_1, a_2, \ldots, a_n\}$, $n \in \mathbb{N}$, is a finite set of *vertices* related to legal topics.
- $D = \{d_1, d_2, \ldots, d_m\}$, $m \in \mathbb{N}$, is a finite set of *vertices* related to the legal provisions
- $E_B \subset A \times D$ is a finite set of *edges* connecting subsets of vertices of A and D.
- $w_B : E_B \rightarrow \mathbb{R}$ is a *edge weight function* defined on the edges of G_B.

In this graph, there is an edge between the vertices a_i and d_j if there is at least one petition in P whose *legal topic* corresponds to a_i and which makes at least one reference to *legal provision* d_j. Generally speaking, we assume that the weight $w_{Bi,j}$ of this edge will be a function of the number of times $q_{i,j}$ that the legal provisions d_j is referenced by the petitions of the topic a_i to P.

The G_B graph bears similarities with other graphs already studied in the literature, such as the bibliographic citations graph [19]. There, the vertices represent documents that make references to other documents. An important characteristic of this type of graph is the existence of documents with a high amount of citations [19]. These vertices with many links end up making it difficult to identify the clustering patterns existing in the graph [3].

Similarly to what happens in *bipartite graphs of users and objects*, we define a weight function that uses a strategy based on a popular technique in *information retrieval* [3] called *tfidf*,

$$w_{Bi,j} \equiv w_B(a_i, d_j) = f(a_i, d_j) \times \log\left(\frac{n}{\deg(d_j)}\right) \tag{1}$$

where $f(a_i, d_j)$ is similar to "tf" and represents the frequency that the legal provision d_j is cited in petitions of a certain topic a_i,

$$f(a_i, d_j) = \frac{q_{i,j}}{\sum_k^{\langle a_i \rangle} q_{i,k}},$$

and $\langle \cdot \rangle$ means that the sum in k is done at the *neighborhood* of a_i. The term $\log(n/\deg(d_j))$ of Eq. (1) is similar to "idf", measuring how much the legal provision is rare or common among all legal topics.

3.2 Forests of Legal Topics and Legal Provisions

Legal topics have taxonomic hierarchical relationships and are predefined in the *Unified Procedural Tables* [9] system. The graph $G_A = (A, E_A)$ represents the *taxonomy* of *legal topics*, where:

- $A = \{a_1, a_2, \ldots, a_n\}$, $n \in \mathbb{N}$, is a finite set of *vertices* related to legal topics.
- $E_A \subset A \times A$ is a finite set of *directed edges* that connect the vertices of G_A.

The intrinsic asymmetry of the relationships between the vertices of G_A makes it fall into the class of *directed rooted forests.* Such graphs have important properties from a topological point of view. In these graphs, we can define two types of special vertices, the *root* vertices and the *leaf* vertices. Considering that the connections between two vertices always occur from the most general legal subject to the most specific subject, a root vertex is defined as a vertex that only has connections "going" from it. Complementary, a leaf vertex can be defined as a vertex that only has links "entering" it.

From the point of view of legal topics, we can interpret leaf vertices as the most specific topics existing in G_A and root vertices as the most generic ones in G_A. Furthermore, the topological structure of *directed rooted trees* allows us to define for each leaf vertex a_i a unique corresponding root vertex $\text{root}(a_i)$, that is, for each specific subject a unique corresponding generic subject.

Similarly to what happens with topics, legal provisions have hierarchical relationships that allow us to define a graph $G_D = (D, E_D)$, where

- $D = \{d_1, d_2, \ldots, d_n\}$, $m \in \mathbb{N}$, is a finite set of *vertices* describing legal provisions.
- $E_D \subset D \times D$ is a finite set of *directed edges* that connect the vertices of G_D representing the *part-of* relationship between the pieces that make up a law.

Although semantically different from the previously described taxonomy, the asymmetry of the relationships between the vertices of G_D also classifies it as a *directed rooted forest.* Legal provisions are hierarchically organized as a *partonomy*, where the vertices of the root type represent complete legal provisions, such as Laws, Decrees and Precedents that are composed of parts such as sections, articles, paragraphs and items. For each leaf vertex d_i there is, also, a single corresponding root vertex $\text{root}(d_i)$.

3.3 Weighted Graph of Legal Topics and Provisions

Given a set of initial petitions P, we can define a *bipartite graph of topics and legal provisions* $G_B = (A, D, E_B, w)$ as shown in Sect. 3.1. Furthermore, the background knowledge of legal topics allows us to build a *forest of legal topics* $G_A = (A, E_A)$ as shown in Sect. 3.2. Thus, we can define a new bipartite graph $G'_B = (A_{\text{leaf}}, D, E'_B, w'_B)$ as the subgraph of G_B, where:

- A_{leaf} is the subset of *vertices* that are of *leaf type* in G_A representing specific topics.
- D is the set of *vertices* of G_B representing the legal provisions.
- E'_B is the subset of *edges* of G_B where $w_{i,j} \geq \tau$.
- $w'_B : E' \to \mathbb{R}$ is an *edge weighting function* defined in E' as $w'_{Bi,j} = w_{Bi,j}$.

Here, τ corresponds to a threshold of *tfidf* below which we consider that an edge does not justify being created in G'_B. It is important to note that, unlike to what happens in G_B,, in G'_B there are only vertices related to legal topics of leaf type. Now, we can define the *weighted graph of legal topics and provisions* $G = (V, E, w)$, through the junction of G'_B and the undirected versions of the forests $G'_A = (A, E'_A)$ and $G'_D = (D, E'_D)$, where:

- $V = A \cup D = \{v_1, v_2, \ldots, v_{n+m}\}$ is the set of *vertices* of G', defined as the union of the vertices related to topics and legal provisions.
- $E = E'_B \cup E'_A \cup E'_D$ is the set of *edges* connecting the vertices in V.
- $w : E \to \mathbb{R}$ is an *edge weighting function* defined in E as:

$$w_{i,j} = \begin{cases} w'_B(v_i, v_j) & \text{case edge } (v_i, v_j) \in E'_B \\ \tau & \text{otherwise} \end{cases}$$

In order to the condition $(v_i, v_j) \in E'_B$ to be satisfied, necessarily, the conditions $v_i \in A$ and $v_j \in D$ must also be satisfied, since G'_B is a bipartite graph. As defined before, the graph G is a simple *weighted graph*. This allows us to generate vector representations for the vertices of this graph, as we will see in the next subsection.

3.4 Generating Embeddings

In this section, we describe the process of generating the vector representation of legal citations from *weighted graph of topics and legal provisions* and the embeddings of petitions that, in turn, use these representations.

Generating Vertices' Embeddings of G. In this step, we use the graph presented in the previous section to generate the embeddings of its vertices. For this, we define the mapping function

$$g_{[G]} : V \to \mathbb{R}^d,$$

which associates to each vertex $v_i \in V$ a vector $\mathbf{v}_i$ of d dimensions. The intuition of this approach is that, in this space, the vector $\mathbf{v}_i$ will be close to vertices belonging to the *neighborhood* of v_i. Then, legal topics will be "placed" close to legal provisions that occur concurrently in petitions and/or next to more general topics in the topic hierarchy. The citation embeddings aforementioned can be seen as an alternative way, other than the natural language text, of representing the arguments used by the petitioner. In this sense, legal provisions are cited to represent and support the matter of a petition and are relevant to determine the legal topic the petition refers to. In order to generate the embedding for G, we used the *node2vec+* algorithm implemented in *pecanpy* [16] library, which was chosen for its ability to capture conformal communities [15].

Generation of a Petition's Embeddings. From the vector representation of each legal citation made in a given petition, the petition's embedding can be calculated. The representation $\mathbf{p}_j$ for a petition from the legal provisions it cites is defined as

$$\mathbf{p}_j = \frac{\sum_k^{\Omega_j} \mathbf{v}_k}{L(\Omega_j)},$$

where the sum is performed on the set Ω_j of all representations corresponding to the legal provisions mentioned in the petition of the process j and $L(\Omega_j)$ is the total number of elements in Ω_j.

3.5 The Classifier

The petition set P is the main input of the classifier, as illustrated in the architecture of our model in Fig. 2. The text of the initial petition goes through a textual standardization phase with the objective of standardizing legal terms (for example, art. 1 of Law No. 11,419 of DECEMBER 19, 2006 to art. 1 of Law 11,419/2006), removing excess spaces, repeated lines and special characters and extending acronyms of agencies and states of the federation. Next, the legal provisions cited in the petition are identified using regular expressions and then we proceed as described in the previous section in order to generate the petition embedding.

The automatic classification is done according to a Hierarchy-aware Prompt Tuning architecture [34]. At a high level, the standardized text of the application is sent to the BERT module, whose architecture is structured in a multi-layered bidirectional Transformer encoder [10]. The input text is encoded in a sequence of tokens generated using WordPiece embeddings [35]. The textual features are then passed to a stack of encoder layers, after augmentation with positional encoding. Positional embeddings are added to each layer of the transformer's encoder to help the transformer learn dependencies between tokens. After applying the transformer coding layers, the BERT module results in a linear pooling layer.

The combination module receives as input the text resources emitted from the Transformer BERT model and the numerical resources coming from the "vector representation of legal citations". From there the module generates a combined multimodal representation. We use the combination module proposed by [34], which receives as input the textual features generated from a Transformers model and numerical features and generates a combined representation, through different methods that combine the different representations in their respective spaces of features.

To incorporate the label hierarchy, we propose a layer-wise prompt (according to [34]). Since the label hierarchy is a tree, we construct templates based on the depth of the label hierarchy. The hierarchy constraint only introduces depth of labels but lacks their connectivity. To make full use of the label hierarchy in an MLM-manner, we further inject the per-layer label hierarchy knowledge into template embedding. A K-layer stacked Graph Attention Network (GAT) [12] is adopted to model the label hierarchy.

Since hierarchical text classification is a multi-label classification problem, in this paper, instead of calculating the score of each label separately, we expect the scores of all target labels to be greater than those of all non-target labels. We use a multilabel cross entropy (MLCE) loss [33], introducing an anchor label with a constant score of 0 in MLCE and hope that the scores of the target labels and non-target labels are all greater and less than 0 respectively. Thus, we form a zero-bounded multi-label cross entropy (ZMLCE) loss. To be consistent with the hierarchy constraint, we adopt ZMLCE at each label hierarchy layer for the layer-wise prediction.

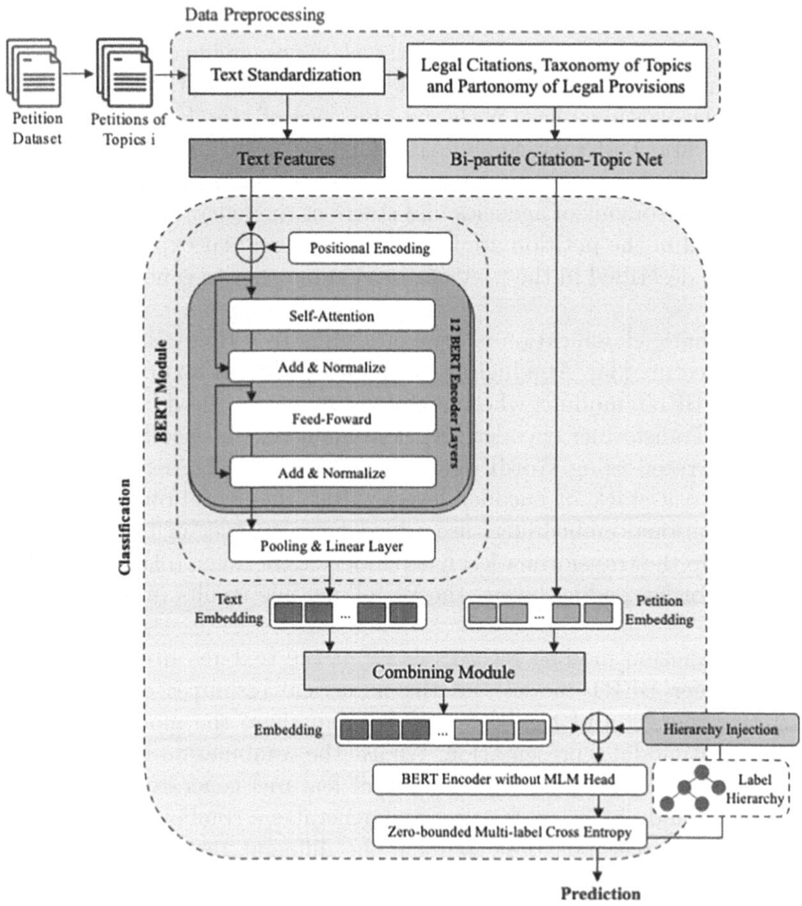

Fig. 2. Architecture of our model. The architecture is divided into two stages: (i) data preprocessing: the text of the initial petition is pre-processed to standardize legal terms and to identify the articles and laws cited in the text, and (ii) classification: combines the input data (text and bi-partite citation topic-net) and uses the architecture of Hierarchy-aware Prompt Tuning (HPT) [32] during training. HPT transforms hierarchical text classification (HTC) into a hierarchy-aware multi-label masked language model (MLM) problem, incorporing the label hierarchy knowledge. HPT transforms HTC into a multi-label MLM task with a zero-bounded multi-label cross entropy loss.

4 The Classifier in Use

The performance evaluation of the model was made from its application in real data of 300,000 petitions of legal processes coming from Labor Courts, Federal Regional Courts and Courts of Justice of the Brazilian states.

4.1 The Petition Dataset

As already stated in Sect. 3.2, legal topics are categorized according to a specific table of topics called *Unified Procedural Table* TPU [9] developed by the *Brazilian National Council of Justice* CNJ. The hierarchy present in these tables allows us to aggregate specific topics into generic ones. We used this strategy to separate our dataset into very distinct topic groups, which allowed us to test our approach at different levels of complexity.

In Table 1, we show the percentage distribution of initial petitions for each generic subject in the dataset of 300,000 petitions.

Table 1. Percentage distribution of lawsuits by legal topics. The codes used in this table are defined by the TPU and are the same ones that will be used in the results section.

Generic Topic	Lawsuits (%)
14 - Tax Law	4.4
195 - Pension Law	4.7
287 - Criminal Law	5.7
864 - Labor Law	11.1
899 - Civil Right	29.0
1156 - Consumer Rights	25.2
1209 - Criminal Procedural Law	0.4
8826 - Civil Procedural and Labor Law	13.0
9985 - Administrative Law and other matters of Public Law	6.3

The set of 300,000 petitions was then divided following a ratio of 80% for training and 20% for testing. We took care that the dataset only contained petitions with at least 50 words and that more than 80% of valid words were valid[1].

4.2 Training Process

The graph G was created from the legal provisions identified in the set of petitions P of the training data. Then, the vector representation of the petition

[1] The verification of valid words was performed using the lexicons proposed by [18].

citations was generated as described in Sect. 3.4. The value $\tau \approx 8.48 \cdot 10^{-3}$ was empirically adjusted by identifying the value that best separated the topics from G_B. This was done by analyzing the *assortativity coefficient* [20] of the one-mode projection [19] relative to the *legal topic* partition.

A fine-tuning of the pre-trained BERTimbau [31] model was done, where all parameters were tuned using the specific topics of the petitions used in the training. In the mesun_etal_2020rge module, we use the gated summation merge method in the transformer output for the petition text and the vector representation of the citation, before the final classifier layer[2]. The model was trained with batch size of 16 examples, 30 epochs max and a early stopping of 5 epochs. Adam Optimizer was chosen with learning rate of 10^{-5} and 5-fold cross-validation. Random oversampling was applied in order to equalize number of examples of each class.

4.3 Comparative Evaluation and Deployment

For a detailed evaluation of our approach, we compare our model (descibed in Sect. 3.5) with the original hierarchical classification method [34] (that uses only text as input). The results for each specific topic were calculated in terms of Precision, Recall and F1-score, obtained from the actual and predicted labels on the test dataset. In Table 2, we show the results of the classifiers for the different generic topics shown in the previous section.

The results in this table show that the classifier that uses information from the citations performs better in some situations. Precisely, it shows better results in "864 - Labor Law", "899 - Civil Right" and "1209 - Criminal Procedural Law" generic topics. We believe that, in these cases, our approach is able to better capture the relationships that exist between legal provisions and specific topics due to the existence of legal provisions that detail several specific situations in these branches of Law.

An example of an application where this approach is very useful occurs when a petitioner only informs a generic topic when submitting the petition. In this case, a meta-model would be capable of identifying which approach is most suitable for classifying the specific subject of that petition based on the generic topic informed.

Indeed, we incorporated a version of a hierarchical classification model into the CNJ's SINAPSES platform [8] in the form of a micro-service. This platform aims to share, among Brazilian judicial institutions, models of artificial intelligence developed by the public and private sector. In Fig. 3, we shows the interface of the judicial process registration system, PJe, after the implementation of the leaf topic prediction functionality with the call of the classification micro-service implemented in the SINAPSES platform.

[2] This merge method is inspired by [26].

Table 2. Precision, Recall and F1-score results for different generic topics. Here, N is the number of specific topics present in each generic topic. The metrics show the comparison between the classifier that uses only the text of the petitions and the one that uses the text and the vector representation defined by our approach.

Generic Topic	N	Only Text			Text + G		
		Precision	Recall	F1-score	Precision	Recall	F1-score
14	9	**0.899**	**0.835**	**0.859**	0.898	0.833	0.857
195	13	**0.743**	**0.719**	**0.725**	0.709	0.675	0.690
287	12	**0.829**	**0.795**	**0.811**	0.823	0.791	0.806
864	22	0.528	0.487	0.488	**0.556**	**0.538**	**0.541**
899	38	0.749	0.635	0.680	**0.779**	**0.637**	**0.692**
1156	24	**0.664**	**0.599**	0.622	**0.664**	0.597	**0.625**
1209	3	0.694	**0.395**	**0.484**	**0.735**	0.364	0.462
8826	22	**0.823**	0.491	0.603	0.807	**0.521**	**0.624**
9985	10	**0.766**	**0.586**	**0.649**	0.758	0.548	0.622

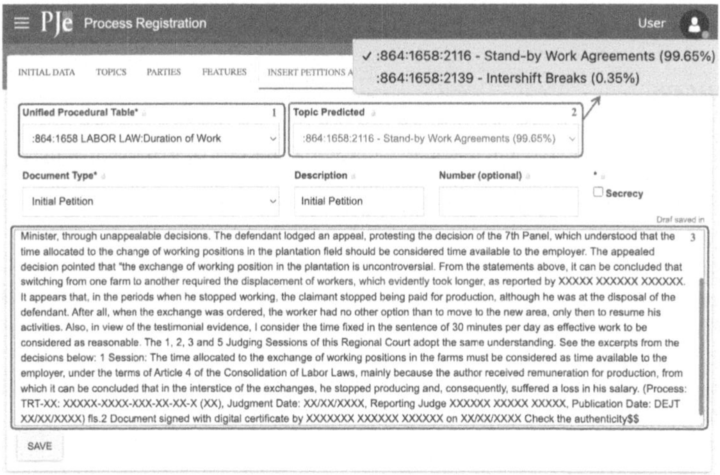

Fig. 3. Screenshot of PJe system with the feature of predicting the topic of a petition. The screenshot is numbered to show: 1) is the main topic informed by the user, 2) is the predicted specific topic and 3) is the text of the petition.

5 Conclusion

Citations to existing legal provisions in lawsuits is an important particularity of this type of document. They add additional information to the text and can be used by automatic natural language processing methods for a variety of tasks. In this article we describe how the automatic topic classification of a legal petition can be enhanced based on these citations. The approach integrates transformer-based methods with a complex network in order to capture, in addition to text features, the relationship between legal citations that are present in the petitions and the topic to which the petition refers to.

This research work characterizes itself by the proposition of a novel method based on the combination of Deep Learning and Complex Networks, and in the application of these ideas in the real world. The implementation of a microservice into the Justice National Platform as well as the implementation of a front-end that uses this service constitutes a novel tool to be used country wide by different institutions of the Brazilian Justice. It is expected bringing productivity augmentation, reduction of costs and more agility to the judicial sentences.

Future work involves improving accuracy of the method via fine-tuning of parameters and techniques used to manipulate the network. For instance, investigations of alternative methods for graph embedding generation for capturing temporal relations may improve the final results. Also, monitoring the use of the classifier by the different institutions may shed lights on the user experience and the real benefits the approach brings to the citizen.

Acknowledgments. We gratefully acknowledge CNPq, CAPES, FUNCAP, BNB, and the Edson Queiroz Foundation for financial support.

References

1. Aguiar, A., Silveira, R., Furtado, V., Pinheiro, V., Neto, J.A.M.: Using topic modeling in classification of Brazilian lawsuits. In: Pinheiro, V., et al. (eds.) PROPOR 2022. LNCS (LNAI), vol. 13208, pp. 233–242. Springer, Cham (2022). https://doi.org/10.1007/978-3-030-98305-5_22
2. Aguiar, A., Silveira, R., Pinheiro, V., Furtado, V., Neto, J.A.: Text classification in legal documents extracted from lawsuits in Brazilian courts. In: Anais da X Brazilian Conference on Intelligent Systems. SBC, Porto Alegre, RS, Brasil (2021). https://sol.sbc.org.br/index.php/bracis/article/view/19093
3. Alupoaie, S., Cunningham, P.: Using TF-IDF as an edge weighting scheme in user-object bipartite networks. arXiv preprint arXiv:1308.6118 (2013)
4. Luz de Araujo, P.H., de Campos, T.E., Ataides Braz, F., Correia da Silva, N.: VICTOR: a dataset for Brazilian legal documents classification. In: Proceedings of the 12th Language Resources and Evaluation Conference, pp. 1449–1458. European Language Resources Association, Marseille, France (2020). https://aclanthology.org/2020.lrec-1.181
5. Carmichael, I., Wudel, J., Kim, M., Jushchuk, J.: Examining the evolution of legal precedent through citation network analysis. NCL Rev. **96**, 227 (2017)
6. Chalkidis, I., Androutsopoulos, I., Aletras, N.: Neural legal judgment prediction in english. CoRR abs/1906.02059 (2019). http://arxiv.org/abs/1906.02059
7. Cheng, J., Danescu-Niculescu-Mizil, C., Leskovec, J.: Antisocial behavior in online discussion communities. CoRR abs/1504.00680 (2015). http://arxiv.org/abs/1504.00680
8. CNJ: Sinapses (2022). https://sinapses.ia.pje.jus.br/. Accessed 9 Aug 2022
9. CNJ: Sistema de gestão de tabelas processuais unificadas (2022). https://www.cnj.jus.br/sgt/consulta_publica_assuntos.php. Accessed 9 Aug 2022

10. Devlin, J., Chang, M., Lee, K., Toutanova, K.: BERT: pre-training of deep bidirectional transformers for language understanding. In: Burstein, J., Doran, C., Solorio, T. (eds.) Proceedings of the 2019 Conference of the North American Chapter of the Association for Computational Linguistics: Human Language Technologies, NAACL-HLT 2019, Minneapolis, MN, USA, 2–7 June 2019, Volume 1 (Long and Short Papers), pp. 4171–4186. Association for Computational Linguistics (2019). https://doi.org/10.18653/v1/n19-1423
11. Fowler, J.H., Johnson, T.R., Spriggs, J.F., Jeon, S., Wahlbeck, P.J.: Network analysis and the law: measuring the legal importance of precedents at the us supreme court. Polit. Anal. **15**(3), 324–346 (2007)
12. Kipf, T.N., Welling, M.: Semisupervised classification with graph convolutional networks. In: 5th International Conference on Learning Representations (ICLR) (2017)
13. Koniaris, M., Anagnostopoulos, I., Vassiliou, Y.: Network analysis in the legal domain: a complex model for European union legal sources. J. Complex Netw. **6**(2), 243–268 (2018)
14. Limsopatham, N.: Effectively leveraging BERT for legal document classification. In: Proceedings of the Natural Legal Language Processing Workshop 2021, pp. 210–216. Association for Computational Linguistics, Punta Cana, Dominican Republic (2021). https://doi.org/10.18653/v1/2021.nllp-1.22. https://aclanthology.org/2021.nllp-1.22
15. Liu, R., Hirn, M., Krishnan, A.: Accurately modeling biased random walks on weighted graphs using node2vec+. arXiv preprint arXiv:2109.08031 (2021)
16. Liu, R., Krishnan, A.: Pecanpy: a fast, efficient and parallelized python implementation of node2vec. Bioinformatics **37**(19), 3377–3379 (2021)
17. Liu, Y., et al.: Roberta: a robustly optimized BERT pretraining approach. CoRR abs/1907.11692 (2019). http://arxiv.org/abs/1907.11692
18. Muniz, M.: A construcão de recursos linguistico-computacionais para o português do Brasil: o projeto Unitex-PB. Dissertação (Mestrado em Ciências de Computação e Matemática Computacional) - Instituto de Ciências Matemáticas e de Computação, Universidade de São Paulo (2004). https://doi.org/10.11606/D.55.2020.tde-19022020-151305
19. Newman, M.: Networks. Oxford University Press (2018)
20. Newman, M.E.: Mixing patterns in networks. Phys. Rev. E **67**(2), 026126 (2003)
21. Noguti, M.Y., Vellasques, E., Oliveira, L.S.: Legal document classification: an application to law area prediction of petitions to public prosecution service. In: 2020 International Joint Conference on Neural Networks (IJCNN), pp. 1–8 (2020). https://doi.org/10.1109/IJCNN48605.2020.9207211
22. Pérez-Rosas, V., Mihalcea, R.: Experiments in open domain deception detection. In: Proceedings of the 2015 Conference on Empirical Methods in Natural Language Processing, pp. 1120–1125. Association for Computational Linguistics, Lisbon, Portugal (2015). https://doi.org/10.18653/v1/D15-1133. https://aclanthology.org/D15-1133
23. Pinheiro, V., Pequeno, T., Furtado, V., Nogueira, D.: Information extraction from text based on semantic inferentialism. In: Andreasen, T., Yager, R.R., Bulskov, H., Christiansen, H., Larsen, H.L. (eds.) FQAS 2009. LNCS (LNAI), vol. 5822, pp. 333–344. Springer, Heidelberg (2009). https://doi.org/10.1007/978-3-642-04957-6_29

24. Pires, R.S., Oliveira, E.A., Almeida, V.F., Neto, J.A.M., Furtado, V.: Análise da ontologia dos assuntos jurídicos e suas respectivas legislações através de redes complexas. In: Anais do XI Brazilian Workshop on Social Network Analysis and Mining, pp. 181–191. SBC (2022)
25. Pires, R.S., Oliveira, E.A., Fernandes, C.G., Neto, J.A.M., Furtado, V.: Mapping landmark cases in the us legal system. CEUR Workshop Proceedings (2021)
26. Rahman, W., et al.: Integrating multimodal information in large pretrained transformers. In: Proceedings of the 58th Annual Meeting of the Association for Computational Linguistics, pp. 2359–2369. Association for Computational Linguistics, Online (2020). https://doi.org/10.18653/v1/2020.acl-main.214. https://aclanthology.org/2020.acl-main.214
27. Ruhl, J., Katz, D.M., Bommarito, M.J.: Harnessing legal complexity. Science **355**(6332), 1377–1378 (2017)
28. Sanh, V., Debut, L., Chaumond, J., Wolf, T.: Distilbert, a distilled version of BERT: smaller, faster, cheaper and lighter. CoRR abs/1910.01108 (2019). http://arxiv.org/abs/1910.01108
29. Shaheen, Z., Wohlgenannt, G., Filtz, E.: Large scale legal text classification using transformer models. CoRR abs/2010.12871 (2020). https://arxiv.org/abs/2010.12871
30. Silva, N., Braz, F., de Campos, T.: Document type classification for brazil's supreme court using a convolutional neural network. In: 10th International Conference on Forensic Computer Science and Cyber Law (ICoFCS), Sao Paulo, Brazil, pp. 7–11 (2018). https://doi.org/10.5769/C2018001
31. Souza, F., Nogueira, R., Lotufo, R.: BERTimbau: pretrained BERT models for Brazilian Portuguese. In: Cerri, R., Prati, R.C. (eds.) BRACIS 2020. LNCS (LNAI), vol. 12319, pp. 403–417. Springer, Cham (2020). https://doi.org/10.1007/978-3-030-61377-8_28
32. Sumner, C., Byers, A., Boochever, R., Park, G.J.: Predicting dark triad personality traits from twitter usage and a linguistic analysis of tweets. In: 2012 11th International Conference on Machine Learning and Applications, vol. 2, pp. 386–393 (2012). https://doi.org/10.1109/ICMLA.2012.218
33. Sun, Y., et al.: Circle loss: a unified perspective of pair similarity optimization. In: Proceedings of the IEEE/CVF Conference on Computer Vision and Pattern Recognition (2020)
34. Wang, Z., et al.: HPT: hierarchy-aware prompt tuning for hierarchical text classification. arXiv preprint arXiv:2204.13413 (2022)
35. Wu, Y., et al.: Google's neural machine translation system: bridging the gap between human and machine translation. CoRR abs/1609.08144 (2016). http://arxiv.org/abs/1609.08144
36. Yang, Z., Dai, Z., Yang, Y., Carbonell, J.G., Salakhutdinov, R., Le, Q.V.: Xlnet: generalized autoregressive pretraining for language understanding. CoRR abs/1906.08237 (2019). http://arxiv.org/abs/1906.08237

Author Index

A. Paes and F. A. N. Verri (Eds.): BRACIS 2024, LNAI 15413, pp. 491–492, 2025.
https://doi.org/10.1007/978-3-031-79032-4

The manufacturer's authorised representative in the EU is Springer Nature Customer Service Centre GmbH, Europaplatz 3, 69115 Heidelberg, Germany. If you have any concerns regarding our products, please contact ProductSafety@springernature.com

Printed and bound by CPI Group (UK) Ltd, Croydon, CR0 4YY
30/09/2026
02225597-0020